Jan Klein · Naoyuki Takahata

WHERE DO WE COME FROM?

Springer

Berlin
Heidelberg
New York
Barcelona
Hong Kong
London
Milan
Paris
Tokyo

Jan Klein · Naoyuki Takahata

WHERE DO WE COME FROM?

The Molecular Evidence for Human Descent

With 124 Figures

Springer

Professor Dr. JAN KLEIN
Max Planck Institute for Biology
Department of Immunogenetics
Corrensstrasse 42
72076 Tübingen
Germany

Professor Dr. NAOYUKI TAKAHATA
The Graduate University for Advanced Studies
Department of Biosystems Science
Hayama
Kanagawa 240-0193
Japan

Cataloging-in-Publication Data applied for

Die Deutsche Bibliothek - CIP-Einheitsaufnahme
Klein, Jan: Where Do We Come From? The Molecular Evidence For Human Descent / Jan Klein and Naoyuki Takahata. - Berlin; Heidelberg; New York: Springer 2002
    ISBN 3-540-42564-00

ISBN 3-540-42564-0 Springer-Verlag Berlin Heidelberg New York

Springer-Verlag Berlin Heidelberg New York
a member of BertelsmannSpringer Science+Business Media GmbH

http://www.springer.de

© Springer-Verlag Berlin Heidelberg 2002
Printed in Germany

Production: PRO EDIT GmbH, Heidelberg, Germany
Cover design: design + production GmbH, Heidelberg, Germany
Typesetting and reproduction of the figures: AM-productions GmbH, Wiesloch, Germany
Printed on acid-free paper   SPIN 10850978   27/3130/ML - 5 4 3 2 1 0

DEDICATION

We dedicate this book to our students and associates who have worked with us over the years. They have come from different countries – Australia, Austria, Chile, Colombia, Croatia, Czech Republic, England, France, Germany, Greece, Hungary, India, Ireland, Israel, Japan, South Korea, People´s Republic of China, Peru, Philippines, Poland, Republic of Tuva, Russia, Scotland, Slovakia, Spain, Taiwan, The Netherlands, United States of America, Yugoslavia – and from many different cultural backgrounds. Through them we have learned much about human diversity and unity, and contacts with them have enriched our lives.

PREFACE

It used to be that books popularizing science were written largely by practicing scientists. Gamow, Jeans, Eddington, Huxley, Haldane … these are some of the illustrious names that immediately spring to mind as authors of memorable opuses, some of which have still not lost their appeal more than half a century after they were penned. The primary aim of these books was to inform and enlighten the reader and in the process, to convey the excitement of discovery. Many scientists today will admit that their interest in science is the direct result of an earlier encounter with one of these books, or with the work of less well known authors. The books were written for intelligent readers who were willing to follow the authors through their arguments, even if it did require a certain effort. For the authors did not step down to the level of the lay reader, they wanted to lift the latter to their own heights. They did not strive to entertain the readers at all costs, they wanted them to comprehend and to wonder.

All this has changed. Popularization of science is now a full-time occupation practiced by a special guild of authors – the science writers – who more often than not are journalists with little or no expertise in the areas they choose to cover. As journalists, these writers have brought sensationalism, superficiality, and ephemerality into popularization of science. Not education, but entertainment of the reader is their primary goal. As a result, more is learned about the character, habits, or social life of practicing scientists than about their discoveries. You are treated to lengthy descriptions of pitching a tent in an Ethiopian desert, of chats with an anthropologist who startles the guests in a Parisian café by whisking a Neandertal skull out of his briefcase, or to the latest gossip about intrigues among the prima donnas of science. Not infrequently, these dilettantes try to force their own views or even hypotheses on their readers, views that a layperson cannot evaluate, but that an expert would have no difficulty demolishing.

This book is written in the spirit, if not with the skills, of an old and now largely abandoned tradition. It strives to provide a picture of human evolution for those who want to be informed rather than be entertained. It does not assume that the readers are so simple-minded that their interest has continually to be whetted by amusing anecdotes. On the contrary, it expects the reader to have a good measure of intelligence and sufficient interest in the topic to continue reading even when it comes to some of the more technical parts.

There are many popular books on the origin of the human species, but the majority are focused on the testimony provided by old bones, while molecular evidence is merely skimmed over. With the present book it is just the reverse: although the archeological and paleontological evidence is summarized, its marrow is the information provided by molecules, first and foremost nucleic acids. The molecular aspect of human evolution may regularly be covered by the media but it is often presented in a grossly distorted way revealing a principal misunderstanding of the underlying concepts. Distorted descriptions of human molecular evolution have even found their way into textbooks. In our text, great emphasis is therefore placed on the description and critical evaluation of methods and concepts, as well as on the hopefully unbiased evaluation of results. It will certainly be heavy going in places, for some of the methods and concepts are based on mathematical and statistical arguments, and we do not spare the reader if the math is essential for grasping the principles involved. Although math has always been a scarecrow in the popularization of science, with authors waxing apologetic when compelled to include even minimal mathematical arguments, we feel its notoriety is not characteristic of the subject itself, but of the way it is explained. Mathematics is not difficult, only mathematicians are: they often seem unaware of what a lay reader needs to be told to follow their arguments. For this reason, although one of us has had solid mathematical training, the mathematical sections have been written by the one who does not. To avoid disrupting the flow of the narrative unduly, however, we have relegated much of the mathematical background information to the appendices. It is our hope that the reader who does require a math refresher will not ignore this part of the book simply because it is dense with algebraic symbolism. In particular, all students among the readership are strongly urged to refer to this part, for it might be the only place in the literature available on molecular evolution in which they will find, for example, a comprehensive derivation of the widely-used Jukes-Cantor formula.

Writing this book has taken much longer than originally anticipated. Indeed, the fact that it has been successfully completed has to be credited to several persons. First and foremost we must name our editorial assistants Ms. Jane Kraushaar and Ms. Lynne Yakes. The former took charge of all formal aspects in preparing a word-processor version of the manuscript, including all the figures. Without her efficiency, dependability, and conscientiousness we would probably have abandoned the project in its early phase. The latter devoted many hours of her free time to polishing and otherwise improving the manuscript. As always, her assistance has been indispensable and as always, it has been a pleasure to work with her. Our colleagues, Drs. Sang-Hee Lee, Werner E. Mayer, Colm O'hUigin, Akie Sato, Yoko Satta, Naoko Takezaki, and Herbert Tichy have helped in many ways by assembling and interpreting material for this book. Whenever help was urgently needed, they could always be relied upon. Finally, representing Springer Verlag, Dr. Rolf Lange's interest in the project has been a continuous source of encouragement during the long months of writing. To all these persons we are greatly indebted.

Tübingen and Hayama
November 2001

JAN KLEIN
NAOYUKI TAKAHATA

C O N T E N T S

How to Read this Book

Instead of a glossary at the end of the book, we have italicized *terms* in the text where they first appear and are defined. *Abbreviations* are explained where they first appear in the text and then again in Appendix Five. *Appendices* One through Four provide a brief review of the elementary concepts underlying the *mathematical parts* of the text. A math-phobic reader may skip these parts without losing the thread of the remainder of the text. *Publications* describing studies mentioned specifically in the text are listed in the *Sources* for each chapter at the end of the book. In the text, the studies are referred to by the name of the first author of multi-authored publications. Where no author is mentioned in the text, the relevant publications are nevertheless easily identifiable from their titles in the Sources. *Further Reading* in the reference section lists books or review articles in which the reader can find further information on the topics covered by the particular chapter.

CHAPTER 1

A Tahitian Prelude

Art, Myth, and Science

The present life of men on earth, O king, as compared with the whole length of time which is unknowable to us, seems to me to be like this: as if, when you are sitting at dinner with your chiefs and ministers in wintertime, ... one of the sparrows from outside flew very quickly through the hall; as if it came in one door and soon went out through another. In that actual time it is indoors it is not touched by the winter's storm; but yet the tiny period of calm is over in a moment, and having come out of the winter it soon returns to the winter and slips out of your sight. Man's life appears to be more or less like this; and of what may follow it, or what preceded it, we are absolutely ignorant.

The Venerable Bede: *Historia Ecclesiastica Gentis Anglorum*

Gauguin's Testament

On July 2, 1895, Paul Gauguin boarded the *Australian* at Marseilles on a voyage that would take him to the ports of Corfu and Aden in the Mediterranean, then across the Indian Ocean to Sydney and Melbourne in Australia, from where he would proceed to Auckland, New Zealand, and finally, for the second time, to Tahiti. He arrived in Papeete on September 9, full of hope that he would be able to pick up the pieces of his shattered existence and begin a new life.

At first, it seemed he might succeed. He rented a piece of land at Punaania, a few kilometers south of Papeete, had a big Tahitian hut built on it *in the shade at the road-side, with a view of the marvellous mountains**, and took a new *vahine*, a female companion. He also befriended the new governor, with whom he undertook trips to the neighboring islands.

But things soon took a turn for the worse. An old ankle injury flared up, he began to cough up blood, and throbbing pain kept him awake at night. He ran out of money and was forced to live on rice and water. His companion deserted him and when his landlord died, he was evicted from his hut. Shunned by the Europeans, dejected, and exhausted, he lost interest even in painting. *I am on my knees and have to put all my pride aside. I am nothing but a wreck*, he confessed in a letter to a friend in Paris. And then in April 1897 the devastating news from Denmark broke upon him: his beloved daughter Aline had died of pneumonia. *I cry out: God Almighty, if you exist, I accuse you of injustice and malice*, he exclaimed in another letter. Spiritually, physically, and emotionally drained, he decided to put an end to his misery.

Before taking his own life, however, he resolved to paint one last canvas to summarize his entire artistic and personal experience – experience *of a man who had roamed continents and cultures, whose thinking embraced many philosophies and religions, who had experienced the contradiction of a revolutionary age, who had been rich and poor, lived in mansions and hovels, who had known sexual ecstasy and humiliating failure* (Sweetman 1995). The painting would be the testament of an artist and a human being, and its subject would be nothing less than the meaning of human existence.

"Why am I here?" is a question posed by every sentient being at some point in their lifetime, particularly during a crisis as extreme as that experienced by Gauguin in 1898 when he received the news of Aline's tragic demise. But this question must have tormented him many times before since there had been other moments in Gauguin's life when he pondered the absurdity of existence in an indifferent world. Now, however, he was prepared to deal with it *sub specie aeternitatis*, as a person soon to face eternity.

The form most befitting a momentous subject such as this would have been a mural, a fresco, or a frieze, but how does one paint a fresco on a bamboo partition? Canvas would have to do, but in his current situation, Gauguin was not able to scrape the money together to procure sufficient linen for the size he had in mind. So he tore open several sacks, stitched them together, and stretched them out as far as his studio allowed – full four and a half meters long and one meter seventy high. Then he summoned up all his remaining energy, and began to cover the rough surface, full of knots and creases, with broad strokes of color using large brushes and working directly from a few preliminary sketches. He

---

* The quotations in this section are from *Gauguin's Letters from the South Pacific Seas*, see Sources, as well as from M. Malingue: *Lettres de Gauguin à sa femme et à ses amis*, Paris 1946, translated by J.K.

Fig. 1.1
Paul Gauguin: *D'où Venons-Nous? Que Sommes-Nous? Où Allons-Nous? (Where Do We Come From? What Are We? Where Are We Going?)* (Museum of Fine Arts, Boston.)

worked feverishly, day and night, in a trance-like state. After a month of all-consuming effort, the painting was essentially completed (Fig. 1.1). To enhance its fresco-like qualities, he painted its upper corners in chrome-yellow, creating an illusion of paint peeling off from a golden wall. In the golden patch on the right-hand side he signed his name, in the left-hand corner he inscribed the title of the work: *D'où Venons-Nous? Que Sommes-Nous? Où Allons-Nous?: Where Do We Come From? What Are We? Where Are We Going?*

For a description of the events that followed, we only have Gauguin's word, which some art historians doubt. By Christmas 1987, he retreated into the mountains, where his *body would have been devoured by the ants*, and there he swallowed several capsules of arsenic that he had hoarded for the treatment of his eczema. The dose was too large, however, and induced heavy vomiting accompanied by terrible pains and cramps, but did not kill him. After a night of dreadful suffering, he dragged himself back to his hut, remaining immobilized for several days before he eventually recovered.

Did this attempt at suicide really take place as described or is the story an extension of the myths and legends that Gauguin invoked to enshroud his paintings and indeed his life? We will never know. The experts doubt that the large dose of arsenic he purportedly swallowed would not have been lethal, regardless of its immediate effects. But they could be wrong. Suicide attempt or not, the fact remains that Gauguin lived for another six years, long enough to provide an interpretation of his masterpiece. Although he produced several more works during this time, none of them compared with *D'où Venons-Nous?* He died in 1903, presumably of a heart attack, although there were those who believed that he had been poisoned by his enemies, who numbered many. Thus the death of this lonely, despairing man was woven into the legends that Paul Gauguin himself had nurtured during his lifetime.

What, then, does *D'où Venons-Nous?* represent? Despite Gauguin's vague explanatory notes (contradicted by his own remonstrations that the work has no explicit meaning: *my dream is intangible*, he insisted) and *because* of the numerous interpretations offered by scores of art critics, the painting's message remains obscure. This is perhaps only to be expected from a work rich in symbolic colors (gold, black, white, and red), laden with symbols, and adopting three fundamental, philosophical questions as its title. Although a general guideline to our perception of the masterpiece is available, a specific relation between the work and a viewer is established individually, according to each person's intellectual and emotional disposition.

On the surface of it, the painting – which now hangs in the Museum of Fine Arts in Boston – depicts a fairly realistic scene on a tropical island. It is presumably Tahiti for the mountains looming in the distance can be identified as those of the Moorea Island. In a jungle clearing, on the bank of a river, a scattered group of natives is seen in various leisurely poses. *Mañana*, seems to be their philosophy of life – relax today and postpone any chores for some later date. The scene seems to echo some of the arcadian paintings of Pierre Puvis de Chavannes, one of Gauguin's artist-heroes. In contrast to the transparency of Puvis de Chavannes' allegories, however, Gauguin's rendition evokes in the viewer a feeling of mystery, of the intangible. The presence of an idol standing out *against a clump of trees such as do not grow on earth but only in Paradise*, the eerie bluish glow of the idol and some of the branches in the jungle, as if they have been illuminated momentarily by a flash of lightning, the fluorescent patches of vegetation contrasting with the jungle's dark recesses, the Veronese green of the sunlit sea in the background setting off the adumbrated scene in the foreground, all this conspires to create the impression of viewing a stage on which a mystery play is being acted out. There are 12 figures on the stage – eight women, two men, and two children – together with an assortment of domesticated animals. The key to the entire work, however, is to be found in three of the figures, the newborn on the extreme right, the erect figure in the center, and the withered old woman, a crone, on the lower left-hand side. They epitomize Birth, Life, and Death, respectively – the beginning, the culmination, and the closing of one round of a cycle, the life cycle.

The Myth of Eden

The three symbolic figures are anchored in mythology, Gauguin's brand of mythology, with elements drawn from diverse cultural traditions – Polynesian, Peruvian Indian, Eastern, African, and above all Judeo-Christian. Gauguin subscribed to the notion, popular in his time, that all mythologies have a single common source in Africa. In his youth, Gauguin was indoctrinated with Biblical mythology which he reinterpreted later under the influence of other mythologies with which he had by then become familiar. *D'où Venons-Nous?* is a good example of this. Its symbolism is based foremost on the first book of the Bible, the Book of Genesis, which is the story of creation. In Genesis, God is credited with creating the first man by fashioning a figurine from clay in his own image (the implication being that God is male) and bringing it to life by breathing into its nostrils. The duality of material body and immaterial soul, a notion popular in Western culture, has its origin in this story. As "earth" in Hebrew is *adamah*, the word *adam* came to signify "man" and, more generally, "mankind". It also became the name of the first man.

Next, God created a place for Adam to live, a beautiful garden in the land of Eden which Biblical scholars have now localized in the valley of the Euphrates River, as indeed the Book of Genesis says that a river flowed through the Garden of Eden. Adam was instructed not only to enjoy the Garden but also to take care of it and God created a helper for him by removing one of his ribs (under anesthesia, of course) and fashioning the first woman from it, whom Adam later called *Hawwāh* (presumably from Hebrew *hayah*, to live). This name came to be translated into Greek as Eva and into English as Eve.

In the Garden of Eden, God planted many luscious trees (the word eden means pleasure, luxury) which were beautiful to look at and whose fruit was delicious to eat. And, of course, he also populated the Garden with many different animals. Among the luxurious vegetation were two trees that stood out from the rest. One was the Tree of Life, bearing fruit which bestowed immortality upon the consumer (Fig. 1.2). The other was the Tree of

Fig. 1.2
The Tree of Life (or the Tree of Knowledge) in the
rendition of the Dutch painter Hieronymus Bosch.
A detail from the left-hand panel of the triptych
The Garden of Earthly Delights executed about 1500.
(Prado Museum, Madrid, Spain.)

Knowledge, the nature of which is rather obscure and open to a variety of interpretations.
(Some Biblical scholars argue that in the original version of Genesis there was only one
special tree, but they cannot agree on which.) One interpretation claims that by eating its
fruit, Adam and Eve would have acquired knowledge of how to create life – knowledge of
sex. As creators of life, they would have become God-like. Knowledge of this kind would
have brought disaster to the Garden of Eden, which was apparently limited in its size and
resources: the continuous birth of new beings in the absence of death (for everybody
would obviously want to eat fruit from the Tree of Life and live forever) would rapidly have
led to hopeless overcrowding. Hence it made perfect sense that God allowed the first couple to eat fruit from the Tree of Life, but forbade them on pain of death to consume any
fruit from the Tree of Knowledge. Had they learned how to reproduce themselves, mortality would have been the only possible solution, since death alone could prevent overcrowding. (Here, however, God seems to have underestimated the power of the urge to reproduce, since not even death could hinder Adam's descendants from multiplying to excess.)

For a while, the first couple led a carefree and guiltless existence as God intended them
to do. But God should have known better than to single out one tree from many and hang
a DON'T EAT THE FRUIT! sign on it. The first couple was devoured by curiosity: Why are
we not allowed to eat fruit from that particular tree? What is so special about that fruit?
And what would happen if we did eat it? What knowledge is God hiding from us?
Unfortunately, there was a spoiler in the Garden, who was burning with desire to enlighten
them. It was the serpent, who was *more subtle than any beast of the field* and moreover
could speak with human voice. Why the serpent was so intent on spoiling the idyll is not

clear. Some theologians believe that he just wanted to play a prank on God, while others re-gard him as the incorporation of Satan. Either way, it was an easy task for the serpent to convince Eve that by eating the forbidden fruit she would be like God (which was true in a certain sense). And so Eve plucked an apple from the Tree and shared it with Adam.

The consequences did not take long in coming. After eating the fruit of the Tree of Knowledge, *Adam knew Eve his wife, and she conceived* says the author of Genesis, where-by "knowledge" here refers to sexual intercourse. In other words, upon consuming the for-bidden fruit, the first human pair discovered the secret of creating new life. The whole no-tion of the Tree of Knowledge can thus be regarded as an elaborate metaphor for the act of reproduction. God, as stated earlier, had His reasons for not wanting Adam and Eve to re-produce. By conceiving new life, the couple in essence created their own successors and God now had no other choice than to make them mortal.

The relationship between God and the first human pair is portrayed by the author of Genesis as that between a father and his children. Eating the forbidden fruit in defiance of God's commandment was therefore an act of disobedience or a sin, which God the Father had to punish. The penalty He chose was to associate death with fear, and childbirth – which was to follow nine months after the act of "knowing" – with pain. Fear and pain are nasty things – they are the essential elements of evil. And so Adam and Eve came to appre-ciate the distinction between good and evil. The pleasure they experienced while "know-ing" each other felt good, but the fear and pain of the consequences were evil – hence the full name of the metaphorical plant: The Tree of Knowledge of Good and Evil. Adam and Eve were soon to learn many other forms of good and evil, especially of the latter. Scolded by God for their bad behavior, they also experienced feelings of shame and guilt for the first time. For the author of Genesis, the sexual act was evil and dirty and this notion was later to become one of the basic tenets of Christianity. The loss of innocence and end of the blissful state in which there was no need to make a distinction between good and evil, no reason for feeling shame and guilt, was the most profound consequence of the first, the *original sin.*

After this incident in which Adam and Eve fell from God's grace and which came to be known accordingly as The Fall, they could, of course, no longer remain in the Garden of Eden with its Tree of Life, the fruit of which they would surely have been tempted to eat as a remedy for God's curse. God therefore expelled them into a hostile land and so put an abrupt end to the leisurely life they had been accustomed to in the Garden. Now they had to work the soil to eke out a living and cope with many adversities that the switch from food collecting to farming brought with it. Fear, pain, hard labor, disease, frustration, and defeat became their constant companions. But worse was still to come: envy, wickedness, and murder, followed by a range of other crimes.

Art and Mythos

So much for the orthodox version of the Biblical creation story. Gauguin used various ele-ments of the story in his paintings, but in an unorthodox fashion. In the orthodox version, the original sin stigmatized not only Adam and Eve, but all of their descendants. From then on, all human beings have entered the world already infected with a germ of wicked-ness, as sinners. The human race is, according to this dogma, by its very nature corrupted. Gauguin, however, could not accept such a fatalistic view. He refused to believe that a new-born should be held responsible for the sins of its ancestors, or that people living in a nat-ural state, like the natives in Polynesia, could be innately corrupt. Although he had lost most of his illusions about the nobility of the Tahitian natives by the time he began to work on *D'où Venons-Nous?*, in the painting itself he clung to his conviction that even if they were not noble now, they had been until corrupted by civilization. The sleeping baby symbolizing the beginning of a new life cycle in the right-hand panel of *D'où Venons-Nous?* is at peace with the world, innocent and guiltless, still unaware of good and evil. Gauguin's message seems to be that as long as it remains in its natural state, growing up in harmony with nature and leading an unspoiled, primitive life, the baby will retain its innocence. Gauguin had dreamed of this idyllic setting when he first headed for Tahiti, believing that he could *love, sing, and die* there, free of money worries, in ecstasy, and for art. The right-hand panel of *D'où Venons-Nous?* might represent the tropical Eden of Gauguin's imagination, and the answer to the first of his three questions: we begin our life cycle in a state of innocence, which we either retain or which we lose to corruption, depending on circumstances.

The loss of innocence and the corrupt stage of the life cycle are depicted by the central panel, Gauguin's answer to the second question of the tripartite title, *Que Sommes-Nous?, What Are We?* The loss, Gauguin's version of the Fall, is epitomized by the figure picking the fruit (perhaps an apple, but more likely a mango, since we are in Polynesia) from the Tree of Knowledge which, however, is not visible. Here again, Gauguin's rendition of the Fall departs significantly from the orthodox version. The fruit-plucking figure is often referred to as the "Tahitian Eve", but its gender is in dispute. Its long hair and oval face do indeed suggest femininity, but the chest is that of a man. An androgyne? Not likely, for two reasons. First, the figure is apparently based on a drawing attributed to the School of Rembrandt, which Gauguin copied in the Louvre and in that drawing, the figure is clearly a male. Second, some art historians claim that the folds of the figure's loincloth suggest a powerful erection underneath. The figure therefore cannot be a Tahitian Eve, it must be a Tahitian Adam in a state of sexual arousal! The suggestion that the perpetrator of the original sin was not a woman but a man could have ingratiated Gauguin to modern-day feminists were it not for the many signs of male chauvinism that pervade his life and works. The true implication of the figure is, perhaps, that both man and woman were responsible for the Fall. Modern critics, particularly those with a psychoanalytical bent, attribute great significance to the intimation of a phallus, hidden or not, real or imaginary, in the center of the painting. They argue that a phallus, as a symbol of sexual potency, a seat of pleasurable experience, and an emblem of male generative and creative power, rightfully deserves to be placed in the center of the stage on which a life cycle is being played out. In direct contraposition to Christianity, which strives to suppress sexuality, Gauguin sets it on a pedestal as the fount of creativity.

Gauguin also parts company with Christianity in his interpretation of the Tree of Knowledge. The word "knowledge", in the context of the Tree, should not, according to Gauguin, be understood in its archaic sense of one person knowing another sexually, but rather in its modern connotation as comprising sets of assertions about the nature of the universe, such as those that make up the subject matter of philosophy, science, and technology. The sets are obtained through inquiry by curious and inquisitive minds and the possession of knowledge confers the power to exploit nature and to exploit other people, as well as the capacity to ask questions about the meaning of our existence. The power of knowledge inevitably disrupts the harmony between nature and human beings; it leads to the enslavement of one group of people by another, and creates fertile conditions for the germination and growth of all evils associated with civilized societies. In short, knowledge corrupts – and this, according to Gauguin, is the true message of the metaphor of the Tree of Knowledge, the loss of innocence through knowledge.

An answer to the third question, *Où Allons-Nous? (Where Are We Going?)* is provided by the artist in the left-hand panel of the canvas. It is epitomized by the crouching figure of an old woman modeled on the pose of a Peruvian mummy that Gauguin studied in the Musée d'Ethnologie in Paris. He must have been quite impressed by it since women appeared subsequently in several of his paintings in similar poses, notably in *Misères humaines (Human Anguish)* and *Breton Eve*. In *D'où Venons-Nous?* the closed eyes of the woman, the way she supports her head with her hands, and her crouched posture, all indicate that she is nearing her end. The life cycle, which unfolded from right to left like a sentence in an Eastern text, comes to a close. The figure suggests more than that, however: its poise is that of a fetus in the womb. And this is apparently the message Gauguin wants to convey: one life comes to an end, but another begins. The sequence returns to the right-hand side to begin a new round of the life cycle. Where are we going? To a new beginning!

The newborn, the Tahitian Adam, and the crone are not the only figures in the painting that have a symbolic meaning; perhaps all the figures do. Gauguin only provided explanations for some of them, however, and even these are so vague that they are open to a variety of interpretations. The sculpture *raising her arms mysteriously and rhythmically to indicate the Beyond*, he explained, is not really a deity *but a woman turned into a statue, still remaining alive yet becoming an idol*. A similar idol appears in other paintings by Gauguin, such as *Te Pape Nave Nave (Delicious Waters*, 1898), *Merahi Metua no Tehamana (The Ancestors of Tehamana*, 1893), and *Mahana no Atua (Day of God*, 1894). In these, it represents Hina, the Tahitian goddess of the moon, a deity responsible for regeneration and rebirth (an allusion to the constant renewal of the moon). Her function in the left-hand panel of *D'où Venons-Nous?* may be to underscore life's cyclicity.

Of the strange white bird with a lizard in its claws at the crone's feet, Gauguin said that it represents *the futility of vain words*, but neglected to specify their nature. Is he referring to the hopelessness of seeking answers to questions posed by the painting's title? Or, perhaps, to the ruthlessness of the life cycle? We note also that a lizard symbolically represents the serpent before the Fall, and that in Polynesia, the nocturnal arboreal creatures were believed to be the spirits of the dead rustling in the trees. In the medieval church, the lizard stood for rebirth and resurrection and for hope of life beyond the grave.

For the presence of the dog, the peacock (if this really is the other bird), the cat, and the goat, Gauguin left no explanation. We note again that pure-bred, short-haired dogs were trusted companions of Polynesian navigators on their epic voyages to discover new lands. In this work the breed, which became extinct in Gauguin's time, may symbolize the lost, pre-European order. The peacock in Christian iconography was a symbol of the Resurrection, possibly because after molting it regains its feathery splendor again. A cat was traditionally associated with the moon, which waxes and wanes and disappears from the sky. More often, however, cats were regarded as the embodiment of evil, often representing the devil himself. As for the goat, in St. Matthew's Gospel (Chapter 25), Jesus likens the righteous to sheep and the wicked to goats condemned to burn in the eternal fire.

Some Gauguin scholars have attempted to impose another layer of symbolism on his masterpiece, one related to his own life history. They suggest that the crone is a reminder of his Peruvian ancestry on his mother's side; that the two figures seen in intimate conversation walking in the background are Gauguin and his daughter Aline; and that the little girl seen eating a fruit at Adam's feet in the foreground symbolizes his daughter by Pau'ura, his *vahine* in Tahiti (the child died a few days after birth). In a letter to a friend, Gauguin described the *two sinister figures, shrouded in garments of somber color, conversing intimately near the tree of knowledge about the anguish of knowledge*. The ghostly, shadowy appearance of the figures may indicate that Aline is portrayed after death and that Gauguin is at death's door, resigned to killing himself after completion of the painting. They walk, or rather glide by, from left to right and so symbolically, in the context of the canvas, from Death to Birth. Before them *an enormous crouching figure, out of all proportion, and intentionally so, raises its arms and stares in astonishment upon these two, who dare to think of their destiny.* Has Gauguin, as some art historians suggest, stylized himself into a wiser but sadder Westerner, contemplating the unspoiled primitive life made beautiful by his own creative fantasy?

Whether any of the remaining figures have a symbolic meaning or have merely been placed in the painting for esthetic reasons, we do not know. For example, the two women sitting deep in thought next to the sleeping child on the right have been suggested to be meditating on the meaning of life, and at the same time to symbolize ancestor-descendant relationships in the life cycle. They could, however, have been placed there by Gauguin to watch over the baby: it would have been strange, indeed, to portray a baby asleep on a river bank without any supervision.

Thus, the broad message of Gauguin's painting appears to be that we participate in the circuit of life, each of us representing one turn in the perpetually rotating life cycle. As individuals, we enter the circle from a void at one point and exit it at another. We are like the sparrow in the haunting metaphor of the eighth century English Benedictine monk and scholar, The Venerable Bede, entering, for a short moment, a well-lit hall before disappearing again into the cold darkness of the night. We enter the life cycle in a state of blissful ignorance, but as we progress, we are gradually corrupted by knowledge, until – worn out like the crone – we are ready to face the blankness of death, knowing that our tenure in the circle of life was part of an eternal renewal.

This is a fairly modern message; as biologists, we might only want to strip it of its metaphysical garb. But as eloquent as it is visually, the painting does not *really* answer any of the questions it poses. Nor was it intended to. Gauguin himself pointed out that *for our modern minds, the problem of Where do we come from? What are we? Where are we going? has been greatly clarified by the torch of reason alone. Let the fable and the legend continue as they are, of utmost beauty …; they have nothing to do with scientific reasoning.* Gauguin's intent was to expose us to the beauty of the fable and to the mystery of the riddle, and allow us to respond emotionally to them. This he achieved magnificently. If we want more than vague, deliberately mystifying hints, we have to follow his advice and turn to science for answers. Before doing so, however, we might want to briefly consider the relationship between fable and scientific reasoning and explain why we prefer the latter to the former as a source of answers.

Mythos and Logos

In exploring the nature of the world and the essence of our existence, fable is the oldest source of information. It must have appeared along with the first flicker of cognition in the human mind and the articulation of the first sentences, for curiosity about the world is paramount to humanity. Curiosity creates thirst for answers and the most facile way of providing answers is by fabrication and fabling. A bit of observation combined with a large dose of fantasy can be interwoven into the rich tapestry of a fable by a skillful and imaginative narrator. When humankind was young, ignorant, and gullible, it found such answers satisfying. In this regard it was like an inquisitive child who is told that babies are delivered by Federal Express in the form of a stork. Such an answer cannot satisfy a child for long, however, for the inquirer quickly becomes aware of its absurdity. A featherweight of a bird carrying a six-pound baby? Even if that were possible, where does the baby come from? Who made it? Sooner or later the child is provided with a more sensible answer that squares with its experience and is in accordance with the knowledge already assimilated from other sources. Analogously, at an early age, humankind contented itself with stories that were appealingly simple and poetic but also unsophisticated and naïve. The stories were a blend of natural and supernatural, familiar and unfathomable. Depending on their birth place, people were told that the first human beings were fashioned by gods from the material at hand – clay on the banks of the rivers Eufrat and Tigris, grass on Mindanao, rush on Luzon, a tree-trunk on Borneo, bamboo on the Andaman Islands, bird's eggs on Fiji, and a decaying worm on Samoa. The gods brought the figurines to life either by breathing into their nostrils, incantation, tickling, dancing, placing ginger into their ears, or spitting. As charming as these explanations were, they were rife with inconsistencies, contrary to reason, and incongruent with experience. As humanity came of age, the more inquisitive minds began to search for an explanation that steered clear of these shortcomings and was both rational and consistent with the entire body of accumulated knowledge.

In no other place was the quest for rational explanations nurtured more than in classical Greece, where a clear distinction between *Mythos*, a fable, and *Logos*, a more rational, more objective mode of discourse, was first made. Ancient Greeks applied *logos* to many areas of inquiry, but – strangely enough – not to the question of human origin. In this area they stuck to their numerous mythological explanations, the best known of which claims that one of the Titans (Prometheus, a potter) shaped the first humans out of clay and that the goddess Athena gave the figures life. (The Titans were a race of giant deities who were overthrown to be succeeded by the Olympians.) Prometheus continued to champion the human race, going as far as to steal fire from the gods and bestow it as a gift on humanity. The punishment that Zeus meted out to him for this and other offenses is well known.

Two main reasons for the failure of the ancient Greeks to initiate a more rational inquiry into human origins come to mind. One is their attitude to empirically acquired knowledge. Milesian and Athenian intellectuals seem to have been convinced that they could solve all the world's riddles by appealing to reason alone, without having to resort to observation, not to mention experimentation. They were not at all eager to roam the Peloponnesian mountains and valleys in search of a new variety of plant or beetle; to dissect worms, frogs, or – Zeus forbid! – human cadavers; or to spend long hours watching the mating ritual of a spider. No, they preferred to sit comfortably around a banquet table at a *symposion* or stroll leisurely in the shade of olive trees in a grove, engaged in a mental ping-pong game of arguments, counterarguments, and syntheses – may the best thought win! This outlook worked well as far as math, geometry, and philosophy were concerned, but it certainly had limited value in what would later become known as natural philosophy, and later still natural sciences. (Curiously, the idea that problems of origin can be solved by abstract reasoning has recently been resurrected, only this time the task of reasoning has been delegated to computers.) It definitely failed to provide any reasonable answers to the question of human origin. Only in the philosophy of Anaximander (611-547 BC), a pre-Socratic thinker best known for his map of the world, did it assume a semblance of rationality. Like others before him, Anaximander set out from the assumption that the world is made of a single substance, not to be confused with any known material but one which was responsible for properties such as hot and cold. At the beginning of time, according to Anaximander, opposing properties in the "indefinite" stuff segregated from each other, hot moving outward and so separating from cold, followed by dry parting from wet. The segregation ultimately led to the formation of the earth, the sun, and all the other celestial bodies. Life on earth sprung from the primeval mud heated by the sun, and then spread onto land in the form of sea urchin-like creatures. Humans, he thought, hatched inside fish-like beasts. By barring the supernatural from his philosophy and by presenting an internally coherent and consistent explanation, Anaximander took a big step away from *mythos* and toward *logos*. Nonetheless, his explanation is little more than an imaginative, ingenuous, and in a certain sense prescient speculation, lacking any support from empirical knowledge.

The second reason why the ancient Greeks, despite their attempts to separate *logos* from *mythos*, contributed so little to expounding the origin of humans was religion. Answers to questions of origin are usually incorporated into religion, and are invariably drawn from existing myths. In many religions, notably Christianity, Judaism, and Islam, the espoused answers have become part of the dogma and are thus inaccessible to rational inquiry. Any disregard is treated as a sin to be penalized by punishment. Even in ancient Greece, blessed by a relatively open-minded society (measured by the standards of its time), tampering with concepts believed to belong to the proprietary sphere of a religion was punishable by death, as Socrates was to learn some 200 years after Anaximander. It is therefore hardly surprising that most philosophers steered well clear of dabbling into human origin. As the influence of Christianity – which by its history of persecution must be judged as one of the most intolerant of all religions – slowly spread, so the reluctance of thinkers to deal with issues of origin increased. It is no coincidence that the rational study of human origin only began in earnest in the nineteenth century, when Christianity's grip on the Western world gradually relaxed and the influence of religion waned.

The *logos*-driven or *scientific* study of human origin draws on two principal sources of information, one based on comparisons of living beings and the other on comparisons of the remains of dead organisms. In both cases, the comparisons can be made among humans themselves or between humans and animals. The comparison among living forms may involve groups of organisms or individual organisms – their bodies, organs, tissues, cells, and molecules; and it may concern behavior, function, and structure. The comparison based on dead organisms is predominantly restricted to bones, but under special circumstances it may involve other parts as well, including molecules; it also frequently involves artifacts created by humans. Of all these sources, this book focuses on one only – the molecules.

Of the various scientific disciplines concerned with human origin, that based on the study of molecules is the youngest. Although initial attempts to reveal relationships among organisms by comparing their molecules were made one century ago, the discipline of molecular evolutionary biology first began to flourish in the early 1960s and its greatest advances have only been achieved in the last ten years. Even greater advances are expected to follow, since the potential of the discipline is enormous. Nevertheless, the information revealed by molecules has already changed current views of human past so radically that it should be of interest to anybody who has ever asked: Where do we come from? What are we? Where are we going?

A Bit of Semantics

Of these three questions, the first is truly basic and when it has been answered, facts relevant for the other two may well be revealed. This book is therefore devoted largely to the first question, although implications concerning human identity are also drawn based on the material presented; a contemplation of humankind's future constitutes the book's last chapter. The focus on the first of the three questions is also dictated by the fact that the other two questions extend into the realm of philosophy in which we do not feel comfortable.

Language can easily become a source of maddening confusion. The history of ideas is replete with examples of endless controversies which could have been avoided had the parties defined their terms and usage beforehand. To nip misunderstandings in the bud, we explain keywords and terms as they first appear in the text and a good place to start with is the book's title. To explain its meaning, let us dissect it.

First, who is "we"? Not we, the authors of this book, nor our families, clans, or nations, but we, all the people alive today, all six billion of us, as well as all those who have ever lived. In short, we the human species, *Homo sapiens*. The species' scientific name is hard to translate into English. In many languages, there is one designation for the male sex, another for the female sex, and a general term to encompass all males and all females without further distinction. In Japanese, for example, *otoko* or *dan-shi* is the human male, *on-na* or *josei*, the female and *hito*, the two together. In Czech, the corresponding designations are *muž*, *žena*, and *člověk*; in German, *der Mann*, *das Weib*, and *der Mensch*. In English, on the other hand, the word *man* was, until recently, used for the species. A current tendency is to refer to the species as *human*, which is not the best choice, first, because the word was meant to be an adjective and second, because it still has a chauvinistic ring to it. Since, however, we know of no better alternative, we use it throughout this work. The context distinguishes when the word is used as an adjective and when as a noun. By *species*, we are referring to a group of interbreeding (or at least potentially interbreeding) individuals. The totality of all human beings, which used to be referred to as *mankind*, is described as *humankind*, *humanity*, or *human race*.

Next, what does "where" mean? It could refer to either a place or source; here it signifies both. We are seeking the place in which the human species originated and the source from which it originated.

Finally "come from", in the present context, signifies origin. It implies that there was once a period in time when no humans existed anywhere in the world, and a time when they first appeared. In short, we want to know the circumstances, the manner, and the time in which the human species arose.

Hikers who choose to wander through the countryside along well-marked trails are assisted by orientation boards that, with the aid of a simplified map of the area, show them how they reached their present position and provide them with information on the next section of their itinerary. In this book, we end each chapter with the equivalent of an orientation board so that you can get the lay of the land. It will summarize the chapter, emphasize its main messages, and introduce you to the succeeding chapter.

You have now arrived at the first of these information boards. On your way here, you should have had a leisurely walk with no rough terrain to negotiate and no high altitudes to overcome. All this will come later. The promenade introduced you to the connotations of the question "Where do we come from?" by means of the artistic masterpiece that immortalized it. The description of the masterpiece and its possible interpretation served to expose you to different approaches for dealing with the question. Several trails were open to you, but you have been directed onto one in particular, the trail of science, and specifically the scientific path that extracts information about the origin of living forms by studying their molecules. The trail will become more and more demanding as it leads you into the jungle of discoveries, but you will be rewarded by breathtaking vistas and fascinating insights. But the difficult sections are still far off in the distance and for now, the next chapter promises a comfortable continuation of the hike as it introduces the molecules in which the information about our origin is stored.

CHAPTER 2

Bridging the Generation Gap

The Physical Basis of Ancestor-Descendant Relationship

Dear
Diana,
Hi! and a
white note probably
unlike any you've *evuh*
eyed before. Doodling intertia,
goodbye. This took considerable
time, but it's worth it, for say, a banana.
Yep! a ripe munchy monkey's fruit & a smooch intricately
entwined around my lips would be absolute nirvana.
How's psychology in provincial Philadelphia?
Lie down, yawn to stir up alpha waves
and use the right brain's retina
to visualize a ballerina
lithely *en pointe*
among these lines,
staccato & undying,
amid the bells & fierce fire
of a tintinnabulating typewriter
-- sparks & syncopation, rhythm & rhyme --
The parable of the Phoenix! Everything dies & rises
again, the wanderers strain through long night's
passage, weary as a flicker of light (flame!)
on water, but together they fight (flame!)
against the soft slow silent
inevitable steady slide
asleep til sunrise &
petals & ignite
alive! On
arrival in
Orlando,
Florida, untanned
& disordered, I meandered
through the airport amid thousands
of visitors: some rushing, some stranded
for simply hours with no apparent explanation,
short people with lean, lanky, vociferous demands,
torn sneakers, fur coats, t-shirts, Panama hats & bandanas,
quarreling parents, caged pets, & carefree children hugging panda
or pooh bears, or hidden beneath Micky Mouse hats from the fantasy
world of Disney, smiles & tears, the wheezy propaganda
of a preacher denouncing vice, smelly sandals,
snoring fat men & gabby wives, and I
saw that the colorful banter of
foreign accents seems bland &
boring when a chewy pecan
corn pone lazily expands,
"Shor is grand!"
Oranges,

Andrew Weiman: *Andy-Diana DNA Letter**

Bloodlines

Ever since you stopped believing in the stork, you have known that one human being comes from another: a child from its mother. Anybody who has experienced or has been present at the birth of a child must be aware of this simple fact of life. Less obvious has been the role of the male, the father, in procreation. It has always been known, of course, that a father is necessary to produce offspring. Immaculate conception is acceptable in myths, but in real life, a daughter would have a hard time convincing her parents that she became pregnant through a miracle, nor would a husband returning home after twelve months absence accept a wondrous event as an explanation for the newborn in his wife's arms. The father's contribution to procreation was shrouded in mystery for a long time, as was the nature of procreation. It was a not-too-well kept secret that sexual intercourse was somehow involved, but the nature of this involvement was the subject of the wildest speculations.

The problem of reproduction is not as trivial as it may seem to those of us who were enlightened on this subject in high-school biology classes. The key issue in the puzzle is this: the parents represent one *generation* of individuals and their offspring another. In both generations, all the individuals are mortal; each individual has a beginning and an end. Where the end of an individual lies is obvious, but where is its beginning? Not in the moment of birth, for the nine-month pregnancy must also be considered. The beginning of a human being must be dated back to the moment of conception. But if so, what is the essence of conception? We now know of course the answer. At conception, a sperm cell from the male fuses with the egg cell of the female and this act of *fertilization* initiates new life arising from the fertilized egg.

This answer, however, only transposes our question to another level. A sperm cell is a separate entity which can exist, if only for a few minutes, independently of the body that produced it. It therefore represents a physical discontinuity, a gap, between succeeding generations. How does life bridge this gap? The descendants of one generation resemble the ancestors of the preceding generation quite closely (not only are the descendants human beings, like the ancestors, but they commonly look more like their own parents than any other humans). Resemblance implies some form of physical continuity from one generation to the next. How, then, is such continuity achieved in face of the obvious discontinuity associated with the beginning of a new life?

In a certain sense, the speculations proffered before the true nature of conception became known explained the generation gap better than the facts that were discovered later. For example, one widespread belief maintained that life's formative force resided in the blood. This seemed logical since blood permeates the whole body; severe blood loss leads to death; blood is fluid and mobile, and mobility is one of life's characteristics; and blood's cauldron, the heart, is life's prime mover. Is it not reasonable to conclude that blood not only kept the body alive, but also shaped it during development and was responsible for its characteristics? It also seemed logical to assume that blood provided the continuity between generations. All that was required to explain the generation gap was a small-scale blood transfusion during sexual intercourse. A candidate event for the transfusion was male ejaculation at the peak of sexual arousal. There was only one minor problem with this explanation: semen discharged during ejaculation is white, whereas blood, as everybody knows, is red. Aristotle, the Greek

---

* The beginning of the poem Andy-Diana DNA Letter which the American poet Andrew Weiman wrote while a student in clinical psychology at New York University in 1980. (Nims 1981.) Superimposed on it is the DNA double helix with the two strands twisted around each other.

philosopher who seemed to know the answer to every question, however, solved this incongruity by suggesting that semen is the foam on churning blood.

A tack-on to this proposition, certainly appreciated by the all-male community of philosophers, was the deduction that since semen comes from the male, male blood is the decisive factor. A female merely provides a receptacle for the development of new life. The club of male philosophers cherished the idea that males provided the seeds of life exclusively. (Although, to be fair, some of its members did agree to the participation of the mother's blood in the process.)

Blood was thus thought to provide the physical link between generations. The belief became so entrenched in people's minds that to this day it is echoed in many Indo-European languages. In English, words and expressions such as "bloodline", "blood relation", a "blood horse" (thoroughbred), a "prince of true blood" (one of a royal family), and "it runs in the blood" (i.e., it is inherited as a family character) continue to remind us of a past belief that at one time seemed to make sense. It seemed to be rational and logical, but ultimately proved to be totally false.

The rise of experimental biology in the seventeenth, and particularly eighteenth, century put an end to these male fantasies. The first serious blow to the image of blood flowing from generation to generation like an underground river, now disappearing, now resurfacing again, was dealt by the incomparable amateur scientist, a Dutchman from the house at "Lion's corner" in Deft, Antony van Leeuwenhoek (1632-1723). When not vending buttons and ribbons, hats and neckties, gloves and handkerchiefs in his draper's shop or, later, taking stock of the city's finances, Leeuwenhoek ground lenses to such perfection that they could magnify objects almost 300 times – at the time an achievement unthought-of. He trained the lenses on whatever tickled his fancy – raindrops, pickings from the cavities in his teeth, body parts from insects, plant leaves and blossoms, rabbit ears, and the webbed feet of frogs – and he saw things nobody had ever seen before him. And what he saw, he described both accurately and simply and sent the descriptions to the Royal Society in London. Members of the Society translated and published the reports in the Society's *Philosophical Transactions*.

In one of his letters, written in 1677, Leeuwenhoek described how he had heard from a certain Jan Ham, presumably a medical student at Arnheim (his identity, including his first name and the spelling of his last name, are uncertain), of the presence of *vermiculi* in the semen of a patient suffering from gonorrhea. Leeuwenhoek also found similar "little worms" in the semen of healthy men, as well as dogs and other animals, and so came to the conclusion that the *vermiculi* were not the products of semen decay, as claimed by Ham, but normal constituents of the ejaculate.

The implication of the Ham-Leeuwenhoek discovery was profound. Since the vermiculi, which really did look wormlike with their little tails, were very different from the globular corpuscles Leeuwenhoek had discovered in the blood three years earlier, semen could not be derived from blood. Thus the bloodline hypothesis, cherished for centuries, if not millennia, had to be abandoned. And if semen did not originate from blood, then semen and not blood bridged the gap between generations. Although not everybody was immediately convinced by this argument and the bloodline hypothesis persisted in various modifications for a while, ultimately it was written off entirely. Two questions then arose: Which component of the semen formed the bridge between generations? And where and how did semen originate in the body? Let us take the first question first.

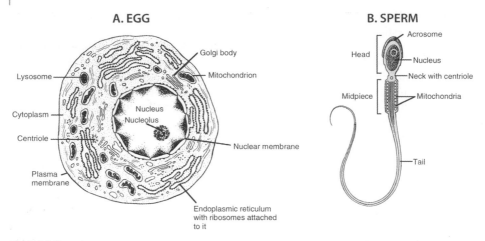

A. EGG

Golgi body
Lysosome
Mitochondrion
Cytoplasm
Nucleus
Nucleolus
Centriole
Nuclear membrane
Plasma membrane
Endoplasmic reticulum with ribosomes attached to it

B. SPERM

Acrosome
Head
Nucleus
Neck with centriole
Midpiece
Mitochondria
Tail

Fig. 2.1 A,B
The bridge between generations. In its structure and components, the human egg (**A**) resembles other cells in the body. The sperm (**B**), on the other hand, is a highly modified cell. The functions of the components shown are as follows. Acrosome: release of egg-penetrating enzymes; centriole: cell division; endoplasmic reticulum: protein synthesis; Golgi body: concentration and secretion of large molecules; lysosome: digestion of large molecules; mitochondrion: energy production; nucleus: storage of hereditary material; nucleolus: synthesis of RNA. Note that in reality the egg cell is much larger than the sperm. (**A** is from Bodmer & Cavalli-Sforza 1976; **B** is from Tortore 1986; both are modified.)

Vapors, Little Worms, and Eggs

Following the Ham-Leeuwenhoek discovery, two principal components of semen were recognized: the vermiculi, which were subsequently renamed *sperm* (from Greek *sperma*, seed, germ; Fig. 2.1) or *spermatozoa* (Greek *zoion*, animal), and the fluid in which the spermatozoa swam. In the seventeenth century, however, a number of scholars argued that a third component should be considered – the vapors rising from the seminal fluid!

It has always been known that a chicken emerges from an egg, but only if the hen has been inseminated by a cock. One has also always been aware of the fact that frogs, fish, and other animals lay similar egg-like contraptions which, after their exposure to semen, develop into animals. Leeuwenhoek himself observed and described how an egg turns into a larva, then into a pupa, and finally into an ant. Most animals would therefore seem to begin their life cycles as eggs fertilized by sperm. Were mammals, including humans, an exception to this rule, or were their eggs hidden inside the females' bodies? Many scientists suspected the latter.

Ultimately, the mammalian egg was indeed discovered, but relatively late because it turned out to be rather small (only about 100 to 150 micrometers in diameter; Fig. 2.1). The human egg was first observed in 1827 by the Lithuanian-born embryologist Karl Ernst von Baer. In the meantime, scientists had realized that animal and plant bodies consisted of cells and that the egg and the spermatozoon were in fact two specialized cells. (A fertilized hen egg, however, is composed of many cells because by the time it is laid, the original fertilized single cell has already developed into a multicellular embryo surrounded by the yolk, the albumen, the membranes, and the shell.) Things thus began to fall into place. What every pupil now learns in high school biology classes was 200 years ago gradually becoming the knowledge of a privileged few, the enlightened experimentalists: the egg must be fertilized by the semen to initiate the development of an individual. But what component of the semen was the actual agent of fertilization?

A group of natural philosophers, their minds still weighed down by Aristotelian metaphysical baggage, maintained that neither the seminal fluid, nor spermatozoa, but an *aura seminalis*, vapors emanating from the semen, was responsible for the fertilization of the egg. This idea may now seem ridiculous because we no longer think in terms of nonmaterial vital forces, souls of inanimate matter, or spiritual form-giving principles, but in the seventeenth century, nothing was thought unusual in vapors initiating new life. A simple experiment by the Italian microscopist Lazzaro Spallanzani, however, disposed of this notion. Spallanzani placed a drop of frog seminal fluid onto the bottom of a glass chamber and then covered it with a lid which had unfertilized frog eggs attached to its inner surface. After five hours of incubation, moisture condensing on the eggs indicated that vapors had risen from the semen, yet no fertilization took place. Ironically, Spallanzani believed that it was the fluid and not the particulate part of the semen that was responsible for fertilization. He was not corrected until 1875 when two German zoologists, Oscar Hertwig and Hermann Fol actually observed the penetration of a sea-urchin egg by a spermatozoon under a microscope, followed by the development of the fertilized egg.

The Homunculus Hypothesis

The bridge between successive generations is therefore formed by two cells, the sperm and the egg. In humans, one of these cells (the sperm) leaves the body in which it was generated, but in many animals (e.g., frogs and fish), both cells leave the bodies of their origin and fertilization takes place externally. (In certain clinics, however, the human egg, too, can be removed from the woman's body and fertilized *in vitro*.) Since the two elements that link successive generations are discrete entities, the bridge metaphor is not really appropriate. Instead of a bridge, it would be better to speak of a ferry that transports its load across a stretch of water – here the gap between two generations. What, however, is the load on the ferry? What is the physical nature of the entity responsible for continuity between generations?

Here we must introduce a term that will be used repeatedly in this book. The Greek verb *phaínesthai*, to appear, is the root of the English word *phenomenon*, one meaning of which, according to Webster's Collegiate Dictionary, is something "that is apparent to the senses and that can be scientifically described or appraised". The scientific term derived from *phaínesthai* is *phenotype*; it signifies the visible characters distinguishing one organism from the others (Fig. 2.2). It pairs with another term, *genotype*, which denotes the entity responsible for the appearance of the visible characters and to which we return later. In the present context, the word *character* is used to signify "one of the attributes or features that make up and distinguish an individual" (Webster). It will be in this sense that the word will be used throughout the book. The question raised by the new metaphor can now be rephrased: do the two ferryboats that cross the generation gap, the sperm and the egg cells, transport the phenotype or the genotype?

In the age of Leeuwenhoek and Spallanzani, many natural philosophers believed that either the sperm or the egg contained a preformed, miniature human body, the *homunculus* ("little human"; Fig. 2.3). The philosophers were *preformists* – either *spermists* or *ovists* (the Latin term for an egg being *ovum*). When he peered through his lenses (microscopes) at a drop of semen on a pin's tip for long enough, even Leeuwenhoek thought that he could recognize the contour of a human body in each vermiculus – a disproportionally large

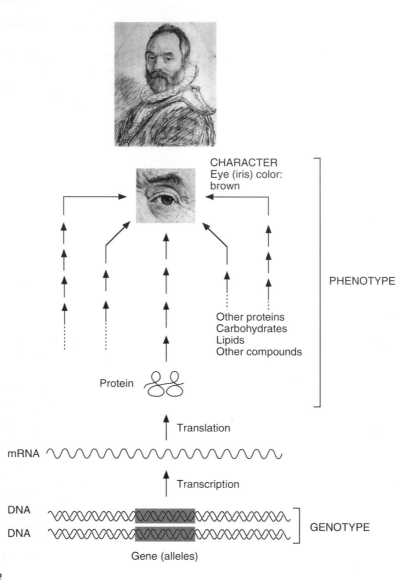

Fig. 2.2

Character, phenotype, genotype. A *character* is any specific, visible, and heritable attribute possessed by an individual which enables it to be distinguished from others. Here, the character is eye color: this man's irises contain small spots of brown pigment, while those of other persons may contain blue or gray pigments. The alternative forms of the character (brown, blue, or gray eye color) constitute *character states*. The form taken by the particular character (or group of characters) is the *phenotype*. The pigment grains responsible for eye color are the products of complex, largely unelucidated biochemical pathways involving intermediate compounds of many kinds. If any of these compounds can be used to differentiate individuals, it, too, becomes a character and its form a phenotype. The characters and the phenotypes are an outward manifestation of a specific *genotype*, a specific form assumed by the factors of heredity, which later in the chapter are identified as genes and their alleles borne by DNA molecules. The brown eyes are those of the Flemish sculptor Jean de Boulogne, who was known in Tuscany, where he worked for most of his life, as Giambologne. He is portrayed here by the Dutch painter Hendrick Goltzius. The painting is from the year 1591 and can be viewed in the Teylers Museum, Haarlem, The Netherlands.

Fig. 2.3
Preformist (spermist) hypothesis of bridging the generation gap: homunculus in a sperm. Each sperm of the homunculus contains an even smaller homunculus, whose sperm cells contain homunculi that are smaller still, and so on *ad infinitum* and *ad absurdum*.

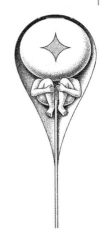

head followed by a much smaller, squatting torso, complete with arms and legs. He was not to be the first, nor the last scientist to see what he was looking for, but what was not there. He did not observe an object, he imagined it. The same thing happened to other microscopists of his time; they, too, fell victim to optical illusions.

There are, of course, no homunculi in either the sperm or the eggs. Cross-sections through the two elements, when observed with modern electron microscopes, reveal them to be modified single cells containing cytoplasm, nucleus, mitochondria, and some of the other structures present in a cell (Fig. 2.1). In the case of the sperm, the illusion of a miniature human body is created by the particular modification of the cell's structure. Here, the cytoplasm is reduced to a thin rim surrounding the nucleus which forms the sperm's head, while the mitochondria and other subcellular structures are packed into the midpiece attached to the head via a narrow neck. At the other end, the midpiece continues into a long, whiplike tail whose contractile fibers are responsible for sperm motility. Leeuwenhoek and his contemporaries had, of course, no inkling of these details; they saw a silhouette that reminded them of human bodies, either male or female. The ovists lacked even this illusion, for the first sighting of a human egg was made, as already mentioned, in 1827. The preformists concluded that the intergenerational ferryboats carried no lesser load than entire humanity, all the people that have ever lived and will ever be born, plus all those that never were and never will be born (man's ejaculate contains more than 100 million sperm, of which usually only one, if any, fertilizes the egg; the rest perish and with them, according to the spermists, also all the homunculi contained within). The homunculus hypothesis assumes that the human sperm (or egg) is organized like a set of boxes, one inside the other. Some of the sperm contain male and other female bodies. Depending on which of these arrives first at a specific site in the woman's body during the sexual act, either a boy or a girl begins to develop, the development consisting largely of growth in size. Combined with the Judeo-Christian creation myth, the homunculus hypothesis implies that Adam or Eve were literally the bearers of the entire humankind.

It is easy to ridicule such a naïve concept and point out its absurdity. But before shaking our heads in disbelief, let us remind ourselves that 300 years ago, people could not draw on the knowledge that we have today; indeed at that time the entire intellectual atmosphere was very different from ours. Let us also remind ourselves that according to a recent survey, some 60% of the population of the United States of America still believes in the existence of guardian angels. These people, unlike our ancestors 300 years ago, certainly *should* realize the absurdity of such superstition.

The preformist explanation of development is, in principle, untenable, if for no other reason than that it requires more bodies to be stored in a single sperm or egg than there are molecules. The direct transmission of a phenotype from one generation to the next can therefore be ruled out; only the genotype responsible for the phenotype comes into question. The development of an individual is not preformistic but *epigenetic*: the specific features of a body always arise anew from an undifferentiated state. The developing body is constructed from molecules just as a house is built from bricks. Although no preformed bodies are present in the sperm or the egg, these two cells – the *gametes* (from Greek *gamein*, to marry) or *germ cells* (because new life germinates from them like a plant rising from a seed) – must possess all that is necessary to specify the bodies in all their intricacies and complexities. So where do the germ cells come from?

The Germ Line

To start a new life and a new generation, the sperm must find the egg and the two gametes must fuse to form a single cell, the fertilized egg or *zygote*. The latter term derives from the Greek *zygon*, which means a yoke, the wooden bar that joins two draft oxen at their heads for working together. In the zygote, too, the two germ cells have been similarly yoked together to function as a single cell. Not only that, and in contrast to the oxen, which remain two separate individuals, the zygote *is* a single cell. Fusion is an unusual thing for cells to do; normally they guard their integrity and protect themselves against its loss by a variety of measures. Fertilization must therefore involve events that overcome these measures. The opposite of fusion is *cell division*, a property most other cells, and especially the zygotes (but not the mature germ cells), possess aplenty.

A mature human body that ultimately arises from the zygote consists of 10^{13} cells. The process that transforms the undifferentiated zygote into an adult body differentiated into many different cells, tissues, and organs, each specialized to a certain function, is referred to as *development*. In humans, development lasts nine months within the mother's body (*embryonic development*) and several years outside of it. It culminates in the formation of germ cells and the ability to *reproduce* sexually, to start a new life. Following the period in which there is a potential for sexual reproduction, the body ages and dies. One *life cycle* has been completed and a new cycle has begun. What Gauguin depicted symbolically and expressively in his painting, biologists present in a diagram:

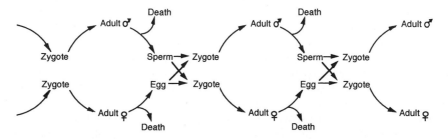

One turn of the cycle is one *generation*, whether it is measured from one zygote to the next, from one adult parent to the adult offspring, or from anywhere else in the cycle to the corresponding point of the succeeding cycle. The three principal components of each cycle are the body (Greek *soma*), the germ cells (gametes), and the fertilized egg (zygote). The gametes form the bridge (the ferryboats) between two successive cycles. They are the only physical links between the generations since all the other 10^{14} cells of the body – the *somatic cells* – perish in each cycle (as do all the germ cells that do not participate in fertilization). All that an individual contributes to the next generation is borne by a single cell, either the sperm or the egg.

As the zygote develops into an adult body, a separation into two principal components takes place: the somatic cells, which do not contribute to the next generation, and the germ cells, which do have such potential. The former cells are mortal; the latter are potentially immortal in the sense that they appear to transmit something from one cycle to the next. But when does the *germ line* – the sequence of cells committed to becoming gametes – separate from the somatic line?

Right at the beginning of development, argued the German zoologist August F. L. Weismann. A superb empiricist, Weismann preferred to spend his time in the laboratory, experimenting and observing. When his eyesight began to fail him, however, and he could no longer peer through the microscope, he turned to organizing the accumulated knowledge into broad concepts, filling the gaps with speculations. One of the concepts, the one for which he is best remembered, was the *germ plasm theory*. Weismann first formulated it in 1883, but during the rest of his life he kept modifying it in the light of new discoveries. In the original version of the theory, Weismann envisioned the cytoplasm (protoplasm) of the zygote partitioned into *germ plasm* (*Keimplasma* in German), destined to give rise to the germ cells, and *somatoplasm,* which is passed onto the somatic cells. As the germ plasm does not mix with the somatoplasm, it can be imagined, according to Weismann, as flowing through the generations like a continuous, uninterrupted stream; in this sense, the germ plasm is immortal. The somatoplasm, by contrast, perishes in each generation with the death of the body after the completion of a life cycle and in each new generation, it is formed anew from the germ plasm.

In some animals, the fruit fly or the clawed frog, for example, a specialized region destined to be included in the germ cells can indeed be recognized in the mature egg (Fig. 2.4), the rest of the cytoplasm being passed onto the somatic cells. In other animals, such as the mouse, and apparently also in humans, no such regions can be identified in the mature egg. The first few divisions of the fertilized egg result in cells which can give rise to the entire body, both somatic and germ cells. Only later, but still relatively early in development (the precise time may differ according to the animal species), does a group of cells – the *primordial germ cells* – segregate from the somatic cells.

Once the primordial germ cells have separated from the rest of the developing embryo, regardless of when this event occurs, they no longer participate in the development of the organism. From this point on, they give rise only to germ cells, but not to somatic cells. Nevertheless, they retain the *potential* to generate the entire body, both the somatic and the germ line parts. This potential is suppressed during development; it is expressed only after the primordial germ cells have differentiated into mature germ cells and the activating stimulus has been delivered during fertilization. During normal development, none of the somatic cells can give rise to germ cells (but see below for a qualification of this statement); they can, however, differentiate into various *types* of somatic cells. In the course of development, their ability to differentiate is gradually restricted, until ultimately, a given somatic cell is committed to differentiation into one type only – an erythrocyte, a melanocyte (pigment cell), or a neuron, for example. In the mouse, and presumably in humans as well, the earliest cells resulting from the first few divisions of the zygote cannot be classified as germ-line or somatic-line cells. They are *totipotent* (pluripotent) because each of them can produce the entire body,

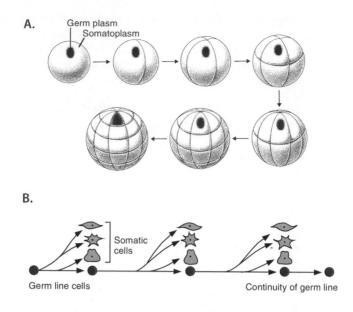

Fig. 2.4A,B
The germ line concept. **A.** In a frog's egg, germ plasm is separated from somatoplasm. As the fertilized egg divides (cleaves), the germ plasm ends up in a group of cells that develop into the germ line and ultimately into mature gametes. **B.** Weissman's germ plasm theory. Germ line cells can give rise to somatic cells, but not vice versa. Any hereditary changes that occur in the somatic cells remain confined to these and are not passed on to succeeding generations. Only changes that occur in germ line cells can, potentially, be transmitted into the next generation. (**B** from Wilson 1896.)

both germ and somatic cells. (Incidentally, in animals the primordial germ cells do not arise within the forming sex organs; they migrate into the testes or the ovaries, which develop from somatic cells, at a later stage. Each primordial germ cell has the potential of differentiating into either sperm or egg cells, depending on the embryonic environment in which it matures.)

Weismann's original theory, therefore, applies to some animals only. Weismann himself felt compelled to modify it radically when, toward the end of the nineteenth century, indications that the phenotype is controlled by factors residing in the cell nucleus accumulated. Specifically, the newly discovered thread- or rod-like structures, the *chromosomes* (literally "color bodies"), made visible by the application of special stains to dividing cells, became attractive candidates for being the bearers of the germ plasm (Fig. 2.5). To bring his theory in line with the new findings, Weismann moved the germ plasm from the cytoplasm to the nucleus and proposed that it might be identical with the material constituting chromosomes. In this, he was right in principle, for the essential component of chromosomes does indeed come close to the concept of germ plasm. The two are not one and the same thing, however.

Weismann believed that as somatic cells gradually specialize to the performance of different functions, they lose chromosomal material, ultimately retaining only that which is necessary for the execution of their particular duties. Primordial germ cells, on the other hand, are spared such losses, and for this reason remain totipotential. Hence the essential difference between somatic and germ cells, according to Weismann, is that the former have lost different amounts of chromosomal material and with it the ability to specify some of the phenotypic characters, whereas the latter, having suffered no such loss, are able to specify the complete phenotype.

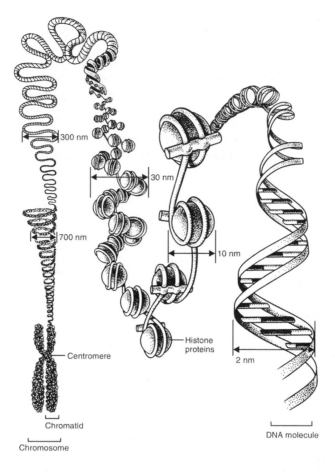

Fig. 2.5
The teleporter. The nucleus of a human somatic cell contains very long DNA molecules wrapped around spherical proteins (histones) to look like beads on a string; they are wound into a hierarchy of coils. Pairs of these molecules are yoked together during most of the cell's life cycle at a single site – the centromere. When the time comes for the cell to divide, the compactness of the packaging of the DNA molecules increases dramatically to give rise to bodies called chromosomes, which can be visualized by the application of special stains. Each chromosome is then seen to consist of two chromatids joined at the centromere, each chromatid containing a single DNA molecule. The width of the various structural elements is indicated in nanometers, one nanometer (nm) equaling one billionth of a meter. (Redrawn and modified from Purves et al. 1992.)

Elimination of chromosomal material from somatic cells has indeed been observed. In certain nematodes (round worms), the commitment of cells to the somatic line is accompanied by the expulsion of up to 90% of the chromosomal material. Alas, such cases are an exception. In the vast majority of animal species, no expulsion of chromosomal material takes place and there is no difference between the chromosomes of somatic and primordial germ cells. Moreover, the nuclei of the somatic and primordial germ cells are functionally equivalent. This has been demonstrated by experiments in which the nucleus of a somatic cell has been used to replace the nucleus of an egg cell and the latter shown to develop into a normal, complete organism. These experiments were first performed on frogs, but recently they have also been successfully carried out in mammals, the notorious sheep Dolly being the first case. The implications of these experiments are, first, that whatever changes might take place in the nuclei when cells gradually become specialized to the performance of specific functions, they are apparently not irreversible; and second, specialization involves the interplay between nucleus and the surrounding cytoplasm. Thus the difference in the cytoplasm of an egg cell compared to a somatic cell would seem to be responsible for the totipotentiality of the nucleus in the former and the unipotentiality of the nucleus in the latter.

The cytoplasm must also determine the difference between germ and somatic cells. In the fruit fly or frog egg, it is not the nucleus but the cytoplasm that is committed to either the germ or the somatic line. As the cell begins to divide after fertilization, the nuclei that find themselves in the germ plasm are set on a path that leads to the formation of germ cells (Fig. 2.4), whereas nuclei in the somatoplasm are prompted to enter the somatic line of development. The components of the germ plasm are manufactured while the egg is still in the mother's body and they are produced according to instructions supplied by the nucleus. Exactly how long the components persist through the successive cell divisions in the germ line is not known, but most – indeed probably all – of them are replaced by embryo-derived components at some point prior to the formation of a new generation of egg cells. So at best, only they can be said to link two successive generations, and this only in some species. The components therefore do not fit into the image of a germ line flowing uninterrupted through an endless succession of generations. At the same time, the chromosomal component that might be fitting (but see below) does not distinguish the germ from the somatic line cells. Under these circumstances, the term "germ line" has relinquished its original meaning and is best abandoned altogether. The question remains, however, whether there really is an immortal substance that flows steadily through generations. For an answer we must turn our attention to the substance in the chromosomes that is the physical carrier of heredity.

From Gemmules to Pustules

In the year in which Weismann published his original version of the germ plasm theory, there was only one person who understood the principles of heredity. His name was Gregor Johann Mendel, an unobtrusive monk living in the sleepy town of Brno, in a backwater province (Moravia) of the Austrian-Hungarian Empire. He made a few insipid attempts to communicate the implications of his hybridization experiments with different varieties of the garden pea, but nobody paid any attention. The audience attending his lectures did not understand him and the distinguished scientists to whom he sent the printed version of his lecture either did not read it or dismissed it as the work of a dilettante. In this case, however, history proved to be just. Mendel came to be regarded, alas posthumously, as one of the greatest figures in the annals of modern science, whereas his arrogant contemporaries are now remembered mostly for their failure to recognize the significance of his work. Because of their failure, science remained in the dark, as far as heredity was concerned, for three more decades, mired in arcane and esoteric speculations.

Born-again preformists abandoned their concept of innumerable homunculi telescoped one into the other in an interminable series, and came up with a new notion, no less crack-brained, in which gametes were imagined to contain the rudiments of individual organs capable of self-assembly following fertilization of the egg. Other scientists were taken by the idea of pangenesis which had been circulating at least since the time of the Greek physician Hippocrates of Cos. In Charles Darwin's version of the hypothesis, the body contained tiny granules, the *gemmules* (from Greek *gemma*, bud), which were responsible for the formation of individual organs and characters – the phenotype. According to Darwin, the gemmules normally circulate through the body, but when the stage of germ-cell production begins, they enter the gametes, one of each kind, and are thus transmitted to the next generation. The most bizarre set of speculations attributed heredity to nonmaterial causes that took the form of mystical, vital forces, which effected the formation of organs in a developing individual.

Mendel approached the problem of heredity in the manner of Alexander the Great who is said to have untied the Gordian knot with one quick slash of his sword. Mendel's experiments, interpreted in modern terminology, revealed that each observable feature (a phenotypic *character*) is controlled by a separate factor or factors (now *genes* at a *locus*) which can occur in multiple versions or *alleles* (Fig. 2.6). Each somatic cell possesses two copies of a gene (two alleles), each germ cell only one; the former is now said to be *diploid*, the latter *haploid*. The conversion from the diploid status of the primordial germ cells to the haploid status of the mature germ cells takes place during the maturation of the gametes. The relationship between two alleles at a given locus can be such that one of them *dominates* the other. In such a case, the presence of the *recessive* allele is revealed phenotypically only when it occurs in two copies in each somatic cell. An organism possessing one dominant and one recessive allele expresses only the character specified by the former. The recessive character can then disappear from one generation to the next, only to reappear again in one of the succeeding generations. During their passage from parents to offspring, genes behave as independent units (this conclusion, however, had to be modified by later discoveries), obeying the laws of probability. Individual characters therefore appear among the progeny of a cross between two different parents in statistically predictable ratios.

Mendel could not know the nature of the "factors", but he assumed that they were chemical entities. Later, toward the end of the nineteenth century, a growing number of scientists, Weismann among them, came to a similar conclusion. According to Weismann, units of heredity were subject to change, the consequences of which depended on whether the change occurred in the somatic or in the germ cells. Since somatic cells do not communicate with germ cells, and since they perish in each generation, the changes they incur cannot be transmitted from parent to offspring. Only changes taking place in the germ cells are inherited.

As the notion of a physical basis of heredity gradually attained respectability, naturally the unit's identity became of major interest. There were only three candidates to be taken into account: proteins, sugars, and fats. In 1871, however, a fourth was added to the list, a totally new substance, whose existence until then was not even suspected. It was discovered by Johann Friedrich Miescher, a young Swiss biochemist who, after receiving his M.D. degree from the University of Basel, came to the University of Tübingen in southern Germany to study under Ernst Felix Hoppe-Seyler, at that time a well-known and powerful personality in biochemistry. The task Hoppe-Seyler assigned to Miescher was to determine the chemical composition of the cells present in pus, which the local surgical clinic could provide in large quantities. After obtaining the cells by washing bandages of patients with infected wounds, digesting their proteins with pepsin-containing extract from a pig stomach, and a few other steps, Miescher was able to isolate material that behaved like an acid in certain tests and contained a surprisingly high proportion of phosphorus. There were only two or three organic compounds with a high phosphorus content known at that time and when these were ruled out, Miescher knew that he had something very interesting on his hands. The cells in the pus were mostly white blood cells (leukocytes), many of

A. GENOME

B. GENE, LOCUS

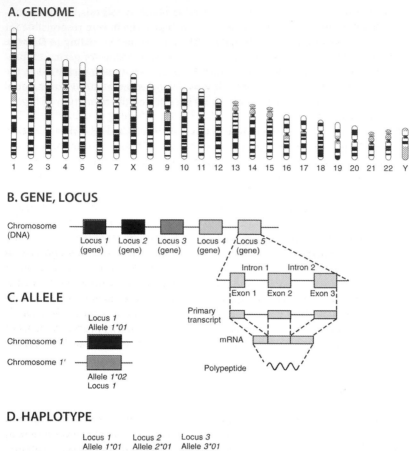

C. ALLELE

D. HAPLOTYPE

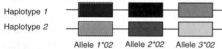

Fig. 2.6A–D
Genome, gene, locus, allele, haplotype: important terms used throughout this book (some will be introduced later in the text). The total DNA of a nucleus (*nuclear DNA*) of a gamete (sperm or egg) constitutes the nuclear *genome*. The DNA is partitioned into *chromosomes* (23 in the case of a human gamete as shown here), each chromosome containing a single DNA molecule. Somatic cells have double the amount of DNA compared to the gamete, i.e., they are *diploid*, while the gametes are *haploid* (somatic cells have two copies of each of the 23 chromosomes shown). Researchers often focus on a particular segment of the DNA molecule, which is then called a *locus*. Loci that are copied (*transcribed*) into RNA (the *primary transcript*) and have a specific function are referred to as *genes*. The primary transcripts of protein-specifying genes contain some parts (*introns*) that are removed (spliced out), while the remaining parts (*exons*) are spliced together into a single *messenger RNA* (mRNA). Parts of the mRNA are then translated into a polypeptide chain (protein). Genes in corresponding positions in chromosomes of a given pair (i.e., at the same locus) are *alleles*. The constellation of alleles at two or more loci of a single chromosome is the *haplotype*. (**A** from Strachan & Read 1996.)

which have large nuclei and rather scanty cytoplasm. Since the pepsin digestions destroyed most of the proteins of the cytoplasm, Miescher's procedure constituted a crude method for the isolation of nuclei. From this, Miescher concluded that the new substance was a component of the nucleus and so he named it *nuclein*. Later research by Miescher, who subsequently became a professor at the University of Basel, as well as by others, revealed nuclein to be a complex of a protein and a phosphorus-containing compound named by one of Miescher's students *nucleic acid*. Miescher completed the characterization of nuclein in 1869, but the paper describing the finding did not appear until 1871. The reason for the delay was that Hoppe-Seyler, the editor of the journal to which Miescher submitted his manuscript, insisted on repeating the experiments himself to make sure that no flaw had crept into the study. Only when he obtained the same results as Miescher, did Hoppe-Seyler proceed with publication, appending his own confirmatory note to the paper.

Miescher belonged to a group of scientists who were firmly convinced that there was a chemical basis to heredity, but he did not consider nuclein, and even less nucleic acid, to be a suitable candidate for this function. Instead, along with other chemists, he believed that proteins with their multitude of building blocks were probably involved. Although several of Miescher's contemporaries did entertain the possibility of nuclein's participation in heredity, the prevailing opinion favored proteins. It was not until 1944 that the first convincing evidence was obtained to establish that nucleic acids, whose chemical composition had in the meantime been elucidated, were the physical carriers of heredity. Miescher's discovery of nuclein must therefore be rated among the greatest scientific achievements of the nineteenth century. Curiously, it is rarely presented as such by historians of science.

From Miescher to Escher

There are two principal chemical types of nucleic acids, the *ribonucleic acid* or RNA and the *deoxyribonucleic acid* or DNA. Both DNA and RNA can function as physical carriers of heredity. In humans, however, as well as in all animals and plants, this function is entrusted exclusively to DNA molecules, which will therefore be the main focus of our attention. To explain the main features of nucleic acids, we resort to the use of two analogies, each of which will help to describe different aspects of their structure. First, we can compare a nucleic acid molecule to a train. Essentially, a train is a linear array of railroad cars, each car hooked up to the one in front and the one behind (except for the first and the last cars, of course). Each car has three main parts, the couplers to hitch it to the adjoining cars, the chassis, and the body mounted on the chassis. The couplers and the chassis are the same in all cars of a train, but the bodies of the cars differ. There are freight and passenger cars, and there are three basic types of freight cars (open-top, boxcar, and flatcar), as well as various passenger cars (restaurants, sleepers, compartments). The length of a train can be varied by adding or removing cars at either end or anywhere else along the train.

Analogically, a nucleic acid molecule is a linear array of many chemical units, the *nucleotides*, which are equivalent to the railroad cars; it is a *polynucleotide chain* or *strand* (Fig. 2.7). Each nucleotide consists of three components, a *phosphate group* (four oxygen atoms linked to a single phosphorus atom) which hitches it to its neighboring nucleotide (comparable to the coupler); a *sugar* containing five carbon atoms (corresponding to the chassis), and a *nitrogenous base* (one or two rings of carbon atoms with one nitrogen atom) attached to the sugar and so corresponding to the body of the railroad car. (Here, "base" refers to the tendency of the nitrogen to take up charged hydrogen atoms deprived of their electrons

A.

B.

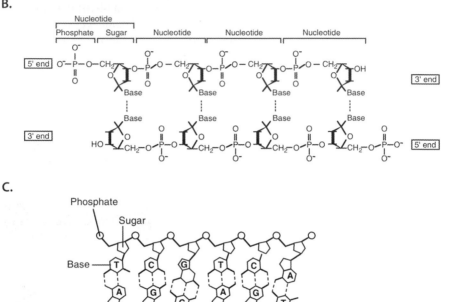

C.

Fig. 2.7A-C
The structure of the DNA molecule. **A.** A single DNA strand can be compared to a train and its nucleotides to the individual railway cars. **B.** The backbone of the DNA strand is the repeating unit consisting of a sugar (deoxyribose, the chassis of the car) and a phosphate group connecting it to the next sugar (the coupler in the train analogy). There are two strands to each DNA molecule. **C.** The sugars bear nitrogenous bases (the car's body mounted on the chassis), which are of four kinds: adenine (A), guanine (G), cytosine (C), and thymine (T). Counterpositioned bases of the two strands are complementary in their structure and are therefore able to form hydrogen bonds holding the two strands together in the form of a double helix. The ends of the strands, which are distinguished by the terminal chemical groups are referred to as 5' (five-prime) and 3' (three-prime) and correspond to the locomotive and the caboose of a train. The two strands run antiparallel to one another: if one strand terminates with a 5' end, its partner terminates at the corresponding position with a 3' end, and vice versa. They are like two trains on parallel railroad tracks moving in opposite directions. (**A** from *The World Book Encyclopedia.* Vol. 16. 1996. **B** from Klein & Hořejší 1998.)

from the solution.) Two types of sugars and five types of nitrogenous bases occur in nucleic acids. The sugars are called *ribose* and *deoxyribose*, the latter differing from the former by having one oxygen atom less (the Latin prefix *de-*, from, away, is used in chemistry to indicate removal). It is from ribose and deoxyribose that the RNA and DNA molecules, respectively, derive their names. The five nitrogenous bases, commonly referred to merely as *bases*, are adenine, guanine, cytosine, thymine, and uracil and are abbreviated to A, G, C, T, and U, respectively. The first four of these occur in DNA and the first three are also present in RNA, in which the thymine is replaced by uracil. (Other nitrogenous bases are also known, but these only occur infrequently in nucleic acids or under abnormal circumstances.) Adenine and guanine molecules each comprise two conjoined rings, one consisting of six and the other of five atoms: they belong to a group of compounds called *purines*. Thymine, cytosine, and uracil, the three *pyrimidines*, each have a single ring of six atoms. The bases differentiate the nucleotides just as the bodies distinguish different railroad cars. It is therefore common practice to refer to the different nucleotides by the single-letter abbreviations of their bases and to use the term "base" when referring to the entire nucleotide. The length of a nucleic acid molecule (its "size"), for example, is commonly expressed in terms of the number of *bases* (in the case of single-stranded RNA molecules) or *base pairs* (bp; in the case of double-stranded DNA molecules). Like a train, a nucleic acid chain can be extended or shortened by the addition (insertion) or removal (deletion) of units, here nucleotides.

The train analogy draws attention to three important features of nucleic acids – their linearity, the repetitive nature of their structure, and their potential for unlimited growth. To explain additional fundamental features we now resort to our second analogy. The

Fig. 2.8A-C
Copying (replication) of DNA molecules. **A** and **B**. The principle of complementarity illustrated by motifs from Maurits Cornelis Escher. **C**. Unwinding of a single DNA molecule and synthesis of two new molecules using the strands of the original as a template for the assembly of complementary nucleotides (indicated by letters specifying different nitrogenous bases). (**C** from Beck et al. 1991.)

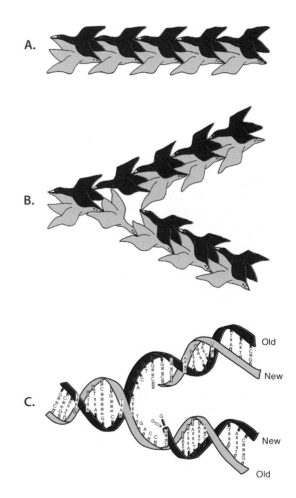

drawing in Figure 2.8A is based on the *Study of Regular Division of the Plane with Birds* executed in 1938 by the Dutch artist Maurits Cornelis Escher. We use it here to illustrate the *principle of complementarity* on which Escher based many of his works. In the drawing, we see two flocks of birds, probably geese, flying in straight lines but in opposite directions. They are arranged in such a way that they are "mutually supplying each other's lack", which is the dictionary definition of complementarity. By matching the wing contours of a bird in one row with the head-and neck contour of a bird in the second row, Escher succeeded in bringing the two so close together that no free space remains between them.

Complementarity is an extremely important property of biological molecules because it allows them to come so close to each other that interactions between them can take place. Many life processes are critically dependent on these interactions, including those that involve the DNA molecules. Although atoms of a molecule are largely composed of a void, their boundaries are delineated by force fields that guard the molecule's territory and prevent it from being penetrated by another molecule. The surface of a molecule can therefore be visualized as that of a flying bird in Escher's drawing, with a distinct relief. If the surfaces are complementary, the molecules can snuggle up closely like two birds from opposite rows. The close fit brings the molecules within range of the weak attraction forces that may emanate from some of their atoms (this feature, however, has no analogy in Escher's drawing). One of these forces leads to the formation of the *hydrogen bond*. Two atoms can become strongly bound together by sharing electrons of their outermost shells, each atom donating one electron to the union. In this *covalent bond*, each contributing electron can move not only around its own atom's nucleus, but also around the nucleus of the other atom. Outer shell electrons that do not participate in a covalent bond may form "lone pairs". Since hydrogen atoms that are covalently bound to nitrogen or oxygen atoms of one molecule acquire a slight positive charge, they are attracted to the negatively charged lone pairs of electrons in the nitrogen or oxygen atoms of other molecules. This weak attraction between a negatively charged nitrogen or oxygen atom and hydrogen that is covalently bonded to another atom is the hydrogen bond. For the bond to form between two molecules, these must not only possess hydrogen and nitrogen or oxygen atoms in the right position and configuration, they must also come very close together. Although weak, the hydrogen bond is so essential for the interaction between such a large number of molecules (including molecules of water) that life, at least in the form we know it, is unimaginable without it. Without hydrogen bonds, the DNA would also be unable to attain the structure and properties that have made it the physical carrier of heredity.

In the DNA molecule, conditions for hydrogen bond formation are provided by the nitrogenous bases (Fig. 2.7). Complementary molecular surfaces exist between thymine and adenine, as well as between guanine and cytosine – in both cases between a single-ring and a double-ring base. And the bases have nitrogen and oxygen atoms with lone electron pairs, as well as hydrogen attached to nitrogen atoms. The distribution of these atoms is such that adenine can bind to thymine via two hydrogen bonds, whereas cytosine can pair with thymine via three. The complementarity of the bases enables pairs of polynucleotide chains to form molecular duplexes which the hydrogen bonds hold together. In the duplex, one strand is oriented in one direction and the other in the opposite direction, like the configuration of birds in Escher's drawing. The nucleotides of the two strands are paired like ballet dancers in a *pas de deux*: where there is an A in one strand, its partner is a T in the opposite strand, and vice versa; and likewise, the G of one strand is always matched with a C in the other. The distribution of atoms compels the duplex to assume a specific configuration, the *double helix*, in which one strand winds around the other (Figs. 2.5, 2.7; see also the *Andy-Diana DNA Letter* on page 13).

Omnis DNA e DNA

The complementarity of nucleotides in the two strands of a DNA molecule enables the molecule to perform an act unique among biological molecules, an act of *reproducing* itself. In biology, the term "reproduce" means to give rise to a life-form of the same kind. The "life-form" is normally an organism, but it can also be a cell of a multicellular organism, or a DNA molecule. Reproduction is the most fundamental process of life, and one that has been responsible for life's persistence on earth for more than three billion years. Demise is an inevitable component of the life cycle, since in the beginning, the end is indeed already programmed. Reproduction is a means of replacing one life cycle with a new one. In one's end is one's beginning, not the spiritual beginning which Mary, Queen of Scots prayed for*, but a biological beginning as implied in T. S. Eliot's *Four Quartets*** and also by Gauguin's masterpiece. Nevertheless, the true function of reproduction is not to keep the cycles revolving, but to provide an opportunity for change – for the evolution of life. And this opportunity is realized originally at the DNA level and only secondarily at the cellular and organismal levels. At the DNA level, reproduction amounts to copying, and copying involves making mistakes. And mistakes, as we shall learn, are the cause of evolution.

When life began, the first organic molecules arose from inorganic matter and the first cells from acellular organic matter. Ever since this initial phase of life's evolution, however, cells arise from cells and DNA molecules from existing DNA molecules. The first of these certitudes is most concisely expressed by the adage *Omnis cellula e cellula* originally coined in 1855 by the German pathologist Rudolf Carl Virchow; the second can be expressed by *Omnis DNA e DNA*. For an organism to reproduce, it must first reproduce its cells; for a cell to reproduce, it must first reproduce its DNA molecules. Of the three levels of reproduction – organismal, cellular, and molecular – the molecular level is, therefore, the most fundamental. Some biologists even argue that the two other levels evolved merely to render DNA reproduction more efficient and less prone to failure.

The reproduction of DNA molecules is referred to as *DNA replication*. (The Latin verb *replicare* means "to fold back, to repeat" and a *replica* is a close copy of an original.) DNA replication is therefore the ability of DNA molecules to produce copies of themselves. The two essential steps in DNA replication are the separation of the two strands of the double helix and the use of each strand as a template for the assembly of a complementary strand. Starting from particular sites in a DNA molecule and proceeding in one direction, the hydrogen bonds between the complementary bases break and the two strands come apart. Immediately after this takes place, however, free nucleotides that are jostling around aimlessly in the surrounding fluid find their way to the divorced strands and assemble on them according to the complementarity of their surfaces. Where there is a T on the separated strand, it is joined by an A, G pairs C, and so on. Hydrogen bonds spring between the paired bases and by means of covalent bonds, special proteins join the assembled nucleotides to form a growing chain. In the metaphorical depiction in Figure 2.8B, the double row of flying birds is separating into two double rows. When the process is completed along the entire length of the parental DNA molecule, the two daughter molecules come apart: the original has produced two copies of itself, it has *reproduced*.

At long last, we have reached a point in our narrative at which we can answer the question of continuity between generations. We have come to realize that continuity is effect-

---

* *En ma fin est mon commencement* (In my end is my beginning) was the motto embroidered on the chair of state of Mary Queen of Scots.
** *In my beginning is my end* is the opening verse of East Coker, the second of the *Four Quartets*.

ed by the germ cells, the sperm of the father and the egg of the mother, for these are the only physical components which derive from the bodies of the parents and participate in the development of offspring. Since it is only the nucleus of the sperm cell that penetrates the egg at fertilization, continuity must reside in the nucleus. Now, continuity presupposes permanence of that which is being continued. A semblance of permanence in both germ cells is provided by the DNA. Everything else, the proteins, sugars, lipids, and the vast array of small molecules comprising the fertilized egg, has a very ephemeral existence. This material is used up, broken down into smaller components, reused for manufacturing new components, "burned" to produce energy, excreted, replaced, turned over – in short, *metabolized*. Some of the components may persist through a few cell divisions after fertilization of the egg, but by the time the new individual is born, not a single molecule remains that could be claimed to have been produced by the parents. Without exception, they have been replaced by molecules manufactured by the cells of the new individual. Goods "made in parents" have been substituted by goods "made in offspring". This is true not only for the components of the somatic cells, but also for those of the germ cells. Even in organisms such as the fruit fly, which produces eggs with distinct regions of germ plasm, it has been demonstrated that the components of the germ plasm are synthesized during the maturation of the egg. These components are therefore handed down one generation, but no more than one.

The only component appearing to be permanent is the DNA, which owes its privileged status to its ability to replicate. But even in this case, it is not a physical permanence. To explain, let us follow the fate of one particular DNA molecule passed from a mother to her offspring. Before the fertilized egg cell divides, the DNA molecule replicates and it does so prior to every cell division. The egg cell has a pool of nucleotides which were provided by the mother and this supply may last for a few divisions. Ultimately, however, the supply is exhausted and from then on, the developing embryo relies on nucleotides synthesized from its own resources. When this happens for the first time, at the next round of DNA replication, each of the two daughter molecules will possess one strand of nucleotides contributed by the mother and one strand of nucleotides produced by the embryo (Fig. 2.8). The following division will result in one cell with a DNA molecule made exclusively of embryo-produced nucleotides, and another cell which will again be half-and-half. As the divisions continue, the number of cells with DNA molecules made exclusively of embryo-produced nucleotides rapidly increases, while the number of cells containing half-and-half molecules at best remains the same. It is more likely however, that some or all of the cells with these molecules will, by chance, be eliminated and the overwhelming majority, if not all the cells, will contain DNA molecules made of embryo-produced nucleotides exclusively. This will be true of both somatic and germ cells. For as long as the half-and-half DNA molecules persist, they can be considered as representing a physical link to the preceding generation. In reality, however, even this tenuous, short-term link does not exist in any truly physical sense of permanence. In a living cell, all the constituents are continually turning over, atoms and molecules being lost and replaced by new ones. After a while, none of the mother's atoms may be present in the nucleotides considered above as being maternally-derived. So, we must conclude that any physical link between generations that might exist is short-lived, probably bridging no more than two generations.

Yet, something *is* being passed down the generations. If it were not so, progenies would not resemble their parents, which they obviously do, if in nothing else than in their species characteristics. Not only do humans bear humans and chimpanzees bear chimpanzees, there are also more subtle similarities between a particular human offspring and its parent, and the same is true for chimpanzees. The "something" is information.

The Informer and the Replicator

What is information? Consider the following array of letters:

MNNTRFOOAII

Try as hard as you might, consult dictionaries in as many languages as you can find, you will fail to find a meaning. The array has no meaning. And now consider this array:

INFORMATION

It consists of the same letters as the first array, but arranged in a different order. In this arrangement, the collection of 11 letters has a meaning, not only in English, but in several other languages as well. It is a word, whereas the first array was just a random string of letters. The word may have a different meaning to different people, but it means something to everybody familiar with it. We say that the specific sequence of 11 letters, in contrast to the first, meaningless array, contains *information*. The difference between the two arrays is in that in the array INFORMATION the identity of the letter appearing at each of the 11 positions or *sites* is specified, whereas in the array MNNTRFOOAII it is not. In the latter, any letter can appear at each of the sites and it will have no effect on the meaning of the array. The 11 letters of eight different kinds can be arranged in 8^{11} or 858,998,492 different orders, only one of which has our particular meaning. One particular sequence of letters alone brings forward the specific idea or concept in our minds. For the present purpose, we can therefore define information as the attribute inherent in one of two or more alternative arrangements of something that produces a specific effect. Here, "alternative arrangements" refer to the different orders of letters and the specific effect produced is the acquisition of meaning. This definition is close to the original meaning of the Latin verb *informare*, "to give a form to", from which the word "information" is derived.

One strand of a DNA molecule can also be viewed as a very long array of letters of four kinds, A, T, G, and C, the four different nucleotides. Scanning the array, one passes long stretches that, like MNNTRFOOAII, do not make any sense. From time to time, however, one hits upon sequences of letters that do make sense because they contain *genetic information*. They do not, of course, specify ideas or concepts like words in a printed text. Instead, they specify the order in which amino acid residues should be hooked together when a protein is made to render a molecule capable of carrying out a particular function. Other stretches specify the order of nucleotides of various kinds of RNAs that participate as auxiliary molecules in protein synthesis. The stretch of DNA that contains genetic information is the *gene*.

To achieve continuity between generations, the information stored in the nucleotide sequences of the DNA molecules must be passed on from parents to offspring. The double-stranded structure of the molecules, the complementarity of the nitrogenous bases in the two strands of the double helix, and the molecules' manner of replication ensure just such a passage. Of the two strands comprising a gene, always only one, the *sense strand*, contains a nucleotide sequence that specifies a protein. (For the sake of simplicity, here and in most of the following text we disregard the genes that specify the different RNA molecules.) The other, the *antisense strand*, bears no protein-specifying information, but it does contain information on the order in which nucleotides of the protein-specifying strand are to be assembled. In the DNA molecule, some protein-specifying sequences are on one strand, others on the second strand. Replication of the DNA molecule therefore amounts to information transfer: the assembly of a sense, protein-specifying sequence on the template of the antisense strands produces a daughter molecule with the same or nearly the same informational content as the parental molecule. Because of this arrangement, the offspring receives genetic information that is very similar to the information their parents possessed.

And so we have finally arrived at the true nature of the bridge spanning the generation gap. Neither blood, nor immaterial vital force, nor germ plasm flows under it. As generations pass, blood forms anew in each individual, cell structures come and go, and immaterial forces never materialize. Even the DNA, the informer and replicator, continually sheds and replaces its nucleotides just as a pine tree sheds its needles, and it only passes on Xerox copies of itself. But the genetic information flows uninterrupted from cell to cell and cascades down from generation to generation. It is genetic information that connects and unites us all, not only those alive today, but all those who have ever lived. If we want to know where we come from, we must follow the stream back in time and trust that it will lead us to the source.

The Teleporter

Now that we have bridged the generation gap, we should move on to the next chapter. Before we do, however, we ought to touch upon an issue tangentially raised by this chapter's theme. The genetic information that is transmitted from generation to generation specifies, as stated earlier, proteins and some RNA molecules necessary for protein synthesis. The proteins then set into motion all the chemical reactions associated with life. They catalyze the synthesis of other organic compounds, including the DNA molecules; they participate in energy production, distribution, and consumption; they are involved in the cellular import-export trade; they serve as construction material for a variety of cellular and extracellular structures; they organize cellular reproduction; and they coordinate, regulate, and control these activities. All this hustle and bustle that keeps a cell alive is determined, directly or indirectly, by the genetic information stored in the DNA molecules. If, however, the life of a cell is specified by the genetic information contained in its DNA molecules, what specifies the development of a multicellular body from a single cell, the fertilized egg?

During development, the egg divides to give rise to trillions of cells; the cells differentiate into hundreds of different cell types such as neurons, muscle cells, erythrocytes, and pigment cells, each cell type specialized in morphology and function; the cells organize themselves into distinct tissues and these into highly complex organs such as the spleen, the eye, or the brain; and the organs assume specific positions in the body. The entire process takes place in a precisely timed and coordinated sequence. Since the course of development is the same for all individuals of a given species but specific for different species, it can be presumed to be genetically determined. But does this mean that in addition to protein- and RNA-specifying genes, there are other sequences in the DNA which are concerned exclusively with coordination of development? Are there sequences that specify the point at which a cell should stop dividing, the shape it should assume, or the position in the embryo that it should move to? Or sequences that coordinate the organization of tissues into organs? Development is often compared to the realization of a computer program in which each step is conditioned on preceding steps. If so, is the program actually imprinted on the DNA?

To the best of our knowledge – which, admittedly, is still fragmentary – the answer to all these questions is "No". Although there are sequences in the DNA that do not specify proteins or RNA molecules, yet are vital for development, they are associated with and considered to be part of the genes. They are referred to as *regulatory regions* because they determine whether the adjoining protein (RNA)-specifying sequences are used, how effectively they are used, and how much of the protein (RNA) is produced. Setting aside regulatory regions, no other development-specifying DNA regions have been identified, not even in organisms whose complete DNA sequences have been determined. There are no regions that store information responsible for a developmental program: apparently, no such program is inscribed into the DNA.

In another frequently used analogy the DNA is said to contain a "blueprint" for the construction of an organism, but this, too, is misleading. A blueprint contains information for the assembly of a structure from its components, but here again, the DNA contains no such information. There is, for example, no gene that instructs certain lipids to arrange themselves into the two layers of the plasma membrane that envelops all cells. At a higher level, there are no detailed instructions impressed onto the DNA in the form of a specific sequence for the pattern of assembly of organs in an embryo.

What, then, actuates development? The best that can be deduced from the present state of knowledge is that once crank-started by fertilization, development unfolds on its own. The principle by which this takes place can be described rather simply, although its application to the development of even a very simple multicellular organism is extremely intricate. To explain, we must return to the gene's regulatory region.

To construct a protein that is based on information from a DNA sequence is a complex process which begins with the *transcription* of the sense strand into a complementary strand of *messenger (m) RNA*. The transcription initiation involves many proteins which must first gain access to the gene. The proteins that normally coat the DNA must be removed, the multiple-folded DNA duplex partially unfolded, and the cleared area seized by *transcription factors* – the proteins that initiate and regulate the transcription of a gene into mRNA. To act, the transcription factors must bind to a short sequence of the DNA duplex, each respective factor to a different sequence (Fig. 2.9). Different genes have distinct sequences in their regulatory regions and thus are able to bind different combinations of transcription factors. Only when the full set of factors has ensconced itself in the regulatory region can transcription begin in full force. The transcription factors themselves are products of genes which, like all other genes, must be activated by other transcription factors. Each transcription factor may participate in the activation of different genes since the specificity and effectiveness of activation depends not on individual factors, but on their combination. Thus, an extremely elaborate interdependence of genes exists, which is mediated by transcription factors (Fig. 2.10). A gene can be activated and transcribed only at the point in development at which the appropriate set of transcription factors has become available. These factors, in turn, become available only when another set of transcription factors has been produced which is necessary for the activation of their genes. And so it continues, a cascade of factors racing through development.

Greatly simplified, fertilization, the initiator of development can be assumed to provide the original impetus for the production of the first battery of transcription factors. These then activate genes to produce a second battery of factors, some of which activate genes involved in early steps of development; the other factors activate genes specifying a third battery of transcription factors (Fig. 2.10). This process then continues and involves not only the activation of additional genes, but also silencing of previously active genes. The sequence in which genes are activated and silenced is not imprinted on the DNA; it unfolds automatically according to the changing constellations of transcription factors available at any particular stage of development. The sequence has been established gradually by evolution over many millions of years.

Development is a realization of a phenotype from a genotype, the conversion of one form of information (sequence of nucleotides in DNA) to many other forms (sequence of amino acids in proteins, the organization of molecules into subcellular structures, the organization of cells into tissues and organs, and so on). The realization can be compared to

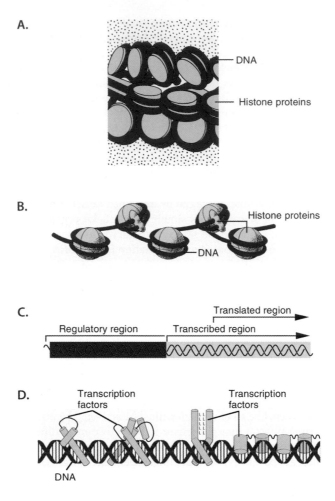

Fig. 2.9A-C
Activation of a gene. **A.** In an inactive part of a chromosome, the DNA wound around its protein is tightly coiled, closely compacted, and inaccessible to activating proteins (transcription factors). **B.** The first step in the activation of a gene is a local loosening (unwinding) of the chromosome structure which opens the inactive part to regulatory proteins. **C.** The exposure of the regulatory region at one end of the gene paves the way for binding of regulatory proteins. **D.** Different types of transcript factors bind to specific nucleotide sequences of the DNA double helix with their specialized DNA-interaction domains. The assembly of a particular combination of regulatory proteins activates the gene: a specialized protein complex initiates transcription of the adjacent part of the gene into an RNA molecule (the primary transcript). (**D** from Klein & Hořejší 1998.)

"teleportation" popularized by the *Star Trek* movies. In Nature's version of "Beam me up, Scotty", the information teleported from one generation to the next brings a copy of the body from which it originated into existence. If the copy is not exactly like the original, it is only because there is not one original, but two. The mixing of information from these two sources leads to the appearance of a new phenotype which resembles the originals, but is at the same time distinct from them. The implication of the mixing for the interpretation of human origin is the subject of the next chapter.

Fig. 2.10
The principle of sequential gene activation during the development of an embryo. Here development is envisioned as a self-propelled system set in motion by the fertilization of the egg. The system is simplistically portrayed as consisting of different tiers of gene sets activated by specific combinations of transcription factors (TFs). In each tier, part of the proteins produced by the activated set is used to activate a new set of genes. The other part participates in the various activities associated with the growth and differentiation of the embryo.

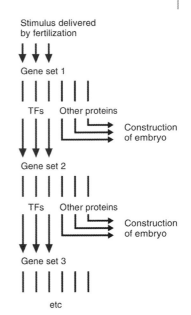

In this chapter we began our search for human origin by examining the most basic fact of life, the paradox that life is mortal and immortal at the same time. Individuals are mortal, but life itself has endured for a period of time that to the human mind is on a par with immortality – for over three billion years (by). The resolution of this paradox lies in the turnover of generations, the process in which a new life cycle begins before an old one has come to an end. We then looked for the bridge spanning two successive life cycles only to realize that in reality no such bridge exists. Instead we had to conclude that the link between two generations can be compared to ferryboats transmitting information of a particular kind – genetic information recorded in the sequence of nucleotides composing nucleic acids. In a broad outline we learned how the information is stored, transmitted from one generation to the next, and how it is unfolded during the development of a new organism. The generation turnover represents a chain of ancestor-descendant relationships in which each link is both a descendant of that preceding it and an ancestor of the link succeeding it. In the next chapter, we examine the nature of the ancestor-descendant relationship and its utility for tracing the origin of organisms.

CHAPTER 3

Crane's Foot

Biological Meaning of Descent

MACBETH

*Thou are too like the spirit
of Banquo. Down!
Thy crown does sear mine eyeballs.
And thy hair,
Thou other gold-bound brow,
is like the first.
A third is like the former. Filthy hags!
Why do you show me this?
A fourth? Start, eyes!
What, will the line stretch out
to th' crack of doom?
Another yet? A seventh? I'll see no more.
And yet the eighth appears,
who bears a glass
Which shows me many more;
and some I see
That twofold balls and treble
sceptres carry.
Horrible sight! Now I see 'tis true;
For the blood-bolter'd Banquo
smiles upon me
And points at them for his.
[Apparitions vanish.] What?
Is this so?*

William Shakespeare: *Macbeth*, Act 4, Scene 1

A Show of Eight Kings

In this famous scene, Macbeth calls on the three witches to learn whether Banquo's descendants will ever reign in Scotland. To his horror, among the spirits conjured up, that of Banquo appears first, followed by a show of eight kings whom Shakespeare's contemporaries could identify as the eight Stuart Kings of Scotland, King Robert I and II and King James I-VI. Last in line is James VI, in Shakespeare's time the King of England and so the playwright's lord and master, as well as benefactor. The procession is a unique theatrical depiction of a line of descent, a genealogical *tableau vivant*. Viewing Shakespeare's play, King James VI must have felt extremely flattered for three reasons. First, for seeing his roots displayed so vividly on the stage. Second, for the playwright's tact in leaving out Mary, Queen of Scots from the procession, who preceded James VI in Scotland, but for obvious reasons was not popular in England. And third, for closing the show of kings with James VI holding a mirror in his hands in which many more figures are reflected. The kings in the mirror suggest continuation of the dynasty beyond James VI. (The mirror was a neat trick on Shakespeare's part since he could identify James VI's predecessors, of course, but not his successors.) Macbeth's exclamation *What, will the line stretch out to th' crack of doom?* even suggested that the line of Stuart successors would be long – a reasonable and accurate prediction, as it turned out.

Parading apparitions on the stage is a somewhat unusual form of presenting a *genealogy* – an account of descent from a progenitor (from Greek *genea*, race, family; hence also "family history"). The more conventional forms include oral recitations, inscriptions on stone slabs (stellae), or records that list the ancestors in a table or a diagram on parchment or on paper. A diagram is the most common form of presenting genealogy, a small cutout of which may look like this:

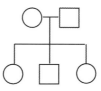

The diagram, when read from left to right and from top to bottom, informs us that a female (O) and a male (□) wed (—) and that three children emerged from the union (⊓), two girls and one boy. The sign for procreation can also be drawn diagonally (⋏), in which case it resembles the foot of a crane, *pié de grue* in French. Thus the diagram came to be referred to as a *pedigree* or alternatively a *family tree*. A more complex pedigree – that of Charles Darwin, of whom we will have more to say later – is displayed in Figure 3.1. Each row of "O" and "□" symbols in a pedigree represents one *generation*. In the layout of generations, the one at a lower level is a descendant of the generation above it, which is ancestral. "Descendant" and "ancestral" are therefore relative terms. A succession of ancestors spanning several generations constitutes a *line of descent*.

Interest in ancestry extends as far back as human consciousness. Some of the oldest literary documents and works, the List of Kings of Sumer, the Turin papyrus from Ancient Egypt, the great Hindu epic poems *Ramayana* and *Mahabharata*, the *Old Testament* of the Hebrews, the *Iliad* and *Odyssey* of the Ancient Greeks, and the Nordic sagas, all contain lengthy descriptions of lines of descent. Since most of these works existed as oral traditions before being committed to paper, we must assume that the genealogies are much older than the written word. In all these cases, however, the genealogies concern a few privi-

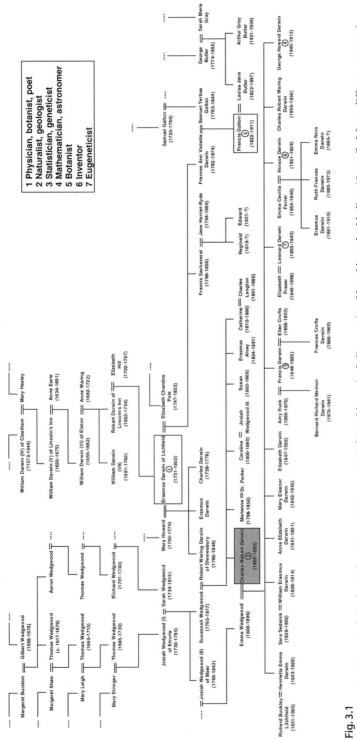

Fig. 3.1
Family tree (pedigree) of Charles Robert Darwin. Descent is indicated by a single line, conjugal bond by a double line. (Compiled from different sources.)

leged individuals – rulers, prophets, heroes, aristocrats, and others who succeeded in distinguishing themselves from their fellow human beings. Common people used not to have genealogies. Even today, if you are not "blue-blooded"* and yet still attempt to trace your ancestry, the chances are that you will not reach far beyond the sixteenth century. In that century, namely, sovereigns in many European countries introduced mandatory registration of all births – not because of their overwhelming interest in genealogy of common people, but to get a better idea of the taxes they could expect to collect.

The privileged had two compelling reasons for their interest in their own genealogy. One was biological inheritance, firmly entrenched in the concept of bloodlines, or more specifically, male bloodlines. Biological inheritance was exploited to rationalize claims to the same social status as an ancestor. Anyone who proclaimed: "In my veins flows the blood of an illustrious forefather" expected others to believe that his ancestor's exceptional qualities had been handed down to him. He who could extend his genealogy to a divine ancestor, such as the goddess Venus in the case of Julius Caesar, or the pagan storm-god Woden in the case of Saxon Kings of Wessex, was a made man. The second reason was social inheritance. To inherit a title, property, and other assets, it was imperative to prove a blood relation to the ancestor who had amassed them.

This was thus the ancient function of pedigrees – to rule the roost and to inherit. But what does a pedigree really represent? What *kind* of relationship does it depict? And more important for our purpose: can it really lead us to the ancestor of the human species?

Adam's Descendants and Darwin's Ancestors

One of the oldest pedigrees is to be found in Chapter Five of the Book of Genesis. It begins thus:

And Adam lived an hundred and thirty years, and begat a son in his own likeness, after his image; and called his name Seth:

And the days of Adam after he had begotten Seth were eight hundred years: and he begat sons and daughters:

And all the days that Adam lived were nine hundred and thirty years: and he died.

And Seth lived an hundred and five years, and begat Enos:

And Seth lived after he begat Enos eight hundred and seven years, and begat sons and daughters:

And all the days of Seth were nine hundred and twelve years and he died.

And Enos lived ninety years, and begat Cainan …

The text continues in this vein all the way to Noah of the Flood and thus to the first great decimation of the human population. This mythological pedigree illustrates the purpose of such a roster of successive begettings: to connect an individual of a later generation with one of an earlier generation genealogically, here Noah with Adam. It also illustrates the most effective procedure for establishing this connection: one takes only male descendants into account and ignores wives, daughters, and also any sons that are not on the male line of descent.

This practice has been followed by generations of genealogists. Think of a medieval genealogist commissioned by his sovereign to prepare a pedigree showing the monarch to be

---

* The expression arose in Spain in the Middle Ages, where the veins of pure-blooded aristocrats with no Moorish taint were bluer than those of people with mixed ancestry.

a remote descendant of King Arthur. How does he go about it? He examines all the relevant records registering births, deaths, and marriages, starting with the sovereign's parents, his grandparents, and then continuing through all the earlier ancestors, generation after generation, until he finally reaches King Arthur. (If he doesn't, and runs instead into a void, he can try to fill it with imaginary ancestors to preclude displeasing his master.) Alas, by using this method, the genealogist has to face the problem that every human being is the product not of one, but of two parents. Consequently, when traced back in time, the lines of descent split first into two, then in the next generation into four, into eight, and so on (Fig. 3.1). In each generation he therefore has to decide which of the two lines to follow. The proper procedure would be to follow them all, but this is impractical not only because the numbers rapidly become unmanageable, but also because the spoor in the female line of descent soon grows cold. In the case of our medieval genealogist, however, it would not even enter his mind to follow maternal lines of descent, first because he would be convinced that biologically, the male alone is responsible for procreation; and second, because in his male-chauvinistic society, records are often only kept of male descendants. By limiting his search to the male line, the genealogist thus saves himself a lot of trouble, just like the author of Genesis.

But times have changed and we have come to realize that the two presumptions behind the method of pedigree-making in the old days are invalid. We now know that both sexes contribute equally to the genetic endowment of offspring, and we treat, in theory at least, both sexes as equal in a growing number of societies. Nevertheless, pedigrees continue to be constructed and so we must determine whether they still meet their original purpose and if not, try to pinpoint the purpose they do fulfill.

Since there is no justification for tracking the male lines and ignoring those of the female, both must be followed. The consequence is easy to grasp. Let us designate the generation to which our person of interest (Charles Darwin in Figure 3.1) belongs as G_0 and use it as a starting point for counting generations retrocessively (i.e., G_1, G_2, G_3, G_4, etc.). Furthermore, let us assume a simple relationship between generations and the number of ancestors of our particular descendant: in generations $G_1, G_2, G_3, G_4, \ldots, G_n$, the descendant had $2 = 2^1, 4 = 2^2, 8 = 2^3, 16 = 2^4, \ldots, 2^n$ ancestors, respectively. In most of the pedigrees from the last 1000 years, the average time span from one generation to the next (i.e., the *generation time*, also defined as the mean age of parents at which their middle child is born) is about 20 years. A pedigree spanning the entire 1000 years would therefore cover $1000/20 = 50$ generations. A person in the current generation could then be expected to have some 2^{50} ancestors in generation G_{50}, which is a number (10^{15}) represented by 10 followed by 14 zeroes. For comparison, the present size of the human population is 6 billion or 6×10^9 (one billion being 10 followed by eight zeroes).

We thus reach the absurd conclusion that 50 generations ago, the number of ancestors of each person alive today exceeded the current population size, which is known to be the largest in human history. The reason for this paradox is that in the calculations, we assumed independence of all ancestors, whereas in reality, in any finite population, individuals are necessarily related to one another. Marriages between related individuals lower the number of different lines of descent in a pedigree and thus the number of ancestors. In Figure 3.1, the marriage between Sarah Wedgwood and Josiah Wedgwood, Darwin's grandparents on the maternal side, is an example of such a union between kin, as is Darwin's own marriage to his cousin, Emma Wedgwood. However, even after taking this reduction through kinship into account, the numbers of expected ancestors remain high for distant generations.

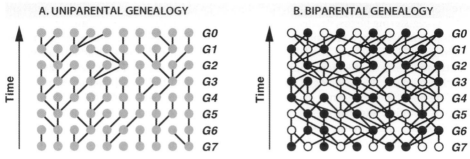

Fig. 3.2A,B
Uniparental (A) and biparental (B) genealogies. Gray circles in **A** represent asexually reproducing individuals; open and closed circles in **B** represent mothers and fathers, respectively. Connecting lines are lines of descent. In **A**, each individual has 0, 1, 2, or more descendants, but is itself derived from a single ancestor (parent). In **B**, each individual can contribute to the production of 0, 1, 2, or more descendants, but is derived from two ancestors (parents).

Grand Unifications

The expected extent of ancestor sharing and its consequences are difficult to ascertain from real pedigrees. They are too complicated on the one hand and too incomplete on the other – in short, *too* real. To glean the general features of genealogy, theoreticians resort to model populations. They work either with model genealogies that they themselves have created or with virtual pedigrees generated by a computer. In the former case, rules of descendancy are derived by deductive reasoning of the kind employed by mathematicians to deduce their theorems (= *analytical approach*). In the latter method, computer programs (*algorithms*) are written that feign real genealogies. The computer is allowed to run through the genealogies many times over and is then instructed to extract general patterns of behavior from a comparison of individual runs (= *computer simulation approach*). In both methods, the genealogy is idealized – trivial variation, which complicates the ancestor-descendant relationships without influencing the overall characteristics of the genealogy, is erased.

A simple model genealogy is depicted in Figure 3.2. Here, each circle represents an individual, each row of circles a generation, and connecting lines ancestor-descendant relationships between individuals of successive generations. In a departure from conditions existing in a real genealogy, here the generations are non-overlapping, the size of the population (number of individuals, N) remains constant through the generations, and ancestor-descendant relationships are determined by chance: each descendant picks its ancestors randomly from the preceding generation. Randomness is achieved by casting a die or by programming a computer to select a random number. Figure 3.2A depicts a one-parent genealogy in which each descendant has a single ancestor; Figure 3.2B shows a two-parent genealogy. Humans have, of course, a two-parent genealogy, as do all sexually reproducing organisms. One-parent genealogy is restricted to asexually reproducing forms. In the woods around Aspen, Colorado, for example, most of the aspen trees are related to one another by this form of genealogy. One-parent genealogy is easier to study and more easily apprehended than two-parent genealogy.

In Figure 3.2A, two descendants in a given *G0* generation share a common ancestor if their lines of descent *coalesce* in a single individual in some earlier generation. All the ancestors of this common ancestor are, of course, also common ancestors of the two *G0* individuals. It is therefore necessary to differentiate between the *most recent common ancestor* (MRCA), which occurs in the generation closest to *Go*, and all the other *common ancestors* which occur in earlier generations. After identifying the MRCA of two *Go* individuals, we may want to join the line of descent of a third individual with the lines of the first two, and find the MRCA of all three. The third line may coalesce with one of the first two before they coalesce, and in this case the MRCA of the first two is also the ancestor of the third *Go* individual. Alternatively, the third line may coalesce with one of the MRCA's ancestors so that the MRCA of all three individuals will precede the MRCA of the first two individuals. In this manner, we can coalesce the lines of one *Go* individual after another until ultimately, we have identified the most recent common ancestor of *all* the individuals in the *Go* generation.

Two characteristics of this coalescence process deserve to be pointed out. First, the lines of all *Go* individuals, regardless of the population's size in that generation (as long as it is finite), ultimately coalesce in a single MRCA. Second, under the specified conditions of the model, the MRCA is never alone in its generation. It has many contemporaries and its only claim to distinction is that it was followed by a line of descendants from whom all the individuals of one of the later generations ultimately arose. All the lines derived from the non-MRCA individuals died out one after the other, they became *extinct*, in the succeeding generations. Which line is to be perpetuated and which is to face extinction is determined by chance alone – again under the specified conditions of the model – so that no special status should be attributed to the MRCA in terms of any exceptional feature or merit. Which of the ancestors becomes an MRCA in one of the succeeding generations is a matter of chance; but that one becomes an MRCA is a matter of necessity. It always happens and after it has happened once, it will happen again. In one of the generations that follow *Go*, one of the individuals again becomes the ancestor of all the individuals in a generation that lies even farther into the future. And so it continues. Time and again an individual emerges from whom all individuals in one of the future generations will be derived.

Later on (see Chapter Eleven), we show that there is a simple relationship between the population size and the average number of generations separating any given generation from the MRCA of all individuals. For a population of *N* individuals, the number is about four times the population size.

In a situation in which each individual has two parents (Fig. 3.2B), lines of descent coalesce into two MRCAs. Under certain simplifying assumptions, a simple formula can be deduced for this situation, too, which estimates the number of generations separating a given generation from the most recent common ancestors of all its individuals. Since the choices of the two parents are made independently, at random, and without regard for the gender of the selected individuals, the same ancestor can be chosen twice as a parent for the same descendant. To overcome this artifact, one can assume that each "individual" is in reality a pair of individuals and that the population in each generation consists of *N* monogamous couples.

Joseph T. Chang used this model to reach the conclusion that the number of generations T_N, going back to the most recent common ancestor couple of all the present-day individuals, depends on the population size, *N*. Provided that the population is very large, T_N is approximately equal to *lgN*, where *lg* is the logarithm to base 2. Applied to the present-day human population of one billion*, $T_N = lg10^9 = 29.9$ generations. In other words, according to

---

* Although the current population size is six billion individuals, we assume a size of one billion per generation. The remaining five billion belong to one or two generations before or after the particular generation.

this model, all the people alive today are derived from a single couple who lived 30 generations ago. Assuming that one generation equals approximately 20 years, this ancestral couple lived a mere 600 years ago! This surprising result with its unrealistically short time period to the MRCA is apparently the consequence of the model's simplifying assumptions. Nevertheless, even if the estimate is wrong by a factor of two or more, it still points to the very recent descent of large groups from a single ancestral couple. This result is in stark contrast to that of the one-parent model, which predicts a much longer time to coalescence.

The two-parent model, however, has more surprises in store. Consider a situation in which all Go individuals have been traced back to a pair of MRCAs in any particular generation Gn, not too distant from Go. The parents of the two MRCAs are also common ancestors of all Go individuals, as are the grandparents, the great-grandparents, and so on. As we follow the track backward through the generations, the number of common ancestors increases until in some distant generation Gx, each individual is either a common ancestor or its line of descent has become extinct at some point between this generation and Go. Going back further from the Gx generation, the same situation repeats itself forever in that every individual whose line has not become extinct is a common ancestor. Chang has deduced that this situation is reached in $1.77 \, lgN$ generations. For the one billion humans composing one generation today, Gx is reached in $1.77 \, lg10^9 = 53$ generations. So, if you meet someone who claims to be a descendant of Charlemagne, you need not be too impressed. You can reply with confidence: "So am I."

Should anyone be tempted to conclude that Chang is substantiating the Book of Genesis, we hasten to reiterate: side-by-side with the MRCA couple were many other individuals. A person attempting to identify the MRCA pair with Adam and Eve either does not understand the term "common ancestor" or is urgently advised to read the Bible more carefully.

From this theoretical analysis, it must be concluded that when correctly constructed and interpreted, pedigrees do not serve the purpose for which they were originally intended – to uphold the VIP status. If every peasant can claim descent from the same illustrious ancestor as an aristocrat, then the latter obviously has nothing to brag about. Pedigrees do not justify usurpation of power and privileges. And as if they had anticipated that the time for devaluation of pedigrees was drawing nigh, the aristocrats set down the inheritance of their privileges in a law and so rendered their special status incontestable.

Unfortunately, standard pedigrees are not suited for our purposes either – to gain an insight into the origin of the human species. If the MRCA pair of all present-day persons was only one of many thousands or millions of biologically equivalent pairs of their generation, and if the pair's ancestors were as human as the pair itself, then the identification of the MRCA does not bring us any closer to our origin. Nonetheless, we have not wasted our time by familiarizing ourselves with the principles of genealogy. If not the pedigrees themselves, then the principles on which they are based will allow us later to study characteristics of past generations which would otherwise have been beyond our grasp. Even more important, in the next chapter we describe how the application of genealogical principles has led to the development of methods which prove suitable for the study of our origin. To understand this application, however, we must first examine the biological underpinning of genealogy. In the preceding chapter, we concluded that the essence of procreation is the transfer of genetic information. Now we must ask what happens to the genetic information of a given ancestor as it passes through a succession of descendants and specifically, how much of it can be expected to filter through the generations from a common ancestor to a group of descendants.

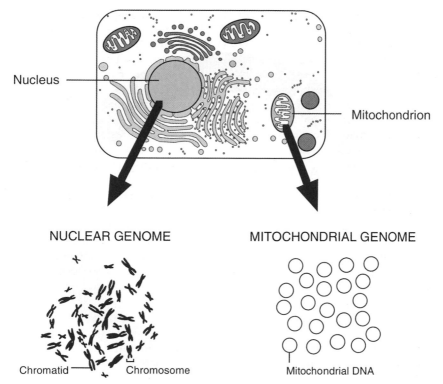

Fig. 3.3
Nuclear and mitochondrial genomes. The human somatic cell shown here contains two editions of
the nuclear genome in its nucleus compacted into 23 pairs of chromosomes, one edition derived from
the mother and the other from the father. (The chromosomes shown here are from a dividing cell and
since they have just duplicated their DNA, each contains two DNA molecules, one in each chromatid.
Hence at this stage, the genome contains four copies of the nuclear genome present in a gamete.) In
its numerous mitochondria the same cell may contain some 80,000 copies of the mitochondrial
genome – 80,000 virtually identical, circular mtDNA molecules. (Modified from Linder 1983 and
Alberts et al. 1989.)

The Floppy Discs of Heredity

The term *genotype* denotes either all the genes of an individual or a few selected genes
which, for one reason or another, are the focus of attention (Fig. 2.2). Perhaps because of
this ambiguity, molecular biologists prefer to use another term, *genome*, to encompass all
the genetic material present in a single cell, genes as well as other parts (Fig. 2.6). By far the
largest proportion (more than 99 percent) of the human genome is found in the *nucleus*
and only a tiny part is present in subcellular structures called the *mitochondria* (Fig. 3.3).
The *nuclear genome* exists in a single (*haploid*) dose in each germ cell and in a double
(*diploid*) dose in each somatic cell. The human haploid genome is partitioned into 23
linear DNA molecules which range in length from 53 to 250 million nucleotide pairs,
the combined length of the molecules being approximately three billion nucleotide
pairs. Returning to the train analogy of Chapter Two and taking 26 meters as the length of

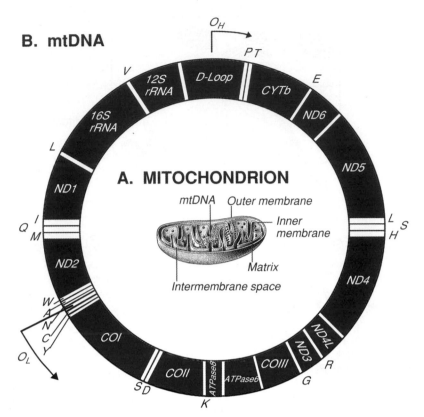

Fig. 3.4

Mitochondrion (**A**) and mitochondrial DNA (**B**). Food taken up by a human being is degraded into fatty acids and glucose which enter the blood stream and are delivered to the cells. In the cytosol (the fluid phase of the cytoplasm), fatty acids and glucose are degraded further and some of the degradation products enter mitochondria, granules of the size and shape of bacteria. In the mitochondria, final degradation of the products takes place in the presence of oxygen. The energy released during the degradation is stored in adenosine triphosphate (ATP) molecules (a nucleotide with three phosphate groups) and used in energy-requiring cellular reactions. Each mitochondrion is enclosed by two membranes, a smooth outer and an elaborately folded inner membrane. The space bounded by the inner membrane is filled with a matrix containing a highly concentrated mixture of enzymes, DNA molecules, and other components. The mitochondrial DNA is a closed circle, present, on average, in ten presumably identical copies in each mitochondrion. It contains genes coding for ribosomal RNA (12S and 16S rRNA, see Chapter Six), transfer RNA (*P, T, E, L, S, H, R, G, K, S, D, Y C, N, A, W, M, Q, I, L,* and *V*; the letters stand for the different kinds of amino acids that the various tRNA deliver to the site of protein synthesis), and enzymes (*cytb, ND1* through *ND6, COI, COII, COIII, ATPase 6* and *8*). The mtDNA replicates independently of the nuclear DNA molecules, the replication of the two strands ("heavy" and "light", H and L, respectively) beginning at different sites (O_H and O_L, respectively; O stands for "origin"; arrows indicate the direction of the replication). The replication of the heavy strand begins by a strand displacement in a segment called the D-loop or the control region, *CR*. The designations "heavy" and "light" refer to the fact that the two strands have a different nucleotide composition and hence different molecular weight (mass).

a railroad car, one strand of the longest human DNA molecule corresponds to a train 6.5 million kilometers long. The real length of the extended molecule is 82 centimeters (the thickness of one nucleotide being 0.327 nanometers). It fits into the nucleus only because it is folded into loops, the loops into larger loops, and these folded again at several levels (Fig. 2.5). The folding is facilitated by specialized proteins which, together with other proteins, the DNA, and certain other materials, constitute the *chromosome*. A human gamete therefore contains 23 chromosomes and each chromosome contains a single DNA molecule.

The exact number of genes present in a haploid genome is still not known, with estimates now generally not exceeding 40,000. The chromosomes are numbered in the order of their decreasing length, with one exception which will be detailed later. The average gene length is 10,000 to 15,000 nucleotide pairs, so that the genes only take up about 3 percent of the DNA sequence. The function of the remaining 97 percent of DNA sequence is not known, but it is generally believed that much of it has no function at all. It is estimated that there is, on average, one gene per 40,000 nucleotide pairs of DNA, but some DNA segments are gene-denser than others. The average number of genes per chromosome is estimated at 1500. Each somatic cell contains twice the amount of DNA compared to a germ cell, since it contains two copies of each DNA molecule (and two copies of each chromosome). The order of nucleotides (their *sequences*) of the two DNA molecules of each pair is very similar, but not identical. On average, one in every 1000 nucleotide pairs is different between the two molecules.

Mitochondria (singular mitochondrion; Fig. 3.4A) are energy-producing granules in the cytoplasm of animal and plant cells. They are the remains of microorganisms that invaded the then emerging single-celled organisms with nuclei 1.5 billion years ago and were ultimately adopted by their hosts to become one of their own components (see Chapter Six). Like the original invaders, mitochondria still have their own DNA molecules, which are not linear but circular, and largely free of associated proteins. The circle has since lost many of the genes it originally contained (some of them moved into the nucleus and were integrated into the chromosomal DNA), but has retained those genes necessary for energy production, as well as a few others. The human mitochondrial (mt) DNA is 16,569 nucleotide pairs long and contains 37 genes (Fig. 3.4B). Each mitochondrion has approximately ten virtually identical circular DNA molecules, and each human cell contains, on average, 8000 mitochondria. (The maturing egg cell, however, contains an estimated 100,000 mtDNA molecules.) How the 99.9 percent identity of the mtDNA molecules in all cells of a body is achieved is still unclear. The favored explanation hypothesizes the derivation of nearly all molecules from a single DNA molecule during the maturation of an egg cell. Of course, if the mature egg cell contains only one kind of mtDNA molecule, then all the derived cells will also contain these molecules. The supposition that sperm cells do not contribute any mtDNA molecules to the adult body and its germ line is supported by most observations, although there are some that contradict it. The sperm cell contains mitochondria in its midpiece and these do enter the egg's cytoplasm during fertilization, but are subsequently destroyed, either in the zygote or in the cells derived from it by the first few divisions.

The chromosomes in the nucleus, together with the DNA circles in the mitochondria, contain the entire genetic information that must be teleported across the gap between two succeeding generations. In this context, the DNA molecules can therefore be compared to floppy discs that store information and enable it to be transferred from one computer to another.

Urdur, Verdandi, Skuld

According to ancient Icelandic sagas, a person's fate is determined by three witches, Urdur, Verdandi, and Skuld. Their counterparts in classical mythology are the three Fates, Clotho, who spins the thread of life, Lachesis, who holds it, and Atropos, who cuts it off. The genetic destiny of an individual is also determined by three events in which blind chance is a decisive factor. All three events are connected with the behavior of chromosomes (DNA molecules) during cell division. To explain, let us denote the molecules present in a mature egg M1 through M23 (M for maternal, the numerals denoting different molecules) and those found in the sperm cell P1 through P23 (P for paternal, the numerals indicating correspondence to M molecules). At fertilization, the M and P sets come together in a single cell, the zygote, thus doubling its DNA content:

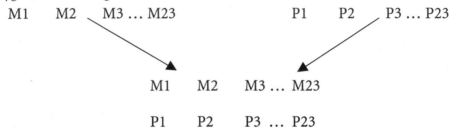

Prior to each of the 10^{14} cell divisions that follow and generate the 10^{14} cells of the human body, the DNA molecules replicate, each producing a copy of itself:

$$
\begin{array}{cccc}
\text{M1} & \text{M2} & \text{M3} \dots & \text{M23} \\
| & | & | & | \\
\text{M1} & \text{M2} & \text{M3} \dots & \text{M23} \\
\\
\text{P1} & \text{P2} & \text{P3} \dots & \text{P23} \\
| & | & | & | \\
\text{P1} & \text{P2} & \text{P3} \dots & \text{P23}
\end{array}
$$

The copies differ from the original only by the uncorrected copying errors (to be discussed later). The vertical lines indicate that the DNA molecules in the chromosome remain physically connected, glued together by specialized sticky proteins, *cohesins*, along their entire length.

During cell division (*mitosis*), the DNA molecules participate in a ritual that culminates in a molecular jamboree, a gathering in the equatorial plane of the spherical cell (Fig. 3.5A). Once assembled in the plane, M pairs commingle with P pairs and each of the molecules is roped by a cordage emanating from the cell's poles. The ropes tie themselves to a special attachment site, the *centromere* (see Fig. 2.5), which has a characteristic, fixed position in the chromosome and also a characteristic nucleotide sequence. The cohesins that hold the two sister molecules together help to orient the attachment sites in such a way that a rope extending from one pole is fastened to one molecule of each pair and another rope extending from the opposite pole is fastened to the other molecule. Once the ropes are in place, a protein is activated that degrades the cohesins, allowing the sister molecules to separate from each other and form distinctly recognizable *chromatids* (see Fig. 2.5). As the ropes then begin to contract, they pull one molecule to one pole and its sister to the opposite pole:

A. MITOSIS   B. MEIOSIS

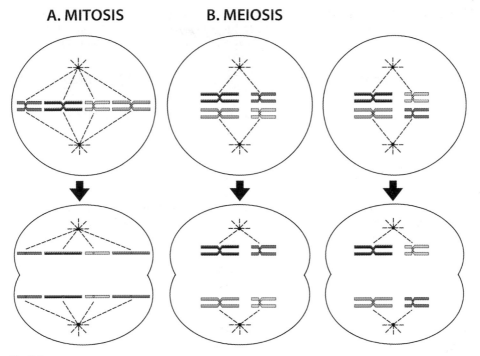

Fig. 3.5
The fate of maternal and paternal chromosomes in mitosis (**A**) and meiosis (**B**). In a somatic cell, one chromosome of each pair is of maternal and the other of paternal origin. Here, only two chromosome pairs are shown, a long (L) pair is distinguished from a short (S) one by color; the maternal (M) from the paternal (P) chromosomes by shading. In mitosis, the division by which somatic cells multiply, each daughter cell receives both the maternal and paternal chromosomes. In meiosis, by contrast, each gamete receives either a maternal or a paternal chromosome. Which chromosome of a pair a given gamete will receive depends on the chance orientation of chromosomes when they come together to form pairs in the equatorial plane. Because of this feature, different gametes end up with different combinations of maternal and paternal chromosomes. Here, only two orientations and the resulting combinations are shown. In the first case, one set of gametes receives the combination LM, SP and the other LP, SM, whereas in the second, the combination in one set is LM, SM and in the other LP, SP. For simplicity, only the relevant stages of the cell divisions are shown. The first meiotic division is then followed by a second (not shown) which resembles mitosis in that it leads to the separation of the two chromatids of each chromosome and their distribution into the daughter cells so that four gametes result from each primordial germ cell. Meiosis serves to shuffle genetic material of the parents before it is passed on to the offspring.

↑	↑	↑	↑	↑	↑	↑	↑
M1	P1	M2	P2	M3	P3 ...	M23	P23
M1	P1	M2	P2	M3	P3 ...	M23	P23
↓	↓	↓	↓	↓	↓	↓	↓

(Here M and P molecules of the same denumeration are shown together in the plane, but in reality they are arranged randomly.) Once the two moving formations have cleared the equatorial plane, a partition arises to split the single cell into two daughter cells, each of which ends up with one M and one P set of DNA molecules:

First daughter cell:	M1	M2	M3 ... M23
	P1	P2	P3 ... P23
Second daughter cell:	M1	M2	M3 ... M23
	P1	P2	P3 ... P23

The M molecules of the same denomination are identical in the two cells, except for the copying errors that arose during replication, and the same is true for the P molecules. Before the next division, the DNA molecules replicate again and the process repeats itself division after division. As a result, all the somatic cells and all the primordial germ cells are genetically identical except for the copying errors.

When, however, the time comes to generate gametes from the primordial germ cells, the ritual changes and so does the outcome. The main change involves the addition of a tête-à-tête between the M and P molecules during their equatorial jamboree (Fig. 3.5B). As in the other form of cell division (mitosis), in the altered form (*meiosis*), the replicated sister DNA molecules stay together, glued to each other by cohesins, but now a new type of bond also forms between the maternal and paternal molecules – M1 pairs with P1, M2 with P2, M3 with P3, and so on:

$$
\begin{array}{cccc}
\text{M1} & \text{P2} & \text{M3} \ldots \text{P23} \\
| & | & | \qquad | \\
\text{M1} & \text{P2} & \text{M3} \ldots \text{P23} \\
|| & || & || \qquad || \\
\text{P1} & \text{M2} & \text{P3} \ldots \text{M23} \\
| & | & | \qquad | \\
\text{P1} & \text{M2} & \text{P3} \ldots \text{M23}
\end{array}
$$

(The pairing is indicated by double vertical lines.) The nature of the bonding between M and P chromosomes remains obscure, but at one stage of the process, the chromosomes are seen to be zipped up by a structure that has formed between them, and at another stage they are physically connected by bridges in the shape of the Greek letter χ (chi), the *chiasmata*. In contrast to mitosis, in which one type of bond temporarily holds the molecules together in pairs, in meiosis two types of bond hold the molecules in quartets, each quartet consisting of one M and one P pair. Then, as the M and P pairs are roped by the cordage from the poles, the chiasmata mysteriously disappear but the cohesins keep holding the two molecules in the M and P pairs together. The ropes are fastened to the same centromere as during mitosis, but now when they begin to contract, they each pull not one, but two DNA molecules to the poles. The first division is immediately followed by a second without an additional round of DNA replication. During the second division, the cohesins holding the two molecules of each pair together are degraded and each member of a pair is pulled to a different pole. The two divisions thus produce four germ cells, each of which contains a single set of DNA molecules (chromosomes), instead of the double set present in somatic cells or primordial germ cells. Since the quartets orient themselves randomly relative to the poles in the equatorial plane of the first division, in some quartets the M pair faces the "south" and the P pair the "north" pole, while in others the opposite is the case. And since the segregation of the M and P molecules into the forming gametes (the pairs are pulled to those poles that they faced from the equatorial plane) depends on the arrangement of the quartets, different gametes receive different combinations of M and P molecules. As there are 23 quartets in the equatorial plane of a human meiotic division, 2^{23} or 8,388,608 different arrangements of quartets are possible. In other words, more than 8.3 million different sperm or egg cells can theoretically be generated. One gamete may end up with the combination M1, M2, M3, P4, P5, P6, M7, M8, P9, M10, P11, P12, P13, P14, M15, M16, P17, P18, P19, M20, M21, P22, P23; another cell may inherit the combination P1, P2, P3, M4, P5, M6, M7, P8, P9, P10, M11, P12, M13, M14, M15, M16, M17, M18, P19, P20, M21, M22, M23; and so on. And since M and P chromosomes of the same denomination may carry different alleles, the single choice that is made from the more than 8.3 million possibilities is Clotho's effect on the genetic endowment, the first of the three lotteries that determine the fate of an individual. The first person to realize it, precociously as it turned out, was Johann Gregor Mendel.

The Greek χ Conundrum

Although chiasmata were first noticed and recognized for what they were almost 100 years ago, the mechanism by which they arise stubbornly resisted elucidation and is not fully understood to this day. From their appearance, it could be deduced that they were either the cause or the consequence of one DNA molecule (expressed in modern terms) from an M-pair crossing over to one molecule of the P-pair and vice versa, and the two molecules breaking at the crossover point and rejoining reciprocally (Fig. 3.6). This deduction was supported by genetic experiments which indicated consistently that genetic exchange did indeed occur between maternal and paternal DNA molecules during the formation of gametes. When electron microscopy later brought to light the existence of the proteinaceous zipper, the *synaptonemal complex*, which aligns the maternal and paternal pairs of DNA molecules along their entire length, the question arose whether it is the process of chiasma formation or the zipper that brings the two pairs together. A definitive answer to this question is still not available, but in the human cells, the former possibility is favored.

Our knowledge of the mechanism by which the maternal and paternal molecules exchange segments is still scanty (Fig. 3.6). It appears, however, that at a certain stage of gamete formation, proteins are made which, when they come into contact with DNA, break both strands of the double helix. The distribution of breaks does not seem to be entirely random. Not only are there certain "hot spots" in the DNA molecules, there also appears to be a tendency for the breaks to occur in the regulatory regions of genes. At the site of the break, specialized enzymes remove nucleotides from one end of each stand, while other proteins protect the other ends, and turn them into flexible tendrils in search of an attachment site. A tendril invades the partner DNA molecule, displaces the noncomplementary strand, and pairs with the complementary strand, forming a *heteroduplex* (a structure in which one strand is derived from one DNA molecule and the other from another). The displaced strand loops out, invades the donor of the tendrils and pairs with its complementary strand. A crossover thus arises between the two DNA molecules which then becomes visible in the microscope as a chiasma.

Ultimately, the chiasmata break at the crossover points, the broken ends are resealed by another set of proteins, and each DNA molecule regains its full integrity. Alas, the molecules now differ from their original state at the beginning of the process: they are no

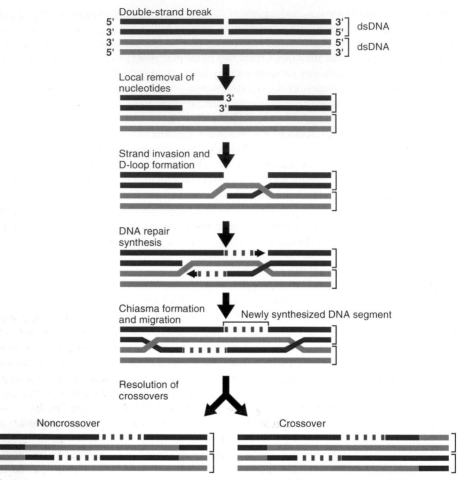

Fig. 3.6
Genetic recombination: shuffling of genetic material between partner chromosomes of a pair. At meiosis, when corresponding chromosomes of a pair come together and align along their length, batteries of specialized proteins effect the exchange of segments between DNA molecules of the partners in the manner shown and described in the text. At the stage in which the exchange takes place, each of the two chromosomes has two DNA molecules so that there are four per pair. Of these, however, only one from each chromosome is shown here. ds, double-stranded. (Based on Roeder 1997.)

longer of the maternal or paternal type but are pastiches of alternating maternal and paternal segments. The number of segments, their length, and their distribution depend on the number and distribution of the breaks in the molecules, the migration of the chiasmata, and other factors. Since, however, each DNA quartet must form at least one chiasma if the division is not to be aborted, in each quartet there are always some composite DNA molecules – molecules that have exchanged or *recombined* their segments by the process of *crossing over*.

The consequence of the exchange of segments is that each gamete receives not only a unique assortment of maternal and paternal DNA molecules, but also molecules with a unique nucleotide sequence composed of both maternal and paternal sequences. This is where the second of the three Fates, Lachesis, leaves her mark on the life of the future individual. The differences between the sequences of the maternal and paternal molecules and the possibility that these differences may translate into dissimilarities in the appearance, physiology, or behavior of an individual means that the length and distribution of the exchanged segments have a significant impact on the individual's life. Since the breaks leading to the exchanges are distributed randomly along the DNA molecules, the number of possible outcomes is enormous and Lachesis distributes them with her eyes covered among the forming gametes.

The third lottery, by which Atropos leaves her mark on the genetic endowment, takes place just before and at the time of fertilization. Atropos selects one of the thousands of eggs for maturation and arranges a potential rendezvous with its fertilizer; she also watches over the marathon involving millions of sperm in which only the very best participant emerges as victor and claims the right to smelt with the waiting egg. For it is at the fateful moment of fusion between a genetically unique sperm and a genetically unique egg cell that the decision is made whether the developing embryo will at a later point die in the uterus because of a genetic defect, or whether it will live and develop into an individual.

Time Travel of a Carolingian Chromosome

Keeping in mind the lotteries preceding the conception of an individual, what can a descendant expect to inherit, in terms of genetic information, from its distant ancestor? If the descendants always wedded an unrelated partner in every generation, the arithmetic of inheritance would be simple. First-generation descendants would each inherit 1/2, second-generation descendants 1/2 of 1/2, that is 1/4, third-generation descendants 1/2 of 1/4, that is 1/8, and the n-th generation of descendants $(1/2)^n$ of the ancestor's genetic information. A present-day descendant of Charlemagne, separated from the King of the Franks by ~50 generations, can therefore be expected to possess $(1/2)^{50} = 8.8 \times 10^{-15}$ of the emperor's genetic information, which is a negligibly small amount. Nuptials between related persons, as must certainly have occurred in the finite central European population and especially in royal families, would have increased the proportion, but certainly not to the degree of justifying any person to blow their own trumpet about it.

We know, however, that human genetic information is partitioned into 23 pairs of packages, the 23 pairs of chromosomes (DNA molecules), and that the packages in each pair exchange parts. It is therefore more instructive to follow the fate of the packages or their parts through the generations rather than the individuals themselves. Take as an example Charlemagne's chromosome number 6. It is highly unlikely, for the reasons described earlier, that Charlemagne passed the same chromosome 6 on to his sons. The sperm that participated in the conception of, say, Louis the Pious more likely carried a composite chromosome 6 derived by exchanges of segments between the two chromosomes 6 present in each of the Emperor's cells. When the time came for Louis the Pious to beget Charles the Bald, there was a fifty-fifty chance that chromosome 6 of Charlemagne's wife would be bequeathed to his son. If that happened, chromosome 6 of Charles the Bald would bear no information whatsoever from his grandfather. But let us assume that this was not so and that the exchange between the two chromosomes 6 of Charlemagne and his wife during the formation of the sperm responsible for the conception of Charles the Bald occurred as depicted in Figure 3.7. In this case, Charles the Bald inherited one chromosome 6 that had only part of the information derived from chromosome 6 of Carolus Magnus. In any succeeding generation in the male line of descent, chromosome 6 containing at least a rem-

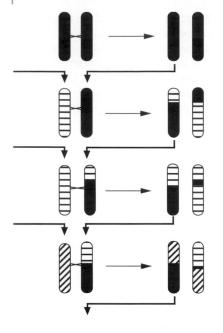

Fig. 3.7
Squandering of inheritance through recombination. The diagram shows how the unique genetic information originally contained in the "black" chromosome is gradually reduced in length over the generations of offspring because of recombination with different partner chromosomes encountered as a result of mating taking place with unrelated individuals. The recombination sites are chosen randomly. A different choice of sites would have produced a different pattern of reduction, but it would nevertheless still have been a reduction. Rods indicate chromosomes, different shading genetic differentiation, "x" recombination, and arrows the origin or fate of chromosomes.

nant of Charlemagne's information might not have been passed on at all, but even if it was, the amount of information would have gradually diminished over the generations of Louis the Stammerer, Charles the Simple, Louis IV "d´Outre-Mer", and all the others that followed them. It probably would not have been so straightforward as the depiction in Figure 3.7 suggests because everything depended on the random positioning of the breaks preceding the exchange of segments, but ultimately the process would have reached a stage in which all that remained of Charlemagne's chromosome 6 in his remote descendant was a single nucleotide pair.

A single nucleotide pair may seem too small a unit of genetic information to follow through generations, but in fact, when the information content of DNA changes, the alteration often consists of the substitution of one nucleotide pair by another. The change is not so different from one that alters the meaning of a word when one letter is substituted by another. The word with the substituted letter often loses its meaning: many single-letter substitutions in the word INFORMATION have this effect – BNFORMATION, INFKRMATION, INFOMBATION, etc. More rarely, it may acquire a new meaning, in which case the change usually involves more than one letter (e.g., CONFORMATION). Changes of nucleotide pairs occur most commonly as a result of DNA replication errors. As the two strands of the DNA duplex separate from each other to serve as a template for the assembly of complementary strands (see Fig. 2.8), from time to time a noncomplementary nucleotide sneaks into one of the sites. This typographical error is picked up by the molecular proofreading machinery and corrected. Now and again, however, the correction may be carried out on the mispaired nucleotide in the template rather than in the newly assembled strand. The result is the substitution of the original nucleotide pair by another, a *mutation*. The frequency with which mutations of this type occur varies depending on a number of factors, but in human cells it is usually to the order of one substitution per 10^9 nucleotide incorporations.

Most of the observed sequence differences between maternal and paternal DNA molecules of a pair are the result of substitutional mutations. The substitutions are usually of no consequence to the individual, either because they occur in the noncoding parts of the DNA or because the change in the coding part is not translated into an amino acid change of a protein. But even mutations that lead to the replacement of one amino acid residue by another often do not affect the protein's function and are therefore phenotypically invisible. Some mutations, however, not only influence a protein's function, but also have an observable effect, and can therefore be easily followed through generations. If the consequence of the mutation is that a protein critical for keeping an organism in a healthy, functional state becomes defective, the result is a *hereditary disorder*. Because most functional genes are of vital importance for a healthy organism, there are as many potential hereditary disorders as there are essential genes. The trail of many of them through generations can be exploited as a source of genealogical information about a single nucleotide pair. One example of a hereditary disorder as a source of genealogical information is hereditary hemochromatosis.

A Celtic Heirloom

One of the genes on human chromosome 6 specifies a protein that is called HFE because it is involved in the regulation of iron (Fe) metabolism. Iron ions* interact with other charged atoms and these interactions play an important part in many of the body's vital functions. The most familiar of these functions is the transport of oxygen by iron ions present in the heme, the nonprotein part of hemoglobin in the red blood cells. There are, however, also several enzymes and other proteins that depend on iron for their function. Iron is absorbed from the ingested food by cells in the upper portion of the small intestine. It is then picked up by the blood plasma protein transferrin and delivered to the various tissues of the body. Since an excess of iron is toxic to the tissues, the uptake of this element must be regulated rigorously – and this is where the HFE protein comes into the picture. It regulates iron uptake by associating itself with another protein on cell surfaces, the transferrin receptor, which in turn binds iron-laden transferrin, internalizes it, and releases iron into the cytosol. Mutations that lead to the replacement of certain critical amino acid residues in the HFE protein abolish the protein's ability to associate with the transferrin receptor. With the regulator incapacitated, the uptake of dietary iron runs out of control and the body begins to accumulate more of the element than it can possibly utilize. The excess iron is deposited in the soft tissues of organs such as the liver, heart, pancreas, and skin. The deposits interfere with the organ's functions and cause the disease *hereditary hemochromatosis* (from Greek *haima*, blood, and *chroma*, color). The name reflects two features of the disease – the fact that its symptoms can be relieved by bloodletting and the characteristic bronzing effect that iron deposits have on the skin. The amount of deposited iron can be enormous: when one of the first autopsies was carried out on a patient who succumbed to the disease, the liver literally creaked under the scalpel. If left untreated, patients with hereditary hemochromatosis die of liver cirrhosis (hardening of the liver caused by excessive formation of connective tissue) or heart failure.

Defective *HFE* genes are a rather common occurrence in certain populations. Among Caucasians, one in every ten individuals has such a gene, which is a very high frequency,

---

* Ions are atoms or groups of atoms that carry an electric charge as a result of having lost or gained one or more electrons.

indeed. (The possession of a defective *HFE* gene does not automatically mean that the individual will develop the disease. Normally both copies of the gene – one inherited from the mother and the other from the father – must be defective for the disorder to manifest itself.) Even more remarkable, however, is the fact that >85 percent of the defective genes have the same mutation, a switch from the G to the A nucleotide at exactly the same site. The substitution translates into the replacement of the amino acid cysteine by tyrosine at position 282 in the HFE protein. The mutation is therefore commonly referred to as Cys282Tyr or C282Y, Cys or C and Tyr or Y being the abbreviations for cysteine and tyrosine, respectively.

Now, mutations are rare events and when they occur, every site in the gene usually has the same probability of being hit. It is therefore highly improbable that among the millions of individuals carrying the Cys282Tyr mutation, the G→A change occurred independently. It is much more likely that in many, if not all of the carriers, the mutation is ultimately derived from the same source, from one individual in whom it originally appeared. Common origin of the mutation is indeed indicated by several studies of hemochromatosis patients in different regions. In one study conducted at the county hospital in Östersund, Sweden, the physicians were able to trace the inheritance of the defective gene through a pedigree encompassing some 9000 individuals, 400 of whom had emigrated to the USA or Canada. The researchers discovered that all their patients had derived the defective *HFE* gene from a common ancestor who lived in early 1700. Similarly, a common origin of the Cys282Tyr mutation could also be assumed for other groups studied by different research teams. Is it possible that the ancestors of these various groups all derived their defective *HFE* gene from a common ancestor? If they did, the ancestor of the ancestors must have lived such a long time ago that no archives, parish registers, or any other source of genealogical information would contain any record of their existence. To find out we must turn from cultural to biological sources of information.

Every individual is genetically unique, the sequence of his or her DNA molecules being different from the corresponding molecules of another individual. The differences are small between individuals of the same species, but they exist and this makes each DNA molecule the one and only in the totality of all the corresponding molecules of all individuals. The uniqueness of DNA molecules is the result of two processes, mutation and recombination. Mutations differentiate corresponding sites in DNA molecules and recombination creates new pastiches from existing sequences. When the Cys282Tyr mutation arose, it emerged in the midst of a unique chromosome 6 sequence of the founder. In subsequent generations, the chromosome disseminated via replication of its DNA molecule into the founder's progeny and, by the union with unrelated partners, also through the whole population. While this was taking place, the segment with the original founder's sequence flanking the Cys282Tyr mutation was progressively becoming shorter through successive recombinations with the partner's molecules, as depicted in Figure 3.7 for the Carolingian chromosome 6. Although the process responsible for shortening is random, it is not unpredictable because it is governed by the laws of probability. These specify that the probability of a break occurring between two sites of the DNA molecules depends on the distance between them. The probability is higher for sites farther apart and lower for sites that are closer together; it is lowest between adjacent sites. They are the same laws that explain why more meteorites fall into the ocean than hit one of the continents. Since the surface area of the oceans is larger than that of the continents, the probability of a meteorite hit is correspondingly higher for the former. Of course, since meteorites keep on falling, they will ultimately fall on continents as well. Similarly, the break will ultimately also occur between specified adjacent sites in the DNA; it just takes, on average, many more generations before it happens.

Suppose we know the rate with which the segment of the original sequence flanking the Cys282Tyr mutation is shortened with time. By comparing chromosomes from different individuals bearing the mutation and determining how much of the original sequence still flanks it, we should be able to estimate how much time has elapsed since the Cys282Tyr mutation arose. Both the rate and the time are measured in generations, because the breaks responsible for the shortening precede generation changeover and because in this case, the clock ticks in terms of the number of opportunities for recombination (i.e., frequency of recombination).

Comparisons of Cys282Tyr-bearing chromosomes from different groups of individuals have revealed that most of the chromosomes share a segment of about six million nucleotide pairs flanking the mutation. Since the frequency of recombination in this chromosomal interval is known to be one recombination per 72 opportunities (divisions of maturing germ cells), it can be estimated that the Cys282Tyr mutation arose ~70 generations ago. Assuming that the generation time in recent history was 25 years (as indicated previously, it was actually shorter in earlier times), we come to the conclusion that the mutation emerged about 2000 years ago. Furthermore, since the mutation occurs most frequently in northwestern Europe or in individuals whose ancestors came from this region, we can infer that it originated at this site. Finally, since 2000 years ago northwestern Europe was still inhabited by Celts (who shortly afterwards were ousted by Romans from the south, Germanic tribes from the north, and Slavs from the east), the ancestor of ancestors was very likely of Celtic descent.

This conclusion does not imply, however, that all people now carrying the Cys282Tyr mutation are of Celtic descent. It only suggests that a piece of DNA, which was originally present in a Celtic person, entered the genome of Germanic people. For this to happen, a single union between persons of these two extractions would have sufficed. Cys282Tyr thus exemplifies how mutations, genes, or other parts of the genome can assume an existence of their own and have their own history. Why this particular part of the human genome has such a long and, in terms of distribution, successful history remains unexplained. By adversely affecting the individual's quality of life, we might expect the mutation to have been lost from the population shortly after its appearance, but in reality just the opposite was the case: the mutation spread. The mutation is therefore thought to bring benefit to its bearers in some important way that more than compensates for the harm it causes. Suggestions for this beneficial effect have ranged from prevention of iron deficiency (and thus protection against anemia caused by hookworm infestation, malaria, multiple pregnancies, or an iron-lacking diet) to protection against infectious diseases caused by pathogens that use the HFE molecule as a hook for attaching themselves to a cell. (The *HFE* mutants have no hook because the altered molecule fails to reach the cell surface.) Evidence to support any of these proposals still has to be put forward, however. It is also interesting that despite the low frequency of mutations, the same mutation apparently arose independently in Sri Lanka. This observation further supports the hypothesis that its possession may well be advantageous.

Taking a Dip in a Gene Pool

The example of hereditary hemochromatosis illustrates the fact that genetic information, like property, is not inherited as a whole, but in pieces, like heirlooms, which get progressively smaller over the passage of generations. The larger pieces are rapidly squandered among the numerous heirs, while the smaller ones have a longer lifetime. If we want to study long-term genealogy, we are therefore well advised to focus on small chunks of genetic information which we can reasonably expect to remain in one piece through many generations. A convenient chunk is the *gene* which in humans is, on average, about 15,000 nucleotide pairs in length and represents a well-defined functional unit specifying one polypeptide or RNA chain.

The study of genes in lines of descent is the subject of *gene genealogy*. To avoid all the complexities that an organism brings with it, it is expedient to put aside the organism as such and concentrate only on the genes. A population of individuals then becomes a population of genes, a *gene pool*, in which each individual is represented by two alleles at each locus, or even two alleles at a single locus only. Producing a new generation amounts to drawing genes from the old pool at random, copying them, returning the originals to the old pool, and placing the copies in a new pool.

The advantage of working with abstract gene pools rather than real groups of individuals is that the behavior of the former is more amenable to mathematical analysis. It can also be simulated by programming a computer to generate new pools over any number of generations, and to keep track of ancestors and descendants during the creation of new generations. At any time during the process, the computer can then provide complete pedigrees of all the genes and extract any information the geneticist might be interested in from them. Both mathematical (analytical) analysis and computer simulation can be used to make predictions about the behavior of genes in real populations, which can be checked by population sampling and testing. Geneticists use the two approaches to define the rules of the coalescence of genes to a common ancestor (the one gene from which they all originated by replication), to follow the fate of new mutations, to determine the consequence of gene migration from one pool to another, and for many other purposes, some of which are discussed later in this book.

A piece of genetic information longer than a gene is called a *haplotype* (Fig. 2.6). Its upper size limit is undefined; it can comprise two or more loci, and also an intergenic region with several mileposts or *markers* in the form of characteristic substitutions along its length. The segment that contains the Cys282Tyr mutation, as well as several other genes in addition to *HFE,* is an example of a haplotype.

In the case of hemochromatosis, fragmentation of genetic information by recombination is exploited to make inferences about a fragment´s origin. Most of the time, however, recombination is a nuisance because it obscures genealogy and complicates its interpretation. Recombination does not stop at the gene's border; intragenic recombination does occur and often frustrates attempts at genealogical reconstruction. Fortunately, there are DNA molecules in the human genome that rarely, if ever, undergo recombination, and it is hardly surprising that these have become popular objects of study for genetical genealogists. They are the mitochondrial (mt) DNA and a large part of the Y-chromosome.

The Circle and the Y

Compared to the 55 million nucleotide pairs of the shortest DNA molecule in the nucleus of a human cell, the mtDNA molecule, with its 16,569 nucleotide pairs, is a Lilliputian. In addition to its small size, mtDNA has four other unusual properties, only one of which, however, is noncontroversial – its circularity (Fig. 3.4B). But the fact that, like any circle, it has no beginning and no end is unusual only in comparison with nuclear DNA. In organisms that have no nucleus – and these are in the overwhelming majority in terms of both numbers and total mass – circularity of DNA is a rule. The three controversial properties

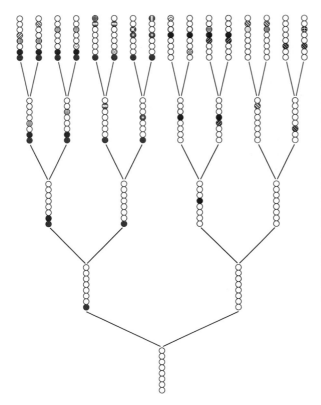

Fig. 3.8
Divergence of DNA molecules by mutations, without recombination. The nature of the circles is unspecified: they could represent nucleotides or genes, and a column of circles could be a single DNA strand or chromosome. The change in the contents of the circles indicates mutations. Connecting lines represent replications of DNA molecules (chromosomes), but also genealogical origin. If viewed from the bottom upwards, the diagram records the gradual accumulation of mutations in successive generations (one mutation per replication). When viewed from the top down, it depicts the reconstruction of genealogy on the basis of mutations being shared between molecules.

of mtDNA are its alleged *homoplasy* (are the thousands of mtDNA molecules in each cell and in all the 10^{14} cells really identical?); its matrilineal inheritance (are all the father's mtDNA molecules delivered by the sperm into the egg cell really destroyed so that only those provided by the mother remain?); and the lack of recombination. Mitochondria, as stated earlier, are vestiges of small bacteria that 1.5 billion years ago invaded a cell with a nucleus and have stayed there ever since. But bacteria with their circular DNA molecules do recombine, so is it really true that mtDNA has lost this ability? There are those who answer this question in the negative, but they are in the minority. It appears that if human mtDNA molecules do barter pieces with one another, they do so with *much* less regularity than nuclear DNA molecules.

If mtDNA is inherited almost exclusively through the female line and recombination takes place at a very low frequency, it allows us to ignore the male parent and thus avoid the problem of progressively increasing numbers of ancestors in a genealogical tracing. In terms of her mtDNA, each daughter has a single mother, a single grandmother, a single great-grandmother – in short, she has a single line of descent. The line ends when it encounters the line of a daughter of another mother. A coalescence of lines is possible because a woman can pass her mtDNA on to several daughters and thus initiate several lines of descent; she becomes the most recent common ancestor of the women at the tips of these lines. In this manner, the lines of descent of all the females in a given generation – for example, all the females alive today – can be traced back to their most recent common ancestor, the notorious "mitochondrial Eve". We say "notorious" because the designation, although widely used, is misleading and has been responsible for a great deal of confusion. We have more to say about it later.

The lack of recombination simplifies ancestry-tracing even further. When there is no shuffling of DNA segments, diversification by mutations proceeds in an orderly manner depicted schematically in Figure 3.8. In this drawing, each circle can be interpreted as one nucleotide pair, one gene, or a longer (entire) segment of the mitochondrial genome (haplotype), depending on the scale one might wish to apply. No matter what we take them for, their change of color or shade signifies that a mutation has occurred. We begin with a single molecule at the bottom of the Figure, with all circles white. The molecule replicates and the copies passed on to the next generation each undergo one mutation. This process is repeated with one mutation taking place in each replication, until we end up with 16 molecules in a generation designated Go, the generation that is accessible to us.

To reconstruct their genealogical history, we note that although the 16 molecules are all unique, they also share mutations (circles of the same color or shade) and that some molecules share more mutations than others. Since mutations occur at a low frequency, we can assume that each of them is unique and that sharing of mutations is an indication of common descent rather than the result of their independent emergence. Furthermore, molecules sharing a larger number of mutations can be assumed to be more closely related by descent than molecules sharing less. Applying these assumptions to the 16 molecules of the Go generation, we first pair them off into eight dyads, each pair sharing a mutation that is absent in all other dyads. We then group the dyads into four quartets and finally the quartets into two octets, always on the basis of shared mutations. To transform the groupings into a pedigree, we connect the members of a group in each tier by lines of descent:

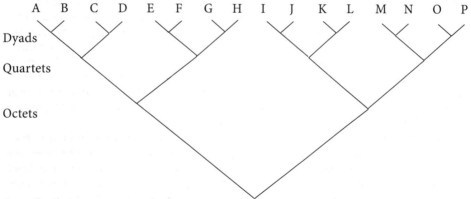

Now you can appreciate how recombination would have scrambled this genealogy. Moving mutations arbitrarily from one line of descent to another would have resulted in groupings that did not reflect genealogy of the mutations.

The above diagram differs from a standard pedigree in that it names only the descendants at the tips of the lines; all of their ancestors remain anonymous and their existence is merely implied by the points of line divergence. Furthermore, the tiers in the diagram do not necessarily represent successive generations, as they do in a standard pedigree. The diagram depicts only those generations and those ancestors in which changes occurred in the character that is being traced (here a mutation, the substitution of one nucleotide pair by another). As depicted, the diagram gives the impression that a mutation occurred each time a DNA molecule replicated. This may be a valid assumption in the case of very long

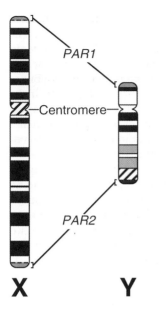

Fig. 3.9
Similarities and differences between human sex chromosomes, X and Y. The chromosomes are depicted as they appear when stained with special dye. Alternating light and dark bands are presumably the result of the dye's binding to regions of differential composition in the chromosome. PAR, pseudoautosomal region in which recombination between the X and the Y chromosome can take place.

DNA molecules, but it does not apply to shorter ones. In the latter, numerous generations may elapse between successive occurrences of two mutations, but these are omitted from the depiction. For these and other reasons, the diagram is no longer referred to as a "pedigree", but is called a *tree* instead, because that is what it resembles in a vague and abstract way – the crown of a tree (see Fig. 1.2). (The term is probably derived from *Stammbaum*, the German word for pedigree, in which *Stamm* is a tribe and *Baum* a tree. Indeed, German evolutionists of the nineteenth century used to depict relationships among organisms or their "tribes" in the form of highly realistically drawn trees.) Correspondingly, the lines of the tree are termed *branches*, the points at which they begin or end *nodes*, and the point or line into which all the lines of descent converge the *root* of the tree.

The mtDNA is not the only part of the human genome that does not recombine or only recombines rarely. A large portion of one of the DNA molecules in the nucleus does the same thing. Recall that there are 23 pairs of chromosomes in each human somatic cell and that in each chromosome there is one DNA molecule. In each pair of chromosomes, with one exception, the DNA molecules are of the same length and of similar nucleotide sequence. The exception is pair No. 23 in male somatic cells, which is an odd couple consisting of a long X chromosome and a short Y chromosome (Fig. 3.9). Female somatic cells each have two X chromosomes with DNA molecules of identical length and similar nucleotide sequence. Because the chromosomes of the 23rd pair distinguish female from male cells, they are referred to as *sex chromosomes*, and all the other 22 pairs are collectively termed *autosomes*.

The Y chromosome differs from the X chromosome not only in length, but also in sequence and gene content. The DNA molecule of an X chromosome is ~165 million nucleotide pairs long, that of a Y chromosome ~60 million nucleotide pairs. The X chromosome contains an estimated 3000 genes, the Y chromosome probably not much more than 30 (26 have thus far been identified), the rest of its DNA lacking genetic information. The known Y-chromosome genes are of two kinds: some have related counterparts on the X chromosome and are expressed in a wide range of tissues, while others are unique to the Y chromosome and their expression is either restricted to or predominant in the testes. The latter include genes that determine the differentiation of a primordial gonad into testes in the developing fetus.

Originally, X and Y were probably a pair of equal chromosomes (autosomes), but at some point in the evolution of mammals, one member of the pair began to specialize in the determination of maleness. This consisted of adapting some of the genes to a new, testes-determining function, inactivating most of the other genes, and converting long stretches into tracts dissimilar in sequence from the original partner. Dissimilarity was achieved through degeneration of the inactivated genes and expansion of repetitive sequences – the amplification of short, meaningless sequence motifs. The consequence of the growing sequence dissimilarity of the two chromosomes was that over long stretches they could no longer exchange segments with each other. In the present situation, recombination between human X and Y chromosomes is limited to two short intervals at the ends of the Y chromosomes, the *pseudoautosomal regions 1* and *2*, *PAR1* and *PAR2* (Fig. 3.9). These two regions have retained sequence similarity with the X chromosome over stretches of ~2.6 million (*PAR1*) and 320,000 (*PAR2*) nucleotide pairs. (They are called "pseudoautosomal" because the inheritance of characters encoded in genes residing in them is not – in contrast to genes in the other parts of the sex chromosomes – linked to the sex of the individual. The regions behave as if they were autosomal although in reality, they are part of a sex chromosome.) *PAR1* engages regularly in exchanges with the corresponding region of the X chromosome; in fact, if at least one exchange per chromosome pair does not take place, the maturation of the sperm cell is aborted. *PAR2* undergoes recombination less regularly and the rest of the Y chromosome, all 57 million nucleotide pairs of it, does not recombine with the X chromosome at all.

If a mutation occurs on the nonrecombining portion of the Y chromosome, it is never passed on to any of the female progeny; it is inherited exclusively through the male line. Like mutations in the mtDNA molecule, those in the Y chromosome have a uniparental mode of descent, but in contrast to the matrilineally inherited mutations in mtDNA, the Y chromosome mutations have a patrilineal inheritance. Except for this difference, the genealogies of the mtDNA and the Y-chromosome mutations are in principle the same and if its meaning were not specified, Figure 3.8 could represent either one. An important difference between the two genealogies is, however, that mutations occur more frequently in mitochondrial than in Y-chromosome DNA but why this should be so is not known. It is possible that the highly active forms of oxygen produced in the energy-generating chemical reactions of mitochondria damage the mtDNA. Another reason could be that mtDNA undergoes many more rounds of replication than nuclear DNA and therefore has a greater opportunity to misincorporate a nucleotide. Also, the proofreading mechanisms might be less efficient in mitochondria than in nuclear DNA. Finally, the mechanisms of mtDNA and nuclear DNA replication are somewhat different and might therefore be prone to errors to a different degree as well.

Mitochondrial DNA and Y-chromosome DNA are two extremes in terms of the speed with which they accumulate mutations, the former being very fast, the latter very slow. Most genes on the autosomes lie somewhere between these extremes, although individual genes vary considerably in this regard. There is thus a wide range of options for a molecular genealogist to chose from, the choice depending on the expected depth of the genealogy. A shallow genealogy, in which lines of descent diverge (or coalesce, depending on one's standpoint) within a short time interval, can best be studied using mtDNA. Particularly suited to this purpose is the part that does not specify any proteins or RNA molecules, the *control region* (it controls replication of the mtDNA molecule) or the *displacement (D)-loop* region (replication begins in this region and proceeds unidirectionally using one of the two old strands as a template and displacing the other; see Fig. 3.4). If the genealogy is expected to be of considerable depth, with lines of descent extending over long time intervals, the choice falls on the slowly diverging, autosomal or Y chromosome DNA. In this respect, the mitochondrial DNA and the Y-chromosome DNA are like a microscope and a telescope, respectively, each more appropriate for a different use, one for "nearsight" and the other for "farsight".

We conclude this chapter with an obligatory stopover at the "orientation board". The chapter's principal message is that there does indeed exist a genealogical record in the DNA molecules which, if correctly used, should lead us to the origin of the human species and beyond. The record is inscribed on the molecules by mutations that change the sequence of nucleotides and thus the molecules' information content. Mutations are responsible for the divergence of DNA molecules in terms of their sequence and this divergence is greater the longer ago the molecules arose by replication from a common ancestor. In this genealogical paradise, however, there is a spoiler in the form of recombination. Exchange of segments between DNA molecules can scramble the genealogical record and so render it unusable. The two ways of avoiding this problem are either to focus on short segments of the genome (such as genes) with a low probability of being broken up by recombination; or to use segments that do not recombine – mitochondrial DNA for tracing female lines of descent and Y-chromosome DNA for tracing male lines. In the next chapter, we investigate ways of exploiting the genealogical information written in the DNA molecules to deduce relationships among species. To this end we must abandon the preoccupation with the traditional genealogy that centers on individuals and focus instead on groups of individuals (populations) and groups of genes (gene pools). We will learn that history of populations and species is often not identical with gene genealogy, not even with the non-recombining portions of the genome. To reconstruct the former, we must study many independent regions of the genome – regions that do not exchange segments by recombination.

CHAPTER 4

Klados and Phylé

The Molecular Nature of Evolution

*T*he Tigris and Euphrates flow
from a single source in the
Achaemenian rocks, where the
Parthian warrior turns in his flight
to shoot his arrows in the pursuing
enemy, but they quickly flow apart
in separate streams. If they should
join their waters again in one
channel, all that each stream car-
ries would come together; the boats
that sail on each, the floating trees
torn up by floods, and the waters
too would mingle by chance. But
steep channels and the downward
flow of the current govern these
seemingly random events. Chance,
too, which seems to rush along
with slack reins, is bridled and
governed by law.

Anicius Manlius Severinus Boethius:
The Consolation of Philosophy

And Everything has Changed so Much in Me

No, this is not going to be a love story from Ovid's *Metamorphoses*. Unlike Philemon and Baucis or Pyramus and Thisbe, Klados and Phylé is the story of two cherished icons of evolutionary symbolism – *klados*, the branch and *phylé,* the tribe. To relate it, let us pick up the thread where we left off in the preceding chapter.

We start with a few simple calculations. The development from a fertilized egg to a human adult body entails, as was already mentioned, approximately 10^{14} cell divisions. Every division is preceded by the replication of all the DNA molecules in the cell and in each replication, uncorrected copying errors occur at a rate of one error per 10^9 incorporated nucleotides. Since a human cell contains 3×10^9 nucleotide pairs, on average 12 copying errors occur during the replication of the DNA before each cell division. Mutations preceding most of the 10^{14} divisions are of no consequence for evolution and are therefore irrelevant for our purpose. They occur in somatic cells (and hence constitute *somatic mutations*) which die before or at the time of the individual's demise and the mutations disappear with them. Somatic mutations may profoundly influence the fate of an individual, for example by contributing to the development of cancer, but since they are not passed on to the progeny, their effect is restricted to a single individual. Mutations in the germ cells or in their progenitors, on the other hand, *are* of interest to us, because they are potential genealogical markers.

Female and male germ cells differ in the number of mutations they accumulate before they participate in fertilization. From the human zygote to a mature egg cell, 24 cell divisions are necessary. Since cells accumulate mutations at a rate of six per division, the mature egg cell gathers $(24)(6) = 144$ mutations that were not present in the zygote. Mutations that occurred in earlier divisions are shared by the different egg cells, whereas those that occurred just before maturation distinguish one egg cell from another. In contrast to eggs, which arise from progenitors that stop dividing at the time of birth, progenitors of sperm cells continue dividing throughout the entire lifetime of the male. The number of mutations a sperm cell has accumulated therefore depends on the age of the male. A sperm cell produced by a 25-year-old male, for example, is separated from the zygote by 312 cell divisions and has therefore accumulated $(312)(6) = 1872$ mutations. A sperm cell of a 50-year-old male is the product of approximately 887 cell divisions and hence has accumulated $(887)(6) = 5322$ mutations. Because, however, only about 3 percent of the total DNA contained in a cell is believed to code for proteins, only four of the 144 mutations borne by an egg cell and 56 of the 1872 mutations borne by a sperm cell of a 25-year-old male change the protein-specifying information. Furthermore, since only approximately two-thirds to three-quarters of the mutations in the coding sequence can be expected to change the corresponding protein sequence (the remaining mutations are of the synonymous type), the egg and the sperm cell of a 25-year-old male can be expected to contribute $(56)(2/3) = 37$ to $(56)(3/4) = 42$ protein-changing mutations to the zygote. When Guillaume Apollinaire wrote in his poem appropriately entitled *Mutation*

>> *Et tout*
>>> *A tant changé*
>>>> *En moi**

---

* *And everything has changed so much in me. In Guillaume Apollinaire: Calligrammes. Poems of Peace and War (1913-1916).* Translated by Anne Hyde Greet. University of California Press, Berkeley, 1980.

he was referring to mental changes inflicted by his war experiences. But he could just as well have been referring to biological mutations his body had undergone.

At fertilization, therefore, the two germ cells introduce more than 2000 new mutations into the zygote. By the time the individual is ready to reproduce, it will contribute its own share of new mutations to the second generation, endowing it with some 4000 mutations that were not present in the zygote two generations earlier. In a direct line of descent, the number of mutations should increase from one zygote to the next by an increment of about 2000 mutations. At this pace, it would take less than one million successive zygotes in a line of descent to scramble the entire DNA sequence, and for the sequences of the first and last zygotes to appear unrelated to each other. Since in many organisms the interval between two successive zygotes in a line of descent is less than one year, any two lines that originated from the same zygote one million years ago should have totally dissimilar DNA sequences. That is not what happens, however. Mutations do not accumulate from zygote to zygote in this simple manner; certain factors slow down the accumulation.

In Tyche's Realm

To understand these factors, we must follow the fate of individual mutations after they have arisen. To lighten our task, we disregard chromosomes, cells, individuals, and groups of individuals, and consider the genes as leading a life of their own – we examine a *gene pool*. In such an abstract situation, the act of reproduction is reduced to a random drawing of genes from one pool to yield a new one. The successive pools represent generations *G0, G1, G2*, etc, starting from the original pool. To ensure that the genes are drawn at random, we use a paradigm of fortuity – a fair die. To accelerate the process, we first consider a ridiculously small gene pool consisting of six genes only. We number the genes from *1* to *6* to keep track of them, and write the *G0* generation thus: {*1, 2, 3, 4, 5, 6*}. To produce *G1*, we pick six genes at random from *G0*, while maintaining a constant pool size in successive generations. To prevent depletion of the sampled pool, and to give each gene an equal chance of being selected in every one of these *random samplings,* we do not actually remove the chosen gene from the pool but copy it and place the *copy* in the next generation pool. We use the die to choose the genes. We throw it once and, *alea jacta est*, take the gene corresponding to the number that comes up; we repeat this process six times, and obtain, for example, the numbers {*2, 3, 4, 6, 1, 3*}, which constitute the *G1* pool. Note that gene 5 has not been sampled at all, whereas gene 3 has been sampled twice, all other genes having been sampled once. In other words, gene 5 has been lost from the pool; it cannot reappear in any of the succeeding generations – it has become *extinct*. The reason for its extinction is by no means inferiority to other genes, but plain bad luck! As with every lottery, random sampling has its winners and its losers, and gene 5 just happened to be on the losing side. Gene 3, on the other hand, made a lucky strike – it was chosen twice.

To produce the next generation, *G2*, we can no longer use the standard die; we have to replace it with a die lacking the ⚄ face and having two ⚀ faces instead, to reflect the gene representation in *G1*.* Throwing this die, you might obtain {*6, 1, 2, 4, 2, 3*} in *G2* and

---

* If you think you can use the standard die and simply ignore throws in which 5 comes up, forget it. Doing so would give all 5 numbers in G1 an equal chance, which would be false. Since 3 is represented twice, it has a better chance of being thrown than the other numbers; this fact must be provided for by using a die with two number 3 faces.

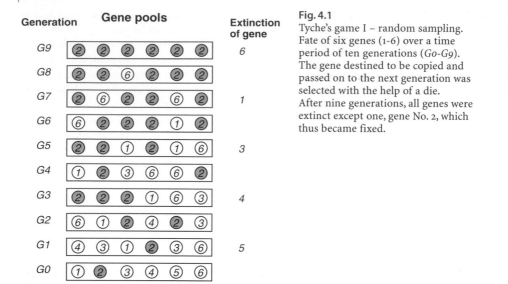

Gene pools

Generation							Extinction of gene
G9	②	②	②	②	②	②	6
G8	②	②	⑥	②	②	②	
G7	②	⑥	②	②	⑥	②	1
G6	⑥	②	②	②	①	②	
G5	②	②	①	②	①	⑥	3
G4	①	②	③	⑥	⑥	②	
G3	②	②	②	①	⑥	③	4
G2	⑥	①	②	④	②	③	
G1	④	③	①	②	③	⑥	5
G0	①	②	③	④	⑤	⑥	

Fig. 4.1

Tyche's game I – random sampling. Fate of six genes (1-6) over a time period of ten generations (G0-G9). The gene destined to be copied and passed on to the next generation was selected with the help of a die. After nine generations, all genes were extinct except one, gene No. 2, which thus became fixed.

{2, 2, 2, 1, 6, 3} in G3. Now gene 4 has also been lost, whereas gene 2 has been selected three times; gene 3, despite its numerical advantage in G1, appears only once. Such are Tyche's caprices – Tyche being the goddess of chance in ancient Greece, our Lady Luck! Continuing in this manner, we gradually lose one gene after another until we are left with one gene only, in our case No. 2 (Fig. 4.1). We say that gene 2 has become *fixed*.

If we were to repeat the experiment, we would observe once more that after a certain number of generations, all the genes except one become extinct. The fixed gene could be the same as in the first experiment, or it could be one of the five other G0 genes, depending purely on chance. The same holds true for any other round of the experiment. Although we cannot predict which of the six genes will become fixed in any specific experiment, we can be certain that one definitely will and that all the other genes will be lost.

We can depict the behavior of the genes graphically by calculating their frequencies in the individual generations. A *frequency* gives the number of elements (events) in the total number of elements (events). Since in G2 there are two copies of gene 3 in the total number of six genes, the frequency of gene 3 is 2/6 = 0.33 = 33 percent. By plotting the frequency of a given gene on the y axis against successive generations on the x axis of the xy coordinate system, we obtain a line that depicts changes in the numerical representation of the gene in succeeding generations (Fig. 4.2). Each of the six genes zigzags through the graph until it either becomes extinct (frequency of 0) or fixed (frequency of 1 or 100 percent). The movement resembles the drift of a leaf that has fallen on water: ultimately the leaf either sinks to the bottom of the sea or is washed up on shore. The random changes in the frequency of a gene over generations are therefore referred to as *random genetic drift* or simply *(genetic) drift*.

Although the six genes in G0 were assigned different numbers, we assumed that they were all identical. In this case, it did not matter which gene became fixed and which was lost; the composition of the gene pool remained the same. If, however, one of the genes mutates, the fate of the genes *is* of consequence. If the mutation had occurred in gene 5 of

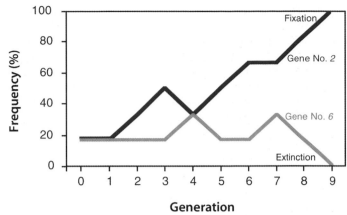

Fig. 4.2
Random walk of two genes from Figure 4.1: the frequency of gene No. 2 increased until, in the ninth generation it replaced all other genes – it became fixed. Gene No. 6 varied in frequency over the generations until it became extinct in the ninth generation.

our example, it would have been lost in *G1*; if it had occurred in gene *4*, it would have been carried over two generations and would then have become extinct. But if the mutation had occurred in gene *2*, it would have become fixed and the entire *G9* gene pool would differ from the *Go* pool. We would then say that the gene pool has *evolved*. And so, after three chapters and over thirty-two thousand words, we finally introduce terminology that, according to the book's title, the reader probably expected to find in the opening paragraph.

The Evolution of Evolution ...

Like many other words, "evolution" has several meanings. An art historian may pronounce that Gauguin's manner of painting evolved from an imitation of the Impressionists to a more subjective, richly colored style of his own. An elderly aficionado of comics may give you a lecture on the evolution of Mickey Mouse's character through the seven decades of his existence. An astronomer may write a book on stellar evolution, from protostars, through the main sequence stars and red giants, to white dwarfs and supernovae. And a politician may agitate for peaceful evolution as opposed to violent revolution, or the other way around. In each of these contexts, "evolution" is used in a different sense, but the different meanings nevertheless all have one thing in common: they imply a continuing process of change.

In biology, an evolutionary change constitutes a *modification* of a previous state. The modification may alter a single attribute or *character* of an organism, or it may affect several characters at once. With time, the same character may change again and again, passing from one *character state* to another, so that it or the affected body part can be said to be evolving. Alternatively, characters may change within a group of organisms (a population, a species) and then it is the group that is evolving.

Change underlying biological evolution has two characteristics that distinguish it from all other forms of evolution. The first of these is its tight link to *descent*. Once it has occurred in an individual, the biological evolutionary change is passed on to the descendants, to be handed on further along the line of descent. The change is *hereditary* and evolution can be described as *descent with modification*, that is, modification along a line of descent. The element of descent is lacking in all nonbiological usages of the word "evolution". Evolution of a painter's style, of a person's character – real or comic, or of a star, is limited to a single individual or an object and all the accumulated changes vanish with the demise of that individual or object. Gauguin's style of painting began to evolve in the summer of 1873 while he was still working as a stockbroker in Paris, and ended on May 8, 1895 when he died on Hiva-Oa, one of the Marquesas Islands in the South Pacific. He did not and could not bequeath it to his children. Similarly, when a star dies, it may give rise to new stars, but it cannot endow them with the changes it has accumulated through its period of existence. An evolutionary change, on the other hand, is perpetuated from generation to generation until it is superseded by a new change.

The second characteristic of a biological evolutionary change distinguishing it from nonbiological changes is its potential for *dispersion*, for spreading through a gene pool (population). Dispersion is the process that leads from the state in which the mutated gene is one of many, to the state in which there are no other genes in the pool. The frequency of the mutated gene straggles through the generations like a drunken sailor thrown out of a pub, now going up, then down, then up again. Biologists say that the mutated gene takes a *random walk* – a process consisting of a sequence of steps each of which is determined by chance. The random walk of both the sailor and the frequency terminates either in a ditch (the mutation becomes extinct) or at the home base (the mutation becomes fixed). Before the mutation occurred, all the genes in the pool were of the same form – the locus was *monomorphic* (Greek *monos*, single; *morphe*, form). While the mutation was drifting through the pool in the succession of generations, and its frequencies were undertaking a random walk, there were two forms (more if additional mutations arose) of the gene – the locus was *polymorphic* (Greek *polys*, many). Fixation of the mutation restored monomorphism once more, but this time at a qualitatively different level: the original gene has been supplanted by its mutated allele. The replacement of one allele by another in a population is termed *substitution* (not to be confused with the form of mutation in which one nucleotide changes to another, which is also call "substitution"). It is this change of states that is referred to as *evolution*. In this sense, *evolution is the change of gene frequencies over the generations*. Evolutionary changes at the molecular level constitute *molecular evolution*. The concept of frequency makes sense only in reference to a *set* of objects and, in the present context, also over a period of time; it cannot be applied to a single entity or to an instant. Obviously, therefore, it is not possible to speak of the evolution of an individual; evolution always refers to a population of individuals or a gene pool.

Of course, genes do not diffuse through a pool like a splash of dye in water. In many organisms, humans included, dispersal is achieved by sexual reproduction. Sex also delimits the pool through which a mutation can disperse. Think of two groups of individuals, *A* and *B*. If, regardless of the reason, group *A* individuals do not seek sexual partners among group *B* individuals, or if they do, but their mating does not result in offspring, or if it does, but the offspring is infertile, then a mutation that occurred in *A* cannot spread into *B*, and vice versa. Biologists say that the two groups are *reproductively isolated*. The only sexual act that counts with respect to evolution is catholic sex, with the goal of producing fertile offspring. Only in a *population of interbreeding individuals* and the corresponding gene pool can there be dispersion and fixation of mutations. We are all well familiar with such groups of interbreeding or potentially interbreeding individuals and call them *species*. Human, common chimpanzee, mountain gorilla, and rhesus macaque are four different species: we do not expect a human to mate naturally, or successfully, with a chimpanzee, a chimp with a gorilla, or a gorilla with a rhesus macaque. We expect humans to mate with other humans, chimps with chimps, gorillas with gorillas, and macaques with macaques.

Because of this reproductive isolation of species, a mutation that has arisen in one species cannot normally (there are exceptions) spread to another species. Recognition of this fact raises the next question: How do species arise?

... and the Evolution of Revolution

This question is viewed very differently in the Eastern and Western cultures. In Eastern thought, humans are considered to be an integral part of nature, and nature itself is wholeness in which transition between forms, including species, is rather fluid. The origin of species is therefore given no close attention, or if it is considered, then devoid of the emotional associations with which it is laden in the West. In Western thought, dominated in the last 2000 years by Judeo-Christianity, by tradition man has not only set himself apart from nature, he has also elevated himself to the position of nature's overlord and conquistador. Both notions are rooted in the Book of Genesis in passages that attribute a singular standing to man ("God created man *in his own image*" implying that he has nothing to do, in terms of origin, with the beasts which were created independently) and giving him a special right to subjugate nature ("... replenish the earth, and subdue it: and have dominion over the fish of the sea, and over the fowl of the air, and over every living thing that moveth upon the earth"). Traditionally, the Book of Genesis has also been credited as the source of the belief that God created all the species now in existence and that species have not changed since their creation. As late as the eighteenth century, "God's Registrar", the Swedish naturalist Carolus Linnaeus stated laconically but authoritatively that *Tot numeramus species, quot creavit ab initio infinium ens* – "there are as many species as were created at the beginning by the infinite being." Although this great classifier of nature apparently began to experience serious doubts about the validity of this proclamation later in life, he was by then too old and too famous to risk endangering his reputation. As the son of a pastor, he knew very well that the church would immediately rise to any challenge that threatened its dogma, the immutability of species. (Incidentally, as we read the Bible, we find no passage that could be interpreted in any way as supporting species immutability.)

Because it is dangerous to tamper with a dogma, any opposition to the notion of species constancy has always been rather meek. Nonetheless, alternative points of view were put forward from time to time and it is probably no coincidence that these were expressed in France and England, two European countries in which the influence of the church had been weakened considerably in the nineteenth century. Furthermore, it is also not surprising that the two men who finally overthrew the dogma, Charles Robert Darwin and Alfred Russell Wallace had both traveled widely and possessed first-hand experience of the tremendous diversity of life in different parts of the world. In the tropics they could not help but notice the fierce competition among living forms, which was much less conspicuous in the manicured agricultural landscapes of countries such as England or France. Charles Darwin is the second of the unlikely revolutionaries in our story, the first being Gregor Mendel. With his distaste of controversy, confrontation, and publicity, and his yearning for a peaceful life of observation, experimentation, study, and writing, Darwin was ill-fitted for overthrowing dogmas. Fortunately, there were others prepared to fight for his cause when in 1859, he finally agreed to the publication of his work *On the Origin of Species by Means of Natural Selection, or the Preservation of Favoured Races in the Struggle for Life*. In this book, Darwin accomplished two things, both of which can be classed among the greatest achievements in biological sciences. He amassed so much evidence for the transmutation of one species into another to change a mere speculation, the *transmu-*

tation hypothesis, into a solid theory. And, concurrently with Wallace, he identified a mechanism – natural selection – which is responsible, in part at least, for the transmutation.

In the first edition of *On the Origin of Species* the word "evolution" does not appear. Although it already existed at that time, it was used in reference to the embryological *development* of an individual, which was interpreted, in a distant echo of the preformation days, as unfolding or rolling out (Latin *evolvere*) of a preexisting design. It was, however, still during Darwin's lifetime that the two terms underwent an inversion: what was previously "evolution" became "development" and what Darwin called "descent with modification" and others referred to as the "developmental hypothesis" became "evolution". Darwin acknowledged this inversion by using the term "evolution" in the sixth edition of *On the Origin of Species* published in 1872.

A Tale of Two Pools

Both Darwin and Wallace argued that extant* species arose from species that are now extinct, and furthermore suggested how this may have taken place. Instead of presenting their argument, we provide here its modern version applied to the molecular level. Imagine that a gene pool has been split and the two pools have become isolated from each other so that no *gene flow* (movement of genes from one pool to another) can take place. A frequent reason for isolation is geographical change. For example, a brook flowing through the jungle can swell to a majestic river which thwarts any attempts to cross it. A monkey population that previously did not heed the brook is suddenly divided into two and the genes of one cannot flow into the other. From this moment on, the populations and their gene pools evolve independently of each other. In each pool, mutations arise and some of them sweep through the pool to become fixed. Because the probability that the same mutation will arise independently in both pools is fairly low (there are so many sites that can mutate and mutations are so rare!), chances are that different mutations become fixed in the two pools. As a result, the pools gradually diverge from each other in the sequences of their genes. Some of the fixed mutations (substitutions) will change the amino acid sequences of the encoded proteins and some of these alterations will change the proteins' function.

Because all genes have the same probability of mutating, it can be expected that at some point, genes that control the monkeys' reproduction will also be changed. There are probably dozens of genes which specify different aspects of reproductive behavior, from courtship and mate selection, through compatibility of sex organs and germ cells, to development of the embryo and fetus in the mother's body. If any of these genes change differentially in the two pools, the populations not only become physically but also reproductively isolated. If the river separating the two populations were then to diminish, allowing individuals of the two populations to intermix freely once more, they would not be able to interbreed. They might no longer recognize each other's courtship signals, their germ cells might be physically incompatible, or the hybrid fetuses might be aborted. No longer separated by a physical barrier, the two populations are nevertheless isolated by a *reproductive barrier*. They have become two distinct species.

The process of species formation is *speciation*. The mode of speciation just described is *cladogenesis*, literally "coming into being by branching". In cladogenesis a single species

---

* *Extant*, as opposed to *extinct*, are those forms whose representatives are still alive today. Their alternative designations are "recent", "modern", or "living".

splits into two, which is why the icon of this process is the letter Y, a stem splitting into two branches on a tree. Theoretically, speciation could also occur without splitting, by *anagenesis* (literally "upward coming into being"). If we could follow a gene pool over many generations, we would witness the fixation of one mutation after another and with each fixation, a further divergence from the initial state. Here, too, we might expect mutations to lead ultimately to reproductive isolation of the later-generation pools from the initial one. The problem is that we cannot compare the two pools because the initial one is long gone.

The principal difference between the two modes of evolution is that in anagenesis, evolution proceeds along a single lineage, whereas in cladogenesis, an ancestral lineage splits into two. Since the single, anagenetically evolving lineage has no counterpart to which it could be compared, the substitutions it accumulates go unrecognized, as if it has not evolved at all. By contrast, each of the cladogenetically split lineages can be compared to its partner and so most of the substitutions are recognized as differences between the two. For these and other reasons, cladogenesis is normally the only mode of speciation accessible to molecular analysis. We have more to say about cladogenesis and anagenesis in Chapter Eleven (see Fig. 11.3).

Since speciation is such an important aspect of evolution, the definition of evolution set out earlier should perhaps be amended. In the amended version, *evolution is the change in gene frequency of a population over the generations that in time produces new species.* Indeed, in Western civilization, most people do not give a hoot about gene frequency changes, but they can become hysterical when the discussion turns to the origin of species.

The Elegant Euphemisms of Algebra

The trademark of Tyche's realm is randomness. But contrary to popular belief, randomness is not akin to unpredictability. The pools – regardless of whether they retain their singularity or keep on splitting, and despite the random walk of their constituting elements – behave in a predictable manner. This behavior is most concisely described when – in the words of the poet W.H. Auden – "couched in the elegant euphemism of algebra". The couching is not always gentle but, delivered in small doses, it may be amiable enough, we hope.

Let us, then, take a closer look at the behavior of the six genes in the pool that we introduced earlier, and let us describe their behavior in terms of probabilities. In mathematics, *probability* is defined as the ratio of the number of specific outcomes of a trial to the total number of possible outcomes. Casting a die is such a trial. It has six possible outcomes and so the probability of one particular face, say $\boxed{\bullet}$, of coming up is 1/6. Similarly, drawing a gene from a pool of six also has six possible outcomes – either gene No. 1, 2, 3, 4, 5, or 6 is picked. Hence the probability that a particular gene will be chosen, say gene No. 1, is also 1/6. The probability that gene No. 1 OR gene No. 2 will be chosen is then 1/6 + 1/6 = 2/6 = 1/3. The probability that gene No. 1 OR No. 2 OR No. 3 OR No. 4 OR No. 5 OR No. 6 will be chosen is 1/6 + 1/6 + 1/6 + 1/6 + 1/6 + 1/6 = 6/6 = 1; it is a certainty. Since one of the genes will be chosen for sure, the probability that no gene will be chosen is 0. Hence probabilities are expressed on a scale from 0 to 1.

To produce the *G1* pool, we pick a gene from the *G0* pool (population), copy it, return the original to *G0*, place the copy into *G1*, and repeat this act six times. Statisticians call a procedure in which an item is picked from a parent population *sampling*, since not all the items but rather only a *sample* is taken from the parent population. And since, in this case, the original is returned to the parent population, they speak of *sampling with replacement*.

If the identity of the gene drawn from the *Go* pool is determined by chance, we speak of *random sampling*. The generation of the *G1* pool from the *Go* pool therefore constitutes random sampling. Assuming randomness, each gene of the parent population has the same probability of being sampled and since it is returned to the *Go* pool, it can be picked up again when the second gene is drawn from the *G1* pool. Theoretically, a gene, say No. 3, may by chance not be included in the *G1* pool at all, or it may be included in 1, 2, 3, 4, 5, or 6 copies, depending on how many times it is picked in the six draws. Each of these possibilities has a certain probability and it is this probability we now would like to define.

The expression "picked once in six draws" can be written as 1/6, in which form it represents a *frequency* and so the entire row of possibilities (0/6, 1/6, 2/6, 3/6, 4/6, 5/6, 6/6) represents a *distribution* of frequencies or simply *frequency distribution*. Since each of the possibilities has a certain probability corresponding to the frequency distribution, there is a *probability distribution* – an array of probabilities of the individual possibilities (see Appendix One).

Rolling a die or letting a computer do it for you is one way of finding out what kind of results you can expect from a random sampling of a population or a gene pool. We can call it *empirical* because it is based on experiment, observation, and experience. Mathematicians, however, prefer to deal with a problem such as population sampling in an *analytical* way by striving to describe a process and express its expected results by using methods of algebra. In our case, the analysis involves probabilities, specifically the probability that a gene will or will not be sampled, and if the former, how many times it can be expected to be sampled in a given number of trials. It is a case in which a mathematician would like to specify the probability that an event will happen (= *success*) or will not happen (= *failure*) and if it does occur, would like to know how many times it can be expected to take place (= number of successes) – to determine the probability distribution of 0, 1, 2, 3, ... k successes.

In a probability distribution describing random gene sampling one would like to have a series of formulae specifying the probability that a gene will be drawn (sampled) 0, 1, 2, 3, ... k times. One would also like to have a formula specifying how to obtain the series of individual probabilities from a general *expression* (i.e., a mathematical symbol or a combination of symbols). Mathematicians call such a formula a *theorem* and the process of converting the general mathematical expression into a series such as that specifying the individual probabilities an *expansion of a theorem*, the individual members of the series being its *terms*. In Appendix One we reach the conclusion that the distribution of probabilities applicable to random gene sampling is specified by the *binomial theorem*. It is so designated because it specifies how to expand the expression $(p + q)^n$ which consists of two terms (symbols) connected by a plus sign (*binomial* means literally "having two names"). The expression instructs the user to carry out the multiplication $(p + q)(p + q)(p + q)$ ... n-number of times. The multiplication is cumbersome, however, especially when n is large. The binomial theorem specifies a procedure which leads to the same result more conveniently:

$$(p+q)^n = \sum_{k=0}^{n} \binom{n}{k} p^k q^{n-k}$$

Here the expression on the right-hand side of the equation is the binomial theorem. The expression $\binom{n}{k}$ is the *binomial coefficient*, which can be rewritten in the form:

$$\binom{n}{k} = \frac{n!}{k!(n-k)!}$$

where ! stands for "factorial", a number obtained by multiplying all the positive integers less than or equal to a given integer (see Appendix One). The expansion of the theorem gives the result of the binomial $(p + q)^n$:

$$\sum_{k=0}^{n}\binom{n}{k}p^k q^{n-k} = \binom{n}{0}q^n + \binom{n}{1}pq^{n-1} + \binom{n}{2}p^2 q^{n-2} + \ldots + \binom{n}{n-2}p^{n-2}q^2 + \binom{n}{n-1}p^{n-1}q + \binom{n}{n}p^n$$

Applied to probability problems the expansion is the *binomial probability distribution* specified by the binomial theorem:

$$B(k;n,p) = \binom{n}{k}p^k q^{n-k}$$

where p is the probability of success (a given gene is drawn) in a single trial; q is the probability of failure (a given gene is not drawn) in a single trial; n is the number of trials; k is the number of successes in n trials (the number of times a gene is drawn in one series of draws); and $B(k;n,p)$ is the distribution of probabilities of 0, 1, 2, 3, ... k number of successes. Hence the probabilities of a gene being sampled 0, 1, 2, 3, ... times in one series of draws (e.g., in the formation of the G_1 pool by sampling of the G_0 pool) is:

Number of times gene has been sampled (k)	0	1	2	3
Probability	$\binom{n}{0}q^n$	$\binom{n}{1}pq^{n-1}$	$\binom{n}{2}p^2 q^{n-2}$	$\binom{n}{3}p^3 q^{n-3}$ $\cdots$

To return to our example, the probability that two copies of gene No. 3 will appear in G_1 is

$$\binom{6}{2}\left(\frac{1}{6}\right)^2\left(\frac{5}{6}\right)^4 = 0.201$$

because p, the probability of success in a single trial, is 1/6; q, the probability of failure, is 1 $- p = 1 - 1/6 = 5/6$; and

$$\binom{6}{2} = \frac{6!}{2!4!} = \frac{(6)(5)(4)(3)(2)(1)}{(2)(1)(4)(3)(2)(1)} = \frac{(6)(5)}{(2)(1)} = 15$$

Compared to the probability that No. 3 will be chosen only once, which is

$$\binom{6}{2}\left(\frac{1}{6}\right)\left(\frac{5}{6}\right)^5 = 0.402$$

the probability of being chosen twice is lower (0.201). Once No. 3 has contributed two copies of itself to G_1, the probability that the following generation (G_2) will receive two copies of No. 3 will be higher than in the preceding generation. Although each of the two copies of No. 3 in G_1 will have the same probability of 1/6 of being chosen, the probability that one OR the other copy will be chosen is 1/6 + 1/6 = 1/3. Since the probability of choosing No. 3 in G_1 is 1/3, the probability that G_2 will contain two copies of No. 3 is

$$\binom{6}{2}\left(\frac{1}{3}\right)^2\left(\frac{2}{3}\right)^4 = 0.329$$

Now imagine a pool composed not of six, but of 6000 genes. In this more realistic case, all probabilities will be lower than in the six-gene pool. The probability of a gene being chosen to contribute a copy of itself to the next generation

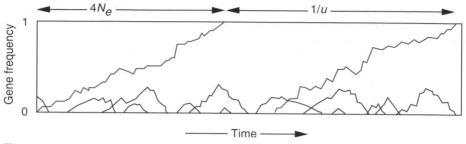

Fig. 4.3
Frequency changes of mutant genes in a finite population over generations of random sampling (time). Each line represents a new mutation. The fate of genes that became extinct is indicated by black lines, the fate of those that became fixed by red lines. The average time of mutations destined to be fixed from their appearance to their fixation is $4N_e$ generations. The average time between two successive fixations is $1/u$. Here, N_e stands for effective population size and u for mutation rate. (Based on Kimura 1983.)

becomes

$$\binom{6000}{1}\left(\frac{1}{6000}\right)\left(\frac{5999}{6000}\right)^{5999} = 0.368$$

instead of 0.402. The probability of the same gene being drawn twice is

$$\binom{6000}{2}\left(\frac{1}{6000}\right)^{2}\left(\frac{5999}{6000}\right)^{5998} = 0.184$$

instead of 0.201; and so on. With such a pool size, the odds against a mutation taking a foothold in the population are enormous and so, as a consequence, the great majority of new mutations become extinct immediately after they have arisen. Nevertheless, given enough time, even in very large gene pools a few lucky mutations will assert themselves against all odds and become fixed. Because they have so many more genes to replace than mutations in small pools, however, it takes them much longer to reach the summit – to achieve the frequency of 1. The larger the pool, the longer and slower the climb of a lucky mutation toward fixation.

The curve representing the ascent of a lucky mutation is in reality a graphic depiction of changes in the gene frequency (here frequency of the mutant gene) over succeeding generations (Fig. 4.3). Since gene frequency is given by the number of copies of a particular gene divided by the total number of genes in the pool, it is obviously related to probability, defined in a similar manner. To give a frequency of a thing or an event is another way of expressing the probability of its occurrence. The probability changes of a mutation over a successive series of lotteries therefore also represent frequency changes of that mutation over successive generations. The amount of frequency change per generation depends on the population size N: the larger the N, the smaller the change. If we knew the precise relationship between the N and the frequency change, and if we had a way of measuring the change per generation, we should be able to estimate the population size without counting heads. Taking a census may seem to be a simpler method of obtaining N, but what if the population does not exist anymore, as would be the case with any past human population?

An underlying assumption of any deduction about the relationship between population size and gene frequency changes is that each gene has an equal chance of contributing its copy to the next generation. If this condition is not fulfilled, the estimate of N will deviate from the real population size. The condition may not be fulfilled under a variety of circumstances. In our earlier example of three individuals (six genes), for instance, one individual would have to be of one sex, say a male, and the remaining two individuals of the opposite sex. Since for the procreation of offspring the participation of both sexes is required, the two genes of the male must make a greater contribution to the next generation than the genes of the females. Continuing with our example, we can also imagine a situation in which the size of the population is reduced to two individuals in one generation and expanded to four in the next. This fluctuation in population size again influences the sampling process. At the stage of the reduction ("bottleneck") to two individuals, only the genes of these two individuals are sampled, the other genes present in earlier generations have no chance whatsoever of passing on their copies through the bottleneck. Or imagine a large population divided into an array of smaller subpopulations. If breeding were to take place predominantly within each of these subpopulations, as would very likely be the case, then genes of a given small population would have a much greater chance of forming pairs among the offspring than genes from different subpopulations. All these situations do indeed occur in nature. Sexes *are* often represented in different proportions; population sizes *do* fluctuate from one generation to the next; and large populations *are* divided into smaller ones.

In these and other similar situations, only a fraction of the total number of individuals is able to produce progeny and so contribute their genes to the next generation. To a geneticist, the genes in a population appear to fall into two categories – the grain and the chaff, the former being necessary for seeding a new generation, the latter being good for nothing. It is, naturally, the grain that a geneticist is interested in – the stockpile of the seeding genes. The number of individuals who bear the genes of the stockpile constitutes the *effective population size*, N_e. It is the size of an ideal population in which each gene has an equal opportunity of contributing its copies to the next generation and which yields the gene frequency changes observed in a given real population. In a randomly mating population, it corresponds roughly to the number of breeding individuals and is usually smaller than the census population. Here again, we have more to say about effective population size later (see in particular Chapter Eleven).

The mathematics of drifting genes has been worked out by two geneticists in particular, the American Sewall Wright and Motoo Kimura from Japan. Paradoxically, neither Wright, nor Kimura had any formal training in math; both were mathematical autodidacts. Wright discovered his talent when, as a postdoctoral fellow, he witnessed a friend struggling with a mathematical interpretation of genetic results. When he could no longer listen to his friend's never-ending curses over pages and pages of calculations, Wright leaned over his shoulder to see what it was all about and then solved the problem within a couple of hours. Similarly, when he discovered his vocation for theoretical genetics and realized that he could not pursue it without a detailed knowledge of mathematics, Kimura made high math his hobby. His other hobby was breeding orchids, Lady Slippers in particular. For Kimura, the two hobbies merged into one when his mathematical deductions acquired the beauty and the elegance of the blossoms he was cultivating on his patio. Kimura died recently. Ironically for a man who had devoted his career to the study of chance, he passed away on the seventieth anniversary of his birth.

Some of the drifting-gene mathematics is simple enough to be understood even by those who toiled their way through primary school math. Other parts, however, are too complicated to be presented here and so a description of their main conclusions will have to suffice. In a population consisting of N breeding individuals, there are $2N$ genes in the pool because each individual carries two copies of a given gene in each of its cells. If each of the 2*N* genes has the same chance of contributing a copy of itself to the next generation, then the probability of a given gene being chosen is 1/(2*N*). Similarly, if each gene has the same opportunity of becoming fixed, the probability of a newly arisen mutation becoming fixed is also 1/(2*N*). If new mutations arise at the rate *u* per gene per generation and if there are 2*N* genes in a generation, the total number of new mutations appearing in each generation equals the total number of genes multiplied by the mutation rate, or 2*Nu*. Each new mutation has the probability 1/(2*N*) of becoming fixed; therefore, the probability (*k*) that any of the mutations will become fixed is the probability that one mutation will become fixed multiplied by the total number of mutations:

$$k = \left(\frac{1}{2\,N} \right) 2\,Nu = u$$

This is an important insight discovered by Kimura: the probability of fixing one of the mutations that has arisen per generation is equal to the mutation rate. We denote this probability the *fixation rate* because it represents the frequency of an event (fixation) per unit of time (Fig. 4.3). Its other names are *substitution rate* and *evolutionary rate*. The former refers to the fact that in the pool, all genes are replaced by the one that suffered a mutation; the latter to the definition of evolution in terms of gene frequency change. Fixation rate is generally very low, to the order of 0.001 fixations per generation, which equals one fixation per 1000 generations. A rate of 1/1000 means that when one fixation occurs, the next fixation can be expected to occur, on average, 1000 generations later. The interval between two successive fixations is, in this example, 1000 generations, which is the reciprocal of 1/1000 (i.e., of the fixation rate). Generally, since the fixation rate is *u*, the average interval between two successive fixations is 1/*u* (Fig. 4.3).

As already mentioned, the length of time a mutation spends climbing toward fixation (the *fixation time*) depends on the size of the gene pool, 2N_e. You can get a fairly good idea about the relationship between fixation time and N_e by comparing the results of random sampling from gene pools of different size, for example 6, 60, and 600. To throw a die for this many genes becomes tedious, but it is relatively simple to program a computer to do so. If you repeat the sampling process several times for each pool size, you realize that the average fixation time is roughly four times the effective population size, or 4N_e generations (Fig. 4.4). This conclusion can also be deduced analytically, albeit with the help of higher math, by using one's brain instead of a computer. The march of a gene toward fixation somewhat resembles Brownian motion as displayed, for example, by a particle of India ink or a pollen grain in water. The particle is bombarded by molecules of water appearing from different directions and with different velocities. At any particular moment, the combined action of the molecules imparts a certain direction and velocity to the particle until in the next moment, the bombarding water molecules conspire to move the particle in another direction and with a different velocity. The particle thus zigzags randomly through the fluid. Since the movement of the particle is determined by random hits, the direction the particle will take and the speed with which it moves at any particu-

Fig. 4.4
Tyche's game II: the experiment depicted in Figure 4.1 was repeated 100,000 times with the help of a computer. In each repetition the generation in which one of the six genes became fixed was recorded and the number of times that fixation occurred in the different generations was plotted. In some experiments, fixation occurred in *G2*; at the other extreme, fixation did not take place until *G40*.

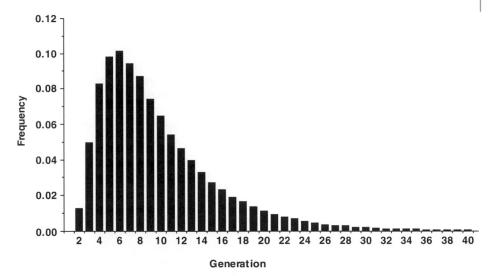

Generation

Therefore the *range* of the fixation time (the difference between the largest and smallest value of the set) was 40 – 2 = 38 generations. The *mean* (average) of the distribution is obtained by taking each value of the variable x (here the fixation time), multiplying it by its frequency (f), adding up the individual products thus obtained, and dividing the sum (Σ) by the total number of observations (n). In a mathematical shorthand description, these instructions are expressed as $\Sigma f_i x_i / n$, where Σ (sigma) means "add up starting with the ith value", $f_i x_i$ is the ith value of the variable multiplied by its frequency, and n is the number of replications. Hence $\Sigma f_i x_i = (1)(0.0001) + (2)(0.0124) + (3)(0.0495) + (4)(0.0826) + \dots /100{,}000 = 9.68$. To evaluate the spread of the individual observations from the mean, we calculate the *variance*, s^2 (v, var), which is the sum of squared differences between each value of x and the mean ($\bar{x}$), divided by the total number of observations, n:

$$s^2 = (x_1 - \bar{x})^2 + (x_2 - \bar{x})^2 + (x_3 - \bar{x})^2 + \dots (x_n - \bar{x})^2 / n$$

or in the shorthand formulation

$$\sum_{i=1}^{n} (x_i - \bar{x})^2 / n$$

For the data shown, we have $s^2 = 34.15$. The square root of the variance, the *standard deviation*, *s.d.* or σ (sigma), calculated for the data shown is

$$\sigma = \sqrt{\sum_{i=1}^{n} (x_i - \bar{x})^2 / n} = 5.84$$

The expected mean fixation time for a gene pool containing six genes ($N_e = 3$) is $4N_e$, i.e. $(4)(3) = 12$ generations. The observed mean fixation time in this computer simulation is, however, 9.68 generations. One reason for the discrepancy is that the $4N_e$ formula was originally derived by Kimura in 1969 under the assumption that the initial frequency of a gene destined for fixation is close to zero. This, obviously, is not the case in a pool containing six genes, each of which has an initial frequency of 1/6 = 0.17. Kimura has also shown that if the initial gene frequency is high, the probability of fixation increases and the expected mean fixation time becomes correspondingly shorter. In the example shown here, the expectation is not $4N_e$ but 91 percent of this value, which comes to 10.92 generations. Eleven generations, however, is still one generation longer than in the simulation result. The likely reason for this discrepancy, according to a study published in 1969 by Kimura and Tomoko Ohta, is the difference in the models used. In the derivation of the $4N_e$ formula, Kimura assumed that time increased continuously, whereas we simulated time increase in discrete intervals. In the continuous time flow model, the genealogy is such that a gene of a preceding generation can pass two copies of itself at most on to the next generation. In the discrete model, on the other hand, a gene can pass on more than two copies of itself, and this feature facilitates the fixation process and shortens the fixation time.

lar moment cannot be predicted. The overall movement of the particle can, however, be foretold with a certain probability. The challenge of determining this probability was taken up by some of the best minds in physics, including Max Planck and Albert Einstein. Planck and A. D. Fokker developed a set of equations for exactly this purpose and these were then used, first by Wright and later in a different way by Kimura, to determine how long a newly arisen mutation takes to reach fixation. Both geneticists came to the conclusion that it requires approximately, and on average, $4N_e$ generations.

Assuming an effective population size of a species as approximately equal to 10,000 individuals (see Chapter Eleven), the fixation of a mutation can take a very long time – 40,000 generations, in each case. For a generation time of 20 years, the fixation time amounts to (40,000)(20) or 800,000 years. (Here, *generation time* refers to the period beginning with a zygote in one generation and ending with the zygote of the following generation. To simplify matters, we assume that generations do not overlap, so that all individuals in a population reproduce more or less at the same time. This assumption obviously does not apply to humans, but it is approximately valid for many other species. Just think of the birds in your garden as spring comes!) In this regard Figure 4.3 is misleading because it gives the impression of rapid frequency changes from one generation to the next, while in reality the changes are quite slow. This impression arises from the compression of generations on the x-axis of the graph. A geneticist, who can only follow genes over a few generations at the most, will hardly notice any frequency change at all. He is like an ant living on a mountain slope: "Which slope?", the ant might ask. To use gene frequency changes over generations to estimate population size is therefore no simple matter.

Let us summarize what we have learned so far about the behavior of mutations in a gene pool (Fig. 4.3). Most new mutations disappear from the pool shortly after their emergence; others linger on for a while but are ultimately lost, too. Only occasionally does a mutant gene continue increasing in frequency until it replaces all other genes in the pool. The probability of a mutant gene being the lucky one is $1/(2N_e)$. The fixation rate is equal to the mutation rate u, the interval between two fixations is $1/u$, and the length of time required by a mutant gene destined for fixation to rise from its initial frequency of $1/(2N_e)$ to the frequency of 1 is approximately $4N_e$ generations.

One point needs to be emphasized as far as these conclusions are concerned: the formulae provide *statistical* estimates; they give an expected *average* value rather than an exact outcome. The values obtained in a specific case may differ from this average considerably – a statistician would say that the individual trials have a large *standard error* (Fig. 4.4; see Appendix One). Thus, for a population of $N_e = 10,000$, the theory predicts a fixation time of 40,000 generations, but the actual value can be 60,000 generations for one mutation and 20,000 generations for another. Statisticians can define the range of standard error and estimate the probability with which a particular deviation from the expected average value will occur; they cannot, however, predict a particular outcome with accuracy.

In the Realm of Antecedent Causes

Up to this point, we have assumed that each gene in a pool has the same probability of being picked to contribute a copy of itself to the forming pool of the next generation. Genes that behave in this manner are said to be *neutral*. All the mutations considered in the preceding section were also assumed to be neutral. Although we encountered genes in the discussion of the effective population size that were excluded from participation in the lottery, this was due to circumstances and not to some intrinsic property of the genes themselves. Had the genes been given the opportunity of joining the lottery, they would have had the same chance of being drawn as all the other genes. In all these cases, chance was the only factor determining whether a gene would be drawn or not. It was the same factor as that responsible for the scatter of bullets fired at a target by a gun with fixed sights. Because the Greek word for a target is *stochos*, behavior with chance as the determining factor is said to be *stochastic*. Up until now we have been looking at the stochastic behavior of genes in a pool. Now, however, we must turn our attention to behavior influenced by antecedent causes, behavior *determined* by factors other than chance alone. The mode of evolution in which not only chance, but also additional factors influence the behavior of genes is called *deterministic*.

In a deterministic mode of evolution a gene is chosen more or less often than would be expected if it were chosen exclusively by chance. The reason for this does not lie in the circumstances but in the gene itself. The gene behaves differently because it is different from the other genes due to a mutation it has suffered. The act of choosing an element from a group because of some characteristic the element possesses is *selection*. The act of choosing a gene from a gene pool because of a mutation it carries is *natural selection*. (The term was introduced by Darwin to distinguish selection taking place in nature from *artificial selection* carried out by humans.) Selection in which the probability of a gene being transmitted into the succeeding generation is lower than that of other genes is called *negative*; selection characterized by an increased probability is *positive*. Mutations responsible for lowering or increasing a gene's probability of transmission are referred to as *deleterious* and *advantageous*, respectively.

Although the reason for the altered transmission probability lies ultimately in the genes themselves, the actual increase or decrease in probability is effected indirectly via a change that the mutated gene causes in the structure or function of the body. The nature of the change, the point in time at which it manifests itself in the life cycle, as well as the site in the body at which it becomes evident vary depending on the gene. Ultimately, however, the change boils down to two major effects: it influences the individual's chances of survival or reproduction. A measure of the change is the *selection coefficient (selection intensity)*, s. To define it, let us assume that in a pool there are genes with a higher probability and genes with a lower probability of transmission to the next generation, and let us designate the transmission probability as *fitness*, w. Next, let us take the gene with the highest transmission probability (e.g., gene A) as a standard and for the sake of argument assign to it the transmission probability (fitness) of 1 or 100 percent. The transmission probability (fitness) of a less successful gene (e.g., a) is then $w = 1 - s$, where s is the selection coefficient. We can therefore define the selection coefficient as the quantity that indicates the degree of reduction in the transmission probability of a gene compared to the standard. A selection coefficient of 0.01, for example, represents a reduction in transmission probability of 1 percent compared to the 100 percent transmission probability of the standard gene. If the fitness of gene A is $w = 1$ (100 percent), then the fitness of gene a is $w = 1 - 0.01 = 0.99$ or 99 percent. Note that both w and s are defined in relative terms by taking one value as a standard and comparing all other values to it. Because of this relativity, the genes can be viewed as if they were competing with one another for transmission to the next generation. In reality, of course, the genes do no such thing, but the individuals that carry them do, and competitive success manifests itself ultimately in the number of progeny they produce – in their *fecundity*. In terms of fitness of an individual all that counts is the number of offspring. Since, however, reproductive success in nature is predicated on *survival*, and survival on physical fitness, reproductive success depends to a large extent on physical fitness. It was in this sense that the word "fitness" was introduced into evolutionary biology

A. NEUTRAL: NO SELECTION

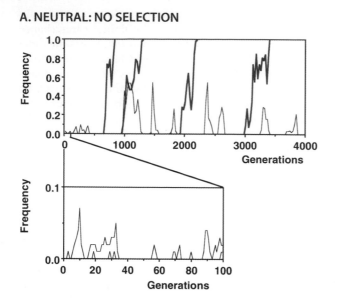

B. DELETERIOUS: NEGATIVE SELECTION

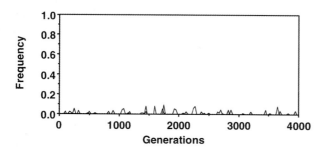

C. ADVANTAGEOUS: POSITIVE SELECTION

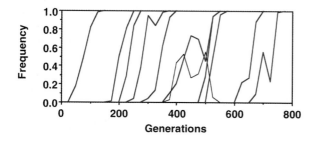

by Darwin when he coined the phrase "survival of the fittest". Because the interactions between genes leading to their differential transmission are acted out among the individuals of a population, the concepts of fitness and selection coefficient are more commonly applied to the level of the phenotype than to that of the molecule. Moreover, since diploid individuals have two genes at each locus, and since it is their combination that determines the fitness of an individual rather than the genes themselves, the two terms are usually applied to genotypes.

The mathematics of gene behavior under natural selection was worked out primarily by two Englishmen, London-born and Cambridge-educated Ronald Aylmer Fisher and John Burdon Sanderson (J.B.S.) Haldane, who was born in Oxford and entered the University there. In character, these two men could not have been more different. Fisher, trained in mathematics and theoretical physics, had narrowly focused interests in theoretical genetics and statistics, and was both withdrawn and conservative. Haldane was trained in mathematics and biology, and in contrast to Fisher had a broad spectrum of interests that ranged from physiology to genetics, and even biochemistry. He was outspoken, liberal, and colorful. Fisher's *The Genetic Theory of Natural Selection,* published in 1930, and Haldane's *The Causes of Evolution,* published in 1932, are still worth reading. To a "mere" biologist, Wright, Fisher, Haldane, and Kimura may appear as mathematical wizards, who have their heads high above the clouds and are often out of touch with reality. It is probably for this reason that their contributions are not always fully appreciated. It is certainly not always easy to follow their argumentation, especially Fisher's. His severe myopia forced him to carry out mathematical derivations in his head without resorting either to blackboard or paper. He was therefore in the habit of excluding derivations in his

◄——

Fig. 4.5A-C
Computer simulation of neutral (**A**), deleterious (**B**), and advantageous (**C**) genes in a pool consisting of 100 genes ($2N_e = 100$) over 4000 generations. In **A**, to form the pool of the next generation, 100 independent random numbers between 1 and 100 were generated and used to determine which genes (numbered from 1 to 100) will be contributed to the next generation. Since it is assumed that each gene has an equal probability of contributing a copy of itself, all 100 genes are equivalent in terms of reproduction. A mutation in a gene creates a new allele. To specify this allele, each gene is assumed to consist of 50 sites, each site occupied by one nucleotide. Genes are assumed to mutate at the rate of u per site per generation. At generation G0, all the genes were assumed to have been in the standard (nonmutated) form (the 50 sites of each gene were labeled by 50 zeros). A mutation at a given site replaced 0 by 1. The frequency of each new mutation was plotted on the ordinate against generations on the abscissa. The lower part of the Figure depicts a blown-up segment of the upper part. Only by a change of scale (the lower part shows frequencies in all the first 100 generations, the upper part every 25 generations only), do the many low-frequency mutations become visible. The upper part shows only alleles that would be considered polymorphic (frequencies of >0.01 and <1.0, black lines) or that ultimately would become fixed (red lines).
In **B**, the simulation scheme is the same as in **A**, except that all new mutations are deleterious. Here, no fixation occurs and no mutant attains a frequency greater than 1 percent. The simulation demonstrates that deleterious mutations do not contribute to molecular evolution because they are eliminated shortly after their appearance in the gene pool. Not all the mutations that have arisen are shown; low-frequency mutations do not show up on the scale used.
In **C**, the simulation scheme is the same again as in **A** except that all new mutations are advantageous over earlier ones. In this set-up, ten mutations became fixed during the first 800 generations. The rate of fixation was 10/800 = 0.012, which is 12 times higher than that (4/1000 = 0.001) in **A**. Furthermore, the time until fixation is considerably shorter than $4N_e$ (200) generations. Here again, not all the mutations are shown. Black lines indicate mutations that are ultimately lost; red lines mutations that became fixed.

publications, understandably to the chagrin of his readers. Wright, perhaps because his mathematical abilities were largely self-taught, was somewhat more considerate of his readers.

There are several varieties of natural selection, each with its own theoretical idiosyncrasies. Here, we focus on one in which both of the two different alleles at a locus have an effect on the phenotype. Haldane and Kimura deduced mathematically that in populations undergoing this form of natural selection, the probability of fixation, f, is $2sN_e/N$ or roughly twice the selection coefficient. The formula applies to cases in which the initial frequency of mutation is $1/(2N)$; in other words, in which the mutation is present in a single copy in a population consisting of N diploid breeding individuals. For a mutation with a selection coefficient of 0.01, for example, the fixation probability is 2 percent, which means that it has a 98 percent chance of being lost. Earlier, we defined the fixation (substitution) rate as the probability of a mutation being fixed times the mutation rate. For advantageous mutations, therefore, the fixation rate, k, is $k = 2Nu2sN_e/N$, which simplifies to $k = 4N_esu$, where u is the rate of occurrence of advantageous mutations. Thus the substitution rate of advantageous mutations depends on the effective population size, the magnitude of the selection coefficient, and the mutation rate.

An important take-home message from all these algebraic deductions is that selective advantage by no means guarantees the fixation of a mutation. On the contrary, the great majority of advantageous mutations, just like the majority of neutral mutations, are predestined for extinction (Fig. 4.5). The only difference between the two is that the advantageous mutation has a somewhat greater chance of becoming fixed than neutral mutations. The lucky advantageous mutations that reach the summit do not do so in a straight line; they drift just like the neutral mutations, but their random walk is of shorter duration. To return to our earlier simile, the advantageous mutations can be compared to a drunken sailor supported by an M.P. The two will still stagger, but the sailor's path home (or the route to jail) is shorter than if he were reeling alone. Deleterious mutations can only become fixed under special circumstances (e.g., in a very small population); normally they are all lost, on average in a shorter time than neutral mutations.

Sports on a Neutral Territory

Charles Darwin knew nothing about genes, gene pools, changes at the molecular level, or random drift, and his view of heredity was hopelessly antiquated. Although a reprint of Mendel's *Versuche* was readily available to him, he almost certainly did not read it (probably because it was written in German) and so was blissfully ignorant of the true nature of heredity. He was acquainted with mutations, but only in the form of *sports*, a term used by animal and plant breeders for rarities that appeared suddenly and deviated strikingly in their appearance from the norm. What he did not know, of course, he could not take into account in his formulation of an evolutionary theory, which he based exclusively on observations at the phenotypic level. He recognized only two types of modification (today we would say mutation) – deleterious and advantageous, and only one kind of evolutionary mechanism determining the fate of the modifications – natural selection. In his view, natural selection weeded out the deleterious mutations and promoted the advantageous ones. Essentially, this was the same view adopted by virtually all knowledgeable evolutionary biologists when genes, gene pools, and the rest, became known. If anybody considered the possibility of neutral mutations also existing, the thought that they may participate in evolution was certainly inconceivable. Sewall Wright was well aware of the existence of random genetic drift, but he did not consider it as a mechanism on a par with natural selection. He believed that in small populations, genetic drift was responsible for rapidly creating new gene combinations from which natural selection then picked the fittest for fixation. But this was as far as he would go in challenging the orthodoxy.

In 1968, however, Motoo Kimura published a short article in *Nature*, innocently entitling it "Evolutionary rate at molecular level". With its seeming innocuousness, however, came a blow under which the monolith of evolutionary theory began to crack. Prompted by certain inconsistencies between predictions based on the theory of natural selection and the degree of genetic variability observed in natural populations at the molecular level, Kimura put forward two revolutionary theses. First, at the molecular level, neutral mutations are far more common than advantageous mutations; and second, neutral mutations are fixed by random genetic drift, fixation being the prevailing mechanism of evolution at the molecular level. The implication of both theses was that in addition to natural selection, there is a second principal mechanism of evolution – fixation of neutral mutations by random genetic drift. The Americans Jack Lester King and Thomas H. Jukes, who reached a similar conclusion at about the same time and published their deductions in *Science* one year after Kimura, expressed their challenge to orthodoxy more bluntly by entitling their article "Non-Darwinian evolution". The soft-spoken, shy, and modest Kimura thus became the third in our gallery of unlikely revolutionaries. The publication of his two theses in *Nature* had an impact on the biological community comparable to Martin Luther posting his 95 Theses on the door of the Wittenberg Castle Church.

The reaction to the Kimura-King-Jukes *neutral theory of molecular evolution*, as it came to be known, was swift and savage. It cannot be true, proclaimed one eminent evolutionary biologist, without offering specific reasons for the rejection. Random drift does occur, but has nothing to do with evolution, said others. Later, as so often happens with revolutionary ideas, there were those who claimed that the neutral theory idea was meritorious but not truly novel. The choice of the expression "non-Darwinian" in the title of the article by King and Jukes in particular was equivalent to poking a stick into a hornets' nest – and the hornets were *not* amused! Much of the initial critique revealed a lack of understanding on the part of the respondents. Some opponents thought, for example, that the neutral theory denied the existence of natural selection, which of course it did not. Kimura made it clear from the very beginning that his theory applied to the major part, but not to all of the mutations occurring in a population. Specifically, he excluded all mutations responsible for the emergence of adaptive characters from his considerations. (*Adaptive*, in this context, refers to any inherited character that makes an organism more adjusted to the conditions of its environment.) These mutations are very rare, but when they do occur, they are rapidly driven toward fixation by Darwinian natural selection. Because of their rare occurrence and rapid fixation, such mutations contribute very little to variability in populations, which Kimura was trying to account for. Kimura also had to explain that he did not deny the existence of disadvantageous mutations and that he was not claiming that all mutations were neutral. (Beyond their initial salvo, King and Jukes did not actively participate in the controversy surrounding the neutral theory, the defense of which was left largely to Kimura and his followers. Kimura later brilliantly summarized his position in his book entitled – how else? – *The Neutral Theory of Molecular Evolution*, a masterpiece of scientific writing.) Kimura chose to ignore the deleterious mutations because they contributed neither to the observable variability, nor to molecular evolution.

The real reason for the ferocity of the attack, however, was probably philosophical. For many people it came as a terrible shock when Darwin proclaimed chance and randomness, in the form of what today are known as mutations, to be the source of evolution. It was, however, far worse when Kimura also made chance responsible for the actual process of evolution at the molecular level. As perceptively pointed out by the philosopher Daniel C. Dennett in his book *Darwin's Dangerous Idea*, the strong opposition to Darwin's theory arose because of the realization that it corroded the cherished notion of a purpose in human existence. Some semblance of purpose could possibly have been retained had there been directionality in natural selection. Although patently false, assuming directionality could have provided the proverbial straw for the drowning concept of purpose. But Kimura seemed set on purloining the very last straw (which was not really true because he did not actually deny the existence of natural selection) and replacing it by total randomness. All that was left was arbitrariness, haphazardness, a world without design or aim. Human existence in a purposeless world is difficult to stomach for many people. Only a philosophical aversion of this kind could explain the acrimony with which the battle over the neutral theory of molecular evolution was fought.

The dust has not yet settled after the battle, but it has cleared to the extent that the positions of the entrenched armies are now visible. The mode of fighting has changed, too. Gone are the days of inflammatory rhetoric; attention is finally focused on designing tests that can pit explanations based on the neutral theory against those based on the selection theory. The two theories differ in their predictions, some of which can be subjected to experimental verification. A major prediction of the neutral theory is that although genes and the proteins they encode may differ in their evolutionary rates, the rate at which a particular gene evolves should remain constant. No such constancy is predicted by the selection theory. So, is the evolutionary rate of genes constant? Before attempting to answer this question, we must point out or reemphasize a few salient features of the neutral theory.

According to their effect on the fitness of their bearers, mutations are divided into three categories, disadvantageous, advantageous, and neutral. *Disadvantageous mutations* have a negative effect because they lower the bearers' chances of survival and reproduction relative to that of other individuals, or they reduce the size of its progeny. These mutations are eliminated from the gene pool by negative (purifying) selection soon after they have arisen. *Advantageous mutations* have a positive effect because they improve the bearers' chances of survival and reproduction and thus have a greater probability of being passed on to the following generations. Relative to other mutations, advantageous mutations have a greater chance of becoming fixed: they are under positive selection. *Neutral mutations* have no effect on the fitness of their bearers; they are not influenced by either negative or positive selection, but some of them can become fixed by drift. Geneticists agree that deleterious mutations, although they constitute the majority in terms of frequency of occurrence, do not contribute to evolution. They also agree that most evolutionary changes of the phenotype are effected by the fixation of advantageous mutations. They disagree, however, on the relative contribution of advantageous and neutral mutations to the divergence of nucleic acids and proteins at the molecular level. The "neutralists", the proponents of the neutral theory of molecular evolution, maintain that the majority of sequence differences observed between nucleic acids (proteins) of two species are the result of fixation of neutral or nearly neutral mutations by random genetic drift. The more traditionally minded biologists, the "selectionists", believe that not only at the phenotypic, but also at the molecular level most evolutionary changes are the result of natural selection. Although the controversy continues, it has become clear that if natural selection influences the fixation of mutations, most of the time its effects are very difficult to demonstrate. The debate has also led to the realization that drift is a factor that cannot be ignored in any explanation of evolutionary changes at the molecular level. Finally, the neutral theory, backed by elegant

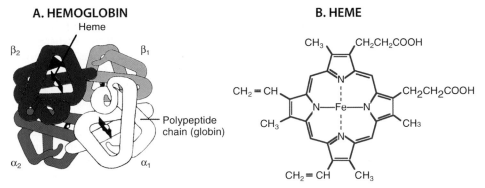

A. HEMOGLOBIN

Heme

β₂

β₁

Polypeptide chain (globin)

α₂

α₁

B. HEME

Fig. 4.6A,B
Structure of a hemoglobin molecule (**A**) and a single iron-containing heme group (**B**). In **A**, the four folded polypeptide chains (2α and 2β chains) are visible. (**A** from Micklos & Freyer 1990.)

mathematical constructions, provides the foundations on which more complex explanations of molecular evolution can be built.

Lest the theory be misunderstood, we emphasize once again that it does not deny the existence of natural selection and its importance in evolution. Nor does it in any way denigrate Darwin's contribution and ideas. Darwin viewed natural selection as occurring at the level of the outward appearance of an organism, at the level of morphology, physiology, and behavior. The neutral theory, by contrast, has little to say about the evolution of the outward appearance of an organism; it deals exclusively with genes at the molecular level. Neutralists freely admit that phenotypic characters and complex organs have evolved by natural selection, without which these organs could never have achieved their marvelous adaptation to the functions they execute.

According to the neutral theory of molecular evolution, most mutations that are not disadvantageous are not advantageous either – they are neutral. But genes control characters and most characters are beneficial to the organism living in a particular environment. Here, however, lies a paradox: how can neutral genes effect the appearance of non-neutral characters? No satisfactory answer to this question has yet been provided.

For reasons we need not go into, it has proved very difficult to come up with evidence for the existence of advantageous mutations at the molecular level. Of the few examples that have been reported, we mention here one – that of the hemoglobin molecule. Hemoglobin is the pigment responsible for the brilliant red color of blood. As the major constituent of red blood cells, hemoglobin functions as an oxygen-delivery system to tissues and as a system of carbon dioxide removal from the tissues. The hemoglobin molecule is composed of a nonprotein part – the *hemes* (from Greek *haima*, blood) and a protein part – the *globins* (a designation derived from "globules", an earlier name of red blood cells; Fig. 4.6). Each heme is a complex of four atomic rings associated with an iron atom (Fe^{2+}) which is capable of binding a single molecule of oxygen. The globin part of the molecule consists of four polypeptide chains, each of which has one heme bound to it. The chains are of two types, each type specified by a different locus (for further details, see Chapter Eight). Oxygen binds to the iron of the hemes, carbon dioxide to the globins.

It has been reported that camel and llama hemoglobins differ in their ability to bind oxygen: llama hemoglobin binds oxygen more easily (has a higher affinity for it) than camel hemoglobin. This makes sense because the llama lives at high altitudes where there is less oxygen in the air than at lower altitudes where the camel is at home. The less oxygen there is, the more difficult it becomes for normal hemoglobin to bind it. Most of us know this from our own experience: at high altitudes we must breathe faster to supply our tissues with enough oxygen than we do at low altitudes. Hemoglobin with high affinity for oxygen makes it possible for the llama to breathe normally.

Similarly, hemoglobin of the bar-headed goose, which spends the winter in India but flies over the Himalayas at an altitude of 9000 meters to its summer feeding grounds, binds oxygen more readily than hemoglobin of the greylag goose, which resides in India all year round. Here again, the difference makes good sense, since the high-affinity hemoglobin enables the bar-headed goose to cope with oxygen shortage at a high altitude.

Finally, the deer mouse *Peromyscus maniculatus*, a common North American rodent, has two types of hemoglobin, one with high and the other with low affinity for oxygen, each controlled by a different gene. And guess what? Deer mice living at higher altitudes have predominantly high-affinity hemoglobin and those living at lower altitudes predominantly low-affinity hemoglobin.

Remarkably, in all these cases, the high- and low-affinity hemoglobins differ in a single amino acid, a different one in each example, and in each case the difference is the result of a single mutation in the hemoglobin gene. It is assumed that the different amino acids are indirectly responsible for the difference in affinity. (Although it is the iron that binds oxygen in the hemoglobin molecule, the protein component may indirectly influence the strength of the binding.) These are thus possible examples of advantageous mutations adapting the hemoglobin molecule to specific environmental conditions. A few advantageous hemoglobin mutations have also been found to exist in the human population, but along with these, over 1000 other mutations are known for which there is no evidence that they provide any selective advantage or disadvantage to their bearers. They may be neutral.

To summarize this part of the chapter: because of the difficulties in demonstrating selection, the neutralist-selectionist controversy will undoubtedly continue for some time. We reiterate, however, that a consensus is slowly emerging according to which most of the DNA evolves predominantly under the influence of random genetic drift. In the small proportion of DNA in which selection may *potentially* effect changes in gene frequencies, the selection intensity is generally so low that under most conditions, the majority of mutations are effectively neutral. The neutrality of mutations is one condition for constancy of evolutionary rates; the other is constancy of mutation rates. But why would one expect a random event, such as a mutation, to occur at a constant rate?

The Clock Symphony

It is our day-to-day experience that events always have a cause. A satellite orbits the Earth because of gravity; we smell a dead rat because volatile substances stimulate odor receptors in our noses; a baby cries because it is hungry, sick, or unhappy. Every effect has its cause. So ingrained is this notion in our brains that we have great difficulty imagining a world in which things happen without a cause. Yet, we live in such a world: at the atomic and subatomic levels, events are acausal. One manifestation of acausality is radioactivity. In a large collection of the same type of carbon atoms, ^{14}C, once in a while one of the atoms disintegrates and turns into one type of nitrogen atom, ^{14}N. There is no identifiable reason why the particular atom has decayed at that precise moment. At best, the physicists assume that the disintegration is the result of random fluctuations in the inner structure of the atom, fluctuations that are unpredictable at the level of individual atoms, but predictable for a large group of atoms. So reliable are these predictions that disintegration of atoms is commonly used to measure geological time and to determine the age of fossils (see Chapter Nine).

Mutations, which underlie evolution, behave similarly to radioactive atoms and for similar reasons. An earlier description of the origin of mutations in the course of DNA replication may have left the reader with the impression that the misincorporation of a nucleotide is comparable to a typist hitting the wrong letter on a keyboard. The real reason for misincorporation is actually quite different. The nitrogenous base of the nucleotide, like any other molecule, is not a rigid structure existing in one fixed form. Rather, it is a dynamic entity oscillating between various forms – isomers or *tautomers*, which differ in the arrangement of some of the atoms and of the interatomic bonds. Some of the hydrogen atoms (protons) of the bases in particular can, from time to time, change places and by doing so, affect the binding properties of the molecule. The alternative forms are relatively rare, however, and exist only fleetingly, the molecule remaining most of the time in the generally stable isomeric forms. If the *tautomeric shift* occurs at the time of DNA replication, either in the base of the template strand or in an incoming nucleotide available for incorporation, the tautomer seeks out the wrong partner. The tautomerically shifted adenine mimics guanine in its binding properties and binds cytosine instead of thymine (and the tautomer of C binds A instead of G), while the tautomer of T binds G instead of A (and the tautomer of G binds T instead of C). The binding "freezes" the abnormal base in its tautomeric form and holds it in this state until the next round of DNA replication, by which time the partner of the nucleotide in the disguised form pairs with its normal counterpart in one of the daughter DNA molecules and thus effects a mutation. Other nucleotide substitutions may take place because of different kinds of tautomeric shifts.

The tautomeric shift, like the disintegration of radioactive atoms, or the flipping of a coin for that matter, is a stochastic process: although entirely random and unpredictable at the level of individual outcomes, its average outcome can be prognosticated with reasonable accuracy. The shifts can therefore be expected to occur at regular intervals, as can the mutations that result from them. The rate at which neutral mutations occur should therefore be constant. And since, according to Kimura, the neutral mutation rate equals the substitution rate, DNA molecules should accumulate substitutions at a stable pace.

In biology, it is impossible to augur the point in time at which a mutation will occur at a particular nucleotide site, or the time at which a mutation will become fixed, but the rates at which mutations appear and become fixed are ascertainable. Fixations of neutral mutations can be expected to occur at more or less regular intervals. And since the partitioning of time into intervals of equal length is an essential feature of a timepiece, the term *molecular clock* was introduced in 1965 for the postulated progressive, steady-pace accumulation of amino acid replacements in proteins during their evolution. The term was coined by Emile Zuckerkandl and Linus C. Pauling, the latter being better known for his research on the nature of the chemical bond, for which he was awarded the Nobel Prize in 1954, and for his advocacy of a vitamin C-rich diet in the treatment of cancer and the common cold. Some of our readers may still recall the steady tick-tock of the grandfather clock in the tranquillity of their grandparents' house. But grandfather clocks are now a thing of the past; they have been replaced by electronically pulsed clocks with silently circling hands and wristwatches with digital displays. So if you want to hear a steady "tick-tock", listen instead to the slow movement of Haydn's Symphony No. 101 in D major:

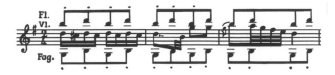

Much more slowly, and perhaps not quite so regularly as a grandfather clock or the wood-winds in Haydn's Symphony, the genes and their derivatives, the proteins, too, are tick-tocking, according to Kimura. The molecules mark evolutionary time by accumulating nucleotide substitutions or amino acid replacements. But the clock – so Kimura – can only tick if random genetic drift decides the fate of mutations. If natural selection were involved, the clock would be expected to run erratically, sometimes very fast and then again more slowly, depending on the changing environmental conditions.

Since its proposal in 1965, the molecular clock hypothesis has remained controversial. Nevertheless, a certain regularity in the accumulation of amino acid replacements can no longer be disputed. Proteins differ, however, in the rate at which they diverge. Some, such as fibrinopeptides, which are fragments of proteins participating in blood clotting, evolve rapidly, while others, such as histones, which form the spools for winding DNA strands in a chromosome, evolve very slowly. The reason for these differences is that proteins are differentially constrained in their evolution by their functions. Some, such as the fibrinopeptides, can change considerably and yet still remain functional, while most amino acid replacements in others, such as histones, would endanger the protein's function. In the former, therefore, many more mutations have a chance of being fixed than in the latter, in which virtually all mutations are eliminated.

Controversial is also the speed of the molecular clock in different evolutionary lineages. Some geneticists have obtained evidence that in certain lineages, for example rodents among the mammals, the clock runs faster than in others, such as primates. Other geneticists contest these findings. On the other hand, it is now generally accepted that at the nucleotide level, some regions of the genome and some sites within a given region evolve faster than others.

A definitive answer to the question of constancy of the molecular clock at the DNA level therefore cannot be given at this time (see also *The Clock in the Molecule* section in Chapter Nine). Enough is known, however, about the clock to warrant – after the necessary precautionary steps have been taken – the application of the clock hypothesis in attempts to solve certain problems in phylogenetic investigations, including the origin of the human species.

So, where do we stand now? In this chapter, we have reduced life to a concatenation of generations, to a sampling of genes from an old pool to create a new pool. If sampling is carried out entirely at random, each gene has the same chance of being chosen to contribute its copy to the next generation. Under these conditions, the sampling process inevitably leads to a gradual replacement, over the generations, of all genes in the pool by one gene from one of the preceding generations, and this process is repeated over and over again. The choice of gene for replacement is unpredictable, but the rate at which these substitutions take place is determinable. During the succession of generations, the genes change – they mutate. Here again, the identity of the gene that will change and the moment at which it will mutate are unfathomable, but that one gene of a specified number will change in a certain number of generations can be prognosticated. Because of the random sampling process, a mutation which originally occurred in a single gene may either become extinct or increase in frequency until it substitutes all other genes. When a pool splits into two and these are prevented from exchanging genes with each other, different mutations become fixed in them and the pools diverge genetically. If the isolation lasts long enough, the two populations represented by the pools may become reproductively isolated species. The change in gene frequencies leading to the emergence of new species is evolution. Genes that behave in the manner just described are said to be neutral. The behavior of neutral genes, although governed by chance, can be described by a set of mathematical equations. Many mutations, however, are not neutral. They are either deleterious (most often) or advantageous (rarely) and as such are subject to natural selection, either negative or positive, which increases the chances of their extinction or fixation, respectively. The proportion of neutral and advantageous mutations is contentious. According to the neutral theory of evolution, the majority of nondeleterious mutations are neutral. The theory posits that at the molecular level genes accumulate mutations with the regularity of a clock.

In the next chapter, we explain how this information can be used to determine the position of the human species in the community of living forms on earth and present the results of such determination. But before closing this chapter, we once again point out the philosophical implications of the presented material. Molecular evolution documents, perhaps better than any other field of study, how much our existence owes to chance. Three crucial events during molecular evolution are determined by chance – the sampling of genes during the formation of a new generation, the generation of mutations, and the fixation of mutations to produce differences between species. The notion that our origin is so deeply rooted in chance may be unpalatable to some, but this is no reason for denying it. To cope with the notion, it may help to realize that chance is not synonymous with chaos and unpredictability. We tried to show in this chapter that chance events have their order, too, but at a different level than that of the events whose immediate causes are apparent. Chance also has an element of predictability which, however, is of a different type than that of causal events. To quote Boethius: *Chance … which seems to rush along with slack reins, is bridled and governed by law.* The philosopher wrote this in prison a few months before he met a most horrible death at the decree of King Theodoric, to whom he remained loyal to the end.

CHAPTER 5

The Painted Tree

Methods of Phylogenetic Reconstruction

*I*t is worse than vain for men
to leave the shore and fish
for truth unless they know the art;
for they return worse off than they
were before.

Dante Alighieri:
The Divine Comedy. Paradiso

Molecular Archives

The nucleus of a cell resembles national archives, a rich source of genealogical information. To retrieve information from archives requires familiarity with methods of studying and interpreting documents. One cannot just walk into the building, pull out the first dossier, and expect to find the sought information. There is so much information stored in the different rooms of the building that without knowing where to look, without understanding the different languages in which the documents are written, and without knowing something about the circumstances surrounding the origin of the documents, the dossiers lie beyond one's grasp.

And so it is with molecular archives. To retrieve genealogical information from DNA molecules, one has to know how to pry open the cell and the nucleus and how to release the DNA without damaging it too much; how to isolate a specific part of the DNA molecule in an amount sufficient for further study; how to determine the order of the nucleotides in this part; and how to convert the order into phylogenetic intelligence. For each of these steps, methods are now available, some of them relatively simple, others highly sophisticated, so that only a trained person can apply them with success.

This is not the place to describe the retrieval of molecular genealogical information in detail. Nonetheless, a brief overview of the general approaches to informational processing and of the interpretative tools applicable to the retrieved information is necessary to appreciate not only the potential, but also the limitations of the molecular archives. If parts of the chapter appear tedious, no harm will come to the reader who decides to skip them. The reader who bears with us, however, will hopefully be rewarded by a deeper understanding of the entire problematic of phylogenetic analysis.

The first step toward phylogenetic reconstruction is to gain entrance to the molecular archives. Molecular biologists do not enter a cell or a nucleus through a gate or by opening a door, with the permission of the curator. Rather they storm the building and expose the documents by taking down the walls. The demolition is usually achieved by treating the cells with a detergent that solubilizes the lipids and by an enzyme that destroys the proteins. (Miescher used pepsin for this purpose, but present-day molecular biologists commonly resort to an enzyme called proteinase K.) After tearing down the house, the DNA must be extricated from the pile of rubble remaining. One way to accomplish this task is to load the rubble onto a column made of a positively charged resin and flush it down. Because of the many phosphate groups in their backbone, the DNA molecules have a strong negative charge and so stick to the resin, while most of the rubble passes unhindered through the column. Any other substances that may have been retained by the column are washed out (eluted) by a solution of low salt concentration and the DNA itself is then eluted by a solution of high salt concentration, the salt abolishing the electrostatic bonds between the DNA and the resin. Next, the solubilized DNA is converted into an insoluble form (precipitated) by the addition of ethanol, removed, washed, and solubilized again. It is not exactly a gentle retrieval of documents, but it serves its purpose.

In the second step, molecular biologists must locate the particular document of interest. The task is comparable to finding a single frame in a pile containing innumerable film reels, each single reel consisting of millions of frames. Molecular biologists can approach this task in one of two ways. The first method is to cut the DNA molecules into pieces of manageable size and insert each piece into a ring which can then be manipulated at will (Fig. 5.1). The DNA molecules are cut by specialized enzymes of microbial origin, *restriction endonucleases*. Microbes have evolved a variety of these enzymes to protect themselves against foreign DNA, such as that of viruses or of other microbes. To cut a DNA mol-

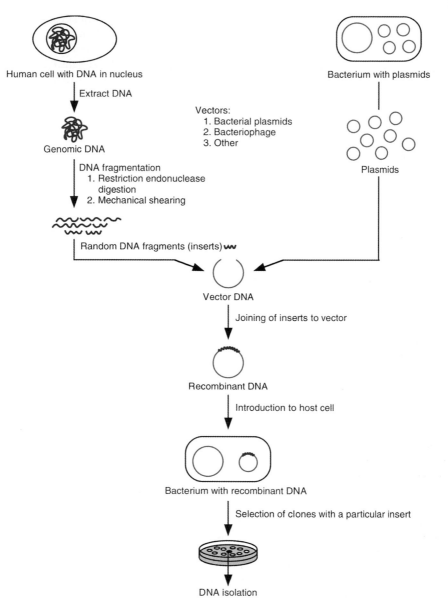

Fig. 5.1

Cloning of DNA molecules with the help of bacteria. DNA extracted from the nucleus of a cell is fragmented either mechanically or by the action of enzymes, purified, and inserted into specifically modified bacterial plasmids (circular pieces of bacterial DNA called *vectors*). The resulting *recombinant DNA* is reintroduced into bacteria and multiplied. When the bacteria are seeded on an agar plate at relatively low density, they form discrete *plaques*, each plaque being a large collection of cells all derived from a single bacterium – the cells constitute a clone. Each clone contains a single type of insert in many copies. The clone (plaque) with the desired insert can be identified by special techniques, isolated, multiplied by regrowth on a new plate and the bacteria then used to isolate the insert DNA in large amounts ecule, the enzyme must find a specific sequence of four or six nucleotides called a *restriction site*, a different sequence for each enzyme (see Chapter Ten and Fig. 10.13). In the microbe's own DNA, these sites are protected from an enzymatic attack, but any unprotected foreign DNA which has invaded the host is immediately fragmented and the fragments degraded further by other enzymes. Molecular biologists have learned to exploit the restriction enzymes for their own purposes. The DNA of a multicellular organism, which is as foreign to a microbe as that of a virus, contains numerous sites for each of the variety of restriction enzymes. Its exposure to a specific enzyme therefore degrades the long molecule into a short fragment whose length depends on the distribution of the restriction sites for that particular enzyme.

The rings (*vectors*) into which the *restriction fragments* are inserted are derived from bacterial or viral DNA modified to serve the specific purpose. When introduced into bacteria, the rings containing the fragments (vectors with *inserts*) are treated by the host cells as if they were their own. As the host cells divide, they replicate the rings and so multiply them. In this manner, numerous identical copies (*clones*) of a given insert are generated – the DNA fragment is *cloned*. In the mixture, the desired clone is identified with a specific *probe*, a short piece of radioactively or otherwise labeled DNA, identical in sequence with a short stretch of nucleotides in the clone. When both the clone and the probe DNA are converted into their single-stranded forms (*denatured*, for example, by heating) and then allowed to reassociate (*reanneal*), the reassociation of one clone strand with a complementary strand of the probe (*hybridization*) identifies the clone. The frame has thus been located on one of the microfilm reels, and the clone can be pulled out from the mixture and recloned to produce a pure preparation of the particular DNA segment.

The second method for obtaining a large amount of a particular DNA segment is based on the *polymerase chain reaction*, PCR (Fig. 5.2). The PCR-based techniques target DNA segments with the help of short, single-stranded stretches of nucleotides (oligonucleotides), the *primers*, complementary to a particular region of the targeted DNA. The primers bind specifically to the targeted DNA and initiate a reaction that produces millions of copies of the particular piece (the DNA is *amplified*) which are easy to isolate. The PCR method therefore enables the investigator to focus on one specific fragment among a blend of different ones.

Since the information contained in the DNA is ingrained in the order (*sequence*) of the nucleotides, to decipher it, the order must be determined – the DNA must be *sequenced*. To this end, the amplified DNA is broken down into a series of fragments whose length increases in increments of one nucleotide (Fig. 5.3). The fragments are separated according to size and their terminal nucleotides are identified by a chemical reaction. Since the fragments are tagged by radioactive atoms or by a fluorescent dye, their distribution can be recorded either on a film or on a screen. The outcome is a sequence, either a ladder of bands or a series of successive peaks (Fig. 5.3C), both of which a trained person can "read", that is, convert into the letters A, G, C, and T symbolizing the four nucleotides. This *DNA sequencing* procedure is now fully entrusted to machines, *automatic sequencers*, which are hooked up to computers programmed to read the order of nucleotides. The sequences are stored in international *databases* where they can be accessed by researchers according to their specific code designation. The databases now contain innumerable DNA sequences from a wide variety of organisms representing all major life forms. Once the sequence of nucleotides in a DNA segment has been determined, it can be analyzed either directly, or alternatively, if the segment specifies a protein, it can first be translated into an *amino acid sequence* and then analyzed. The translation is a straightforward matter because all organisms, with minor exceptions, use the same *genetic code* in which different combinations of

Fig. 5.2A-C
Principle of polymerase chain reaction (PCR).
A, B. A solution containing double-stranded
DNA (dsDNA) molecules, two primers (short
stretches of nucleotides, one complementary to
a DNA segment of one strand, the other to a
segment of the opposite strand, usually less than
1000 nucleotide pairs "downstream" from the
first), the four types of nucleotides from which
DNA molecules are synthesized, and heat-resist-
ant *Taq* DNA polymerase (pol; an enzyme necessary for DNA synthesis, isolated from the hot-
spring bacterium, *Thermus aquaticus*) is heated
to 93°C. This is the temperature at which the
bonds holding the two DNA strands together
break and the single strands (ssDNA) separate
from each other (*denaturation,* d). The temperature is then reduced to approximately 50°C, a
condition at which the primers (but not the separated DNA strands) find and establish bonds
with their complementary sites (*primer anneal-
ing,* a). In the third step, the temperature is
raised to 70°C, an optimum for the activity of
Taq DNA polymerase. The enzyme attaches to
the primer and then moves along the ssDNA
template, adding one complementary nucleotide
after another, all the way to the end of the
strand (*primer extension,* e). Because the same
thing happens on the second strand, two identical dsDNA molecules are produced. The temperature of 70°C is too high for DNA polymerases
of most other organisms to function but not
high enough to denature the DNA molecules.
After completion of the first cycle of the reaction, in the next cycle the temperature is again
raised to 93°C and the steps are repeated. This
reaction cycle is repeated 30 times or more in a
special machine capable of achieving rapid temperature changes (*thermocycler*). C. The number
of copies of DNA molecules increases rapidly
from cycle to cycle as if by chain reaction: the
original single molecule produces two molecules, which produce four molecules, which produce eight molecules, and so on, until the reaction mixture runs out of ingredients. Thirty
reaction cycles produce 3×10^{10} copies of the
target sequence. (Modified from Klein & Hořejší
1997.)

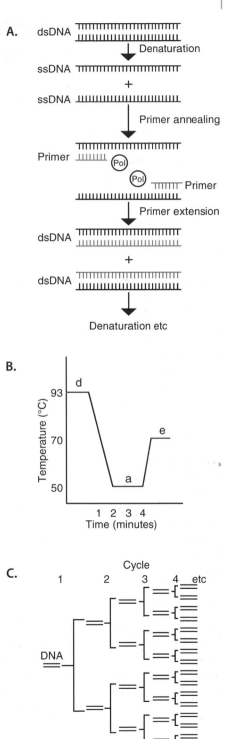

A.

Primer   DNA polymerase

```
...T-C-C-A ●→          Template
...A-G-G-T-A-C-T-C-C-A-G...
```

↓ Strand synthesis

```
...T-C-C-A-T-G-A-G-G ●→
...A-G-G-T-A-C-T-C-C-A-G...
```

B.

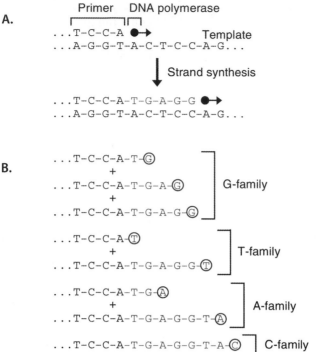

```
...T-C-C-A-T-Ⓖ
       +
...T-C-C-A-T-G-A-Ⓖ          ⎫
       +                    ⎬ G-family
...T-C-C-A-T-G-A-G-Ⓖ        ⎭
```

```
...T-C-C-A-Ⓣ                ⎫
       +                    ⎬ T-family
...T-C-C-A-T-G-A-G-G-Ⓣ      ⎭
```

```
...T-C-C-A-T-G-Ⓐ            ⎫
       +                    ⎬ A-family
...T-C-C-A-T-G-A-G-G-T-Ⓐ    ⎭
```

```
...T-C-C-A-T-G-A-G-G-T-A-Ⓒ  ⎦ C-family
```

C.

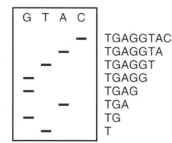

```
G  T  A  C
            —    TGAGGTAC
         —       TGAGGTA
      —          TGAGGT
—                TGAGG
—                TGAG
         —       TGA
—                TG
   —             T
```

or

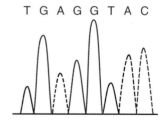

T G A G G T A C

D.

A dideoxynucleotide

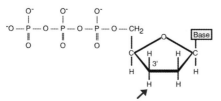

Here the H atom replaces the –OH group of
deoxynucleotide triphosphate

three nucleotides (triplet *codons*) specify unambiguously the various amino acids (Fig. 5.4). The amino acid sequence of a protein can also be determined directly using methods developed specifically for this purpose. DNA sequencing is, however, preferred over protein sequencing because it is faster and cheaper. The latter is now used only in special cases.

In what manner does the sequence of nucleotides or amino acid residues reflect the evolutionary origin of an organism, and how can this information be converted into a genealogical relationship?

Tribal History

As explained in Chapter 3, specifically in Figure 3.8, the reconstruction of evolutionary history is based on the assumption that at the molecular level, the amount of evolutionary change is proportional to elapsed time. In molecular genealogy, DNA molecules that originated from a common ancestor a long time ago have had more time to accumulate mutations (substitutions) than molecules that originated more recently. In other words, genealogically more closely related DNA molecules are expected to be more similar to each other in their sequences than less related ones. Hence, by comparing DNA sequences of

---

Fig. 5.3A-D
Principle of the chain-termination method of DNA sequencing. To use this method successfully, the sequence of a short stretch flanking the DNA segment to be sequenced must be known. This stretch could be located in the vector bearing a given insert or in an adjacent segment sequenced previously. **A.** The two DNA strands of the purified DNA molecules are separated from each other (only one is shown here) and a short oligonucleotide (17-20 nucleotides long) complementary to the template strand is annealed to the flank of the chosen segment. The oligonucleotide serves as a primer that initiates (primes) the synthesis of a complementary strand in a reaction catalyzed by the enzyme DNA polymerase. **B.** The free nucleotides used in the synthesis are of two kinds. The majority are the standard deoxyribonucleotide triphosphates (dATP, dCTP, dGTP, and dTTP). A small proportion are dideoxynucleotides (ddATP, ddCTP, ddGTP, and ddTTP) in which the –OH group normally used to link up neighboring nucleotides in the growing strand is replaced by a hydrogen atom (**D**). The enzyme does not recognize the difference between the types and incorporates the dideoxynucleotide, but the absence of the –OH group then precludes further growth of the strand. Since both types of nucleotide are available, either of them may be incorporated in the growing strands. But whenever the dideoxynucleotide is incorporated, the reaction stops at that site. The result is a mixture of molecules of different length, depending on the site at which strand growth happens to have been terminated. If the proportion of the two nucleotide types has been correctly chosen, each site in the sequence becomes the termination site in some molecules. The mixture therefore contains molecules in which 1, 2, 3, etc nucleotides have been incorporated starting from the end of the primer. **C.** The molecules are then denatured and the single strands analyzed. The newly synthesized strands are arranged according to their size to form bands that increase in length by one nucleotide, each band consisting of strands of the same size. The assortment of the strands is achieved by introducing an electric current into a slab of polyacrylamide gel in which the mixture of DNA strands has been placed in a narrow slot. (Polyacrylamide is a polymer based on acrylamide, which turns into a gelatinous substance in water.) Because of their electric charge, the DNA strands migrate through pores of the gel (= *polyacrylamide gel electrophoresis*), the shorter strands faster than the larger ones. The bands are then visualized by the attachment of either radioactive atoms or fluorescent dyes to the nucleotides of the primer, either one dye to all four nucleotide types or dyes of different colors to the A, C, G, T nucleotides. Depending on the nature of the label and the method of detecting it, the result is either a series of bands which can be "read" in the manner shown (from the bottom to the top of the gel) or a series of peaks of different color. (Since only the newly synthesized strands are labeled, they alone are detected; the unlabeled templates normally do not interfere with the detection.)

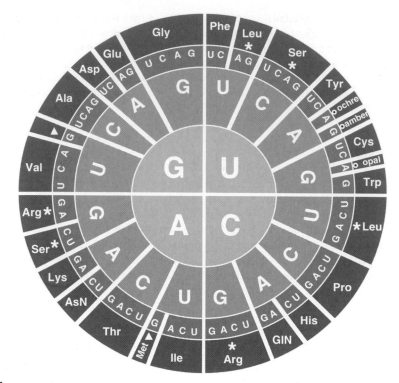

Fig. 5.4

Genetic wheel of fortune: the specification of amino acid residues of a protein by nucleotide *codons* of nucleic acids. Each codon consists of three nucleotides (*triplet*) whose identity and order determine which of the 20 kinds of amino acids the codon specifies. Here the positions of the nucleotides are indicated by the three concentric circles, the fourth, outermost circle representing the amino acids. At the first position of the codon (the first, innermost circle) there can be A, G, U, or C. Each of these can again be associated in the second position (second circle) with A, G, U, or C, and the same is true for the third position (circle), so that altogether 64 different codons are possible. (Note that the DNA sequence is first transcribed into an RNA which has a U at each position in which the DNA has a T. Since the protein is translated from this *messenger RNA*, it is customary to give the *genetic code* with U instead of T.) The genetic code is *degenerate*, which means that in most cases the same amino acid is specified by more than one codon. This is indicated in the wheel by its spokes. For example, valine is specified by codons with G in the first position, U in the second, and U, C, or A in the third position. The AUG codon specifies either the amino acid methionine or the site in the DNA sequence in which transcription begins (it is an *initiation codon*), depending on the nucleotide sequence flanking it. The UAA, UAG, and UGA codons indicate the point at which translation of the transcript stops (they are *stop codons*). Their names (ochre, amber, and opal, respectively) are those of mutants originally described in bacteria and bacteriophages. The names (one-letter abbreviations) of the amino acids are: Ala (A), alanine; Arg (R), arginine; Asn (N), asparagine; Asp (D), aspartic acid; Cys (C), cysteine; Gln (Q), glutamine; Glu (E), glutamic acid; Gly (G), glycine; His (H), histidine; Ile (I), isoleucine; Leu (L), leucine; Lys (K), lysine; Met (M), methionine; Phe (F), phenylalanine; Pro (P), proline; Ser (S), serine; Thr (T) threonine; Trp (W), tryptophan; Tyr (Y), tyrosine; Val (V), valine. The asterisks indicate that each of these amino acids is specified by codons with different letters in the first position. (From Schmitt 1994.)

different organisms, it is possible to infer the degree to which these organisms are related. Genealogy that takes the accumulation of evolutionary changes into account (Darwin's descent with modifications) is *phylogeny*, the history of a tribe. The graphic depiction of phylogeny is the *phylogenetic tree*.

The reconstruction of phylogeny at the molecular level is therefore inseparably bound to the idea of a molecular clock, as Zuckerkandl and Pauling realized when they drew the first molecule-based tree. Regardless of how one views the molecular clock – whether it is regular or erratic, for example – if one denies its existence altogether, one must give up any attempts at reconstructing phylogeny from molecular records. Incidentally, a similar assumption also underlies the reconstruction of evolutionary history at the supramolecular (phenotypic) level. This reconstruction, too, relies on similarity between organisms: generally, the more similar the organisms, the more related they are assumed to be. Since the emergence of dissimilarity requires time, the longer evolutionary lineages are separated from each other, the greater their dissimilarity. The supramolecular clock, however, is much more erratic than the molecular clock, mainly because natural selection not only speeds it up or slows it down, depending on the circumstances, but also creates false similarities. Many of the difficulties that are encountered and the errors that are committed in phenotypic phylogenetic reconstructions are directly attributable to the fitfulness of the supramolecular clock. Nevertheless, a certain time–dependence in the extent of accumulated evolutionary change must be assumed if phenotypic characters are to be used to make phylogenetic inferences. The supramolecular clock may often run haywire, but since it still marks time, it *is* a clock.

To carry out a phylogenetic analysis, we must compare DNA or protein sequences in order to establish and quantify their similarity. Comparisons are easier when objects are placed close to each other – in the case of sequences one below the other. It then becomes immediately apparent that some sites (positions) are occupied by identical and others by disparate nucleotides (amino acid residues). Since any one site can be occupied by one of four different nucleotides, it can be expected that at 25 percent of the sites nucleotides will be shared between any two sequences by chance alone! If, however, the two sequences are related, then as we slide one sequence slowly along the other so that each site of one sequence has a chance of being compared to all sites of the other, we will discover that at a certain constellation, the percentage of matched sites is significantly higher than can be expected by chance alone. We then say that the sequences have been *aligned*. Sometimes, however, an alignment of parts of the sequence becomes possible only when one or more gaps are introduced at specific sites. We then assume that at each of these sites, either an extra nucleotide has been inserted in one of the two sequences or a nucleotide has been deleted in the other. Because it is often not possible to determine whether the former or the latter took place, the gaps are referred to as insertion/deletions or *indels*. One must, however, take care not to get carried away in this regard, because by introducing sufficient gaps, even two totally unrelated sequences can be made to appear related.

But even when the alignment reveals the sequences to be related to each other, the match may still be a case of mixing apples with medlars. To explain, let us assume that gene *A* duplicated in the common ancestor *Z* of two species, *X* and *Y*, and that the copies *A1* and *A2* diverged from each other in their sequences in the two lineages (Fig. 5.5). If, unaware of these circumstances, we compared by chance *A1* of species *X* with *A2* of species *Y*, or vice versa, instead of genes of the same denomination, we might be misled in our deductions concerning the evolutionary relationship of these two species relative to other species. Evolutionary biologists refer to the *A1-A2* type comparisons as *paralogous* and distinguish them from *orthologous* (*A1-A1* or *A2-A2*) comparisons. Either paralogous or orthologous

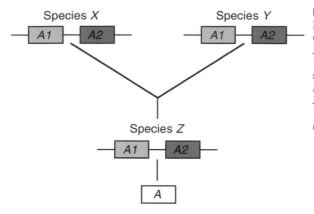

Fig. 5.5
Homology, orthology, paralogy.
Orthologous relationships:
X-A1 with *Y-A1* and *X-A2* with
Y-A2 (here *X* and *Y* indicate the
species of origin of the gene).
Paralogous relationships:
X-A1 with *Y-A2* and *X-A2* with
Y-A1. All four relationships are
homologous.

comparisons are termed *homologous*. If the sequences of both genes from each of the two species are not available, it is often difficult to ascertain whether a comparison is ortho- or paralogous.

What's Hidden Under the Surface?

Assuming that we have aligned our sequences optimally and made reasonably certain that they are from orthologous genes, we can begin to assess the extent to which they are related. Here, however, we are immediately confronted by another ruse which evolution may use to thwart our efforts. Imagine that we compare two orthologous DNA sequences obtained from different species, for example human and chimpanzee, and find identical nucleotides at most of the sites and different nucleotides at only a few sites. We are tempted to conclude that since the separation of the human and chimpanzee lineages, single nucleotide substitutions took place at the differential sites and that no substitutions occurred at all the other sites. But here lies the snare: it may or may not be true. Let us consider first the sites occupied by shared nucleotides (Fig. 5.6). We find, for example, that both the human and the chimpanzee have a G at a particular site and are naturally inclined to assume that the common ancestor of these two species also had a G at this site. There are, however, other possibilities. For example, the ancestor may have had a C which then changed to G in both the human and chimpanzee lineages independently (this would make it a case of *parallel substitutions*). Or the ancestor had a C which changed to G in the human lineage, whereas in the chimpanzee lineage it first changed to T and then to G (a case of *convergent substitutions*, since here, in contrast to parallel substitutions, the identical nucleotides are of different derivation). Or the common ancestor may have had a G at this site, which changed to C and then back to G in one of the lineages (a case of *back substitutions*). These three situations are examples of evolutionary *homoplasy* (from Greek *homos*, alike, and *plassein*, to mold; so "molded alike").

Now to the sites at which the two sequences differ, for example a site at which the human has a C and the chimpanzee a T (Fig. 5.6). We might assume that the common ancestor of the two species had a C at this site, which changed to T in the chimpanzee lineage (or a T which changed to C in the human lineage; both would be cases of *single substitution*).

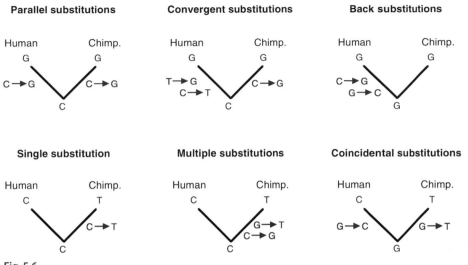

Fig. 5.6
The hidden changes in sequence evolution. In five of the hypothetical examples shown, the evolution at a given nucleotide site from a common ancestor to human and chimpanzee involved changes (indicated by arrows at branches) which the comparison of the two extant sequences does not reveal. Only in the category "single substitution" does the observed change reflect the actual evolutionary pathway accurately. The three examples in the upper section represent cases of *homoplasy*. (Inspired by Page & Holmes 1998.)

Some of the alternatives are, however, that a C of the ancestor changed first to G and then to T in the chimpanzee lineage (which would constitute a case of *multiple substitutions* or *multiple "hits"*). Or that the ancestor had a G which changed to C in the human lineage and to T in the chimpanzee lineage (a case of *coincidental substitutions*).

So, when we align two sequences and find shared nucleotides at some sites and dissimilar ones at others, the comparison may not be revealing all the changes that occurred during the evolution of the sequences from their common ancestor. Under the surface of the observed identities and disparities additional changes may lay hidden which extant sequences do not divulge. Ignoring these hidden changes may distort our interpretation of the genealogy, especially if the genealogy reaches far into the past. Fortunately, our inability to observe some of the changes does not mean that we are compelled to remain ignorant of them. There are certain paths open to us which allow us to make educated guesses about the number, and in some situations also the nature, of the changes that probably occurred but cannot be observed directly. Here, we explain a procedure for estimating the number of hidden changes; in the next section, we describe a method designed to resurrect some of the changes by indirect means.

Horse Kicks and Monsieur le Baron

In the preceding chapter we emphasized that despite their randomness, mutations and substitutions are predictable events. They are predictable not in the sense of forecasting individual events, but rather in terms of statistical averages which can be estimated by the application of appropriate laws of probability. If, however, we can *predict* evolutionary changes, we should also be able to "*post*dict" them and so to use laws of probability to estimate the number of hidden changes.

The two main events underlying the divergence of DNA (protein) sequences are, first, *mutation*, in the simplest case the replacement of one nucleotide by another at a particular site of a DNA molecule; and second, *substitution*, the replacement of all DNA molecules not bearing the mutation by the mutated molecule in the gene pool. Both events are random. Since, according to the neutral theory of molecular evolution, the substitution rate equals the rate at which neutral mutations arise, in what follows we treat the two events conjointly. Mutations (substitutions) either occur (= "success") or do not occur (= "failure") in a series of *n* independent trials, at each of which they have a certain probability *p* of appearing. (Here a trial encompasses each DNA replication that provides a nucleotide with the opportunity to change, and random sampling underlying the creation of a new generation gene pool provides an opportunity for fixation and so for substitution.) From this point of view, therefore, it should be possible to model these two events by the binomial probability distribution which we described in the preceding chapter and in Appendix One. This modeling, however, is not convenient because the number of trials, *n*, is very large (every cell division preceded by DNA replication constitutes a trial) and the binomial expansion is difficult to manage for large numbers. Fortunately, there is another probability distribution which is easy to handle even for a large *n* and which can be used to approximate the binomial distribution.

The distribution bears the name of the French mathematician and physicist Simeón-Denis Poisson who described it in 1837. Poisson received many honors during his lifetime for his various scientific contributions, including elevation to baronage, but curiously the achievement for which he is now best remembered was not among those aggrandized. The *Poisson probability distribution*, which has only one parameter, λ (lambda), is defined by the formula:

$$P(k;\lambda) = \frac{\lambda^k}{k!} e^{-\lambda}$$

where $P(k;\lambda)$ stands for the probability that a rare random event will occur *k* number of times; λ is the product of the number of trials (*n*) times the probability of success (*p*) in *n* number of trials (i.e., $\lambda = np$; it represents the mean of the distribution); *e* is a constant approximately equal to 2.71828; and ! is the factorial symbol. The derivation of the formula and of the number *e* is described in Appendix Two. The probabilities (the relative expected frequencies) of the rare event happening 0, 1, 2, 3, 4, ... , *k* times are obtained by the expansion of the formula:

k:	0	1	2	3	4	...	*k*
$P(k;\lambda)$:	$e^{-\lambda}$	$\lambda e^{-\lambda}$	$\frac{\lambda^2}{2!}e^{-\lambda}$	$\frac{\lambda^3}{3!}e^{-\lambda}$	$\frac{\lambda^4}{4!}e^{-\lambda}$ ...		$\frac{\lambda^k}{k!}e^{-\lambda}$

An example of using the formula to obtain the individual probabilities is also given in Appendix Two, in which, furthermore, we derive the Poisson from the binomial distribution. The Poisson distribution can, however, be interpreted on its own as a description of a distinct *Poisson process*. The latter is defined as a process in which events occur at random and the number of events in any time interval depends only on the length of the interval *t* and is specified by the Poisson probability distribution. To introduce the Poisson process and the distribution, we avail ourselves of an example popularized by Ronald A. Fisher.

In earlier days, the glory of an army used to be the cavalry. Alas, being a cavalryman brought with it certain imperilments to which ordinary soldiers were not exposed. The cavalrymen's companions are often temperamental creatures and when a horse is in a bad mood, accidents, even fatal ones, can happen. Some of the best records of such accidents were kept, to nobody's surprise, by the Prussian army. When they were declassified, statisticians exploited them to illustrate the laws governing random events. One set of these records of ten army corps kept over a period of 20 years was used by Fisher as an example of the Poisson process in action. An important factor in the Poisson process is time. Although continuous, time can be divided into intervals of specified, fixed length and the distribution of events over these intervals determined. Depicted graphically, such distribution may look like this:

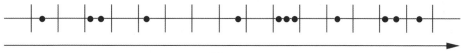

Time

Here the vertical bars delineate intervals and the dots mark events, in our case the deaths of cavalrymen by horse kicks. The diagram explicates the randomness of the process (indicated by the irregularity in the distribution of the dots) and the rarity of the events (signified by the long distances between events). Clearly, the time factor must be taken into account in the analysis of the process and this can be done by relating the process to a fixed time interval. Specifically, in the Poisson formula, the expected (mean) number of events, λ, is replaced by the expected number of events per fixed time interval, λt. The formula thus becomes:

$$P(k;\lambda t) = \frac{(\lambda t)^k}{k!}e^{-\lambda t}$$

The application of the formula to the records of the ten Prussian army corps is shown in Table 5.1. Here the fixed time interval is one year, and since the observation period extended over 20 years, there were (10)(20) = 200 intervals altogether. In more than half of these

Table 5.1
Statistical analysis of data on deaths caused by horse kicks, based on the records of ten army corps over a period of 20 years.

k	N_k	$(k)(N_k)$	$P(k;\lambda t = 0.61)$	NP
0	109	(0)(65) = 0	0.55	(200)(0.55) = 110
1	65	(1)(65) = 65	0.33	(200)(0.33) = 66
2	22	(2)(22) = 44	0.10	(200)(0.10) = 20
3	3	(3)(3) = 9	0.02	(200)(0.02) = 4
4	1	(4)(1) = 4	0.003	(200)(0.003) = 0.6
5	0	(5)(0) = 0	0.0004	(200)(0.0004) = 0
	$N = 200$	$T = 122$		$N = 200.6$

k, number of deaths caused by horse kicks per year per corps; (k) (N_k), number of fixed time intervals in which k number of men were killed (the dimensions of one time interval being per one year per one corps); $P(k;\lambda t = 0.61)$, probability that k number of men will be killed if the expected number of deaths in the fixed time interval is $\lambda t = T/N = 122/200 = 0.61$; NP, expected number of intervals in which k number of men will be killed; N, total number of intervals [(10)(20) = 200]; T, total number of deaths in the 200 time intervals. $P(k;\lambda t = 0.61) = \frac{(0.61)^k}{k!}e^{-0.61}$

(109 to be exact), no accident happened; in 65 of the one-year-long intervals, one man was killed by a horse kick in each; in 22, two men were killed; and so on. The total number of men killed over the entire observation period of 200 intervals was 122 and the mean number of men killed per time interval was λt = 122/200 = 0.61. Applying this value to the Poisson distribution formula, the probability (expected frequency) of 0, 1, 2, 3, ... men being horse-kicked to death can be computed. The values can then be used to calculate the number of intervals in which 0, 1, 2, 3, ... deaths could be expected to occur. As can be seen from the *NP* column in Table 5.1, the numbers are in good agreement with the observed values. The Poisson process thus describes the work of the Grim Reaper among the cavalrymen well. But is it also suited to modeling molecular evolution?

Applied to molecular evolution, the Poisson distribution specifies the probabilities that 0, 1, 2, 3, ... substitutions will occur in a DNA segment of a certain length in a defined time interval. Since substitutions are normally observed by comparing two DNA sequences that diverged *t* years ago, the mean (expected) number of substitutions in a fixed time interval is given by the product $2t\mu$. Here, μ is the substitution rate (the mean number of substitutions per unit of sequence lengths, per unit of time), and *t* is the time elapsed since the two sequences shared a most recent common ancestor. Since the two lineages leading from the most recent common ancestor to the extant sequences accumulated substitutions independently for *t* years each, together they had $t + t = 2t$ years to diverge: *t* years from the common ancestor to one sequence plus another *t* years from the ancestor to the other sequence. The Poisson formula for molecular evolution therefore has the form

$$P(k;2t\mu)=\frac{(2t\mu)^k}{k!}e^{-2t\mu}$$

where $P(k;2t\mu)$ is the probability that *k* number (0, 1, 2, 3, ...) of substitutions will occur at a nucleotide site in a time interval *t* when the substitution rate is μ. The expanded form of the formula then gives the probabilities for the different *k* values:

Number of substitutions *(k)*:	0	1	2	3
Probability [*P(k; 2tμ)*]:	$e^{-2t\mu}$	$2t\mu e^{-2t\mu}$	$\dfrac{(2t\mu)^2}{2!}e^{-2t\mu}$	$\dfrac{(2t\mu)^3}{3!}e^{-2t\mu}$

To give an example: if $\mu = 10^{-9}$ substitutions per nucleotide site per year and $t = 5 \times 10^6$ years, the probabilities that 0, 1, 2, or 3, substitutions occurred at this site in the five million years are:

Number of substitutions:	0	1	2	3
Probability:	0.99	0.0099	4.95 x 10^{-5}*	1.65 x 10^{-7}

---

* A brief refresher on scientific notation based on the decimal system:
$10^0 = 1$ (by definition)

$10^1 = 10$ $10^{-1} = 1/10 = 0.1$
$10^2 = (10)(10) = 100$ $10^{-2} = 1/100 = (1/10)(1/10) = 0.01$
$10^3 = (10)(10)(10) = 1000$ $10^{-3} = 1/1000 = (1/10)(1/10)(1/10) = 0.001$
$10^4 = (10)(10)(10)(10) = 10,000$ $10^{-4} = 1/10,000 = (1/10)(1/10)(1/10)(1/10) = 0.0001$

Hence the *exponent* of each 10, if positive, specifies the number of zeros that follow the 1 if the number is written out. If negative, it is one less than the number of zeros following the decimal point. Hence 4.95×10^5 would equal 495,000 and 4.95×10^{-5} equals $(4.95)(1/100,000) = 0.0000495$ (the decimal point has been moved to the left by the number of places specified by the exponent, i.e., 5).

The problem is, however, that often we know neither the substitution rate nor the divergence time of the two sequences; in fact, in some cases (see Chapter Eleven), we even use the Poisson formula to estimate t. We must therefore find some other way of defining the mean, independently of μ and t. In the horse-kick example, we obtained λt by multiplying the observed number (0, 1, 2, 3, …) of fatalities per year by the number of years in which the specified numbers of accidents happened, adding up the individual categories, and dividing the sum by the total number of time intervals for which records were available. Theoretically we should be able to obtain $2t\mu$ for the case of molecular evolution in a similar fashion – from the observed number of substitutions differentiating the two sequences. The *proportion of differences*, p (the number of observed substitutions, n_d, divided by L, the total number of nucleotide sites compared; i.e., n_d/L; it is also referred to as p *distance*), is in reality a form of a mean. In this case, however, we only have two categories that we can actually observe – "no substitutions" and "substitutions" – and in each of them we have hidden events. For this reason we cannot proceed in the same manner as in the horse-kick example. To obtain the mean, we must take the hidden substitutions into account and so convert the proportion p into an *evolutionary distance*, d. How this is done is explained in Appendix Two. In principle, one considers all the possible changes a particular nucleotide and the changed nucleotides at a given site can undergo. One then calculates the probabilities of the individual changes under the assumption of a Poisson process and estimates the numbers of changes hidden behind those revealed by the comparison of the two sequences. The complicated procedure can be condensed into a mathematical formula which can then be readily applied to any number of sequence comparisons. Several such formulae are now available; they differ in the assumptions regarding the nature of the nucleotide changes. The first and simplest of them was derived in 1969 by Thomas H. Jukes and Charles R. Cantor. It is based on the assumption that the four nucleotides are represented equally in a sequence, and that $A \leftrightarrow G$ and $C \leftrightarrow T$ changes (*transitions*) occur with the same probability as all other changes (*transversions*). Neither of these assumptions is borne out by observations. Other, more complex formulae take into account variation in nucleotide composition, differential probabilities of transitions and transversions, as well as other factors influencing the frequency and nature of nucleotide changes and so presumably provide a more accurate correction for hidden changes. The *Jukes-Cantor formula* is

$$d = -\frac{3}{4}ln\left(1-\frac{4}{3}p\right)$$

where d is the estimated number of nucleotide substitutions per site or *evolutionary distance* ($2t\mu$), ln the natural logarithm (i.e., the logarithm which has as its base the number e mentioned earlier), p is the observed proportion of differences between two sequences: $p = n_d/L$, and L is the number of nucleotide sites compared. (The formula is derived in Appendix Two.) For two sequences, each of which is 500 nucleotide pairs long and which differ at 50 sites (i.e., $p = 50/500 = 0.1$), the number of nucleotide substitutions is $d = 0.1073$ per site. Applied to the entire sequence, this comes to $(0.1073)(500) = 53.66$ substitutions. Hence, the correction formula reveals 3 to 4 hidden substitutions in addition to the 50 observed ones.

The Art of Drawing a Tree

To sum up: the essence of molecular evolution is the gradual accumulation of nucleotide substitutions, which can be modeled by the Poisson process governed by laws of probability. The result of molecular evolution is the appearance of differences in DNA, and secondarily also protein sequences, in which the genealogical record of the evolutionary relationships among the organisms is encrypted. The task we now face is how to convert the differences into a phylogenetic tree.

But why do we expect the depiction of phylogeny to assume the shape of a tree? At the molecular level, DNA replication is a bifurcating process in which each parental molecule produces two daughter molecules, each of which again produces two molecules, and so on, with the accumulation of substitutions superimposed on this process. When depicted graphically (see Fig. 3.8), DNA replication at the molecular level resembles an abstract tree. At the species level, too, populations are presumed to split mostly by bifurcation and the diagram outlining the process again has a tree-like shape. Thus because the two essential processes that underlie evolution have a bifurcating, tree-like geometry, the tree would seem to be an appropriate metaphor for organismal phylogeny (but perhaps not for all domains of life, as becomes apparent in the next chapter).

The conversion of sequence data into a phylogenetic tree requires, first, a general principle for making choices among various options, and second, a specific method of converting the data into a tree. The guiding principle of most phylogenetic reconstructions is the *principle of parsimony,* which has long been known to philosophers as *Ockham's razor.* William of Ockham was a fourteenth century English theologian and philosopher who made the demand for simplicity of arguments his platform. If his colleagues became entangled in a convoluted war of words, he would interrupt the debate with a suitably cutting remark such as *entia non sunt multiplicanda praeter necessitatem* (roughly translated "plurality is not to be assumed without necessity") or "What can be done with few assumptions is done in vain with more". So effectively did he slice through the wrangles that his contemporaries began to speak of "Ockham's razor", and this phrase has been retained as a demand for parsimony (from Latin *parsus,* thrift, economy) in explanations. The theologian Ockham would surely have been tickled pink to learn that his principle became the golden rule of evolutionary thinking!

Applied to evolution, the parsimony principle states that given a choice between a more direct path involving fewer steps and a less direct one involving more, evolution opts for the former. Evolution is thrifty, parsimonious. Why should this be so? If the choice is between several hypotheses, why should we want to choose the simplest? Theologian Ockham had a ready answer: since the world was created and is superintended by God, and since God is, among other things, a Supreme Logician, it is inconceivable that He would tolerate any choice other than the simplest. Modern philosophers of science are less certain about their reasons why evolution should adhere to the parsimony principle. Two arguments are usually put forward to justify its application, philosophical and statistical, but neither of them is entirely satisfying. The philosophical argument reflects Ockham's original one: if the choice is between two hypotheses, one which explains all observations without having to resort to any auxiliary assumptions, and another that is untenable without ad hoc conjectures, it is more economical to give precedence to the former over the latter. In phylogenetic analysis, the assumption of homoplasy is an ad hoc explanation. If fewer assumptions have to be introduced and fewer changes postulated, the phylogenetic hypothesis becomes more likely. This may be all very well philosophically, but why should Nature act according to what philosophers think is best for her? Here, one could perhaps argue that natural selection favors those individuals who achieve an adaptive change before the rest, which usually means in fewer steps. (In other words: Nature does not tolerate bureaucrats. In human societies, bureaucrats thrive because they are exempted from selection pressure. Just imagine what would happen if, upon leaving the city hall, driver's license bureau, or passport office, you were handed a score card to grade the bureaucrat's efficiency and the low scores were then used to dismiss inefficient clerks! Well, its nice to daydream occasionally.) The statistical argument is based on the observation that evolutionary changes are rare and their probability of occurrence is low. Since the probability of multiple events is lower than that of a single event, explanations invoking fewer changes are more likely to be correct than those postulating more numerous events.

The input data necessary for drawing a phylogenetic tree (evolutionary biologists commonly speak of "constructing", "building", or "making" a tree, as if it were actually possible to obtain a tree by any of these activities!) are of two kinds: they comprise either the individual substitutions differentiating the sequences, or the overall differences between sequences in the form of genetic distances. In evolutionary analysis, the site occupied by a nucleotide in the DNA (or a position occupied by an amino acid residue in a protein) is a *character* and the identity of the nucleotide (amino acid residues) is the *character state*. Tree-drawing procedures based on the identity of nucleotides (amino acid residues) at individual sites (positions) are therefore referred to as *character state methods*. They use the character states of extant sequences to infer the optimal tree.

In the *distance methods*, it is of no primary concern whether a site is occupied by A, G, C, or T, but whether it is occupied by identical or different nucleotides in the compared sequences. Also, in contrast to the character state methods, distance methods do not attempt to reconstruct ancestral states; they assess similarity or dissimilarity of two sequences strictly on the basis of the genetic distance between them and use this distance to assign positions of the sequences on the tree. Earlier we defined *genetic distance* as the proportion of different nucleotides (amino acid residues) between two sequences corrected for hidden changes. The term is, however, often used in reference to the uncorrected number of differences between two sequences, or the uncorrected proportion (the number of sites at which two sequences differ, divided by the total number of sites). Graphically, distance can be represented as a line connecting two points, *A* and *B*:

In phylogenetic reconstruction, the distance *AB* between two extant sequences is in reality a composite of two distances, one from *A* to ancestor *X* and the other from ancestor *X* to *B*:

Another way of depicting this relationship graphically is:

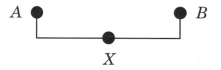

The two depictions are equivalent and both are used interchangeably in phylogenetic iconography. Under the assumption of a molecular clock, the *AX* and *BX* distances should be of equal length because the time that has elapsed since the splitting of the two lines of descent is the same for both. In reality, however, the *AX* and *BX* branches are often unequal, first of all, because molecular evolution is a stochastic process during which, by chance, more substitutions may accumulate on one branch than on the other; and second, because evolution may proceed faster in one lineage compared to another.

The casting of the data into the form of a tree can be accomplished in one of two principal ways. The data can be pitted against all the theoretically possible ready-made trees and the tree that accommodates the data optimally is then chosen as most likely representing the phylogeny of the set of sequences. Or the tree is "assembled" branch-by-branch according to a specified, fixed routine applied to the data set. The former procedure requires certain optimality criteria which, when applied to the data, enable the biologist to choose among the trees. The latter approach is contingent on the availability of an appropriate algorithm, a step-by-step set of instructions for grouping the data.

Depending on the nature of the input data (character state versus distance) and the procedure used to obtain the tree (the assessment of ready-made trees as opposed to the "assembly" of a specific tree), there are four groups of methods commonly used for tree drawing: maximum parsimony (MP), maximum likelihood (ML), minimum evolution (ME), and neighbor-joining (NJ). The first two of these are character-state methods using optimality criteria to select the best fitting tree from those that are theoretically possible. The last two are distance methods, one of which (ME) uses an optimality criterion to choose a ready-made tree while the other (NJ) relies on a clustering algorithm. None of the methods is perfect; each has its advantages and disadvantages. All perform well when the data set is unambiguous and each outperforms the others, depending on the circumstances, when the set permits certain ambivalence.

MP, ML, ME, and NJ

The *maximum parsimony method* strives to reconstruct the pathway of substitutions that may have led to the observed sequence differences. It considers all the possible trees and selects those that require the fewest changes (most thrifty, maximally parsimonious) to account for the differences. To grasp its principle, consider four sequences, W, X, Y, and Z, and four sites, 1-4:

Sequence	Site			
	1	2	3	4
W	G	T	A	G
X	G	T	A	C
Y	G	T	G	G
Z	G	C	G	C

The four sequences can, theoretically, be related to one another in three different ways, I, II, or III:

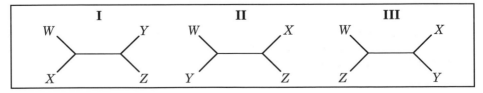

The three diagrams represent the three theoretically possible unrooted trees; any other trees, such as ![diagram], are trivial variants of the three trees. [The total number of theoretically possible unrooted trees is given by the formula: $(2n-5)!/(2^{n-3})(n-3)!$, whre n is the nuber of difrent squences. For four squences, as in the abve example, there are $(3 \times 2 \times 1)/2 = 3$ unrooted trees.]

Let us now consider one site after the other in terms of the minimum changes it involves. Site 1 does not seem to have changed at all and can therefore be ignored. Site 2 must have suffered at least one change, no matter how we arrange the branches:

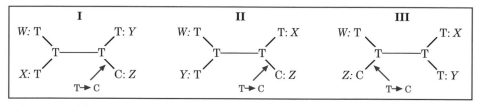

(The site could, of course, have undergone more than one change, but this assumption is not maximally parsimonious and so is not taken into consideration.) As far as site 2 is concerned, all three trees are equally likely and so the site is regarded, like site 1, as being *phylogenetically uninformative* and as such is not considered any further. Site 3, on the other hand, is *phylogenetically informative*, because it enables us to choose the most parsimonious tree of the three:

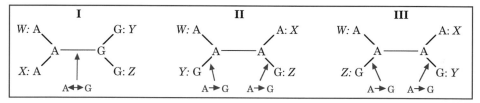

Tree I requires only one change, whereas trees II and III each require at least two changes to explain the pattern of nucleotide distribution at this site among the four sequences. (Note that there are several other possibilities with regard to the identity of the nucleotides in the two ancestors, but no matter which is considered, tree I always requires fewer changes than trees II and III.) Finally, site 4 is also phylogenetically informative but, alas, it identifies tree II rather than tree I as the most parsimonious of the three:

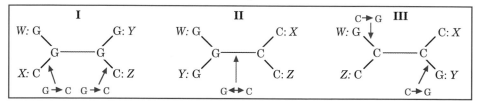

In this case, it is tree II that requires a single change, whereas both trees I and III each require at least two changes. The conflict between the two informative sites can be resolved by the inclusion of additional informative sites, if they are available: the tree identified by *most* sites as requiring the smallest number of changes would then be considered as maximally parsimonious and so as reflecting the actual evolutionary changes most closely. The sites contradicting this tree would be assumed to have undergone hidden changes, giving the tree that is based on them the appearance of being more parsimonious than it is in reality. As in the above case, however, often not one, but several trees are identified as being the most parsimonious. These can then be combined into one *consensus tree* by displaying, in the form of dichotomous splits, only those branches on which either all (*strict consensus*) or most (*majority rule consensus*) of the trees agree. The splits on which the trees disagree are presented as *polytomies* – multiple branches issuing from a single point (node).

For large numbers of sequences, the number of theoretically possible trees is enormous. For example, ten sequences can already be arranged into more than two million unrooted and more than 34 million rooted trees. Given larger numbers of sequences, the number of trees increases to such an extent that methods have first to be applied to restrict the search for the right tree to a small section of the forest in which its location is presumed. The methods are usually hitched to the tree-drawing process itself. A system is employed in which the whole process is divided into a series of pathways, incorporating certain criteria into each pathway that render some more likely than others to lead to the right tree. Since, however, the criteria cannot guarantee that the baby will not be thrown out with the bath water, one can never be sure that the identified tree is truly the best-fitting one of the whole lot. Procedures that aid the discovery of the right tree are called *heuristic* (from Greek *heuriskein*, to discover).

An optimal tree, as defined by the optimality criterion of the MP method, is the one requiring the fewest evolutionary (character state) changes to explain the observed differences among the analyzed sequences. In the second of the four groups of tree-drawing methods, the optimality criterion is based on the principle of *maximum likelihood*. The expression "likelihood" is often used synonymously with "probability" but to a statistician, these two terms mean different things. To explain the principle of maximum likelihood, we use a coin-flipping experiment. In the usual version of the experiment, we assume that we know the probability p of heads coming up in a single trial (and hence also the probability $q = 1 - p$ of no heads, i.e., tails). Since there are two possible outcomes, heads or tails, and both are equally likely $p = q = 1/2$. By using the binomial probability distribution (see Appendix One), we calculate the probability of heads coming up twice, once, or not at all in two coin tosses. In this model, p is the known *parameter* which we use to estimate the probabilities of the outcome of an experiment.

Now we reverse the situation and assume that we know the outcome of an experiment and from it want to estimate the parameter and thus test whether the coin is fair. Suppose, for example, that we flipped the coin 100 times and observed heads coming up in 60 tosses. What is the coin's p in this case? Is it equal to 0.5, as would be expected from a fair coin, in which case the deviation from the expectation is merely chance fluctuation? Or does it have some other value (0.6?) as a result of the coin's idiosyncrasy?

We cannot answer these questions with certainty, but we can calculate the probabilities of obtaining the observed outcome when p assumes different values and then choose the p that gives the observed outcome with the highest probability as the most likely value of p. We thus have two different situations: in one we calculate the probability of an outcome from the fixed parameter and in the other we estimate the parameter from the fixed outcome of an experiment. To distinguish these two types of probabilities, statisticians from the time of Ronald A. Fisher call the former *probability* and the latter *likelihood*. We can therefore define likelihood, L, as the probability of obtaining the observed data given a particular hypothesis (model) specifying it. The *maximum likelihood estimate* of a parameter is therefore the value of the parameter that maximizes the probability of obtaining the observed outcome. Hence to obtain a maximum likelihood estimate of the value of a parameter, we need both a set of observed data and a model that specifies the probabilities of observing the data given different values of the parameter. In the coin-tossing experiment, the model is the binomial probability distribution and the data, in our case, are represented by the observation that in 100 tosses, heads were thrown 60 times. The formula specifying the probabilities of individual outcomes in a coin-tossing experiment is

$$B(k;n,p) = \frac{n!}{k!(n-k)!} p^k (1-p)^{n-k}$$

where n is the number of trials (tosses), k the number of successes (heads coming up), and p the parameter (the probability of heads coming up in a single toss). In this set-up, the only fixed value is that of the parameter ($p = 0.5$), whereas n and k can be varied. In the reverse set-up, in which the values of n and k are fixed and the parameter p is to be estimated, we use the same model but set $n = 100$, $k = 60$, and consider p as the unknown variable:

$$L(p;k,n) = \frac{n!}{k!(n-k)!} p^k (1-p)^{n-k}$$

This expression is the *likelihood function*. If we put the fixed values of k and n in this formula and vary the value of p, we can compute the likelihood of obtaining the observed result of 60 heads in 100 tosses. By plotting p against $L(p;k,n)$, we obtain a curve of the type shown on the left. The curve changes its slope (gradient) continuously (see Appendix Four), but at one point – the peak of the curve – the slope becomes 0 and the value of L is at its highest. This value represents the maximum likelihood estimate of p. (The binomial coefficient in the above formula influences the height of the curve – the fraction – but not the maximum likelihood estimate of p and can therefore be ignored in the calculations.)

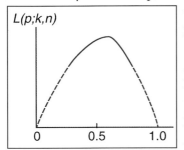

In the case of $n = 100$ and $k = 60$, intuitively we might have expected the most likely value of p to be 60/100 = 0.6, as it was indeed found to be. In general, we would expect the value of the parameter to lie close to the ratio of observed successes to total number of trials in a given experiment. In the coin-tossing experiment, the maximum likelihood estimate of p is clearly apparent, but in more complex situations it is by no means obvious and must therefore be reached by the procedure described. Instead of plotting the changing values of the parameter against the likelihood function and searching for the peak, however, the maximum likelihood estimate is normally obtained by a mathematical procedure called *differentiation*, which is described briefly in Appendix Four.

To apply the maximum likelihood principle to phylogenetic reconstruction from sequence information, our starting kit must contain three things: the sequences whose phylogenetic relationship we want to determine (the observed data), a model of sequence evolution, and a tree whose likelihood we want to compute. The two essential elements of a phylogenetic tree are the branch lengths, and associated with them the branching order (topology). In using the maximum likelihood method we therefore want to find out which branch lengths make the observed differences between the sequences (i.e., the evolutionary distances between them, defined as $d = \mu t$: the product of the substitution rate and the time of sequence divergence) most likely. In this case, therefore, the parameter we want to estimate by the method is the branch length. To compute the likelihood of a given tree, we need to obtain a maximum likelihood estimate of this tree's branch lengths and branching orders, and by examining different trees distinguished by their branch lengths and branching order, we must find the one that gives the greatest likelihood of accounting for the data. The procedure by which this is accomplished is not easy to explain without trivializing it. Although we have restricted the explanation to a simple example of the method's application, we realize that the description that follows may tax the reader's staying power. The method, however, is of such importance that anybody striving to understand how modern phylogenetic reconstructions are carried out should make the effort to understand its principle.

For the sake of simplicity, consider once more four sequences, *1-4*, derived from different extant species. The sequences can be arranged in three ways, as three unrooted trees:

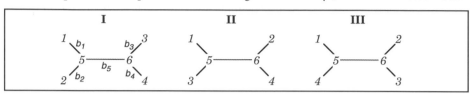

in which the presumed ancestral sequences are designated as *5* and *6* and the branches are denoted b_1-b_5. The branches (distances between adjacent nodes) may be of different length depending on their mutual arrangement and it is their lengths in each of the three trees that we now want to estimate.

Seq. *1*	A	A	G	T
Seq. *2*	A	A	C	T
Seq. *3*	A	A	C	G
Seq. *4*	T	A	G	G
n	20	100	10	5

To this end, we align the four extant sequences and scrutinize the individual sites for their patterns of nucleotide distribution. Suppose we find four patterns in the entire sequence (vertical columns in the matrix on the left), each occurring at the number of sites indicated by *n*. Let us first focus on the pattern AAAT which occurs at 20 sites in the sequences. Theoretically, the ancestral sequences 5 and 6 could have had any of the four different nucleotides (A, C, G, T) at these sites. In other words, in Tree I the distribution could have been

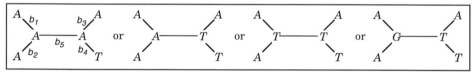

or any other of the 4 x 4 = 16 different variants of tree I (variants 1-16). Which of these variants is most likely to have been the one that occurred? To find out, we must examine each of the variants, branch by branch, and then calculate for each of them the probability that the nucleotide has either changed or remained the same in the pair of sequences connected by the branch. To do this, we need to first specify the model of the evolutionary process, in other words the general probabilities of the change or no change of a nucleotide at a site. One such model – one of the simplest – is the Jukes-Cantor model described earlier in this chapter and in Appendix Two. The model assumes, you will recall, that each nucleotide has an equal probability of changing to any other of the four nucleotides. The probabilities are

$$P_{ii}(b)=\frac{1}{4}+\frac{3}{4}e^{-\frac{4}{3}b}$$

for a nucleotide staying the same and

$$P_{ij}(b)=\frac{1}{4}-\frac{1}{4}e^{-\frac{4}{3}b}$$

for it changing to any other nucleotide. Using these formulae we calculate the probability of the expected outcome for each of the five branches in each of the 16 variants.

Designating the probabilities of the individual branches as $P_{AA}(b_1)$, $P_{AA}(b_2)$, etc (where the letters AA stand for the nucleotides at the branch ends and b_1, b_2, ... are the branch designations), we can write the total probability of variant 1 of tree I (P_1), for example, thus:

$$P_1 = \frac{1}{4}\left[P_{AA}(b_1)P_{AA}(b_2)P_{AA}(b_5)P_{AA}(b_3)P_{AT}(b_4)\right]$$

(The fraction 1/4 is necessary to take into account the choice of one nucleotide out of four.) Each of these probabilities [i.e., $P_{AA}(b_1)$, $P_{AA}(b_2)$, etc] is calculated by the application of the appropriate Jukes-Cantor formula, P_{ii} for P_{AA} and P_{ij} for P_{AT}. Since b, the branch length, is the parameter to be estimated, we vary it and calculate P_{ii} and P_{ij} for each chosen value. Obviously, however, we must place some restrictions on the choice of the b values to be plugged into the formula, because otherwise we would be calculating until the end of our days. Specifically, we choose a particular value as the starting point of our calculations which we have reason to believe is close to the value expected to give the highest P_{ii} or P_{ij} result. This initial value is obtained from the evolutionary distances between the extant sequences. Thus, for example, the initial length of the b_1 branch in variant 1 of tree I can be computed using the equation

$$b_1 = \frac{1}{2}d_{12} + \frac{1}{4}\left(d_{13} + d_{14} - d_{23} - d_{24}\right)$$

(Here d_{12} is the number of nucleotide substitutions between sequences 1 and 2, d_{13} the distance between 1 and 3, etc). This formula has been derived by adding up equations $d_{12} = b_1 + b_2$, $d_{13} = b_1 + b_5 + b_3$, $d_{14} = b_1 + b_5 + b_4$ etc (see diagram of tree I above) and solving the sum algebraically by replacing all b's except b_1 by d's. Starting with these initial b values and plugging them into the P_1 formula above, we then gradually increase them, thereby ascending the curve in a process similar to that of the coin-tossing experiment, until we reach its peak – the point of highest probability. (Here again, however, the probability is determined by differentiation rather than by drawing curves.) The same computation is carried out for all 16 variants, and the probabilities P_1, P_2, P_3 etc are then added up to obtain the likelihood of pattern AAAT:

$L_{AAAT} = P_1 + P_2 + P_3 + ... + P_{16}$
Next, likelihoods of the other patterns (L_{AAAA}, L_{GCCG}, L_{TTGG}) are computed in the same manner and these are then used to obtain the likelihood of tree I:

$$L_I = \frac{135!}{20!\,100!\,10!\,5!}\left(L_{AAAT}\right)^{20}\left(L_{AAAA}\right)^{100}\left(L_{GCCG}\right)^{10}\left(L_{TGGT}\right)^{5}$$

or generally $$L(b, X) = \frac{n!}{n_1!\,n_2!\,n_3!...n_{n_m}!}L_1(b)^{n_1}\,L_2(b)^{n_2}\,L_3(b)^{n_3}...L_{n_m}(b)^{n_m}$$

where $L(b,X)$ is the likelihood function of the branch length (the likelihood that the observed pattern X of differences among the sequences is the result of an arrangement of branches with specific lengths b (i.e., b_1, b_2, b_3,...); n_1, n_2, n_3,...n_m are the numbers of sites with a given nucleotide pattern; n is the total number of sites; $n!/n_1!n_2!n_3!...n_m!$ is the coefficient specifying the number of combinations in which the patterns occur along the sequence; and $L_1(b)^{n_1}$, $L_2(b)^{n_2}$, etc are the likelihoods of observing a particular nucleotide pattern. After computing the likelihood of tree I, the same procedure is applied to trees II and III, and at the end the tree giving the greatest likelihood of producing the observed differences among the extant sequences is chosen.

From this description, it is apparent that the maximum likelihood method of tree drawing is computationally demanding and its application would be unthinkable if it could not be entrusted to computers. But even computers require a long time to carry out all these steps and the method can therefore be applied to a relatively small number of sequences only.

The third category of methods employs the *minimum evolution* criterion to find the optimal tree. In contrast to the MP and ML methods, which are used to scrutinize individual substitutions, the ME methods deal with genetic distances. In common with the MP and ML methods, however, the ME methods also rely on ready-made trees (i.e., they examine all theoretically possible trees). The only directly computable genetic distances are those between two extant sequences. However, as pointed out earlier, each of these distances is a composite of at least two distances – one from one extant sequence to that of the common ancestor and the other from this ancestor to the second extant sequence. Distances between less related sequences can be partitioned into several segments corresponding to the individual branches of the tree. As also stated earlier, the distance between two adjacent nodes is the *branch length* and the sum of all the branch lengths is the *length of the tree*. An unrooted tree of n extant sequences has $(2n-3)$ branches and the lengths of all these branches must be added up to obtain the tree lengths. Different trees drawn from the same set of extant sequences have different lengths, some representing shorter and others longer evolutionary pathways. Because the shortest pathway entails the fewest changes required to produce the observed differences between the extant sequences, and as minimum evolution is the most parsimonious explanation of the origin of these differences, the shortest tree is assumed to explain the evolution of a given sequence set best.

To identify the shortest tree in the collection of all possible trees, each sequence is compared with every other sequence in the set (= *pairwise comparison*) and the differences distinguishing members of each pair are counted. The distances thus obtained are then organized into a *distance matrix* – a rectangular array of rows and columns listing the differences between the two members of all the possible pairwise combinations. The distance matrix of the W, X, Y, Z sequences introduced earlier is this:

	W	X	Y	Z
W	—	1	1	3
X	—	—	2	2
Y	—	—	—	2
Z	—	—	—	—

(Here, for simplicity, the distances are expressed in terms of number of substitutions and have not been corrected for hidden substitutions.) The matrix is then used to obtain the length of each of the theoretically possible trees. Once again, the four sequences of our example can be arranged into three unrooted trees, each involving two ancestral sequences, here designated U and V:

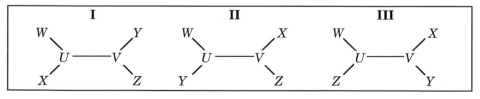

The length of tree I, for example, is the sum of the distance between the individual nodes: $WU + XU + UV + YV + ZV$, none of which we know. We know, however, the $WX, WY, WZ, XY, XZ,$ and YZ distances and from these, using a simple algebraic trick, we can compute the tree's length. We note that:

$$WY = WU + UV + VY$$
$$WZ = WU + UV + VZ$$
$$XY = XU + UV + VY$$
$$XZ = XU + UV + VZ$$

By adding these equations we obtain

$$WY + WZ + XY + XZ = 2WU + 2XU + 4UV + 2VY + 2VZ$$

Since $WU + XU = WX$ and $VY + VZ = YZ$, we have

$$WY + WZ + XY + XZ = 2WX + 4UV + 2YZ$$

so that

$$UV = 1/4[(WY + WZ + XY + XZ) - 2WX - 2YZ]$$

Hence the total length of tree I is

$$S_I = WX + YZ + 1/4[(WY + WZ + XY + XZ) - 2WX - 2YZ]$$
$$= 1/4[WY + WZ + XY + XZ + 2WX + 2YZ]$$

By substituting the symbols in this equation by distances from the distance matrix, we obtain $S_I = 3.5$. For trees II and III, the equations are

$$S_{II} = WY + XZ + 1/4[(WX + WZ + YX + YZ) - 2WY - 2XZ]$$
$$= 1/4[WX + WZ + YX + YZ + 2WY + 2XZ] = 3.5$$
$$S_{III} = WZ + XY + 1/4[(WY + WX + ZX + ZY) - 2WZ - 2XY]$$
$$= 1/4[WY + WX + YZ + ZX + 2WZ + 2YX]$$
$$= 1/4[1 + 1 + 2 + 2 + 6 + 4] = 16/4 = 4$$

This approach identifies trees I and II as the shortest of all three. (Recall that tree II was also identified as one of the two optimal trees by the maximum parsimony method.)

Neighbor joining is, as the name implies, an example of a distance method in which a tree is drawn step-wise, by the successive addition of branches. The neighbors are two sequences on the tree which are interconnected via a single node (ancestor), implying that they are more closely related to each other than either of them is to any other sequence in the set. The neighbor-joining procedure consists of finding these sequences sequentially and segregating them from the rest. Two sequences are identified as neighbors when, after being joined to their most recent common ancestor, the partly-drawn tree is shorter than it would be if any other two sequences were made into neighbors. Once identified as neighbors, the two sequences are treated as if they were a single sequence and the procedure is repeated until all sequences have found their neighbors.

To start the procedure, we arrange the sequences *A* through *H* in a star-like configuration, as if they were all derived from a single common ancestor *X*:

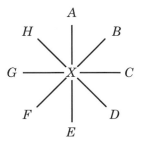

We then choose two sequences from the set at random, for example A and B, make them neighbors interconnected via the ancestor X,

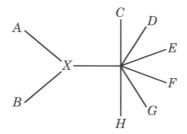

and compute the length of this partly-drawn tree by adding up the lengths of the individual branches. We then choose another two sequences, make them neighbors, and repeat the entire process. We do this for all the possible combinations of sequences in the set and identify the pair that gives the shortest total tree length. These two sequences then become definitive neighbors in the final tree.

Once the first pair of neighbors has been identified and the length of its branches to their shared common ancestor determined, we collapse the pair into a single, composite sequence, and repeat the whole process with the reduced set of sequences. Assume, for example, that we have identified the A, B pair as giving the shortest tree. In the new cycle, we again choose two sequences at random from the set which now consists of (A, B), C, D, E, F, G, and H sequences, say E and F, and make them neighbors:

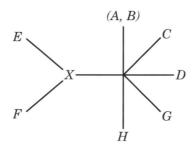

We compute the sum of the branch lengths of this tree, test all possible neighbor combinations, identify the pair giving the shortest tree, and collapse them into a composite sequence. We proceed in this manner, successively joining pairs of neighbors until every sequence has a partner and we end up with a single tree of specific topology. The method, developed in 1987 by Naruya Saitou and Masatoshi Nei, is quite popular because it is relatively simple to use, produces trees very rapidly, and can handle large amounts of sequence data. Unfortunately, it does not guarantee an optimal tree.

Catching a Liar

Once we have obtained a tree, we have to face the question: how confident can we be that it reflects true phylogeny? The differences we base the trees on are of stochastic nature, the result of chance events, so how do we know that Lady Luck is not playing pranks with us? Statisticians have developed a formidable array of tests that evaluate the reliability of data-

Fig. 5.7
Baron Münchausen* pulling
himself and his horse up from
the bottom of a lake by tugging
on his own ponytail. Illustration
by Gustav Doré. (Reproduced
from Forberg)

set interpretations objectively. Surely, one might think, they have come up with statistical tests for objectively evaluating phylogenetic trees by now. Surprisingly, this is not so. Although several methods have been proposed, their value is controversial.

The difficulty in determining confidence in a phylogenetic tree lies in the nature of the data. The examined sequences represent a small *sample* taken from a very large *population* of sequences. Taking a sample from a population is always associated with a certain *sampling error* and so with a potential misrepresentation of the conditions existing in the population. To establish the degree to which the sample misrepresents the population, the sampling is repeated and the observed variation in the result is used to measure the sampling error. The problem with sequence sampling is that sequence determination is expensive and time-consuming, so that extensive resampling is normally out of the question. The commonly used method of determining the reliability of a phylogenetic tree attempts to bypass this problem by a mock sampling: instead of resampling the population, one resamples the sample. This trick, called *bootstrapping,* was introduced into statistics in 1979 by Bradley Efron. According to Efron, the use of the term "bootstrapping" was inspired by the phrase *to pull oneself up by one's bootstrap,* which in turn presumably has its origin in

---

* In German, the name is spelled "Münchhausen", but the book was first published in English and in this edition the name appears with only one "h". It is in this form that the name has entered the English language.

one of the *Adventures of Baron Münchausen* related by the eighteenth century German writer Rudolph Erich Raspe. In this particular story, the Baron, riding a horse, falls to the bottom of a deep lake but can save himself at the last minute by pulling himself up to the surface with the power of his own bootstraps. (In Münchausen editions we consulted, we found a different version of the tale in which the Baron tugs on his own ponytail rather than on his bootstraps, in the manner depicted by Gustave Doré in Figure 5.7.)

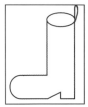 Boots with straps are now somewhat out of fashion, so we have drawn one on the left. The essence of statistical bootstrapping is to obtain information about the characteristics (parameters) of a population by first taking a random sample of n items from it and then sampling the sample with replacements (i.e., by returning each drawn item to the sample after its properties have been recorded) just as the boot is pulled on by the power of its own strap.

The bootstrap method was applied to phylogenetic tree drawing in 1985 by Joseph Felsenstein. To illustrate the principle of its application, consider the following distribution of nucleotides at six sites 1-6 in four sequences W, X, Y, and Z:

	1	2	3	4	5	6
W	C	C	G	A	A	G
X	C	C	G	C	A	A
Y	T	T	C	A	A	A
Z	T	T	C	C	G	G
Favored tree :	I	I	I	II	NI	III

NI, not informative

The six sites of the four sequences represent the sample that we have taken from a very large population of sites in the genomes of the four species. We use this small sample to draw a phylogenetic tree of the four sequences by the maximum parsimony method. Specifically, of the three trees that are theoretically possible, we want to select just one:

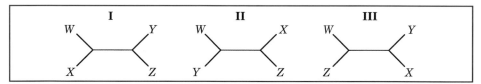

By considering one site after the other, placing the nucleotides into the trees, and counting the number of changes necessary to account for the distribution of the nucleotides at each site in the manner described earlier in this chapter, we come to the conclusion that sites 1, 2 and 3 favor tree I, sites 4 and 6 favor trees II and III, respectively, and site 5 is not informative. The analysis thus favors tree I as providing the most parsimonious explanation for the differences between the sequences. But since sites 4 and 6 favor different trees, we would like to know just how reliable our conclusion is. To find out we resort to the random resampling of the sample.

To this end we roll a die (as a generator of random numbers) six times and register the numbers that come up. Suppose we obtain the following numbers: 5, 4, 1, 6, 2, and 5. They identify the sites for us that we should resample from the sample to produce four pseudosequences:

	5	4	1	6	2	5
W	A	A	C	G	C	A
X	A	C	G	A	C	A
Y	A	A	T	A	T	A
Z	G	C	T	G	T	G
Favored tree :	NI	II	I	III	I	NI

NI, not informative

We then subject the pseudo-sequences to the same maximum parsimony analysis as the original sequences. We find that now two sites support tree I, and one site each trees II and III, so that in this case, too, tree I is favored. If we then repeat the resampling procedure 500-1000 times, we find that 78 percent of the pseudo-sequences support tree I, 11 percent tree II, and 11 percent tree III. These numbers are the bootstrap values that support the different trees. Since in this unrooted tree there are only two pairs of sequences, they receive equal bootstrap support. If more than four sequences were present, bootstrapping would partition them into groups, each bootstrap trial supporting a particular grouping with the exclusion of all other sequences. The different partitions can then receive different amounts of bootstrap support. In other words, the bootstrap support pertains to clusters of sequences and is obtained by counting the number of trees in which a given cluster of sequences has been recovered.

The resampling *randomizes* the data. If there is a clear relationship between two sequences, randomization will not erase it. If, on the other hand, the relationship indicated by the original tree is spurious, it may disappear upon randomization, which changes the frequencies of the individual sites. Bootstrap values of 95 to 100 percent indicate a high confidence level in a given node; values below 95 percent do not necessarily mean that the obtained grouping of sequences is not genuine, but that the available data do not support it convincingly. On the other hand, a node can be supported by a high bootstrap value and yet still be illegitimate.

Fishing for the Truth

The study of molecular evolution only became possible after methods were developed for isolating and sequencing DNA molecules and for analyzing the sequence data statistically. On both these fronts, tremendous advances have been made in the last decade and there is every reason to believe that the rate of progress will not only be sustained, but will even accelerate in the future. The trend toward standardizing and automating laboratory procedures, toward using ever smaller amounts of sampled material, and toward generating sequences more and more rapidly can be expected to continue so that it will be possible to analyze ever increasing numbers of DNA molecules and increasingly longer sequence stretches.

Similarly, on the data analysis front, the tendency toward more objective, quantitative, reliable, and sophisticated methods that take an increasing number of parameters into account will revolutionize our interpretation of evolutionary history. Not only will seemingly unresolvable phylogenies (we give examples of some in the next chapter) be solved, entirely new approaches to phylogenetic reconstruction will be developed. Methods based on the maximum likelihood and similar approaches will attain such a degree of refinement that it will be possible to literally reconstruct ancestor DNA and protein sequences. The first steps in this direction have already been taken. Once such reconstructions become routine, it will then be possible to test the function of the recreated genes and proteins. From this point on, the recreation of crucial phenotypic characters and the ability to test their adaptation to conditions that once existed but now no longer do is close at hand. The tests will enable us to travel the path from molecules to phenotypes in evolutionary reconstruction and infer conditions that existed along the way. A whole new dimension in evolutionary biology will open up to transform the discipline itself and with it, the entire field of biology. Evolutionary biology will become experimental science and biology itself will become far more quantitative than it is at present. In some areas at least, the vision of biology as a quantitative science comparable to physics or chemistry will no longer be intangible. These exciting prospects for the expected advances in evolutionary biology contingent on the development of new methods of fishing for the truth were behind our decision to devote an entire chapter to a topic that some readers may have found vapid. Not only for fishermen, for scientists, too, "know thy art" is the First Commandment. Armed with the knowledge of the methods by which phylogeny can be reconstructed, we can now embark on tracing humanity's origin from the time of appearance of the first organisms on earth – on painting, with broad brush strokes, the Tree of Life.

CHAPTER 6

The Tree of Life

From the Root to the Crown

*I saw in Louisiana
a live-oak growing,
All alone stood it and the moss
hung down from the branches,
Without any companion it grew
there uttering joyous leaves
of dark green,
And its look, rude, unbending,
lusty, made me think of myself.*

Walt Whitman: *Leaves of Grass*

The Loneliness of a Long Perduring Tree

Like the oak tree Walt Whitman saw in Louisiana, the Tree of Life, the community of all living forms on this planet, both extant and extinct, stands alone. There may be trees of life on other planets in the universe and saplings of different trees may once have flourished on this planet; if they did, however, our own Tree of Life must have smothered them all. It soared above them billions of years ago, sank its roots deep into the fertile soil and furcated at an early stage into three beamy limbs, each of which ramified zillions of times over, giving forth to progressively thinner branches that produced a magnificent crown. The extant species are leaves on the terminal branches and like all leaves, they will one day be shed, only to be replaced by new ones on the most recently branched twigs. As on a live oak, every leaf on the Tree of Life is connected to all the other leaves through the pattern of growth – the twigs, the branches, the stem, and the roots. In contrast to a genuine live oak, however, the leaves on the Tree of Life are only connected by imaginary branches. The only real parts of the Tree of Life are the living individuals comprising the species, while the twigs, branches, and so forth are mental constructs conjured up to depict the presumed genealogical relationships among the individual species and groups of species. In other words, the Tree of Life is not a true *arbre* but a pedigree in the shape of a tree. In a pedigree, only the individuals correspond to reality, while the interconnecting lines designate relationships reconstructed from the historical record.

In this chapter we sketch out the Tree of Life on the basis of genealogical information extracted from the molecular archives, point out the changes in perspective occasioned by molecular studies, and specify the position of our species on the Tree. Before we turn our attention to the Tree itself, however, we must answer two questions: first, what makes us so sure that there is only one Tree of Life? And second, what kind of molecular record extends so far into the past that it enables us to dig right down to the Tree's deepest roots?

If all life forms do indeed possess a common ancestor, they should all share certain attributes that arose in this universal forerunner and have been retained ever since. The sharing of such attributes could then be taken as evidence for common ancestry. It would be difficult to argue that groups of organisms that emerged independently of one another opted for the same attribute just by chance. Do such shared attributes exist? They do, indeed. A good example is the genetic code. The "Wheel of Fortune" in Figure 5.4 is one of 10^{70} theoretically possible wheels of this kind in which each of the 64 possible combinations of three nucleotides always specifies one of 20 amino acids. (The immensity of this number becomes apparent when we remind ourselves that the entire Universe is estimated to contain 10^{78} atoms.) Yet this one particular wheel is used by all living forms examined, with very few minor deviations from the standard. How was the wheel chosen?

Some investigators have argued that an as yet unidentified relationship exists between the codons and their amino acids which restricted the number of choices, but nobody has yet come up with convincing evidence for it. Many biologists therefore believe that the code was chosen by chance. Once the choice was made, there was no turning back because changing it in midcourse during evolution would be like shifting into reverse gear in a Porsche racing at over 200 km per hour on a German *Autobahn*. Or, perhaps more aptly, it would be akin to replacing the keyboard on your computer by one with a different arrangement of characters, but keeping the old software. Obviously, any text you typed on your new keyboard would make no sense whatsoever, just as the altered code would specify a heap of useless proteins or no proteins at all. Any evolutionary line in which the code was changed would end up like reckless drivers traveling at breakneck speed on the

Autobahn – extinct. The fact that all organisms use the same genetic code is therefore a powerful argument for the assumption that the code was chosen once and once only and hence that all living forms stem from one group of ancestors. Other examples of attributes shared by all living forms include the system of translation from nucleic acids into proteins, the system used to replicate nucleic acids, the fundamental organization of the cell, and the principal metabolic pathways. Every one of these, and many other features in addition, are the result of one particular strike among many possibilities and they therefore all support the notion of a single Tree of Life.

If all living forms on Earth are derived from a common ancestor, it should be possible to place them on a single phylogenetic tree. Alternatively, if different groups of organisms were derived from distinct ancestors that arose independently of one another, more than one phylogenetic tree should exist and there should be no genealogical connections among the various trees. The existence of a single Tree of Life would therefore be strong evidence in itself for the origin of life from a single root.

This brings us to the second question. To draw the Tree of Life, we must identify a historical record encompassing all living forms and extending all the way back to life's very beginnings. By necessity, the record is molecular in nature, because only molecules could reasonably be expected to have been present at the emergence of life and to have kept a genealogical record extending from that time to the present day. What *kind* of molecules could be expected to have kept such a record? Presumably those involved in the most basic attributes of life – replication of nucleic acids, protein synthesis, protection of the emerging life forms from noxious agents, and the regulation of the essential metabolic processes – attributes that must have been acquired at the onset of life.

Several of these molecules or the genes specifying them have been rigorously interrogated to discover how much they can recall about the history of life, and one of them in particular was found to remember a lot – the ribosomal (r) RNA. Ribosomes are the free-floating or membrane-bound granules in the cytoplasm which provide the physical setting for protein synthesis. They bring the necessary ingredients together and provide the tools, the machinery, and everything else that is required to construct a protein. Each granule consists of two parts; in bacteria these are the small 30S and the large 50S subunit. (The numbers refer to the speed with which the granules sediment in a sucrose-gradient solution spun at high velocity in an ultracentrifuge. The letter "S" stands for units of sedimentation introduced by the chemist Theodor Svedberg.) The small subunit of a bacterial ribosome contains a single 16S rRNA molecule approximately 1500 nucleotides long (Fig. 6.1). The large subunit contains two rRNA molecules, 23S and 5S in size and 2900 and 120 nucleotides long, respectively. The corresponding sizes of animal rRNA molecules are 18S (2300 nucleotides) in the small subunit and 28S, 5S, as well as 5.8S (4200, 120, and 160 nucleotides long, respectively) in the large subunit. To fit into the granule and to function properly, the rRNA molecules must fold in a specific manner by interactions that take place between complementary nucleotides and between nucleotides and ribosomal proteins. Because of these structural constraints, the rRNA molecules tolerate very few changes and so evolve slowly. These properties – their ubiquitous presence in all cellular forms of life, their age stretching right back to the very beginning of life, and their conservative mode of evolution – combined with easy isolation and handling, have turned the rRNAs, and particularly the 16S/18S types, into some of the phylogenetically most avidly studied molecules. The use of 16S/18S rRNA molecules in phylogenetic research was pioneered by Carl R. Woese and his colleagues.

PROKARYOTE

A.

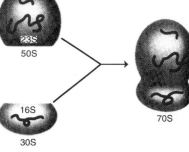

Subunits   Assembled
ribosomes

5S

23S

50S

16S

70S

30S

B.

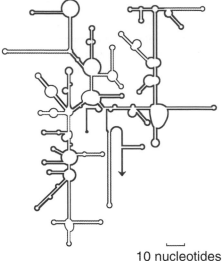

10 nucleotides

EUKARYOTE

C.

5.8S   5S

28S

60S

18S

80S

40 S

D.

Fig. 6.1A-D
Ribosomal subunits, assembled ribosomes, and ribosomal RNA molecules of prokaryotes (**A, B**) and
eukaryotes (**C, D**). Each ribosome consists of a large (50S or 60S) and a small (30S or 40S) subunit
and each subunit is comprised of many different proteins (gray shading) and one or two rRNA molecules (colored chains). The rRNA molecules (**B, D**) are folded characteristically: although single-
stranded, they contain complementary stretches which form double-stranded segments (indicated
by parallel lines); the noncomplementary stretches form loops separating the complementary regions. The three-dimensional structure of the prokaryotic 16S (**B**) and the eukaryotic 18S (**D**) rRNA
molecules are similar (similarities are highlighted in red), but not identical. (Adapted from Darnell et
al. 1990.)

The Woesian Revolution

To sequence a gene in the pre-PCR days, one had first to clone fragments of the genome, identify the fragment carrying the gene, separate it from the rest of the clones, produce large quantities of it, and then extract the segment of interest from it. This elaborate procedure is not suitable for the large-scale sequencing of genes from phylogenetically distant life forms. The rRNA offered a suitable alternative to the standard procedure of DNA cloning. The rRNA is encoded in rDNA genes which are part of the genome in the nucleus. The rRNA itself, however, is present in the cytoplasm, from which it can be isolated in a pure form in large amounts and sequenced directly, without cloning. Once Woese and his coworkers had established a routine for rRNA isolation and purification, they were able to rapidly sequence one rRNA molecule after the other. In this manner, they could accumulate a large number of rRNA sequences, including molecules from phylogenetically distant forms. Their example was followed by many other investigators, and the joint effort has led to the establishment of a database that contains rRNA sequences from thousands of species.

One handicap of the original method was that the starting material had to be isolated from a pure population of a single organism; a medley of different species would give a potpourri of sequences that would prove hard to sort out. This requirement was not difficult to fulfil when dealing with large organisms such as the various vertebrate species, for example, but it presented a major problem where microorganisms were involved. Although techniques for growing pure cultures of microorganisms have been available since the pioneering days of bacteriology, their application to new, previously uncharacterized forms is tedious because it demands the painstaking testing of various conditions until those supporting the growth of the particular organism are hit upon. For many microorganisms, optimal conditions have still not been found – but certainly not for lack of effort. This handicap has been overcome by the development of the PCR technique that enables researchers to pinpoint one specific form among a blend of many different ones. By applying this technique, they can now pick out one type of organism from a mixture and sequence snippets of its genome, without being waylaid by the rest.

As Woese diligently accrued one rRNA sequence after another, a picture emerged that was very different from the traditional view of the living world. At that stage, he must have felt like a man who, after living all his life in a valley protected by high mountains, decides one day to climb one of the peaks. After a long ascent and great exertion he finally reaches the summit and is stunned by the panorama spread out before him: as far as his eyes can see there are mountain ranges, many higher than the only one familiar to him. A whole new, unfathomed world opens up before him!

Before the advent of mass rRNA sequencing, the living world seemed to be partitioned into a few kingdoms, inhabited for the most part by creatures visible to the naked eye. Linnaeus, who was the first to take a census of living forms, recognized only the plant and animal kingdoms, which he considered to represent the highest categories of his classification system (his third kingdom was nonliving – that of minerals and rocks). Later, after the microscope came into wide use, biologists recognized the existence of organisms which were neither plants nor animals, and realized that there were many small organisms, or microorganisms, which were difficult to distinguish from one another because they seemed featureless. Clearly, there were more than two kingdoms, but the actual number remained controversial until the late 1960s, when an agreement was reached to divide the living world into five kingdoms – animals, plants, fungi, protista (protozoans), and monera (bacteria). The first four of these were grouped together as *eukaryotes* (literally, organisms

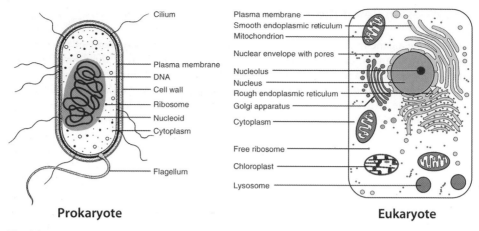

Prokaryote

Eukaryote

Fig. 6.2
Comparison of prokaryote and eukaryote (plant) cell structure. See text for description. (Based on Linder 1983.)

with a "true kernel", i.e., nucleus), the last constituted the *prokaryotes* ("before kernel"). All prokaryotes and some eukaryotes are single-celled organisms, but the cells are organized differently in these two groups (Fig. 6.2). Eukaryotic cells are generally larger and more complex than prokaryotic cells. Each eukaryotic cell has a clearly delineated nucleus bounded by a membrane and incorporating several linear DNA molecules permanently associated with specialized proteins and organized into chromosomes. The eukaryotic cytoplasm contains a number of membranous structures, the *organelles*, specialized in carrying out different functions. A prokaryotic cell, on the other hand, lacks a clearly differentiated nucleus and has one or more nucleoids instead – clear regions containing most of the hereditary material, but not surrounded by a membrane. Each nucleoid contains a single, circular DNA molecule which only associates with proteins transiently. The prokaryotic cytoplasm lacks membranous organelles, but contains ribosomes which are smaller (70S compared to 80S) than those of a eukaryotic cell.

The traditional view gave only a token acknowledgment of the existence of the microbial world and perpetuated the belief that the bulk of evolutionary diversity was present in the kingdoms encompassing large organisms. Monera were seen as a relatively undifferentiated bunch, of interest only because they were known to harbor a number of human pathogens. This sentiment is still reflected in contemporary textbooks of biology which concentrate on the kingdoms of large organisms and pay only lip-service to the monera. But after Woese had peered over the mountains of rRNA sequence files and had used them to draw phylogenetic trees, like the man who finally ascended the mountain peak above his valley, he saw an entirely new vista for the first time. He perceived the microbial world not only to be far more diverse than had been believed up until then, but the diversity of the large organisms actually paled in comparison to the microorganisms. On the Tree of Life that Woese drew (Fig. 6.3), the computers relegated the large organisms to an undistinguished position in the terminal part of one section of the crown, while the microbial world was differentiated into many distinct, ancient lineages with histories extending close to the origin of life. He saw all living forms fall into three major groups which did not cor-

Fig. 6.3
The Tree of Life: an early version based on rRNA sequences obtained by Carl R. Woese and his coworkers, as well as other researchers. (Compare with Figures 6.8 and 6.9.)

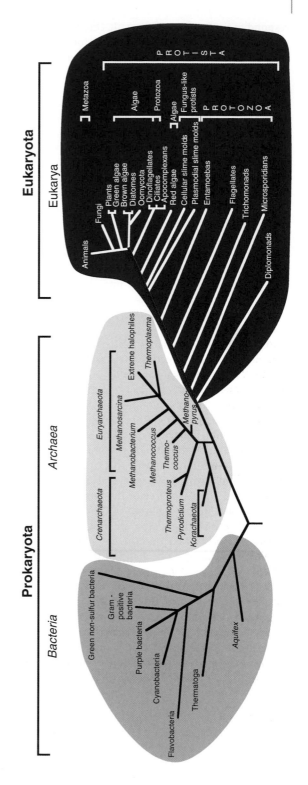

respond to the five kingdoms at all. One of the kingdoms (monera) was scattered between two of the groups, while the remaining four were all in one group, most of them located in an insignificant recess. In short, Woese had laid out an entirely new perspective of the living world.

The Three Domains of Life

Woese called the three groups *domains* and named them *Eubacteria, Archaebacteria,* and *Eukaryota,* but the designations were later changed to *Bacteria, Archaea,* and *Eukarya,* respectively, whereby the first two domains comprise the *Prokarya,* the former prokaryotes. For any one of his claims alone, Woese must have expected to be struck by the wrath of his contemporary traditionally-minded taxonomists, but it was in fact his division of the formerly so undistinguished microorganisms into two domains – the highest taxonomical categories – that set him up as the target of extreme hostilities. Woese persevered, however. He and a small group of his followers presented such a convincing case that ultimately, it could neither be ignored nor derided; the almost universal acceptance of his concepts followed. And with this 'Happy End' the story should come to a close but it does not: no sooner had the three-domain view of life been canonized than the first complications appeared. Before we turn to them, however, we must describe, if only superficially, the domains.

In addition to similarities in their rRNA sequences, the *Bacteria* share several characters. The cell wall protecting their bodies contains a compound, muramic acid, which is found in no other organism. In all other living forms, the first amino acid residue with which the synthesis of every protein chain begins is methionine; in bacteria it is its modified version, formylmethionine. In all organisms except bacteria, protein synthesis is blocked by a toxin derived from *Corynebacterium diphtheriae,* the causative agent of diphtheria. Bacterial, but not archaeal or eukaryotic protein synthesis, is inhibited by antibiotics such as chloramphenicol, streptomycin, and kanamycin. Finally, bacteria possess only one relatively simple enzyme that transcribes DNA into RNA – the RNA polymerase; archaea and eukaryotes possess several such enzymes and each is more complex than that of the bacteria. The rRNA tree of the known bacteria has at least 12 major branches which emerged early in the evolution of the domain. Each of the branches can therefore be regarded as constituting a separate kingdom. Only some of the branches had been recognized previously as such on the basis of phenotypic characters, either morphological or physiological; most can be distinguished only on the basis of their nucleic acid or protein sequences.

Representatives of the *Archaea* were already known in the pre-rRNA times, but they were believed to be just odd bacteria. Their rRNA sequences were first to unmask them as comprising an entirely new domain. Even then, however, an aura of oddity remained, largely because of the strange (from our point of view) places in which they were found. These include blazing volcanic vents* at the bottom of the ocean, terrestrial hot sulfur springs, lakes with a high salt concentration, saltworks and brine (the conspicuous pink color of the salt flats you may have noticed while flying over western parts of the United

---

* On the ocean floor, in areas in which two adjacent plates are forced apart by molten magma issuing from the earth's interior, sea water leaks into the volcanic fissures, becomes superheated, and rises up again in escape passages, the *hydrothermal vents*.

States is caused by carotenoid pigment synthesized by certain archaea), and highly acidic (low pH) or highly alkaline (high pH) lakes. According to their environmental preferences, archaea can be heat-loving or *thermophiles*, cold-loving or *psychrophiles*, salt-loving or *halophiles*, acid-loving or *acidophiles*, and alkali-loving or *alkaliphiles*. The thermophiles grow optimally at temperatures between 50°C and 70°C, but *extreme thermophiles* or *hyperthermophiles* thrive at temperatures between 70°C and 110°C; they die of cold when the temperature drops to 55°C. The psychrophiles, at the other extreme, prefer temperatures between 0°C and -10°C. The halophiles find water almost saturated with sodium chloride amiable and inhabit, for example, the Dead Sea, which is therefore not true to its name at all. The acidophiles tolerate pH values close to zero, which is like wallowing in pure acid, whereas the alkaliphiles inhabit environments with pH values as high as 11.5, which is akin to swimming in ammonia. Some archaea grow optimally under conditions that combine different extremes. *Thermoacidophiles*, for example, grow both at high temperatures and at low pH, and *sulfothermophiles* thrive in vents gushing out hot water saturated with sulfurous gases.

In such conditions, ordinary bacteria would be boiled or dissolved by the acid or ammonia. Archaea are protected against the effects of high temperature, high salt concentration, and extreme pH values by specific modifications that their proteins, nucleic acids, and other essential molecules have undergone, as well as by certain alterations of their membranes. Some archaea have also evolved special ways of obtaining energy. The *methanogens*, for example, use carbon dioxide, hydrogen, or acetate to produce methane and grow in environments deprived of oxygen, such as aquatic sediments, bogs, marshes, tundra, the heartwood of infected trees, the rumen, guts, oral cavities of animals, and deep seas. Other archaea use elemental sulfur or sulfates as their source of energy.

Some biologists regard the adaptation of archaea to extreme environmental conditions and their ability to live without molecular oxygen as a relic dating back to a time some two to three billion years ago, when the waters on earth were hot and full of salt and sulfur, and the atmosphere was devoid of oxygen. Other researchers, however, have come up with evidence that the taste for extreme conditions has evolved independently several times in the history of life, and that the present-day extremophiles are of more recent origin than the earlier phylogenies seemed to suggest. It was once thought that all archaeans were extremophiles, but further research disproved this contention. Most archaea, as it turns out, are adapted to ambient conditions, just like bacteria. They are part of the community of floating organisms at the surface of marine and fresh waters (plankton) and they are well represented in agricultural and forest soils. They are so widely distributed that they may be a major constituent of the global biomass.

Among the characteristics distinguishing archaea from both bacteria and eukarya are unique fatty acid types in their lipids; a unique linkage of fatty acid to glycerol; and characteristic RNA polymerases each consisting of eight to 12 subunits. Archaea do share, however, a number of features with bacteria on one hand and eukarya on the other, in both instances to the exclusion of the third domain. The known archaea fall into three major groups or kingdoms – the Crenarchaeota, the Euryarchaeota, and the Korarchaeota (Fig. 6.3).

The distinguishing characteristics of the *Eukarya* were enumerated earlier, the most prominent among them being the presence of organelles in the eukaryotic cell. Certain similarities to bacteria in some of the organelles led biologists in the last century to propose an *endosymbiosis hypothesis* of the organelle's origin. According to this hypothesis, early in the evolution of eukaryotes, when the latter began to feed on bacteria, some of the bacteria taken up by the eukaryotic cells developed ways of escaping digestion and thriv-

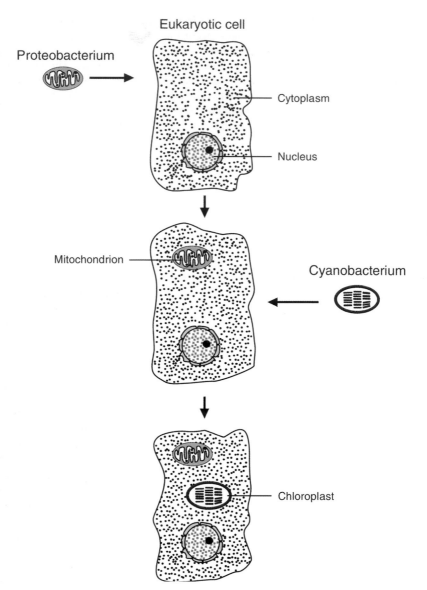

Fig. 6.4
The endosymbiosis hypothesis of the origin of mitochondria and chloroplasts. At some point in their evolution, the eukaryotic cells acquired the ability to engulf smaller prokaryotic cells, destroy them, and utilize their components as nutrients. A few of the engulfed cells escaped destruction, however, and entered into a partnership or endosymbiosis with the host by supplying the host cell with energy and receiving all that is required for survival in return. In the course of evolution, the endosymbionts simplified their structure and also adapted in other ways to life within a cell – they evolved into cellular organelles. Prokaryotes resembling present-day proteobacteria became mitochondria, whereas cyanobacteria-like organisms gave rise to chloroplasts and other plastids.

ing in the host cell cytoplasm (Fig. 6.4). A symbiotic partnership of mutual assistance developed between the guest and the host, each providing certain services to the other. The host protected its guest from being attacked by other eukaryotes and the guest paid for this service by providing new possibilities for energy uptake. Upon entering into the symbiotic relationship, the bacteria had to relinquish a good portion of their freedom and with time they became an integral part of the host, incapable of independent existence. The endosymbiotic origin of mitochondria and plastids is well-supported. (Plastids are pigment-containing organelles which capture some of the energy from sunlight and transfer it through a complex system of chemical reactions into the chemical bonds of organic compounds in the well-known process of photosynthesis. There are several categories of plastids differing in the structure and the nature of the prevailing pigment. In chloroplasts, the prevailing pigment is chlorophyll.) The two organelles are the size of bacteria and like certain bacteria, they are enclosed by two membranes. They contain their own hereditary material separate from the nuclear DNA and the material is of a prokaryotic rather than eukaryotic type (circular, "naked" DNA molecules); they possess independent nucleic acid and protein synthesizing machineries, including RNA polymerases for transcription and tRNA as well as ribosomes for translation; the polymerases and the ribosomes are of prokaryotic type; the synthesis is blocked by the same antibiotics that inhibit bacterial protein synthesis, and so forth. The most convincing evidence for the bacterial origin of mitochondria and chloroplasts is provided, however, by the comparison of rRNA sequences. Instead of clustering with their eukaryotic counterparts, the mitochondrial and plastid rRNA sequences bunch with certain bacterial sequences. The mitochondrial sequences cluster with those of the α-proteobacteria, a group that includes *Escherichia coli*, the well known commensal in the human gut and a favorite experimental model for genetic studies; and the chloroplast sequences cluster with those of the cyanobacteria, a group of photosynthesizing bacteria formerly classified as "blue-green algae".

The evolutionary stage in which the eukaryotic cells acquired the symbiotic bacteria and the number of times this event took place remain controversial. Plastids are restricted to certain eukaryotes, of which plants are the most familiar; mitochondria are ubiquitous in eukaryotes except for the branches that emerge earliest on the phylogenetic tree of the eukaryotic domain. Does this mean that the endosymbiotic event occurred after the emergence of these branches, or that their ancestors originally possessed mitochondria which they subsequently lost? Some observations do support the latter possibility, but their interpretation is not without unambiguity. We shall return to this question later.

The success of the endosymbiotic hypothesis in explaining the origin of mitochondria and plastids has encouraged biologists to seek similar explanations for other organelles, including the nucleus. Some biologists view the eukaryotic cell as a motley of parts acquisitioned mostly from bacteria and as far as organelles are concerned, possessing hardly any of its own constructs. Such a sweeping extension of the endosymbiotic hypothesis, however, seems unwarranted.

The Search for the Root

The three-domain concept raised the question of evolutionary interrelationships: which of the domains is the oldest and where is the root of the Tree of Life? Three possibilities that arose were, first, an early divergence of bacteria from the common ancestor of archaea and eukaryotes; second, an early divergence of archaea from the common ancestor of bacteria and eukaryotes; and third, an early divergence of eukaryotes from the common ancestor of

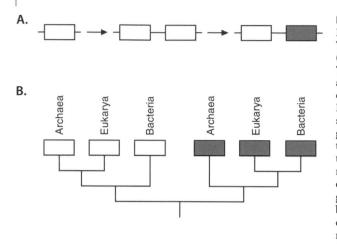

Fig. 6.5A,B
Early attempts at rooting the Tree of Life using duplicated (paralogous) genes. (**A**) Tandem duplication of a gene and the subsequent divergence of the two copies. (**B**) Phylogenetic tree based on the sequences of the duplicated genes. The tree indicates that the duplication occurred before the divergence of the three domains of life so that one copy of the gene serves as an outgroup for rooting the Tree based on the sequences of the other copy. Here, the genes are not specified, but they could be the *EF-Tu* and *EF-G* genes described in the text, for example.

bacteria and eukaryotes. At first glance, it may seem beyond one's grasp to choose from among these possibilities. The standard method for rooting a tree of organisms relies on the existence of an *outgroup* that by common agreement does not belong to the organisms under study. If, however, the Tree of Life encompasses all organisms on Earth, there cannot be an outgroup. Yet, there seemed to be a way of finding the root indirectly, by using genes that duplicated before the divergence of the three domains. An example is the gene *EF*, which codes for the elongation factor, a protein assisting in the translation of nucleotide triplets into strings of amino acid residues. The gene duplicated before the separation of Bacteria, Archaea, and Eukarya to produce two copies, *EF-Tu* and *EF-G,* which are now found in the genomes of all three domains. Since the *EF-Tu* and *EF-G* genes began to diverge from each other before the divergence of Bacteria, Archaea, and Eukarya, the orthologous *EF-Tu* genes of the three taxa should be more similar to one another than any of them is to the paralogous *EF-G* gene (Fig. 6.5). Similarly, the orthologous *EF-G* genes of the three taxa can be expected to exhibit a greater similarity among themselves than any one of them to the paralogous *EF-Tu* gene. The *EF-G* gene can therefore serve as an outgroup for the *EF-Tu* genes of the three domains, and reciprocally, the *EF-Tu* gene can be used as an outgroup for the *EF-G* genes of Bacteria, Archaea, and Eukarya. The rooting of the Tree of Life is therefore based on the use of a paralogous gene as an outgroup for the orthologous genes. The method grouped the Archaea with the Eukarya, separating them from the Bacteria and thus placing the root as issuing from the point at which Bacteria are connected with the common ancestor of Archaea and Eukarya. A similar location was also indicated by another pair of duplicated genes. Paleontological evidence, too, seemed to be in agreement with this placement of the root. It identified the earliest known life-forms in the fossil record as resembling present-day cyanobacteria.

A neat consensus view slowly appeared to be emerging in the early 1990s. Single-celled microorganisms resembling present-day cyanobacteria formed the root of the germinating Tree of Life. Later on, the stem of the Tree split into two major limbs, one giving rise to the bacterial domain and the other to the common ancestor of Archaea and Eukarya. The latter branch subsequently split again, and each of the three domains sprouted multiple branches which became the various kingdoms, some persisting to this day. When the eukaryotes learned to engulf and digest bacteria for food, the stage was set for endosymbiotic events that dramatically changed the eukaryotic cell in its complexity. At least two of these events are known to have occurred, one involving α-proteobacteria and giving rise to mitochondria in the early stage of eukaryotic evolution, and the other involving cyanobacteria and producing, at a later stage, photosynthesizing plastids. For some evolutionary biologists this picture was *too* neat to reflect the whole story, especially since it was based largely on a single molecular system – the rRNA. And they were proved correct, for no sooner had a consensus been reached than observations accumulated to contradict it.

The Pastiched Genomes

The monkey-wrench thrown into the works took the form of the emerging genomic sequences. The genomes of most prokaryotes are on average only a few million nucleotide pairs long; to sequence one such molecule in its entirety takes only about a couple of years for a modestly large research group. Several laboratories in the United States, Europe, and Japan have completed sequencing the genomes of a number of prokaryotes and selected eukaryotes. Each genomic sequence is a gold mine of information as far as the sharing, acquisition, and loss of genes, as well as the relationships among the shared genes are concerned. When the first analyses of the sequenced genomes were carried out, it became apparent that different genes were often telling different stories about the relationships between the three domains. The genomes of the Archaea in particular appeared to be mixed bags containing three categories of genes in terms of their relationships to genes in the two other domains: genes (proteins) exhibiting approximately the same similarity to their counterparts in Bacteria and Eukarya; genes that are significantly more similar to their bacterial than to their eukaryotic homologs; and finally genes that are more similar to their eukaryotic than to their bacterial counterparts. Since, according to the consensus view, Archaea and Eukarya have been separated from each other for a shorter time than either of these two domains has from the Bacteria, the archaeal genes should have accumulated more differences when compared to bacterial genes than when compared to eukaryotic genes. This, however, was clearly not the case, and to make things even worse for the consensus view, the genes in the first two categories were found to outnumber the genes in the third category by far.

Various hypotheses have been proposed to explain these unexpected findings. Among them, the most favored explanation is also the most radical because it bears upon the very essence of phylogenetic reconstruction. The principal tenet of phylogenetic analysis is that once gene pools (species) have become reproductively isolated, gene exchange between them ceases. Since sex in sexually reproducing organisms is normally the only vehicle for the transmission of genomes or parts thereof, the absence of interspecies mating restricts gene travel to intraspecies routes alone. If this were not the case, and genes did indeed travel frequently from species to species, the species concept would collapse. Since the restriction of gene flow is firmly embedded in the classical definition of a species, its violation would undermine the very existence of the classically defined species category. Furthermore, since the current concept of phylogenetic reconstruction is based on the existence of the species category, it would also be jeopardized by widespread interspecies promiscuity. At the molecular level, genes and other genomic segments are used to reconstruct phylogenies of organisms; gene exchange between species would thus violate the bifurcation rule underlying all tree-drawing methods. The diagram of a process in which genes are swapped between species like stamps between stamp collectors would not be a tree, it would be a *network*! Yet, despite all these iconoclastic implications, frequent gene swapping not only between species, but also between members of different domains, now constitutes the preferred hypothesis to account for the observed mosaic nature of prokaryotic genomes.

The concept of interspecies gene travel is not new. Introduced under the name *horizontal (lateral) gene transfer* ("horizontal" to contrast it to the standard "vertical" gene transfer from parent to offspring within a species), it was first put forward as a hypothesis to explain certain anomalous observations, but was later demonstrated experimentally. It has always been assumed to be so rare, however, that its effect on phylogenetic reconstructions could be ignored. The new gene-swapping hypothesis does away with all these assumptions. It posits instead that far from having a negligible influence, horizontal gene transfer has indeed played a major part in some stages of evolution and in certain groups of organisms.

Prokaryotes have developed several means of exchanging DNA fragments (Fig. 6.6). In the simplest case a cell dies, its membrane is dissolved, and DNA spills out to resemble an elaborate bow on a birthday present. The released DNA then fragments into many short pieces, some of which may be lapped up by another cell that happens to possess the appropriate DNA-binding proteins on its surface. If the recipient cell also has a protein in its cytoplasm that protects the ingested DNA from further degradation, the morsel may be given the chance to recombine with the indigenous chromosome. Recombination, which resembles the process described in Chapter Three, integrates the foreign DNA into the recipient's genome at a site of sequence similarity. By doing so, a new gene may be introduced into the recipient's DNA which, when expressed, alters some of the recipient cell's properties – it *transforms* the cell (Fig. 6.6A).

Transformation depends on the chance encounter between a DNA fragment and a suitable (transformable) recipient cell. In *transduction* (Fig. 6.6B), another method of gene swapping, a piece of donor DNA hitches a ride on a virus. Bacteria are parasitized and ultimately killed by specialized viruses, the *bacteriophages* (literally "eaters of bacteria"). Transduction is a form of bacterial retaliation. As a new generation of viruses assembles in the cytoplasm, readying itself for the infection of yet another cell, fragments of host DNA may sneak into the forming particles instead of the viral DNA. Alternatively, as the viral DNA, which in one part of its life cycle integrates into the host genome, exits the bacterial chromosome again, a few bacterial genes may hold on to it for delivery into another cell. In both cases, the virus is duped, because the defective particles are incapable of further reproduction. The bacterial DNA, however, enjoys a free ride into another cell and gets the chance to recombine with the bacterial chromosome of a similar sequence.

The most common form of gene swapping is, however, *conjugation* (Fig. 6.6C), or sex between two consenting prokaryotes (the word is derived from Latin *conjugare*, to unite). When it comes down to essentials, sex is nothing more than a method of handing on DNA molecules from one partner to another. In eukaryotes, the DNA molecules are delivered by sex cells; in prokaryotes they are transported directly from a donor to a recipient cell. With regard to their sexual inclinations, bacteria and other prokaryotes are divided into two types, those that possess the *fertility plasmid* in their cytoplasm and those that do not, F^+ and F^- types, respectively. The F plasmid, like most other plasmids, is a small DNA ring that exists side-by-side in the cell with a much larger ring, the bacterial chromosome. It contains ~200 genes, some of which, when activated under specific conditions, govern the formation of *sex pili*. These are long, hollow, filamentous projections (Latin *pilus* means hair) emerging from the cell membrane (Fig. 6.7C). The pili scout the neighborhood for suitable F^- cells; when they have found them and established contact, the pili begin to con-

A. TRANSFORMATION

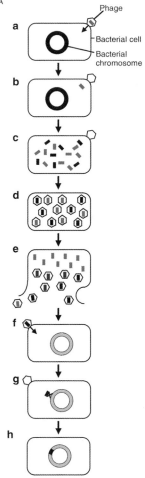

B. TRANSDUCTION

C. CONJUGATION

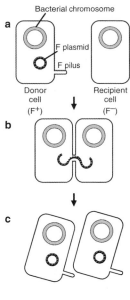

Fig. 6.6A-C

Gene swapping in prokaryotes. **A.** *Transformation.* (a) A double-stranded (ds) DNA fragment derived from another bacterium binds to a receptor protein on the cell's surface. (b) A cell-surface enzyme (nuclease) destroys one of the DNA strands, while the other crosses the membrane and enters the cell. In the cytoplasm, the single-stranded (ss) DNA is protected from degradation by the attachment of DNA-binding proteins. (c) With the help of special RecA proteins, the ssDNA binds to the bacterial chromosome by displacing one of the chromosomal strands. (d) The second strand of the transforming DNA is synthesized and the fragment is thus integrated into the bacterial chromosome. **B.** *Transduction.* (a) Bacteriophage (virus) attaches to a bacterial cell. (b) Phage DNA enters the cell. (c) Phage DNA multiplies in the host cell and the bacterial chromosome is fragmented. (d) New viral particles form and some of them enclose fragments of bacterial DNA instead of phage DNA. (e) Bacterial cell breaks down and phage particles are released. (f) Phage particles containing bacterial DNA (transducing particles) infect another bacterial cell. (g) Bacterial DNA fragment recombines with the host cell chromosome. (h) Recombination integrates the donor cell DNA into the host cell chromosome: the cell is thus transduced. **C.** *Conjugation.* (a) Donor (F+) cell contacts recipient (F−) cell via F pilus. (b) F plasmid of donor cell replicates and one of the DNA strands enters the recipient cell via the F pilus. (c) Synthesis of complementary DNA strands in both the donor and the recipient cells results in the generation of F plasmids in both cells. Some F plasmids may integrate into the bacterial chromosome and then transfer parts of it to the chromosome of the recipient cell (not shown).

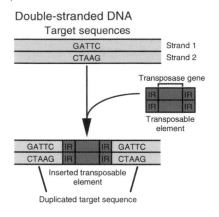

Double-stranded DNA
Target sequences

Fig. 6.7
Insertion of a simple prokaryotic transposable element into a target DNA. The element consists of a transposase gene flanked by inverted repeats (IR). The transposase enzyme encoded in the element is necessary for the integration of the element into the target DNA. The integration duplicates the target sequence (Based on Madigan et al. 1997.)

tract, bringing the donor and the recipient cells together. At the contact site, the membranes of two cells dissolve and the F plasmid begins to replicate. One of its two strands breaks and as the circle begins to roll, the free strand enters the recipient cell, where it serves as a template for the synthesis of a complementary strand. The remaining, unnicked strand in the donor cell similarly templates the synthesis of its complement and the process ends with the formation of two rings, one in the donor and the other in the recipient cell. The transfer completed, the conjugating cells separate, both now being of the F^+ type. If for one reason or another, one of the F^+ cells loses its F plasmid, it converts back to the F^- type. Although plasmids lead an autonomous existence most of the time, on rare occasions they recombine with the chromosome at one of the several sites at which they share sequence similarity, and integrate into it. When that happens and the integrated plasmid is later overcome by urge to have sex, it may assume the function of a pony express delivering genes from one bacterial chromosome to another. Because in prokaryotes sex is not restricted to members of the same species (unlike in most eukaryotes), gene swapping can take place between unrelated bacteria or even between bacteria and archaea.

Have Transposon, Will Travel

In addition to these three mechanisms – transfection, transduction, and conjugation, there is one other process that, although insufficient to effect horizontal gene transfer by itself, nevertheless facilitates gene swapping. It is the faculty of certain genomic segments to move from one position in the prokaryotic chromosome to another. Geneticists refer to these genomic changes of address as *transpositions* and describe the mobile segments as *transposable elements* (Fig. 6.7). To move a genomic segment, two things are necessary: an enzyme, *transposase*, that excises the segment from its original place of residence (but see below) and integrates it into a new domicile; and a tag, an *inverted terminal repeat* that identifies a segment as potentially mobile. The tag is comparable to an estate agent's "For Sale" sign marking property that is available for change of ownership. The simplest prokaryotic transposable elements, the *insertion sequences (gene cassettes),* consist only of the transposase-specifying gene and the inverted terminal repeats. More complex elements, *transposons,* contain additional genes that may have nothing to do with the transposition itself. Some elements move from their original position by excision, while others duplicate and send their duplicated copy to another location, thus increasing their numbers in the genome. When left to their own resources, transposable elements normally take up new residence within the genome of which they are a part. But when caught up in one of the three mechanisms described above, they can jump from genome to genome and take some of their closest genes with them.

How frequent are gene exchanges between prokaryote species? A rough estimate has been calculated for *Escherichia coli*. The bacterium normally lives in harmony with its host, but some of its variants cause urinary tract infections, travelers' diarrhea, and other gastrointestinal ailments. Its close relative, *Salmonella*, is a dangerous pathogen and the causative agent of typhoid fever and "food poisoning". The two genera shared their most recent common ancestor an estimated 100 million years (my) ago. Since that time the *E. coli* genome may have received ~1.16 million nucleotide pairs of foreign DNA on 234 occasions, which is an acquisition rate of 116,000 nucleotide pairs per 1 my. Of the introduced DNA, only ~550,000 nucleotide pairs representing 735 genes remain; the rest has been lost. Since the total length of the *E. coli* chromosome is 4.64 million nucleotide pairs, the genome now contains about 12 percent of foreign DNA.

In general, horizontal gene transfers are thought to introduce as much variation into prokaryotic genomes as that resulting from mutations. Horizontal gene transfer is therefore an important mechanism driving prokaryote evolution. It differs from mutations, which normally change only one or very few nucleotides at a time, by effecting alterations *en bloc*. One single horizontal transfer may introduce an entire cluster of genes into a prokaryotic genome. The cluster often consists of components that are brought together because they jointly specify a set of properties enabling their new owner to exploit new ecological niches. Once assembled, the package assumes an existence of its own, traveling from genome to genome and from species to species. Its lucky recipient may gain an immediate advantage over other forms in a particular environment and so be given the possibility of outclassing rivals or colonizing new niches. Since the recipient's descendants inherit the package and with it, a competitive edge over other prokaryotes, natural selection rapidly sweeps the foreign DNA through the population.

Mutations can have the same effect. In fact, the initial assembly of the block is preceded by the stepwise accumulation of mutations in the individual genes. Without horizontal gene transfer, however, the desired combinations of mutations would have to be assembled independently in different species. The various species would have to reinvent the wheel, so to speak, over and over again. This is indeed the manner in which eukaryotes, lacking widespread horizontal gene transfer, evolve. Eyes, wings, or a streamlined body for swimming in water, all these features evolved independently on several occasions. Prokaryotes, by contrast, can take evolutionary short-cuts by taking advantage of the widespread existence of horizontal gene transfer. They merely assemble the genes governing an advantageous property in a single cluster and then let the cluster tour the genomes of different species. If mutations were to produce the advantageous property *de novo* every time the need for it arose, it would be a tediously slow process. Horizontal transfer can achieve the same effect instantaneously and as many times as necessary, even though it may be as rare as mutations.

The reality of these deductions is demonstrated most convincingly in situations in which horizontal gene transfer has detrimental effects on human health. Bacteriologists have known for some time that harmless prokaryotes such as *E. coli* can suddenly and inexplicably turn into dangerous pathogens, into organisms that harm their hosts and cause disease. Changes in the opposite direction can also occur leading to an oscillation between virulence (ability to cause disease) and avirulence (harmlessness). Comparisons of these two forms have revealed that they often differ primarily in a particular region of the bacterial chromosome, the *pathogenicity island,* present in the virulent form and absent in the avirulent one. The island is occupied by genes that are largely responsible for the pathogenic effects (ability to cause damage) – the attachment of the bacterium to the host cell, secretion of toxins, resistance to the host's immune reactions, and others. Very similar pathogenicity islands are often found in distantly related prokaryotes. Sequence comparisons show that the genes of one island are much more closely related to those of the other island than other genes in the genomes of the two species. The islands have apparently been horizontally transferred into the various hosts by defective bacteriophages and conjugative plasmids. The introduction of a pathogenicity island into a previously harmless prokaryote changes the latter instantaneously into a life-threatening pathogen by providing it, for example, with access to the body's recesses where it interferes with the body's vital functions, or by making it resistant to the body's defense mechanisms.

The most dramatic demonstration of the latter effect in recent years has been the growing bacterial resistance to antibiotics, and indeed to drugs in general. Antibiotics were introduced into clinical practice in the 1940s. At that time, genes determining antibiotic resistance were rare. But in the following 50 years their frequency steadily increased on a global scale as a result of horizontal transfer via gene cassettes, transposons, plasmids, and bacteriophages, without respect for species boundaries. Today, widespread antibiotic resistance of human and animal pathogens presents a major clinical problem throughout the world.

Worlds Apart

It seems only reasonable to assume that mutations and recombinations have been in operation since the beginning of life. Once nucleic acids acquired their ability to replicate, they also began to mutate because replication is inadvertently tied to misincorporation of nucleotides and consequently to mutations. In the early phase of life's history, misincorporations were probably so frequent that they threatened to lead evolution into chaos unless they could be curbed by certain restraining measures. The mechanism that finally did so probably evolved originally to repair lesions inflicted by environmental factors on the poorly protected and highly sensitive nucleic acids. Both repair and proofreading mechanisms relied on enzymes capable of excising defective nucleotide stretches, inserting new sequences into the gaps, and mending the patches to link them up with the rest of the molecule. The early life forms must soon have learned that these enzymes can double in yet another function – the exchange of segments between nucleic acids, or recombination.

An important difference between mutation and recombination is that the former involves a single molecule only, whereas the latter requires two to participate. Consequently, while mutations can transpire at any time and any place, recombination cannot betide two molecules if the circumstances preclude their coming together in a single cell. Not only that, since recombination includes a recognition step based on the complementarity of strands, its occurrence depends on sequence similarity between the two molecules. If the molecules are too dissimilar (a sequence difference of 20 percent or more), the attempted exchange between them is usually aborted.

The two living worlds on earth, the prokaryotes and the eukaryotes, provide similar opportunities for mutations, but not for recombination to occur. In the eukaryotic world, the universe of the DNA molecules is partitioned into a myriad of isolated pools. Ordinarily only molecules of the same pool can recombine; molecules from different pools are excluded because they never come together in a single cell. The isolation of the pools is the consequence of sex. In eukaryotes, sex is restricted to individuals of the same species. Individuals belonging to different species usually do not show any desire to mate with each other, but even if they do, other mechanisms prevent the union from producing fertile offspring. The restriction that is placed on sexual relationships defines a eukaryotic species: a *biological species* is a group of interbreeding organisms. And it is the absence of sex between species that prevents interspecific DNA recombination in eukaryotes.

In the prokaryotic world, the situation is quite different. Although microbiologists refer to *Escherichia coli* as one species and to *Shigella dysenteriae* as another species of a different genus, and even though they use the Linnean system of binomial nomenclature (genus designation followed by a species name) in the same manner as their colleagues studying eukaryotes, the words do not signify the same thing in both cases. The main reason why *E. coli* and *S. dysenteriae* are regarded as different species is that when the bacteriologist Theodor Escherich isolated the former for the first time in 1885, he found it to be a commensal, while *Shigella dysenteriae* was identified as the causative agent of one particularly severe form of bowel inflammation in humans, the bacillary dysentery. This distinctive behavior makes a world of difference to humans, but genetically the two bacteria are so similar that the classification can hardly be justified if any other objective criterion is applied. The major difference between them is that one produces dangerous toxins and the other normally does not. On the other hand, microbiologists include several different strains in *E. coli* that show a similar behavior but are genetically far apart, and certainly more distant from one another than *Escherichia* from *Shigella*.

The main difference between prokaryotic and eukaryotic species is, however, that the concept of reproductive isolation does not apply to the former. Reproductive isolation, as emphasized earlier, is the defining feature of a species and for this reason it is senseless to apply this designation to the various prokaryotic forms. The prokaryotic "species" are discretionary categories encompassing organisms that are united by the sharing of arbitrarily chosen characters. In reality, far from comprising a series of reproductively isolated gene pools, prokaryotes are gene-swapping communities of clones. To a microbiologist a *clone* is a group of organisms descended from a single common ancestor by asexual reproduction – a mode of propagation that does not involve fusion of gametes. To propagate, prokaryotic cells simply divide, each parental cell giving rise to two daughter cells, which then divide again, and so on. In this manner, within a very short time a large number of highly similar cells, a clone, is generated from the initial single cell. The cells in a clone differ only by mutations that may have arisen during their derivation from the common ancestor, and by any elements that may have been introduced by horizontal gene transfer. Since both mutations and horizontal gene transfer are rare events, clones, also referred to as *strains*, may retain genetic similarity for long periods of time. Because clones come and go, however, the structure of prokaryotic populations can be highly dynamic and unstable. Nevertheless, groups of clones (strains), which microbiologists call "species", somehow manage to retain a suite of distinctive characteristics that they have in common. Exactly how the contrasting trends – diversification by mutations and horizontal gene transfer as opposed to retention of "species"-specific characters – square with each other and with the continuing process of clonal expansions and contractions is not well understood.

Clearly, prokaryotes evolve differently from eukaryotes. In global prokaryotic evolution, both mutations and recombination (which underlies horizontal gene transfer) play an important part, and evolution is acted out against the backdrop of forms that are not differentiated into classically defined species. In the eukaryotic world, the global community of organisms is differentiated into discrete units or species that diverge from one another by

Bacteria   Eukarya   Archaea

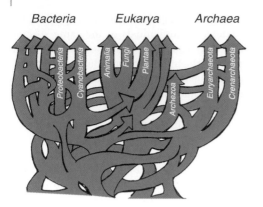

Fig. 6.8
A reticulated Tree (Network) of Life.
(Redrawn from Doolittle 1999.)

mutations only. Here, recombination is essential at the intraspecies, but not at the inter-species level. In terms of their evolution, prokaryotes and eukaryotes are worlds apart.

Viewed from this perspective, the relationships between the three domains and the question of the root appear in a new light. Early on, prokaryotes undoubtedly evolved in the same manner in which they evolve now. If anything, in the earliest stages of life, the swapping of genetic material may have been even more extensive than it is at present. To search for a single root for the Tree of Life may therefore be futile. The phylogeny of the earliest life forms was probably *reticulate*, network-like (Latin *rete*, a net), as a conse-quence of frequent gene exchanges between taxonomically distant groups, and so resem-bled a tree with a root even less than standard eukaryotic phylogenies (Fig. 6.8). Although the global community of organisms may have split at an early stage in evolution into the ancestors of present-day bacteria and the ancestors of present-day archaea, the two protodomains could still have carried on gene swapping more than they do today. A diagram of their phylogenies would therefore show an outline of two massive tree limbs, but interconnected in a most un-tree-like manner by numerous rungs. Because the reti-culation extended most likely all the way down to the beginning of life, the commencement phase would not resemble a root either. Cellular life did not begin as a single organism or a single species, but as a community of diversified, but genetically interacting organisms.

The Prokaryote Success Story

Prokaryotes have been in the catbird seat according to almost any criterion of evolutionary success. In terms of persistence, they include some of the most ancient lineages which can be traced back to the era in which life began. As far as their numbers are concerned, mi-crobiologists like to point out that there are more bacteria and archaea living in the colon of a single human being than there are humans on earth. And with regard to their evolu-tionary diversity, they comprise more kingdoms than existed at any time in the history of Europe. By examining just the crevice between the gum and the tooth of one single person, molecular biologists could tally at least 60 different prokaryotes, and if they hadn't stopped counting, it would undoubtedly have been many more. The genomes of some of these forms ("species") were more dissimilar than the genomes of humans and yeasts. Indeed, the numbers of prokaryotes found at the opposite end of the digestive tract were even greater.

Prokaryotes owe their evolutionary success to their adaptability and inventiveness. Their ability to accommodate themselves to a very wide range of environmental conditions is stunning, especially under such extremes that include boiling water and high acidity at one end of the spectrum, and freezing water and high alkalinity at the other. Prokaryotes are present in every corner of the world, in the waters, in soil, in the air, on the inner or outer surfaces of animal bodies, in cells of other organisms, and even in rocks. It is believed that the Earth's crust is impregnated with prokaryotes up to a depth of several kilometers. To be able to live in various environmental niches, the prokaryotes had to make many specific adjustments to their cell structure, their life styles, and their chemistry.

The seemingly boundless creativity of the prokaryotes is attested by their ability to tap a variety of inorganic and organic sources of nutrients and energy. Not only have they come up with photosynthesis (the sunlight-fueled conversion of carbon dioxide into organic compounds) and respiration (the release of energy by the conversion of organic compounds into carbon dioxide), both in the presence or absence of air; they have also hit upon several more exotic (from the human point of view) methods of energy production. These include oxidation of hydrogen sulfide (H_2S) to elemental sulfur (S^o) and then to sulfates (SO_4^{2-}); oxidation of ferrous iron to ferric iron with the participation of molecular oxygen; oxidation of ammonia (NH_3) to nitrites (NO_2^-), followed by the oxidation of nitrites to nitrates (NO_3^-); oxidation of hydrogen gas (H_2) to water; the conversion of carbon dioxide (CO_2) to methane (CH_4); and the conversion of methane to carbon dioxide. Moreover, the entire system of the body's principal chemical reactions, collectively referred to as *metabolism*, was presumably fashioned by prokaryotes.

In all respects, therefore, prokaryotes have been evolution's *coup de maître*. In all respects but one, that is, because in terms of morphological diversity, prokaryotes – compared to eukaryotes – are still swimming in evolutionary backwaters. Not only are they all basically single-celled organisms, they all look more or less the same under a light microscope. They *do* vary morphologically, of course, both in the appearance of the individual cells and in their cellularity, but their variability is no match for the eukaryotic orgy of shapes and forms. The prokaryotic variation in shape is limited to a few basic forms (spherical, rod-shaped, or corkscrew-like) and some versatility in the projections from the cell surface (whip-like flagella, fiber-like fimbriae, or finger-like pili).

The prokaryotic attempts at multicellularity are nothing more than a few timid steps in that direction. When a prokaryotic cell begins to divide, it produces a small crack in its cell wall with the help of resident enzymes, inserts a patch of newly synthesized wall material into the opening, and mends the borders. It then repeats the process and continues doing so until the cell doubles in size (the process can be compared to producing two sweaters out of one by knitting patches in them). Subsequently, it forms a partition by the inward, centripetal growth of the plasma membrane and the cell wall, which separates the two daughter cells. In some prokaryotes, however, the partition is not cleaved in the middle and the cells remain joined together forming cellular doubles, chains, filaments, cubical packets, or irregular clusters. All the cells in these aggregates are equivalent and, except for physical contact, independent of one another.

In myxobacteria, the attempts at multicellularity have been taken a step further. These organisms pass most of their life cycles as single cells, but when nutrients begin to dwindle or the environmental conditions deteriorate, the cells socialize. Attracted by some as yet unidentified signals, they slide toward one another along the trail of slime they secrete. The aggregates form a mound that then differentiates into a complex structure, the fruiting body, with a stalk composed mostly of slime and a few scattered cells, and a highly cellular "head". The cells of the head differentiate into spores that are resistant to dryness, heat, and ultraviolet (UV) light. When conditions improve again, the spores germinate into single cells and the cycle repeats itself. Some myxobacteria also congregate to surround and attack other bacteria – not unlike a pack of wolves hunting down their prey. Even this form of socializing, however, is a far cry from the genuine multicellularity of eukaryotes.

Why do the prokaryotes then look so drab and why have some of them not become many-celled? Actually, they *have*, but these we now call Eukarya. Nevertheless, the fact remains that enormous numbers of prokaryotes opted to remain morphologically indistinctive and single-celled. Why?

The Advantage of Being a Eukaryote

One of the chief impediments to morphological diversification and multicellularity might have been the *cell wall* – the rigid, rather inflexible casing outside the plasma membrane. There are few prokaryotes that lack the cell wall and those that do either live in protected environments or have strengthened their plasma membranes by the inclusion of special lipids. These exceptions notwithstanding, the possession of a cell wall is essential for the survival of a prokaryotic cell. It protects the cell against enemies on the outside and from a revolution on the inside.

A cell without a wall, enclosed only by a plasma membrane, is like the pig's bladder submerged in water into which your biology teacher poured a sugar solution to demonstrate the effects of osmotic pressure. Since the sugar, but not the water molecules, are too large to cross the membrane, the bag gradually swells as water molecules stream from the region of higher concentration (outside the bag) to the area of lower concentration (inside the bag) in an attempt to equilibrate the difference. In addition to sugars and many other large molecules, a cell contains electrically charged atoms or groups of atoms (ions) at concentrations exceeding those of the aqueous environment in which all prokaryotes presumably once lived, and many still do. If the cell were not protected against the *osmotic pressure*, the water diffusing into it would cause it to burst because the fragile plasma membrane, permeable to water but impermeable to ions and large molecules, would collapse like the walls of the Bastille under the surging crowd of revolutionaries. There are, in principle, two stratagems for upholding a cell under osmotic pressure. One is to build a bulwark sheathing the cell from the outside, a cell wall; the other is to crisscross the cell's interior with ropes and rods fastened to opposite sites on the cytoplasmic face of the plasma membrane – to assemble a *cytoskeleton*. The prokaryotes have stuck to the former stratagem, whereas the eukaryotes opted for the latter.

The advantage of possessing a cell wall is that it not only solves the problem of osmotic pressure, it also shields the cell from external hostilities. Protection, however, has its price. The cell wall is like a suit of armor: for a certain type of warfare, it is great to have the whole body enclosed in chain mail but at the same time, wearing a suit of armor makes it difficult to run after or away from an enemy, to make love, or to socialize. These restrictions were probably one of the reasons why prokaryotes look like a bunch of Dollies, why they have remained single, and why they are skittish about sex.

The transition from life with a cell wall to life without it but with a cytoskeleton was accompanied by the emergence of numerous other innovations. They included changes in the composition of the plasma membrane; an upgrade in the means of intracellular transport; an increase in cell volume; differentiation of the cell's interior into separate compartments delineated by a system of intracellular membranes; acquisition of the ability to change shape and to engulf particles (phagocytosis); an increase in the size and complexity of the genome; the emergence of mitotic cell division; and the appearance of sexual reproduction with germ cells, alteration of diploid and haploid stages of the life cycle, meiosis, and fertilization. The advantages of all these changes are fairly obvious, all except one, that is: sexual reproduction.

Why did sex arise? If you consider this question ridiculous, try – if you can – to forget your hormone-tainted prejudices about sex and think like an economist in terms of costs and returns for investments. Many contemporary evolutionary biologists do indeed think along these lines and seek explanations for biological phenomena by weighing up their potential costs and benefits. The costs of sexual reproduction are heavy for both the female and the male, but more so for the former. In a sexually reproducing species, the female, which produces the egg, supplies it with all the material it will need in its initial development, and frequently also rears the young. The male's contribution, on the other hand, is often limited to the insertion of his genetic material into the egg. Since approximately one-half of the progeny are female, the mother "wastes" her effort on raising fifty percent good-for-nothing males. By contrast, in asexually reproducing species, the parent produces progeny in which all individuals contribute equally to the next generation. In effect, an asexual organism produces twice as many offspring – in terms of contribution to the next generation – as a sexual organism. One would therefore expect natural selection to favor asexual over sexual reproduction and thus prevent the latter from arising or, failing this, to dispose of it entirely. There are in fact even more reasons that can be put forward to argue that sex is not only disadvantageous for the female, but also for the male.

On the positive side, several potential benefits of sexual reproduction can also be listed, only one of which we shall mention here. Suppose two different advantageous mutations have occurred, one in each of two separate individuals. It is faster to bring these two mutations together through sex and produce doubly advantageous individuals than to wait until the two mutations occur in the same individual. This and similar arguments are not entirely convincing, but since sex is so widespread among eukaryotes, its benefits must somehow be greater than its costs. An all-inclusive explanation of why sex arose continues to elude evolutionary biologists.

The Rise of the Eukaryotes

In the original 18S rRNA/rDNA-based trees, eukaryotes were roughly split into two groups, a bunch of long branches issuing from the lower part of the trunk and numerous shorter branches gathered together in the crown (Fig. 6.9). It was assumed that the species on the lower branches were descendants of the earliest eukaryotes and that the "crown group" emerged later in evolution. Seemingly congruent with this interpretation was the observation that the eukaryotes of the lower branches lacked mitochondria. Although the early forms had a nucleus, cytoskeleton, internal membranes, and all the other characteristics of genuine eukaryotes, they did not possess mitochondria, presumably because they lived in an environment that was largely devoid of oxygen (anaerobic).

Eukaryotes emerged, according to this hypothesis, before conditions favoring an aerobic life style and thus the endosymbiosis of α-proteobacteria with archaea prevailed. A re-examination of the phylogenetic position assigned to the "early eukaryotes" suggested, however, that the group might be of far more recent origin than was previously thought and furthermore that their lack of mitochondria might be a derived character. The group includes Microsporidia ("small-spored ones"), Diplomonadia ("doubled ones"), and Trichomonadia ("haired ones"), almost all of which are single-celled parasites adapted to

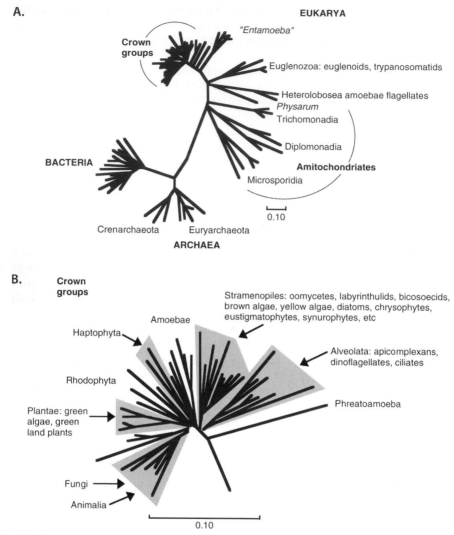

Fig. 6.9A,B
The Tree of Life: a more recent version. **A.** The three domains of life. **B.** The crown groups of Eukarya. This tree, too, is based on rRNA sequences. The lengths of the branches correspond to genetic distances. (Redrawn from Sogin 1991)

life in bodies or cells of other eukaryotes under conditions of limited access to oxygen (Fig. 6.9). Since oxygen is necessary for the normal function of mitochondria, the limited use of the organelle after the transition to the parasitic life style may have led to its loss. At present, all known eukaryotes either have mitochondria or are presumed to be derived from ancestors who had them. Many evolutionary biologists therefore believe that mitochondria, along with the nucleus, cytoskeleton, and internal membranes, were part of the original package that differentiated the emerging eukaryotes from the archaea.

These conclusions, however, have left unanswered the question regarding the circumstances under which certain α-proteobacteria came to be so closely associated with certain archaea that they eventually became part of their cells. Several explanations are currently under consideration. One possibility is that the ancestors of the symbiont were bacteria that produced hydrogen which they supplied, along with carbon dioxide and certain organic compounds, to hydrogen-dependent anaerobic archaea. From living next to each other in this type of bondage it would have been but a small step for one of the organisms to live within the other. Another possibility is that the future symbiont fed on methane produced by the future host. A third hypothesis posits that the future symbiont detoxified the host's cytoplasm by consuming oxygen; when it was later engulfed, it turned into a mitochondrion upon receiving host genes enabling it to produce adenosine triphosphate (ATP). The advantage of these hypotheses is that they explain the origin not only of mitochondria, but also of *hydrogenosomes*, the ATP-generating organelles resembling mitochondria in many respects but differing from them by their ability to produce hydrogen.

The definitive location of Microsporidia, Diplomonadia, and Trichomonadia on the Tree of Life remains unresolved. Microsporidia in particular are homeless at this stage because one set of genes, including the 18S rRNA, supports their original assignment at the base of the eukaryotic section of the Tree, while another set has recently been used to argue that these parasites are in reality weird fungi.

As we climb higher up in the Tree, we pass many branches of mostly single-celled organisms, some of which are amoeba-like, others flagellated, and others still slime-like. Since we are in a hurry to reach one particular species, *Homo sapiens*, we do not stop at any of them. The crown itself is dense with branches, most of them radiating from a narrow area of the trunk, suggesting that they sprouted out almost simultaneously. The most familiar of these are the plants, fungi, and animals. In earlier classifications, each of these three groups was granted the status of a kingdom and all the remaining eukaryotes were lumped together into the kingdom of single-celled organisms. The 18S rRNA-based tree reveals how misleading this classification was (Fig. 6.9). Many branches in the single-celled "kingdom" are not only not directly related to one another, they are also much longer than the plant, fungal, and animal branches. Hence, if plants, fungi, and animals are each granted the status of a kingdom, the same should also apply to all the various branches of the single-celled organisms. The number of eukaryotic kingdoms would then increase from four to several dozen.

The Murmuring of Leaves

In this chapter we climbed the Tree of Life – the only tree of its kind on this planet – from its roots to one particular branch in its crown. The climb enabled us to explore the general topology of the Tree and to reach the conclusion that its shape is very different from the original conception. Early on in our climb, we reached a site from which, looking up, we could clearly see the main trunk splitting into three massive limbs, but looking down, we could not discern the order of splitting with any accuracy. By circumnavigating the trunk's perimeter, we tentatively decided that the limb bearing the bacteria split off first and that the limb bearing the archaea split from that bearing the eukaryotes at a later point. Below this, however, only a dense thicket lay before us in which it was difficult to establish which parts led to the three limbs. And even above the split, the limbs remained connected via many shoots, sprouts, and runners.

We then explored the base of the eukaryotic limb at some length in an attempt to discover the manner in which it separated and distinguished itself from the rest of the Tree.

Among the various innovations enabling the limb's separation, we pointed out two that may have been especially significant – the loss of the cell wall and the development of the cytoskeleton. The latter allowed compartmentalization of the eukaryotic cell and the evolution of organelles. Some organelles apparently evolved from the cell's own resources, while others arose from engulfed bacteria.

We then climbed the eukaryotic limb, expecting to find our own species. Along the way we noticed many long branches splitting off, most of them bearing diverse single-celled organisms. At the end of the limb, we came upon a dense cluster of many branches, all seeming to issue from a single region. We located the branch bearing all the animals and are now perched in the saddle formed by the splitting of the plant, fungal, and animal branches, readying ourselves for scaling the next section in Chapter Seven. Our view, both up and down, is obscured by the thicket of branches and foliage, which is probably just as well, for looking down, we might otherwise be overcome by vertigo, and looking up, we might hesitate to enter the thicket for fear of getting lost. At this moment of repose, Walt Whitman's verses come to mind again:

> *Murmuring out of its myriad leaves,*
> *Down from its lofty top rising two hundred feet high,*
> *Out of its stalwart trunk and limbs, out of its foot-thick bark,*
> *That chant of the seasons and time, chant not of the past only but the future.**

---

* Walt Whitman: Song of the Redwood-tree. From *Leaves of Grass and Selected Prose*. Rinehart & Co., New York 1951.

CHAPTER 7

The Rise of the Metazoan Tribes

The New Phylogeny of the Animal Kingdom

*Y*ou *sound as if you have been playing with R.N.A. It's dangerous, Charlock.*

Lawrence Durrell: *Tunc*

Natural and Unnatural Groupings

If the word "Metazoan" evokes transcendence, it is only because another, more familiar term springs to mind – metaphysics. Yet both words are etymologically devoid of super-natural connotation. Aristotle introduced "metaphysics" simply as a designation for the works in his oeuvre that followed those on physics (*tà metà tà physikà*). And biologists at one point saw the need to split the Linnean animal kingdom into one-celled Protozoa, the animals that come first, and many-celled Metazoa, those that come after them (from Greek *zoon*, animal). The problem with such a division, however, was that among the one-celled eukaryotes, many actually behaved both like animals (in being motile, for example) and plants (in using sunlight energy to convert inorganic into organic matter). To avoid refer-ring to plant-like organisms as "animals", alternative, more neutral designations such as Protista or Protoctista were introduced.

The new names may have solved one problem, but subsequently another, graver one arose. To explain, we have to digress briefly to examine the nature of biological classifica-tion. The essence of classification is the grouping together of objects that have one or more properties in common. The choice of the shared property is dictated by the intended pur-pose of the classification and it, in turn, determines how "natural" the erected categories will become. We can, for example, classify organisms according to their size, but if we do, a giraffe ends up in the same category as the acacia tree on which it feeds. This outcome is hardly satisfactory because although the organisms in the particular category are of simi-lar size, they differ in many other properties which they share with organisms placed by the size criterion into distinct categories. We thus perceive the grouping as unnatural.

We can find a way out of this predicament if we use multiple characters for classification instead of a single one. There is, indeed, a school of taxonomy – *phenetics* – which adheres to the philosophy of clustering organisms together according to the *number* of characters they share. On the basis of overall, superficial similarity, however, we might be tempted to place, for example, the shark and the dolphin in the same category – a grouping which on closer inspection would also turn out to be unnatural. The superficial similarities between the shark and the dolphin would be contradicted by fundamental anatomical differences which place the shark in a very different category (fishes) from the dolphin (mammals).

Most biologists strive to produce a natural rather than an artificial classification of the living world because they are convinced that a natural classification mirrors the historical process by which the similarities and dissimilarities among the living forms arose. Organisms comprising a natural group are similar because they are derived from a com-mon ancestor which possessed the same characters that they now share. Since the time of their origin, the organisms may have acquired unique characters that now distinguish them from one another. Some of these characters may converge on those evolved inde-pendently by another, unrelated group of organisms and so become responsible for super-ficial similarities that can lead to unnatural groupings. Of primary importance, however, are the *shared derived characters* which were present in the most recent common ancestor of a group (but not in earlier ancestors) and which are therefore found in all of its mem-bers, but not in members of other groups. Shared derived characters define a natural group or *clade*, a cluster of taxa* all derived from the same ancestor (i.e., a clade is a *mono-*

---

* *Taxon* (plural *taxa*, from Greek *taxis*, arrangement) is a group of entities sufficiently distinctive to be worthy of a name and a rank in a classification system. Species is a taxon, as are genus, family, or-der, class and other categories. A taxon of a higher rank consists of taxa of a lower rank (e.g., genus consists of species) in a classification hierarchy. The study of the general principles of scientific classification is *taxonomy*.

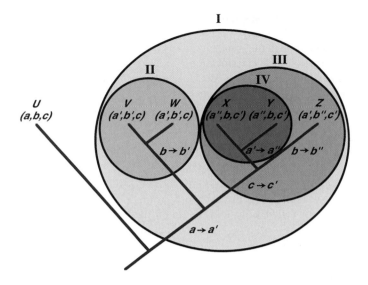

Fig. 7.1
The essence of the Hennigian revolution: Life's history as a series of monophyletic groups (clades) included hierarchically into one another. The cladogram depicts six taxa (*U* through *Z*) defined by three characters (*a*, *b*, *c*). It consists of four monophyletic groups (I through IV, each group encompassing the taxa enclosed by the circle). Monophyletic group I is defined by the shared derived character *a'*, group II by *b'*, group III by *c'*, and group IV by *a"*; *b"* is a unique state. Note that the designations "ancestral" and "derived" are relative: *a'* is a derived character for group I, but ancestral with respect to all other groups; *c'* is a derived character for group III, but ancestral with respect to group IV, and so on. Changes of characters are symbolized by arrows.

phyletic group). The aim of modern biological classification is to produce a system consisting exclusively of monophyletic groups arranged into a hierarchy in which smaller clades are included in progressively larger ones (Fig. 7.1).

This insistence on monophyletic groupings was not always the First Commandment of biological classification as it is now. In the past, taxonomists often relied on their subjective judgment and authority in specifying the characters which were more, and those which were less important for assigning organisms to categories. The revolt against this discretionary practice was led by the entomologist Willi Hennig. Facing the difficult task of classifying an obscure group of New Zealand flies, Hennig developed an entirely new system of classification, complete with precise definitions, a set of objective rules and principles, and a new terminology. Having applied it successfully to the flies, he then extended and generalized it, molding it into a new philosophy of biological classification. His *Grundzüge einer Theorie der phylogenetischen Systematik* (*Principles of a Theory of Phylogenetic Systematics*), published in 1950, had the same effect on biological taxonomy that Immanual Kant's *Kritik der reinen Vernunft* (*Critique of Pure Reason*) had had on philosophy nearly two centuries earlier – and was almost as difficult to read. The author's striving for precise formulations, the ironclad but complex logic, and the penchant of German intellectuals for long sentences, place a heavy demand on the reader of Hennig's *magnum opus*. Not surprisingly, therefore, the book had little initial impact. Only after a brief summary of Hennig's theory was published in English 15 years later, did the cladistic revolution finally take off. Although even now, half a century after publication of Hennig's *Grundzüge*, cladistics in its many variants still has to battle with its detractors, its insistence on defining monophyletic groups by shared derived characters has become the cornerstone of modern biological classification.

From the cladistic point of view, Metazoa but not Protozoa (Protista) are a natural group, a clade. As far as Protozoa are concerned, no single, shared derived character has been found that would unite all one-celled "animals" to the exclusion of all other organisms. Metazoa themselves are only one of several kingdoms in the domain of Eukarya, most of which are one-celled. The decision as to which of these kingdoms should be included in the Protozoa and which excluded is arbitrary because there is no objective criterion on which to base the selection. Nevertheless, the term Protozoa, like so much of the baggage inherited from traditional taxonomy, continues to be used for the sake of convenience and because of inertia. The English word "animals" is now used either to refer to all animals, one- or many-celled, or as a synonym of Metazoa; it is in the latter sense that it will be used here.

From Colonialism to Socialism

Most of the shared derived characters that define the metazoan clade are related to the characteristic of this kingdom – *multicellularity*. All Metazoa are many-celled, but they are not the only multicellular group. Full-blown multicellularity also occurs in fungi and plants, while representatives of several other eukaryotic groups, including slime molds and some algae (the familiar "seaweeds"), have at least taken a few initial steps towards this state. Moreover, various eukaryotes and a few prokaryotes (e.g., cyanobacteria) pass through a *colonial* stage in their life cycle, which is also a multicellularity of sorts. The main difference between the colonial and the multicellular state is that in the former, individual cells retain their full autonomy, whereas in the latter, they specialize to performing different tasks and so become dependent on other cells. Strictly speaking, therefore, multicellularity cannot be regarded as a shared derived character of Metazoa. Or can it? Metazoan multicellularity is very different from the multicellularity of fungi, plants, and the various upstarts. The bodies of many-celled fungi (there are also numerous one-celled forms in this kingdom, baker's yeast for example) are essentially agglomerates of filaments, each filament or *hypha* comprised of a row of cells attached end-to-end. The filaments are either arranged loosely into wads or *mycelia* or are packed tightly into fruiting bodies, the familiar mushrooms. Plants, occasionally referred to as Metaphyta in an obvious allusion to Metazoa, have – like fungi – rediscovered the corset in the form of the *cell wall*, and have made a virtue of its rigidity. The plant cell wall is made of the polysaccharide *cellulose* whose long, extended chains are bundled into fibers and deposited, layer after layer, in an amorphous matrix of other polysaccharides outside the plasma membrane. (The fungal cell wall is made of a different polysaccharide, *chitin*, which the arthropods among the Metazoa use to construct a protective armor encasing their bodies.) The rigidity of the cellulose-based cell wall enables the plant kingdom to create architectural marvels such as the central European oak trees or the redwoods of the Yosemite National Park, California. The corseted plant cells are tightly packed together with only a thin layer of a glue-like substance separating adjacent cell walls. They communicate via a set of tiny tunnels that cross their cell walls. The plasma membrane of one cell connects with the membrane of the neighboring cell via these tunnels. The whole plant thus resembles a system of interconnected lakes – Minnesota in miniature.

Nothing of this sort is found in the Metazoa. Here, each cell is a separate unit surrounded by its own plasma membrane, without any cytoplasmic channels leading to other cells. Further, metazoan multicellularity is distinguished from that of plants and fungi by cellular mobility, the use of buttons for fastening, and the utilization of intercellular spaces. In plants, cells take up their positions and hold them for as long as they live, and often even after they die. Not so the animal cells. Most of them are not only capable of changing their shape to some degree at least, many can also change their position. Some even specialize in cruising through the body, from place to place, like cellular Flying Dutchmen.

In contrast to plant cells, which are held together by glue or mortar, animal cells are linked up by buttons – the *cell junctions* – made of crisscrossing protein strands, glycoprotein aggregates, or proteinaceous cylinders which can open and close. The main distinction between plant (fungal) and metazoan multicellularity, however, is the utilization of intercellular spaces. Plant cells are either stuck together so tightly that there is no space between them except for the glue or if intercellular spaces do occur, they serve as conduits of gases and juices but otherwise remain empty. In animals, on the other hand, intercellular spaces have become an essential part of the body design and have provided an impetus for the creation of a fundamentally new type of tissue. In addition to the *epithelia*, in which closely packed cells largely without intercellular material are arranged into single- or multilayered sheets, metazoans also possess *connective tissue*, in which the intercellular spaces are nearly as important as the cells themselves. The cells of the connective tissue secrete material, the *extracellular matrix,* which consists of proteins, glycoproteins, proteoglycans, polysaccharides, and a suite of low molecular mass substances. The main protein constituent of the matrix and the most abundant protein in the entire animal kingdom is *collagen*. The extracellular matrix and the connective tissue provide the body not only with support and at the same time flexibility, but also with "paths" along which cells travel: they guide the cells, and instruct them on the form they should assume and the tasks they should undertake.

Thy Sister, the Mushroom

Cell junctions, connective tissue, the extracellular matrix, collagen and other specialized molecules are all metazoan inventions, as is their very particular form of multicellularity. When referring to plants, fungi, and other taxa as multicellular organisms, it is nothing more than the use of the very same word for different phenomena. All these features, including the special form of multicellularity, are therefore shared derived characters which attest to the origin of all Metazoa from a single common ancestor and so to the monophyletic status of the group. Having reached this conclusion, we would now like to know the identity of this common ancestor.

Here we have two possibilities: plants and fungi. Both may seem the most unlikely of candidates. What could animals possibly have in common with sunflowers, pine trees, and crabgrasses on one hand and toadstools, fly agarics, and puffballs on the other? Our stereotype of an animal is an elephant, a hare, or a mouse, so much so that a lay person may even hesitate to call an earthworm, a bee, or indeed a blackbird an animal. All these creatures are worlds apart both from plants, which lack the ability to move but are capable of harnessing sunlight energy, and from fungi, which appear to be mere tangles of filaments that occasionally organize themselves into elaborate fruiting bodies. The disparity in the appearance of the present-day forms does not exclude a similarity between ancestors, however. Indications are that the ancestors of plants, fungi, and animals diverged from one another within a short geological time frame so that viewed from our present-day position, they appear to have diverged simultaneously. The divergence of the three kingdoms is seen as a *trichotomy* – the emergence of three branches from a single common ancestor. Since evolutionary biologists have an aversion to trichotomies, efforts to resolve this particular one continue. The assessment of morphological, biochemical, and physiological characters led to the traditional view that fungi are plants, a view now universally rejected. Hopes are therefore pinned on the use of molecular phylogeny analyses.

The four theoretical possibilities are, first, that the trichotomy is real in the sense that the three kingdoms did indeed diverge simultaneously from their common ancestor; second, that Metazoa diverged from the common ancestor of plants and fungi; third, that fungi diverged from the common ancestor of plants and animals; and fourth, that plants diverged from the common ancestor of fungi and animals. The first two are generally discounted, the third is embraced by some evolutionary biologists only, and the fourth is the consensus view.

Several molecular systems have been exploited to examine each of the four possibilities in depth: RNAs of both the small and large ribosomal subunits; various families of ribosomal proteins; and a broad selection of other evolutionarily conserved proteins. The results have varied depending on the system used, the species, and the method of phylogenetic analysis. The consensus view, the hypothesis that fungi and animals are *sister groups*, is largely based on the analysis of conserved proteins involved in the control of protein synthesis (elongation factor 1α), the breakdown of carbohydrates (enolase), and the formation of cytoskeleton (actin, tubulin). The animal-fungus sisterhood is also upheld by the presence of a twelve-amino-acid insertion in the elongation factor and three small gaps in the enolase. All four markers are uniquely shared by the animal and fungal proteins but are absent in the corresponding plant proteins. There are also other characters shared by animals and fungi (but not plants), including mitochondria with similarly reduced gene contents; the use of the UGA triplet to specify the amino acid tryptophan; the use of glycogen for carbohydrate storage; and the use of similar sets of biochemical reactions to synthesize hydroxyproline, chitin, cellulose, and ferritin. We may therefore have to get used to the thought that the cottony mass we see growing on the dead body of an animal or, God forbid, on the food stored in our refrigerators, is in fact a sister lineage of the one we belong to.

Fungi, however, are not the *ancestors* of animals; the two groups arose independently from one-celled organisms. The metazoan ancestor probably either belonged to the slime molds or to the choanoflagellates. For a very long time, *slime molds* were considered a subgroup of fungi until molecular evidence indicated, first, that they branch out independently from fungi, and second, that the three slime mold types (plasmodial, cellular, and oomycotan) represent three independent eukaryotic lineages. Slime molds share characters with both fungi and animals: the filamentous organization of their bodies, the ability to arrange their cells into a fruiting body, and the production of multiple nuclei enclosed by a single plasma membrane are comparable to similar features observed in fungi. Their animal-like characters include locomotion, ingestion of food particles by engulfment, and their passage through radically different developmental stages in their life cycle.

Choanoflagellates are distinguished by cells embellished with a collar (Greek *choanê*, a funnel) and a whip (flagellum). The collar consists of a circle of tiny, finger-like projections emerging from the cell's cytoplasm. The colorless cells feed on organic particles, including bacteria, which are brought to the "mouth" near the base of the flagellum by water currents. Some choanoflagellates lead a hermit-like existence, while others socialize: each time a cell divides, the daughter cells remain together to form a group, a *colony*, that is of-

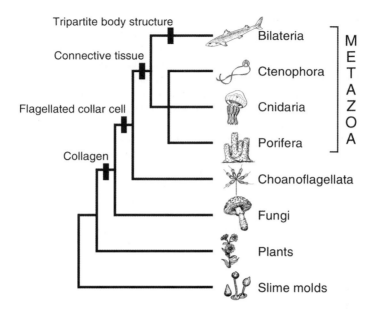

Fig. 7.2
Relationship between animals (Metazoa), choanoflagellates, fungi, plants, and cellular slime molds inferred from rRNA sequences published by Patricia O. Wainright and her coworkers, as well as other investigators. The emergence of critical shared derived characters is indicated for some of the clades by bars intersecting the corresponding branches of the tree.

ten attached to a common stalk. Although the cells may link up with one another by cytoplasmic bridges, each of them exists independently of the rest. The colonies are perceived as an intermediate step in evolution from unicellularity to multicellularity. The collar encircling the flagellum also occurs in cells of animals on the lowest branch of the metazoan clade. A few additional characters unite choanoflagellates and animals, but the strongest indication of a relationship is provided by 16S-18S rRNA sequences. The molecular data place choanoflagellates on a separate branch positioned between the fungal and animal clades (Fig. 7.2).

The Tribes of the Animal Kingdom

The highest category in the classification of organisms belonging to the kingdom of Metazoa is the *phylum*, the Greek word for tribe or race. A phylum is defined by the *body plan*, or *Bauplan*, if you want to sound very learned. The latter term is a throwback to a time some century and half ago when Germany was the hotbed of "evo-devo", in the horrible jargon of researchers studying the evolution of development. A body plan, however, is not what you might take it for – a blueprint for the construction of an animal body. It is a suite of characters that distinguish one group of animals from others possessing different body plans. Thus the body plan of the phylum Arthropoda (which includes insects, lobster-like creatures, spiders, centipedes, and millipedes) comprises a list of characters such as a body partitioned into segments (think of a caterpillar!); segmented appendages (protruding body parts such as legs, antennae, and tentacles; the phylum's name is derived from the Greek *arthron*, a joint, and *pous*, a foot; arthropods are therefore "joint-legged" animals); and a chitin cuticle (a stiff, horny coat secreted by the upper layer of the skin). The body plan of molluscs is characterized, among other things, by the presence of the radula (a horny band that bears chitinized teeth on its backside and is used like a file to tear up food and draw it into the mouth); the mantle (a skin-like fold of the body wall that secretes and lines the shell); and the foot (a retractable muscular underside of the body used for creeping). And the body plan of the chordates, to give one more example, includes the notochord (a rod-like, flexible cord that runs down the back of the body); the hollow nerve cord; and gill slits in the region of the pharynx (throat). All species belonging to a given phylum possess the set of characters defining a particular body plan. The term, however, is also used for categories lower than the phylum. For example, the body plan of vertebrates, a subphylum of chordates, is distinguished by the presence of the vertebral (spinal) column (a skeletal structure that surrounds or replaces the notochord, is segmented into vertebrae, and houses the dorsal nerve cord) and the cranium (the brain case).

There are 35 phyla in the animal kingdom, plus or minus a few. The uncertainty about the total number stems from a disagreement about the phylum status of some groups. The size of the phyla in terms of the number of extant species ranges from one (*Trichoplax adherens* is the only known representative of the phylum Placozoa) to more than one million (the estimated number for the phylum Arthropoda; the exact number is not known because many species remain unidentified while others are continually being lost, mainly as a result of environmental deterioration caused by humankind).

Until recently, the phylogenetic relationships among the metazoan phyla have largely eluded evolutionary biologists. Using traditional methods of classification, many attempts have been undertaken to group the phyla into higher-order categories, the superphyla, but they have all been thwarted by two obstacles: the paucity of shared derived characters and rampant evolutionary convergences in characters. Only recently have molecular evolutionary biologists, who are less handicapped by these hindrances than their more tradition-minded colleagues, scored the first successes in this regard. Before we turn to the new findings, however, we must briefly describe some of the characters conventionally used to group phyla into superphyla. Most of the characters pertain to the different stages of animal development, and these are the focus of the next section; here we sketch out one character that conspicuously differentiates adult bodies – *symmetry*, the arrangement of body parts in relation to an imaginary axis or plane.

There are two principal types of body symmetry – *radial* (pertaining to rays or radii emanating from a center) and *bilateral* (literally "two-sided"). The former is the symmetry of a jellyfish, the latter that of a fish. The parasol of the jellyfish can be slit along the central body axis in an infinite number of ways and always produce two parts, each a mirror image of the other (Fig. 7.3A). By contrast, there is only one way of slicing a fish to obtain two halves that mirror each other – by drawing the knife along a line directly between the eyes and leading it down from the head to the tail (Fig. 7.3B). Radially symmetrical bodies are characteristic of three animal phyla, Porifera (sponges*), Cnidaria (jellyfishes, sea anemones, corals, and hydras, all possessing cnidocytes; Greek *knidé* nettle), and Ctenophora (comb jellies, formerly grouped with Cnidaria into a single phylum Coelenterata). These three phyla are therefore sometimes grouped together into a super-

---

* Only some sponges or certain parts of their bodies show a semblance of radial symmetry; others are asymmetrical. No matter how you slice them, you never end up with two mirror images.

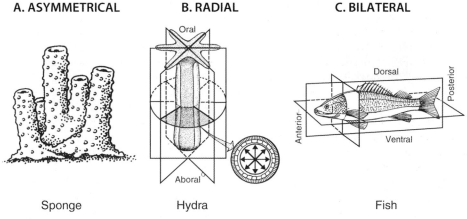

A. ASYMMETRICAL　　　**B. RADIAL**　　　**C. BILATERAL**

Sponge　　　　Hydra　　　　　　　Fish

Fig. 7.3A-C
Symmetry of animal bodies. **A.** Asymmetrical bodies of a poriferan. **B.** Radially symmetrical body of a hydra. A traverse section through the stem (arrow) shows perfect symmetry in all radial directions. **C.** Bilaterally symmetrical body of a fish showing differentiation into anterior and posterior, as well as dorsal and ventral sides. Only the plane in the antero-posterior direction divides the body into two halves that mirror each other. (**B** and **C** modified from Simpson & Beck 1965.)

phylum of Radiata and contrasted with Bilateria, which encompass all or most of the remaining metazoan phyla. Although the adults of some bilaterians (e.g., those of the sea stars) are not bilaterally symmetrical, they are derived from a stage that is. It is therefore believed that animals such as sea stars lost their bilateral symmetry secondarily.

In general, body symmetry reflects more the life style of an animal than its phylogenetic relationship. Sessile animals, which spend most of their life attached to a substrate, tend to be radially symmetrical, whereas motile animals, which move from place to place, are by and large bilaterally symmetrical. In the latter, the front or *anterior part* of the body that enters a new environment first is differentiated from the rear or *posterior part*. Usually the feeding apparatus and many of the sense organs are concentrated in the anterior part, whereas the posterior part often specializes in locomotion, propulsion, and steering. The body thus becomes morphologically differentiated along the *antero-posterior axis*. Similarly, a moving animal faces a different challenge from the environment above than from below, especially if it creeps along the ground or at the bottom of the sea. In response to these challenges, the body has morphologically differentiated its back or *dorsal* from the belly or *ventral* surfaces; it is differentiated along the *dorso-ventral axis*. On the other hand, since the environmental challenge from the left and right side of the body is similar, only limited differentiation of the body parts has occurred along the *left-right axis*. Very few inner organs, such as the swim bladder in fishes or the heart and liver in most vertebrates, disturb the left-right symmetry by occurring singly and on one side only.

Metamorphoses for a Cellular Orchestra

A metazoan is, by definition, a many-celled organism, but it begins its life as a single cell, the fertilized egg. The process that transforms the single cell into a many-celled adult is its *development*. The initial stage of development, in which the germ of an animal, the *embryo*,

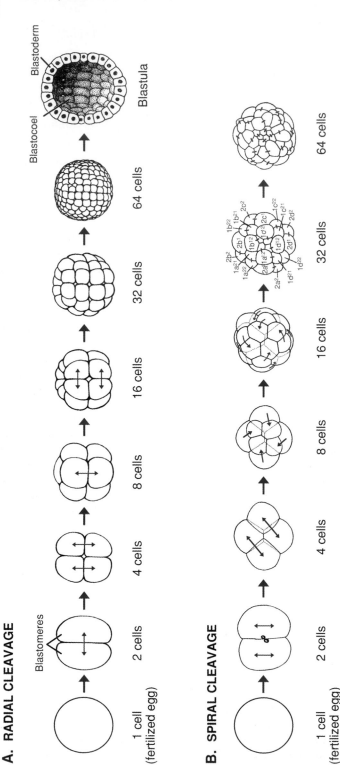

Fig. 7.4A,B

Two paths to multicellularity of animal embryos. A. Radial cleavage such as may take place in a starfish. The fully developed blastula is cut open to reveal that it is hollow. B. Spiral cleavage such as may occur in a snail. Arrows indicate the orientation of the mitotic spindle, numbers and letters in B the origin of the individual blastomeres. In radial cleavage, the cleavage planes all pass either parallel or perpendicular to the axis specified by the egg's poles. In spiral cleavage, the cleavage planes are neither parallel nor perpendicular to the axis; the cleavage involves tilting of the mitotic spindles. (A modified from Simpson & Beck 1965 and Gilbert 1997; B modified from Fioroni 1992.)

subsists on nutrients stored in the egg's yolk or on foodstuff provided by the mother's body, is *embryonic development* or *embryogenesis*. If, at the end of the embryonic phase, the organism resembles the adult, the development is said to be *direct*; if it does not, it is *indirect*. In the latter case, the embryo turns into a *larva* which bears very little or no resemblance to the adult, but is able to feed on its own. In some animals, development passes through several larval stages, in which successive larvae become progressively larger. The transition from one larval stage to the next is accompanied by *molting*, the shedding of the old encasement. The transformation of the larva into an adult, *metamorphosis*, takes place in an immobile, nonfeeding form, the *pupa*.

While certain features of development are common to all Metazoa, others are shared by groups of animals and so can be used to cluster animal phyla. Thus in all animals, after fertilization the egg enters a *cleavage* stage, in which it divides in rapid succession, without pausing between individual mitoses for the growth of the daughter cells, as is customary with other dividing somatic cells. As a result, the mass of the embryo remains the same, but its individual cells become progressively smaller with each division. There are two principal types of cleavage, radial and spiral (Fig. 7.4). If we compare the egg to a globe with the poles, meridians, and the equator, *radial cleavage* (Fig. 7.4A) takes place when the furrow producing the first two daughter cells forms in the plane of one of the meridians. The second furrow arises in the plane of another meridian at a right angle to the first, and the third in the plane of the equator. In subsequent divisions, the furrows form alternately in meridian and equator planes, so that the number of cells grows progressively in a geometric series (2, 4, 8, 16, 32, 64, 128, etc). The cleavage is called "radial" because bisection of the embryo along any meridian results in two halves that are mirror images of each other. Radial cleavage is characteristic of chordates, hemichordates (acorn worms; marine animals with worm-like bodies that live in burrows), echinoderms (sea stars, sea urchins, sea cucumbers, and sea lilies), and a few lesser phyla. In the *quartet form* of the *spiral cleavage* (Fig. 7.4B), the first two divisions of the fertilized egg are meridional, resulting in four large cells. In subsequent divisions, each of these cells then gives rise to many smaller cells in such a way that in each upper tier, one cell is positioned between two cells of the lower tier. Viewed from above the pole, the small cells have a spiral arrangement, which is the consequence of successive divisions occurring in oblique planes. A variant of this mode is the *duet form* of spiral cleavage in which the production of the small cells begins after the first division from the two large cells. The quartet form of spiral cleavage is shared by annelids (exemplified by the earthworms), molluscs, most platyhelminths (flatworms), and several other "worm-like" phyla. The duet form characterizes certain other flatworms (the Acoela).

After this hectic stage, in which one division follows on the heels of the one before, a period sets in which the cells take more time to grow after each mitosis. The rate of divisions slows down and the cells begin to rearrange and to sort out. Distinct regions of the embryo acquire different developmental potentials and become the "germs" of various types of tissue. As these regions often appear as sheets of cells, they are referred to as *germ layers*. The embryos of all metazoans form two such layers of different potential, the ectoderm and the endoderm (literally meaning "outer skin" and "inner skin", respectively). The *ectoderm* is committed to giving rise to the skin and the nervous system, whereas the *endoderm* develops into the gut and its associated organs, such as the liver, pancreas, and the lungs.

The gut is formed by an inward migration of a group of cells originally located in a small, circumscribed patch on the surface of the *blastula*, the hollow ball that emerges in the cleavage stage. In the simplest case, the patch first caves in as a dimple, which then elongates into a sack (Fig. 7.5). The protrusion lengthens across the hollow interior of the

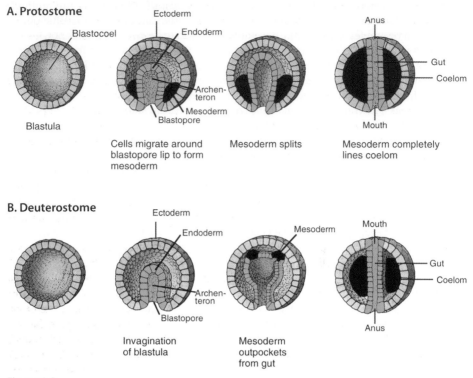

Fig. 7.5A,B
Two ways to open one's mouth. **A.** In protostomes, the mouth develops from the blastopore, the circumscribed area on the surface of the blastula at which a fingerlike projection (archenteron), which ultimately becomes the gut, begins to push into the blastocoel cavity. Cells from the area migrate inside the ball, cluster around the forming gut, and so give rise to the mesoderm. Cavities arising within the cluster develop into the coelom. **B.** In deuterostomes, the mouth forms from an area opposite the blastopore contacted by the projection of the future gut (archenteron). The mesoderm arises by outpocketing of the gut near this area and the coelom develops from the pockets. (Redrawn from Purves et al. 1992.)

blastula until it touches the wall on the opposite side. At the point at which the tip of the ingrowing tube (the endoderm) touches the wall of the embryo (the ectoderm), the tissues at the contact site first fuse and then dissolve to form a small opening. Two openings thus appear in the ball of some metazoan embryos at this stage, one, the *blastopore*, in the area of the original dimple from which cells migrated inwardly to form the endodermal tube, and another at the junction of the invaginating tube with the ectoderm on the opposite side. The process of gut formation or *gastrulation* (Greek *gastér*, stomach) transforms the hollow ball into a tube-within-a-tube structure, for concurrently with the invagination of the endoderm, the embryo elongates along the antero-posterior axis. One of the two openings becomes the mouth, the other the anus. In some metazoan phyla, the mouth arises from the blastopore, in others the blastopore gives rise to the anus and the mouth forms secondarily (Fig. 7.5). The fate of the blastopore is the basis for clustering the phyla into two major groups, the *protostomes* ("mouth-first", i.e., the mouth forms from the first opening that appears in the blastula; Fig. 7.5A) and the *deuterostomes* ("mouth-second", i.e., the mouth forms from the second opening; Fig. 7.5B). The deuterostomes include the echinoderms, the hemichordates, and the chordates, while the protostomes encompass most of the remaining phyla.

Embryos of all the extant phyla pass through both the cleavage and gastrulation stage and so all of them possess two germ layers, the ectoderm and the endoderm. Phyla such as the sponges, cnidarians, and ctenophorans, in which the embryonic tissues are formed from these two germ layers only, are classified as *diploblastic* (Fig. 7.6A). Other phyla are *triploblastic* because some of their tissues (muscle, skeleton, blood, heart, kidney) arise from a third germ layer, the *mesoderm* (the "middle skin"; Fig. 7.6B). The layer forms in two principal ways. In some animals it arises from a single cell (mesentoblast) located near the blastopore. The cell migrates into the interior of the blastula and repeatedly divides to produce a mass of cells between the wall of the gut (endoderm) and the body wall (ectoderm). In other animals, the mesoderm emerges from the wall of the gut (i.e., endoderm) either in the form of pouches or solid sheets, which again insert themselves between the gut and the body wall.

The hollow, fluid-filled space of the blastula is the *primary body cavity* or *blastocoel*. The cavity, which has no cellular lining of its own and no opening to the outside world, persists in various, often greatly reduced forms after gastrulation as the *pseudocoelom* or "false body cavity" (Fig. 7.6C). In many metazoan embryos it is virtually obliterated by the expansion of the mesoderm. In others it is well developed and organized into the blood vascular system or *hemocoel*. In embryos that develop a mesoderm, the third germ layer is also the site that accommodates a new cavity, the *coelom* (Fig. 7.6D). This *secondary body cavity* arises either from fluid-filled pockets forming within clumps of mesodermal cells or from the spaces enclosed by mesoderm pouching out from the gut. The coelom therefore differs from pseudocoelom in that it is lined along its entire surface with mesoderm (the pseudocoelom is either not lined with mesoderm at all, or if it is, then only partially) and often retains a duct for communicating with the exterior.

Zoologists traditionally classify animal phyla into acoelomates, pseudocoelomates, and coelomates. The *acoelomates* lack the secondary body cavity altogether, their primary body cavity between the body wall and the interior organs is filled with loosely connected cells or with mesoderm, and their gut has only one opening. The *pseudocoelomates* have an unfilled body cavity which is believed to represent the blastocoel, the original cavity of the blastula, and their gut has two openings, the mouth and the anus. The *coelomates* have a fully developed body cavity, the coelom. They are divided into two groups according to the manner in which the coelom arises in the embryogenesis: in *schizocoelous coelomates* the body cavity forms by the splitting of the mesoderm, whereas in the *enterocoelous coelomates* it develops by outpocketing of the gut.

Gastrulation, in essence, establishes the body plan shared by all the members of a phylum. Further development results in the appearance of characters that differentiate groups within a given phylum, down to the species level. These characters are therefore of little use in any attempts at establishing relationships among phyla. The one exception among the post-gastrulation characters is the particular form of the larvae in animals undergoing indirect development. There are only a few basic types of larvae, each of which is shared by different phyla. The annelids, molluscs, and four other phyla, for example, share a top-shaped larva, *trochophore*, that bears a tuft of hair at its apex and a girdle of cilia around its waist.

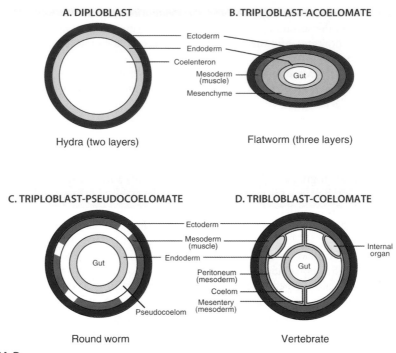

Fig. 7.6A-D
Four ways of hollowing out a body. **A.** In coelenterates (cnidarians and ctenophores), here represented by the hydra, the only body cavity is the coelenteron, an endoderm-lined tube which functions both in the digestion of food and circulation of body fluids. It has a single opening that serves both as mouth and anus. The tube arises by invagination or ingression of cells from the surface of the blastula. The space normally corresponding to the blastocoel is either obliterated by the close apposition of the endoderm to the ectoderm or is filled with extracellular material and, in some species, with a scattering of cells. **B.** Acoelomates, whose bodies consist of three layers, have a digestive cavity (gut), and remnants of the blastocoel, which is usually filled with extracellular material and a scattering of cells, but they have no cavity within the mesoderm. **C.** Pseudocoelomates have a digestive cavity, a cavity (pseudocoel) presumably derived from the blastocoel, but no cavity originating in the mesoderm. The pseudocoel may, however, be partially enclosed by the mesoderm. **D.** Coelomates possess a true coelom that arose within and is entirely enclosed by the mesoderm. The diagrams represent traverse sections through the bodies of the respective animals.

The Testimony of the Cytoplasmic Granules

Body symmetry, mode of cleavage, number of germ layers, the origin of the mouth, and the type of larvae – these are the essential characters available to traditional taxonomists and cladists in their attempts to decipher the relationships among the metazoan phyla. Clearly not enough to reach an agreement, a phylogeny with which all experts would feel comfortable. True, certain groupings do stare one in the face – the fact that the radiate phyla are all diploblasts for example or that the deuterostomes have radial cleavage – but by no means all characters fall so neatly into place. It is not possible to arrange all the phyla into a nested system of characters indicative of a hierarchy of clades without various *ad hoc* hypotheses. Some phyla apparently share characters not because of a common origin but as

A. TRADITIONAL

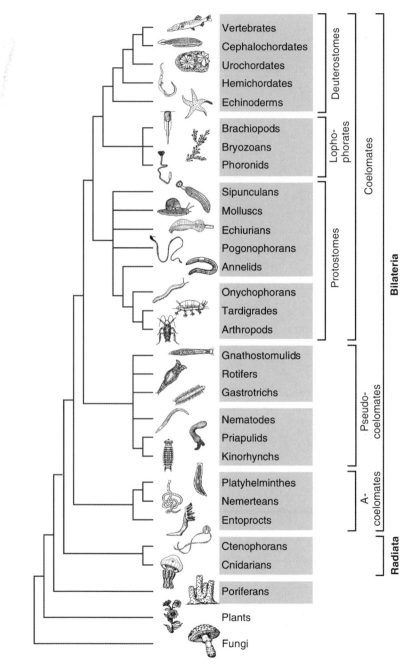

Fig. 7.7A
Traditional (**A**) and molecule-based (mostly rRNA; **B**) interpretations of metazoan phylogeny. Comb-like branching patterns (multifurcations) signify unresolved or uncertain order of divergence. (Modified from Adoutte et al. 2000.)

B. MOLECULE-BASED

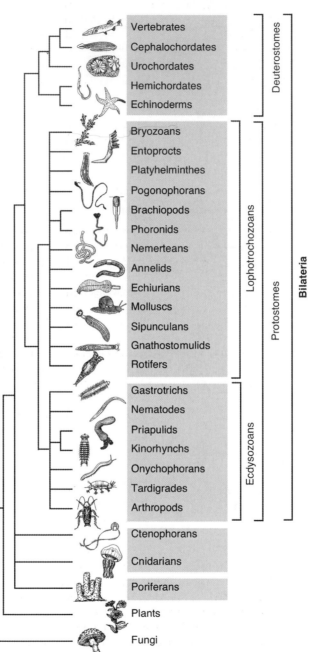

Fig. 7.7B

a result of convergent evolution. The vital question here is which character states are the result of the former and which of the latter process. Unfortunately, this is where the experts don't seem to be able to agree. They each appear to have their favorite *ad hoc* hypotheses and also their preferred phylogenies. The problem would probably have remained insoluble had the doors to the molecular archives not been opened a few years ago. Not that access to the archives has solved all the problems, far from it. But there is now a widely shared sentiment among molecular evolutionary biologists that once the technical snags are overcome, the pieces will fall into place, and a phylogeny will emerge that is acceptable to both morphologists/embryologists and molecular biologists.

One example of the snag is the *long branch attraction*. Occasionally in a collection of sequences, some will appear to evolve faster than others and the more rapidly evolving sequences tend to cluster together even if they are not closely related to one another. Fortunately, tests are now available for measuring whether one sequence evolves faster relative to the others in a sample. Once identified, the rapidly evolving sequences can be left out, thereby precluding spurious groupings.

Technical difficulties of this sort and spurious phylogenies have led to a growing mistrust among some of the more traditional taxonomists as far as the molecular archives are concerned. The impression seems to have been created that molecular methods drag evolutionary biologists into the same mire as the traditional or cladistic methods. As a result, the traditional taxonomists regard molecular taxonomists as *nouveaux riches,* having acquired their wealth by dirty means and being in no position to utilize it properly. Some molecular taxonomists, on the other hand, display a certain cockiness in their attitude toward traditional taxonomy and an inclination to dismiss much of it as bunk. This mutual dearth of respect has occasionally led to friction between the two camps, to the detriment of the discipline. In reality, however, the *nouveaux riches* and the venerable aristocrats are dependent on one another, for their approaches to phylogenetic reconstruction are complementary. Traditional taxonomists and cladists rely on molecular taxonomists to resolve intractable phylogenies; and molecular taxonomists require the help of aristocrats to make sense of the trees by relating them to the phenotypes that differentiate the taxa.

Tradition and the New Look

In the traditional classification, animal phyla are arranged according to an apparent increasing morphological complexity of their bodies. Splitting off first from the base of the traditional tree are the Parazoa, animals with poorly differentiated tissues and no organs, comprising the phyla Porifera and Placozoa. *Poriferans* or *sponges* are immobile animals with radially symmetrical or irregular bodies that have no head, mouth, or nerve cells. A sponge body has neither front (anterior) nor rear (posterior) part; it consists basically of a single wall perforated by a system of canals and chambers through which water containing food particles and gasses is filtered. Embryonic development of the sponge consists of a cleavage stage leading to a hollow blastula that turns first into a free-swimming larva and then into a sessile adult. During the entire embryonic development there is no clear differentiation into germ layers characteristic of gastrulation. The flattened, irregular *placozoan* bodies, by contrast, consist of two sheets of ciliated epithelial-like cells, an upper (dorsal) and lower (ventral) layer, separated by a meshwork of loosely associated, star-shaped cells.

Higher up on the metazoan tree that you may remember from your college biology class, all the phyla represented by animals with tissues and organs branched out, first the Radiata and then the Bilateria (Fig. 7.7A). The bodies of the Radiata (phyla Cnidaria and

Ctenophora) consist of simple tissues, even primitive organs such as ovaries and testes, and possess at one end a mouth leading into a dead-end gut. The radially symmetrical *cnidarians* lack muscle cells but possess muscle fibers in the extended base of their epidermal cells. They develop from diploblastic embryos and so their bodies consist of two layers only, an epidermis covering the outer surface and a gastrodermis lining the gut cavity. The layers are separated either by a membrane or by a layer of jelly-like acellular substance. *Ctenophorans* (comb jellies) are biradially symmetrical (i.e., they are radially symmetrical but flattened somewhat from two sides). They possess muscle cells in the acellular matrix separating the epidermis from the gastrodermis. Because the ctenophoran muscle cells are not organized into a definitive layer as they are in bilaterians, in which they derive from the mesoderm, some embryologists classify the comb jellies as diploblasts. Others argue, however, that the presence of muscle cells alone qualifies the ctenophorans as triploblasts. Others still strive to strike a compromise by calling the two ctenophoran germ layers ectoderm and *endomesoderm*.

The triploblastic phyla have traditionally been divided into three groups, the acoelomates, pseudocoelomates, and coelomates. The *acoelomates* include the phyla Platyhelminthes (flatworms), Nemertea (ribbon worms), Gnathostomulida (tiny animals living in the interstitial spaces of marine sediments), and Mesozoa (ciliated parasites of marine vertebrates). The *pseudocoelomates* comprise the phyla Gastrotricha (strap- and bottle-shaped marine animals with ciliated cells on the belly side), Nematoda (round worms), Nematomorpha (horsehair worms), Rotifera (small animals with saclike bodies and a crown of cilia at the anterior end), Acanthocephala (parasites with a spiny, retractable snout), Kinorhyncha (tiny marine animals which live in sands and muds with bodies divided into 13 segments, a tractable mouth surrounded by a ring of spines), Loricifera (small intestinal parasites with an abdomen encased within a cuticular girdle), and Priapulida (marine animals with bodies ranging in length from less than one millimeter to 20 cm and shaped like the human penis). The *schizocoelous coelomates* include the phyla Sipuncula ("peanut worms" with sausage-shaped, unsegmented bodies, U-shaped gut, and mouth surrounded by tentacles), Mollusca (snails, slugs, clams, oysters, squids, octopuses, etc), Echiura (burrowing, unsegmented marine animals with long, extensible trunk or proboscis), Annelida (earthworms, sandworms, leeches, etc), Pogonophora ("beard worms" living in thin tubes buried in sediments often at great ocean depths), Tardigrada ("water bears", paunchy, eight-legged small animals moving among mosses or in water with a bear-like gait), Onychophora (mostly tropical creatures resembling soft-bodied, unsegmented centipedes), Arthropoda (insects, spiders, centipedes, crabs, lobster, etc), Pentastomida (tongue worms, inhabitants of lungs and passageways of vertebrates), Phoronida (worm-like, marine tube dwellers), Ectoprocta (formerly Bryozoa, "moss animals" forming sessile colonies in marine and freshwater environments), and Brachiopoda (lamp shells with bodies enclosed between a pair of valves). The *enterocoelous coelomates* include Chaetognatha (arrow worms), Hemichordata (acorn worms and pterobranchs), Echinodermata (sea stars, sea lilies, brittle stars, sea cucumbers, sea urchins), and Chordata. The classification was believed to reflect a gradual increase in body complexity, the progression from a single germ-layer organization in the sponges to two layers in the cnidarians, a transition to a three-layer organization in the ctenophores to genuine three layers in the bilaterians. In the bilaterians themselves, the progression was from forms lacking a true body cavity, the acoelomates, through organisms that transformed the blastocoel into a body cavity, the pseudocoelomates, to forms with a true coelom, the coelomates. The neatness of this scheme was its ability to show how the transitions from one group of phyla might have occurred via a series of intermediates. Regardless of the fact that the experts could not agree on the specific manner in which the transitions occurred and indeed very often not even on the groupings of the phyla, the tacit assumption was that these differences of opinion would eventually be ironed out and allow a consensus acceptable to most to emerge. Indeed, the traditional phylogeny described above was close to such a consensus.

The rRNA/DNA studies shatter this paradigm (Fig. 7.7B). Although they leave the sponges, cnidarians, and ctenophores at the base of the tree, they move the placozoans up to a position among the bilaterians and suggest that this phylum's apparent simplicity arose secondarily as an adaptation to a particular life style. Studies have failed to resolve the relationship among the three prebilaterian phyla unambiguously and they also provide no answers as to whether the sponges are really a natural, monophyletic group.

The most profound impact of the rRNA/DNA studies has, however, been reserved for the bilaterians. Here, these investigations disband the two "intermediate" groups, the acoelomates and pseudocoelomates, thought to represent the transition from prebilateria to true bilateria, and scatter them among two of the three bilaterian superphyla, the lophotrochozoans and the ecdysozoans. Of the acoelomates, the flatworms in particular were generally held for the most primitive bilaterians, but molecular studies place them, along with the rest of the group, squarely among the lophotrochozoans, with whom they share the spiral mode of cleavage and a larva resembling the trochophore. The pseudo-coels are unmasked by molecular studies to be an unnatural group, some of its members belonging to the lophotrochozoans and others to the ecdysozoans. Here again, some of the pseudocoels, the nematodes in particular, were long thought to be primitive bilaterians. This was one of the reasons why their representative, the species *Caenorhabditis elegans* or *C. elegans* for short, became the target of an all-out attack on the nature of its embryonic development. But now the new phylogeny perches *C. elegans* and the rest of the round worms much higher up on the tree. The "primitiveness" of the various acoelomates and pseudocoelomates, like that of the placozoans, is thus best explained as a secondary simplification of the structure, an adaptation to a parasitic life style.

Nor have coelomates of the traditional classification been spared from the onslaught of the molecular data, with the result that some of the cherished, long-standing associations between phyla had to be abandoned and the phyla moved to different positions on the tree. Even the Articulata, the group uniting earthworms and their allies (the annelids) with flies and Co. (the arthropods), had to go. The articulates seemed to be tied together by a particularly strong bond – segmentation, the division of their bodies into similar structural modules repeated along the body length. But the rRNA/DNA data are unyielding in this case: they assign annelids to the lophotrochozoans, next to the molluscs, and the arthropods together with nematodes to the ecdysozoans, The annelids may not look much like molluscs, certainly not as far as body segmentation is concerned, but they share other, no less important characters with their new neighbors, for example the spiral mode of cleavage and the trochophore larval stage. Apparently, the evolution of body segmentation has been more complex than originally thought.

The rRNA/DNA-based phylogeny also differs from the traditional classification in the composition of the deuterostome superphylum. It restricts the membership in the superphylum to three phyla, the echinoderms, the hemichordates, and the chordates, while in the traditional classifications deuterostomes also included certain minor phyla ("minor" in terms of the number of extant species they contain). Which minor phyla should be classified as deuterostomes, however, the traditional taxonomists could not agree on. One or more of the following minor phyla were included by the various experts: Phoronida, Brachiopoda, Ectoprocta, and Chaetognatha. The first three of these are distinguished by possession of the lophophore and so constitute the Lophophorata.

You might think that the origin of the mouth from either the primary or secondary opening in the gastrula is something an experienced embryologist should be able to determine unambiguously, leaving no doubts as to which organism is a protostome and which a deuterostome. But in reality it is not as simple as that. Take the chaetognathans as an example. When their embryos begin to gastrulate, they form an opening, the blastopore, at the future animal's posterior end. Then, however, the opening closes and *both* the mouth and the anus form secondarily. So, is this a protostome or deuterostome manner of forming the mouth? It can be either, and correspondingly some taxonomists have assigned arrow worms to protostomes whereas others delegated them to deuterostomes. Now, however, molecular phylogeny assigns all the candidate minor phyla, including the chaetognathans, to the protostomes.

The relationships among the phyla within two of the three great bilaterian groups defined by the rRNA/DNA studies remain unresolved. The phyla within the lophotrochozoans and within the ecdysozoans may have diverged from one another within such a short interval that there was not enough time for the events to be adequately recorded in the archives of the slowly evolving rDNA genes. Whether other more rapidly evolving genes will be able to resolve the order of splitting remains to be seen. By contrast, the order of divergence within the deuterostome superphylum has recently been resolved by rRNA/DNA studies. Here, the record indicates that the chordate ancestor was the first to diverge from the common ancestor of the echinoderms and the hemichordates. The phylum Chordata is traditionally divided into three subphyla – Urochordata (tunicates), Cephalochordata (represented by the amphioxus), and Vertebrata. The evidence available suggests that urochordates diverged first from the common ancestor of the cephalochordates and vertebrates. They then diverged from the other two chordate subphyla to such an extent that some taxonomists promote their elevation to the status of a separate phylum.

The Columnists

Turning our attention to the vertebrates, animals sporting a vertebral column, we must now forgo our reliance on the rDNA genes and cast around for other genes that could replace them. The rDNA genes that served us well where the phylogenetic distances between the compared taxa were large are generally not suitable for intravertebrate comparisons in which the divergence intervals between taxa are often relatively short. We must therefore turn to genes that evolve faster and so have kept a more extensive record of the rapid events than the rDNA genes. Mitochondrial genes, whole mitochondrial genomes, and an assortment of nuclear genes are all being used for this purpose. Nonetheless, if some of the nodes in the vertebrate phylogeny remain unresolved, it is not because the selected genes do not evolve fast enough, but because of another problem: some of these genes apparently diverged from each other before the splitting of the species, and this confounds the interpretation of the species phylogeny. We have more to say about this problem in the next chapter; here we merely point out that the solution to the problem is believed to lie in the use of a large number of genes at independently evolving loci. The phylogeny indicated by the majority of the genes is then taken to be the species phylogeny. Some of the vertebrate nodes remain unresolved because the number of genes sequenced from the representative species is insufficient.

By morphological criteria, the living vertebrates fall into two major groups, those that have jaws and those that do not. According to the fossil record, the jawless vertebrates or Agnatha (Greek *a*, without, and *gnathos*, jaw) were the first to emerge and the jawed verte-

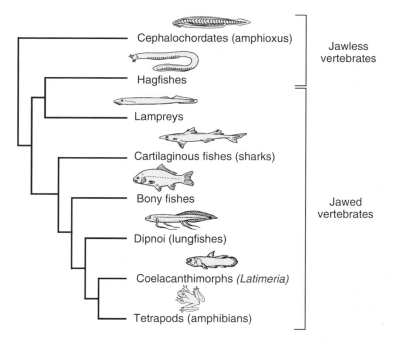

Fig. 7.8
Tentative vertebrate phylogeny deduced from the sequences of multiple genes. The cephalochordate (amphioxus) serves as an outgroup. (Based on data of F. Figueroa, N. Takezaki, and J. Klein, unpublished.)

brates or Gnathostomata (Greek *stoma*, mouth) followed later (Fig. 7.8). The jawless vertebrates were once very abundant and highly diversified, but now they are reduced to two classes only, the 43 species of hagfishes and 40 species of lampreys. Both are eel-like animals lacking scales and paired fins, but possessing a round mouth (for this reason, they are sometimes also referred to as "cyclostomes") and an internal skeleton made exclusively of cartilage. Hagfishes are marine scavengers that live off dead or dying fishes; adult lampreys are fresh or salt-water predators that feed on the flesh and blood of live fishes.

The remaining living vertebrates, the gnathostomes, are traditionally divided into six classes: Chondrichthyes, Osteichthyes, Amphibia, Reptilia, Aves, and Mammalia (Fig. 7.8). The Chondrichthyes or cartilaginous fishes (Greek *chondros*, cartilage, *ichthys*, fish) are characterized by a cartilaginous internal skeleton that never ossifies (turns into bone), paired nostrils, and paired pectoral (chest) and pelvic fins. Members of this class include sharks, skates, rays, and chimeras, altogether ~960 living species. Osteichthyes or bony fishes (Greek *osteon*, bone) have an internal skeleton that contains at least some bone. Osteichthyes are divided into two subclasses, Actinopterygii and Sarcopterygii. Actinopterygii or ray-finned fishes (Greek *aktin*, ray, *pterygon*, fin) have fan-shaped fins supported by radiating bony rays or spines. Most of the fish you are acquainted with are likely to belong to this group. Sarcopterygii or lobe-finned fishes (Greek *sarkodes*, fleshy) possess paired, lobe-shaped fins with a central axis of flesh and bone. They fall into two orders, Crossopterygii and Dipnoi. Crossopterygians (Greek *krossos*, tassels) are survived by two species of the genus *Latimeria*, the coelacanth (Greek *koilos*, hollow, *acantha*, thorn).

Of the Dipnoi or lungfishes (Greek *di*, two, *pnoia*, breath) only five species remain, all of which are able to breathe using both gills and "lungs" – a modified swim-bladder. Some of the species are the only fishes that drown if denied access to atmospheric air.

The rest of the vertebrates are terrestrial animals, most of which have two pairs of limbs – the *tetrapods* (Greek *tetra pous*, four-footed). They consist of the phyla Amphibia, Reptilia, Aves, and Mammalia. Amphibians are so named because most of them return to water to reproduce (Greek *amphi*, double, *bios*, life). They are represented by frogs, toads, salamanders, newts, and caecilians. Reptiles derive their name from their mode of loco-motion (Latin *reptare*, to crawl) and include snakes, lizards, alligators, crocodiles, turtles, and tortoises. *Avis* and *mamma* are Latin terms for bird and breast, respectively. The three classes – reptiles, birds, and mammals – form the *amniotes*, animals carrying the embryo in a fluid-filled sac, the *amnion*, one of the extraembryonic membranes.

Living mammals are traditionally divided into three major groups, monotremes (Prototheria, from Greek *protos*, first and *therion*, wild beast), marsupials (Metatheria, from Greek *meta*, next to), and placentals (Eutheria, from Greek *eu*, true). The monotremes (Greek *monos*, single, *tréma*, hole) resemble reptiles, from which they are indeed derived, in their habit of laying eggs and in possessing a cloaca, a chamber receiving the discharge of the digestive, excretory, and reproductive tracts (hence their name). They are represented by the platypus (duckbill) and echidnas. All other mammals are viviparous: eggs develop within the body of the female, which brings live young into the world. In marsupials, the young are born at an early stage and then complete their development in the marsupium, a pouch formed by a skin fold on the belly side and supplied with mammary glands (Latin *marsupium*, small purse). Representatives of the marsupials include opossums, kangaroos, and wallabies. All other mammals belong to the Placentalia. Their young develop to a rela-tively advanced stage in the uterus, where they are nourished via the *placenta*, a structure formed partly from the mother's uterine lining and partly from embryonic tissues. There are 18 orders of living eutherian mammals, one of which, the Primates, includes *Homo sapi-ens* among its members, the leaf on the Tree of Life that guided our climbing tour.

Although the outline of the vertebrate section of the Tree is uncontested, several nodes still remain unresolved. For example, there are different opinions regarding the relation-ships between the hagfishes, lampreys, and jawed vertebrates. Did the hagfishes diverge first from a common ancestor of the lampreys and jawed vertebrates, or did the jawed ver-tebrates diverge from the common ancestor of hagfishes and lampreys? Neither morpho-logical nor molecular data have thus far provided a definitive answer. Or another example: Is the closest living relative of the tetrapods the lungfish, the coelacanth, or neither of the two? Here again, the molecular information available is insufficient for one to choose be-tween the various competing hypotheses.

A Ribonucleic Hangover?

To recapitulate: we began this chapter by introducing a novel method of classification that partitions the living world into natural groups called clades. All members of a clade are de-rived from the same common ancestor and so constitute a monophyletic group character-ized by shared derived characters. One such shared derived character of the Metazoa is a form of multicellularity in which "naked" cells are fastened together by specialized buttons or separated by large intercellular spaces. Animal multicellularity apparently evolved sep-arately from other types of multicellularity found in plants and fungi. We identified the lat-ter as the nearest relative of the Metazoa. Another character shared by all Metazoa and contrived by them is the specific type of development in which a fertilized egg is transformed into an elaborate body of an adult capable of sexual reproduction. Although every animal species has a unique form, all Metazoa fall into a limited number of phyla, whereby all the members of a phylum share the same body plan. The features that distinguish the three dozen or so phyla from one another include body symmetry, the number of germ layers from which the body is constructed, the presence or absence and the origin of the body cavity, and the embryonic origin of the mouth. Over the years, taxonomists have attempted to establish how the various phyla relate to one another. Although they could not agree on details, they were been able to come up with a consensus view based purely on morphological characters. In this "traditional" classification, the phyla were arranged in the order of increasing complexity in body structure and organization.

The application of rRNA/DNA-based methods to the study of the relationships between the metazoan phyla provides no support for the traditional classification. In the new, molecule-based phylogeny, the metazoans are divided into pre-bilaterian and bilaterian phyla, and the latter are then subdivided into three large groups, the lophotrochozoans, the ecdysozoans, and the deuterostomes. These groups do not correspond to any of the groupings in the traditional phylogeny. The molecular phylogeny does not deny that complexity has been increasing gradually during metazoan evolution, but it disputes the suggestion that the increase took place in the manner implied by the traditional phylogeny. At the morphological level, evolution has apparently independently created similar structures more than once.

The relationships among the phyla in each of the major groups defined by the rRNA/DNA studies remain unresolved or controversial and so are commonly depicted in the form of a comb in the phylogenetic trees. The phyla might have diverged from one another in rapid succession, within a time interval that was too short for the slowly evolving rDNA genes to register the order of the splittings in sufficient detail.

Both the more traditional morphology-based methods and the newer methods based on molecules assign *Homo sapiens* to the deuterostome phylum Chordata, and within the phylum to the subphylum Vertebrata, the class Mammalia, the subclass Placentalia, and the order Primates. In this sense – to quote from Lawrence Durrell's *Tunc* again – *what is our civilization but a ribonucleic hangover, uh?*

But we are not finished with the metazoan tribes yet. One task still remains – to identify the living primate that is our closest relative and so complete the specification of our place in Nature. This is the subject of the next chapter.

CHAPTER 8

Our Place in Nature

The Closest Living Relative

*P*eople are bloody ignorant apes.

Estragon in Samuel B. Beckett's
Waiting for Godot

The First Family

In the *Divine Comedy*, in the sixth cornice near the top of Mount Purgatory, Dante and Virgil are overtaken by a group of emaciated shades. The Gluttonous, whom the shades represent, are being purged of their love of excessive drinking and eating by having to endure thirst and starvation while milling around a pool of running water, in an air scented with the smell of fruit. As they pass by and Dante looks into their faces, reduced to skin and bone, he comments:

Parean l'occiaie anella senza gemme: Each socket seemed a ring without a gem;
chi nel viso de lo nomini lege 'omo' He who reads OMO in man's countenance
ben avria conosciuta l'emme Right plain in these might recognize the M.*

What is he implying? He alludes to the medieval belief that the human face bears an inscription (H)OMO DEI, Man (is) of God; the eyes forming the Os, the line of the cheeks, the eyebrows, and nose the M, the ear the D, the nostrils the E, and the mouth the I:

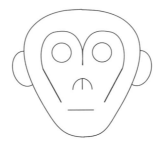

In the wasted faces of the shades, says Dante, the outline of the letter M stood out with clarity.

The belief was part of the doctrine that humans occupy a unique position in Nature, having been created by a special act, separately from all other living creatures. The inscription was thought to be the Creator's signature, just as artists autograph their work. (As the American poet and Dante translator, John Ciardi, has remarked, it was very thoughtful of the Lord to anticipate the Latin alphabet!)

When you look at the sketch above, however, you are likely to be reminded not so much of a human's face as that of a monkey. In reality, it can be either of the two. It points out what is obvious to every visitor of a zoo, that humans and monkeys are astoundingly similar. The similarity is so striking that "God's Registrar", Carolus Linnaeus, keen observer of nature that he was, could not do otherwise than group *Homo sapiens* together with monkeys and apes in his classification system. The son of a pastor must have been well aware that this classification would not sit well with church authorities and so he tried to ameliorate its impact by naming the group *Primates*. The Latin word *primus* means "the first" and is generally used to indicate highest authority or rank. The Archbishop of York, for example is the "primate of England" and the Archbishop of Canterbury the "primate of *all* England." (These two gentlemen were probably not

---

* *The Comedy of Dante Alighieri the Florentine.* Cantica II *Purgatory*. Translated by Dorothy L. Sayers. Penguin Books, Baltimore, MD 1955.

amused by Linnaeus' misappropriation of the word for so lowly a purpose as naming an order in which humans *and* monkeys were placed together.) In Linnaeus' system, therefore, primates were given the highest ranking possible: they became the First Family of the Animal Kingdom. The "secundates" in the system were the remaining mammals and the "tertiates" the rest of the animal kingdom. The implication of the naming was that monkeys and the rest of the order are superior to all other creatures and that among the primates, man stood at the pinnacle as the most perfect product of creation, the image of God himself (who was, of course, assumed to be male).

As a defining character of the primates, Linnaeus laid down the possession of one pair of mammary glands in the chest region of the body (as opposed to several pairs possessed by other mammals). On this basis, he classed the human species together with the various monkeys and apes, lemurs, flying lemurs, and bats. Later, taxonomists removed the flying lemurs and bats from the primate group and assigned them to groups of their own.*

As stated in the preceding chapter, extant mammals are divided into three main clusters – monotremes, marsupials, and placentals. The placentals are also called Eutheria; the alternative names of the monotremes and marsupials are Prototheria and Metatheria, respectively (the latter and the Eutheria forming the Theria).

When more than 230 my ago, a certain group of reptiles began to adapt to environmental niches different from those occupied by other reptiles, the process gradually transformed the descendants of this group into mammals. Among the major adaptations acquired by the early mammal-like reptiles was the morphological and functional differentiation of the original, rather uniform dentition. The teeth lost their conical, peg-like shape and ultimately evolved into incisors, canines, premolars, and molars characterizing recent mammals (Fig. 8.1A). They developed multiple roots and grinding surfaces that occluded so that the opposing surfaces of the teeth in the upper and lower jaws fit together in the closed mouth. At the same time, the grinding surfaces became rippled with pointed projections, the *cusps* (= *cynodont* or *dog-like teeth*; Fig. 8.1B). The earliest mammals had three main cusps aligned from front to back (= *triconodont* teeth; Fig. 8.1C). Later, the three cusps became arranged in a triangular pattern in such a manner that the upper molars fitted into the lower ones like a pestle in a mortar (= *tribosphenic teeth*; Fig. 8.1D). This innovation enabled the mammals not only to tear food into pieces, but also to shear (*sphen*) and grind (*tribein*) it, thus allowing a switch to a tougher diet. The innovation may have contributed to the evolutionary success of mammals in the last 65 my.

Fossil tribosphenic teeth were originally found in mammals classified either as marsupials or placentals and only in regions that were part of Laurasia, the northern continent at the time when all the landmasses were segregated into two supercontinents. Paleontologists therefore believed that during the period from ~150 my to ~65 my ago, the fauna of the two supercontinents evolved in complete isolation and that tribosphenic mammals evolved exclusively on the northern continent. More recently, however, fossil tribosphenic teeth have also been discovered in Australia and Madagascar, which were once part of Gondwana, the southern supercontinent. Consequently, some paleontologists now believe that tribosphenic molars evolved twice independently, once in mammals living on the southern continent and the second time in mammals of the northern continent. The

---

* Interestingly, in some recent phylogenetic trees based on molecular data, flying lemurs split the primate group into two, as if they were indeed themselves primates; see Figure 8.2B.

A.

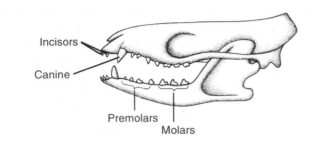

Incisors

Canine

Premolars

Molars

B. CYNODONT   **C. TRICONODONT**   **D. TRIBOSPHENIC**

Cusp

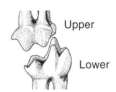

Upper

Lower

E.

Monotremes   Marsupials   Placentals

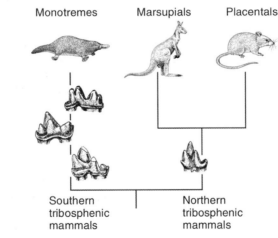

Southern
tribosphenic
mammals

Northern
tribosphenic
mammals

Fig. 8.1A-E
Teeth and the origin of mammals. **A.** Morphological differentiation of teeth in the descendants of early mammals, the insectivores. **B.** Tooth of a cynodont, a representative of mammal-like reptiles which lived some 200 my ago. The surface of the tooth, a molar, is differentiated into cusps which helped the reptiles to tear off bite-sized pieces of meat. **C.** A tooth of a triconodont mammal with three cusps aligned anteroposteriorly (from front to back). **D.** Occlusion of an upper with a lower molar of a tribosphenic mammal. Here the cusps are arranged in a triangular pattern and the opposing grinding surfaces are complementary – they fit together very closely. **E.** Divergence of mammals into three main groups. According to a modern view, the tribosphenic tooth pattern evolved independently at least twice, once in the Southern and a second time in the Northern Hemisphere. (**A** and **C** from Dorit et al. 1991; **B** from Strickberger 1996; **D** from Pough et al. 1989; **E** based in part on Stokstad 2001.)

former are now survived by a few monotreme species only, whereas the latter are represented by marsupials and placentals (Fig. 8.1E).

The extant placentals are subdivided into 18 orders (Fig. 8.2A). Traditional taxonomy, based on fossil record and morphological comparisons of approximately 4356 extant species of placental mammals, has great difficulty in determining unambiguously either the sequence in which the 18 orders emerged during evolution or the relationships among the orders. Various phylogenies of placental mammals and different superordinal groups (i.e., groupings above the order level) have been proposed, but no consensus view has emerged. One such proposal, possibly reflecting a view shared by the majority of experts, is shown in Figure 8.2A. As long as traditional taxonomists arbitrarily gave greater weight to some characters rather than to others, the disagreements could be attributed to this fact alone. Now, however, when not even the objective cladistic methods utilizing sets of shared derived characters are sufficient to unify the experts in their interpretation of placental phylogeny, it appears that the reasons behind the differences in opinion extend beyond subjectivity of judgement.

Initially, molecular taxonomists experienced similar difficulties as their more traditionally minded colleagues. Phylogenies based on individual genes or proteins yielded discordant indications of the relationships among the 18 orders of placental mammals. When, however, researchers turned to the use of groups of genes instead of individual genes, the discrepancies began to disappear one by one. In the last few years, several groups of molecular taxonomists working with different gene sets have arrived at similar phylogenies, which offer a very different view of the relationships among the placental mammals than the schemes proposed by traditional taxonomists. The molecular data cluster the 18 orders of placental mammals into four clades, at least three of which are monophyletic, each being derived from a single ancestral species (Fig. 8.2B). Surprisingly, the grouping does not reflect the distribution of any of the characters that traditional taxonomists use for classification purposes. It does, however, seem to reflect the places of origin of the clades. According to the traditional view, placental mammals originated in the Northern Hemisphere, where the divergence of most or all of the 18 orders also occurred. The molecular data suggest, by contrast, that not only did the initial divergence of each of the four clades take place on different continents, but also that the common ancestor of all the clades originated in the Southern Hemisphere.

Three of the four clades are tentatively designated as Afrotheria (group I), Xenarthra (group II), and Laurasiatheria (group IV). The fourth clade, whose monophyly is still uncertain, does not yet have a name; it is simply referred to as group III. The first two clades diverged in the Southern Hemisphere: Afrotheria in Africa at the time it was still isolated from Eurasia, and Xenarthra in South America before it became connected to North America. Laurasiatheria and possibly also group III orders originated on the continents that originally constituted the Laurasia landmass in the Northern Hemisphere. In addition to creating four new clades that have not been recognized by traditional taxonomists, the molecular data bring species together that up until recently were assigned to distinct orders. Thus, you may have learned in your college zoology course that whales and dolphins belong to the order Cetacea, whereas pigs, peccaries, hippopotamuses, camels, deer, giraffes, and bovids (cattle, antelopes, sheep, and goats) are all members of a separate order of Artiodactyla or even-toed ungulates. Molecules, however, identify whales and dolphins as close relatives of hippopotamuses and indicate that Cetacea and Artiodactyla are in reality a single order now referred to as Cetartiodactyla. On the other hand, molecular phylogeny splits groups that have long been regarded as firmly established. The most striking case of this kind is the order Insectivora. In traditional taxonomies the order included the

A. TRADITIONAL

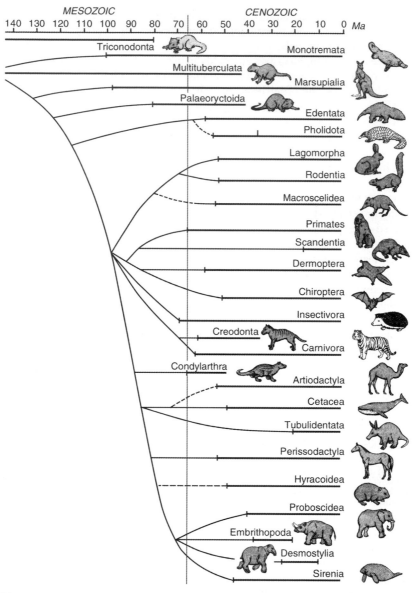

Fig. 8.2A

Two interpretations of mammalian phylogeny. **A.** The traditional interpretation is based on the comparison of visible body structures of both living and extinct mammals. In the version shown, it incorporates elements of cladistic analysis. The thick horizontal lines indicate time spans documented by fossil record; thin solid lines represent an inferred existence and splitting; dashed lines indicate ambiguous relationships. On the time scale, Mesozoic and Cenozoic are two major eras of geological history (see Chapter Nine). The Mesozoic Era was characterized by the presence of dinosaurs, flying reptiles, and ammonites; the Cenozoic Era, which extends to the present time, was marked by the rapid evolution of mammals and birds. (From Novacek 1992)

B. MOLECULAR

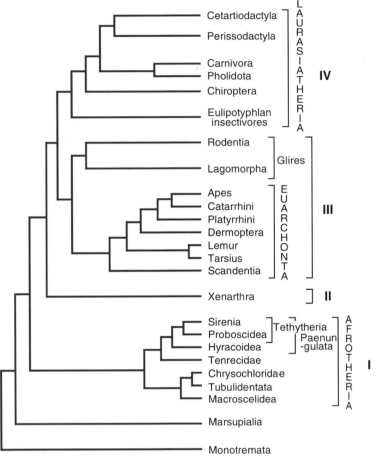

Fig. 8.2B
Two interpretations of mammalian phylogeny. **B.** The molecular interpretation is based on DNA sequences of sets of nuclear genes. Note the division of some classical orders (e.g., Insectivora) and their dispersal into different clades (eulipotyphlan insectivores in clade IV and Tenrecidae, as well as Chrysochloridae, in clade I). (Based on data of Murphy et al. 2001; Madsen et al. 2001; and others.)

families Erinaceidae (hedgehogs, moonrats), Soricidae (shrews), Talpidae (moles, shrew moles, desmans), Tenrecidae (tenrecs, otter-shrews), and Chrysochloridae (golden moles). The new molecular data, however, assign the first three of these families to the Laurasiatheria clade and the last two families to the Afrotheria clade. The morphological similarities between the two groups of families that led traditional taxonomists to lump them into a single order are apparently the result of convergent evolution.

The Afrotheria clade brings together six orders (families) which until recently were not even suspected to be related to one another and which appear to have little in common morphologically: Proboscidea (elephants), Sirenia (sea cows, manatees, dugons), Hyracoidea (hyraxes), Tubulidentata (aardvarks), Macroscelidea (elephant shrews),

Chrysochloridae (golden moles), and Tenrecidae (tenrecs). Of these, elephants, sea cows, and aardvarks had been regarded as close relatives of other hoofed mammals (ungulates), but the molecular data indicate that this similarity, too, stems from convergent evolution. The Xenarthra (Edentata) include American anteaters, sloths, and armadillos.

Group III consists of Rodentia (squirrels, flying squirrels, gophers, mice, rats, beavers, hamsters, mole-rats, voles, gerbils, porcupines, guinea pigs, capybaras, chinchillas), Lagomorpha (rabbits, hares, pikas), Primates, Dermoptera (flying lemurs), and Scandentia (tree shrews). Originally, some molecular data were interpreted as suggesting that guinea pigs are not rodents, but the new data do not support this contention. Similarly, some taxonomists have questioned whether lagomorphs are really the closest relatives of rodents; the new data indicate that they are, the two orders forming a superoder of Glires.

The Laurasiatheria include the orders Certartiodactyla and Perissodactyla mentioned earlier, as well as Carnivora (dogs, jackals, foxes, wolves, bears, raccoons, weasels, badgers, skunks, civets, mongooses, cats, hyenas), Pholidota (pangolins), Chiroptera (bats), and the eulipotyphla group of insectivores (hedgehogs, shrews, and moles).

Morphological analyses suggested a clustering of primates with tree shrews (Scandentia), flying lemurs (Dermoptera), and bats (Chiroptera) to form a group called Archonta. (Note that two of these orders – bats and flying lemurs – were originally classed as primates by Linnaeus; a third order, the tree shrews, were considered to be primates by other taxonomists.) Tree shrews are neither shrews (which are eulipotyphlan insectivores in the new classification) nor do most of them live in trees. They are small mammals from Southeast Asia resembling squirrels with either a bushy tail or a tail tipped with feather-like arrangements of hair. (Their alternative name, tupaias, is derived from the Malayan word for "little fast animals of the trees".) Similarly, the two species of flying lemurs are neither lemurs, nor do they fly. They are tree-dwelling Southeast Asian mammals with skin flaps stretched between their limbs used for tree-to-tree gliding. (The scientific name of the order, Dermoptera, is a combination of the Greek *derma*, skin and *pteron*, wing.) Bats are the only mammals that can fly by flapping their wings derived from the forelimbs, with the hind legs included in the wing membranes. (The designation Chiroptera is derived from the Greek *cheir*, hand and *pteron*, wing.) Close relationship of primates to tree shrews and flying lemurs is strongly supported by molecular data (Fig. 8.2B), the three orders, together with rodents and lagomorphs constituting group III of the most recent classification system. The details of this relationship are, however, uncertain. Molecular data do not support the notion of a close relationship between primates and bats. Phylogenies based on multiple nuclear and mitochondrial genes place the two orders into different clades – primates in group III and bats in Laurasiatheria. They also fail to support the claim that one group of bats, Megachiroptera (fruit bats and flying foxes), are in reality not bats at all but flying primates. The new data indicate that bats – Microchiroptera and Megachiroptera – are monophyletic (Fig. 8.2B).

There are 4629 living mammalian species: three monotreme, 270 marsupial, and 4356 placental species. The order Primates comprises 233 species* and ranks fifth in the total number of species after rodents, bats, eulipotyphlan insectivores, and carnivores with 2015, 925, 385, and 271 species, respectively. It is distinguished from the other placental orders by certain evo-

---

* This is the number given in Wilson & Reeder (1993). Other authorities give different numbers which range from 185 to 252. The discrepancies result from a lack of agreement as to whether certain forms should be regarded as species or subspecies. The numbers given here for the other orders are also based on Wilson and Reeder.

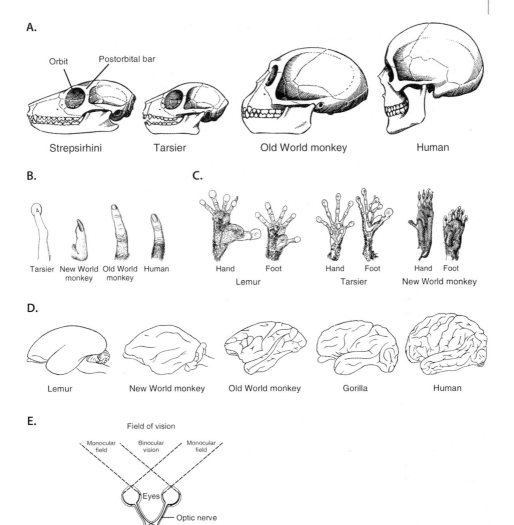

A.

Orbit Postorbital bar

Strepsirhini Tarsier Old World monkey Human

B.

Tarsier New World Old World Human
monkey monkey

C.

Hand Foot Hand Foot Hand Foot
Lemur Tarsier New World monkey

D.

Lemur New World monkey Old World monkey Gorilla Human

E.

Field of vision

Monocular Binocular Monocular
field vision field

Eyes

Optic nerve

Fig. 8.3A-E
Trends in the evolution of primates. **A.** Shortening of the snout and the development of the orbital bar as part of the closing of the eye socket (orbit). **B.** Transformation of claws into nails. Tarsiers' nails are small flakes; in New World monkeys they are narrow and curved; in Old World monkeys and apes they tend to be flat. For strepsirhine nails see Figure 8.7. **C.** Grasping hands and feet and opposable thumbs. **D.** Enlargement of the brain and growth in the complexity of its surface layers. **E.** Binocular vision. (**A** from Portman 1948; **B** and **E** from Napier & Napier 1985; **C** from Vallois 1955; **D** from Ankel 1970.)

lutionary trends (Fig. 8.3). Primates have gradually become less dependent on their sense of smell (olfaction) and have reduced their snout length, shortened their skull, and flattened their face accordingly. Concurrently, they have become increasingly dependent on three-dimensional (stereoscopic) vision as an adaptation to active life in tree canopies, where judging distances and depth is essential for survival. This has led to the development of forward-projecting eyes ("binocular" vision) and the formation of the postorbital bar (the bony strut at the side of the eye socket), the closure of the bony ring encircling the eye, and the formation of the postorbital plate behind the eyes. Little-by-little the grasping and manipulative capabilities of hands and feet have been extended by improvements to the mobility of their digits, the development of an opposable thumb (i.e., a thumb that can rotate along its long axis so that its tip can touch the other finger tips), and the replacement of claws by flat nails on all or nearly all digits. And bit-by-bit, the brain size relative to body size has increased, accompanied by an enlargement of its "higher" centers, notably certain areas of the neocortex – the upper layers of gray matter in the brain hemispheres. According to the stage to which they have progressed in these trends, the living primate species are classified into six groups: lemurs, lorises, tarsiers, New World monkeys, Old World monkeys, and apes (Fig. 8.4).

Lemurs are mouse- to cat-sized primates with fluffy fur, fox-like snouts (Fig. 8.5A), and long furry tails. Most live in trees much of the time, visible from the ground only as shadows behind the dense foliage – hence their name from the Latin word for ghosts, an expression used by ancient Romans to designate household gods and spirits. They are restricted in their distribution to Madagascar off the southeast African coast and include true lemurs, indris, avahi, sifakas, mouse and dwarf lemurs, and the bizarre aye-aye.

Lorises and the closely related *galagos* or bush babies are nocturnal, arboreal primates of tropical Africa and southern Asia. The tailless lorises derive their name from the Dutch word *loeres* or "slow-witted", applied in reference to their slow movements. There is, however, no evidence that the loris is less intelligent than many other mammals and its limited mobility is an adaptation to stalking insects and small birds. The designation "galago" probably stems from the Wolof word for "monkey". Galagos have large eyes, long ears, long tails, and long hind legs used for leaping from branch to branch in the treetops.

Tarsiers are rat-sized nocturnal animals with round heads which they can rotate like owls to look backwards over their shoulders. They have large, forward-directed eyes, big ears, long hind legs adapted to frog-like leaping, and long, nearly hairless tails tipped with a tuft. Their legs have been lengthened by the elongation of their tarsal (ankle) bones and it is this character that has given tarsiers their name. The five tarsier species live in the tropical rainforests of the Malayan Archipelago.

New World monkeys or *Platyrrhini* are, as the name implies, primates of the tropical rainforests in Central and South America. They have a short, broad nose with widely spaced nostrils (Greek *platyrrhis*, broad-nosed; Fig. 8.5C), long, raking limbs, and a prehensile tail (adapted to grasping and wrapping around objects) by which they can hang from branches and which they use as a fifth limb. They live exclusively in treetops and feed on leaves and fruits. New World monkeys include marmosets, tamarins, as well as howler, capuchin, squirrel, spider, and woolly monkeys.

Old World monkeys and *apes*, or *Catarrhini* were once distributed across Europe and Asia, but their present range now extends through the African continent, Gibraltar, the Himalayas, northern Japan, and Southeast Asia. They differ from New World monkeys in that their nostrils are close together (Greek *catyrrhis*, narrow-nosed; Fig. 8.5D), facing downward and outward, their thumbs are fully opposable, their tails are not prehensile, their buttocks have pad-like cushions that facilitate sitting on branches or on the ground, and their shoulder joints are "locked" in position hindering them from rotating their arms and consequently not allowing them to hang or swing. Some Old World monkeys are tree-dwellers, others prefer to live on the ground, and others still exploit both types of habitat; they are mainly quadrupeds who run on all four legs even on beamy branches. They include macaques, baboons, guenons, mangabeys, langurs, drills, mandrills, and colobus monkeys.

Apes are tailless and have flexible shoulders. To a biologist, the term "ape" signifies gibbons and siamang (the *lesser apes*), orangutan, gorilla, and chimpanzee (the *great apes*). A

DERMOPTERA

Flying lemur
(*Cynocephalus volans*)

SCANDENTIA

Lesser tree shrew
(*Tupaia minor*)

PRIMATES - STREPSIRHINI

Ring-tailed lemur
(*Lemur catta*)

Loris
(*Loris tardigradus*)

PRIMATES - HAPLORHINI

PLATYRRHINI - NEW WORLD MONKEYS

Spectral tarsier
(*Tarsius spectrum*)

Santarem marmoset
(*Callithrix humeralifer*)

Red uakari
(*Cacajao rubicundus*)

White-throated capuchin
(*Cebus capucinus*)

CATARRHINI - OLD WORLD MONKEYS

Rhesus macaque
(*Macaca mulatta*)

Baboon
(*Papio*)

CATARRHINI - APES (Asian)

Gibbon
(*Hylobates lar*)

Orangutan
(*Pongo pygmaeus*)

CATARRHINI - APES (African)

Gorilla
(*Gorilla gorilla*)

Common chimpanzee
(*Pan troglodytes*)

Pygmy chimpanzee
(bonobo)
(*Pan paniscus*)

Human
(*Homo sapiens*)

Fig. 8.4
Portrait gallery of primates and their closest relatives. (From Vallois 1955; Le Gros Clark 1959; and Mazák 1977.)

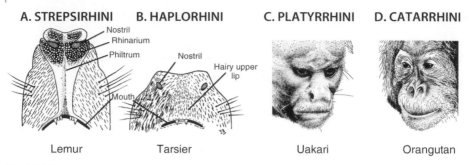

Fig. 8.5A-D
Primate noses. (**A** and **B** modified from Passingham 1982; **C** and **D** modified from Kent & Miller 1997.)

layperson may often use the terms "monkey" and "ape" indiscriminately. Gibbon, siamang, and orangutan are Asian apes; gorilla and chimpanzee are African apes. Gibbons, siamangs and orangutans live on high trees, but can also walk on the ground. Gibbons use alternate arms to swing along a branch or from branch to branch (= *brachiation*). Gorillas and chimpanzees live mostly on the ground, walking as quadrupeds by supporting their bodies with the folded fingers of their hands – a form of locomotion referred to as *knuckle-walking*, even though it does not actually involve knuckles as such. There are 11 species of gibbon, one species each of orangutan and gorilla, and two species of chimpanzee (the common chimpanzee, *Pan troglodytes* and the pygmy chimpanzee or bonobo, *Pan paniscus*).

Clash of Philosophies over Noses

The division of the primates into these six groups is largely uncontested with one important exception: the position of the human species in this scheme. We return to this particular point later; here we address the controversial issue of arranging the six groups into higher categories. There are two major views on this subject, which each reflect different philosophies of biological classification. It is generally agreed that primates should be divided into two higher categories, with lemurs and lorises in one category and New World monkeys, Old World monkeys, and apes in the other. The bone of contention is the tarsier, which assumes a position between the two categories. Taxonomists of the traditional school cluster the five tarsier species with lemurs and lorises into the Prosimii (literally "before monkeys"). And they group the New World monkeys, Old World monkeys, and apes into the Simii, the monkeys, or Anthropoidea (Greek *anthropos*, human being, *eidos*, resembling; Fig. 8.6A). Taxonomists of the cladistic school, on the other hand, insist on clustering lemurs and lorises into one category, the Strepsirhini (Greek *strepsis*, a twisting, turning; *rhis*, nose) and tarsiers, New World monkeys, Old World monkeys, and apes into another category, the Haplorhini (Greek *haploos*, single or simple; Fig. 8.6B). The categories erected by the traditional taxonomists represent *grades*, groups of organisms distinguished by the level of their evolutionary advance, whereas those laid out by the cladists are *clades*. What, then, is the difference between grades and clades? To explain, we must digress somewhat from our main theme.

First let us remind ourselves of the distinction between two types of characters, primitive and shared derived, and of the different types of ancestors. A group of monophyletical-

A. TRADITIONAL

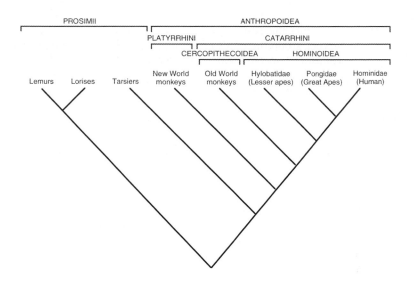

B. CLADISTIC

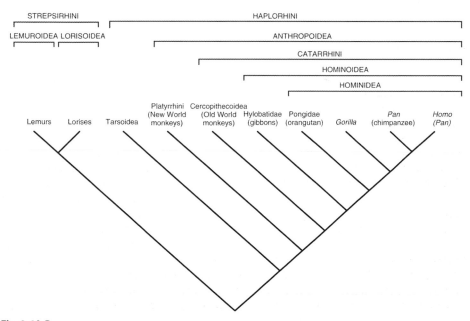

Fig. 8.6A,B
Traditional (**A**) and cladistic (**B**) classification of living primates.

ly related species (or more generally taxa) has one direct (immediate, most recent) common ancestor and numerous indirect (more distant, earlier) ancestors. A character common to members of a group and their direct ancestor, but absent in the indirect ancestors is a *shared derived character* that defines the assembly as a *monophyletic group* or *clade*. A character shared by the members of a group, their direct common ancestor, as well as some of the indirect ancestors is a *primitive character*. Cladists ignore primitive characters and use shared derived characters to resolve the order of branch splitting. Thus, if a character is shared by all primates, but is also found in, for example, the dog and other carnivores, then it obviously cannot be used to define the point in phylogeny at which the primate branch split off from the mammalian stem. Since, however, the definition of shared derived characters is always relative, a primitive character in the context of one group can be a shared derived character of another, larger group. In our example, it may be suited to identifying the point at which the branch bearing both the carnivores and the primates split off from the stem.

Traditional taxonomists have few scruples about the use of primitive characters. On the contrary, they exploit suites of primitive characters to differentiate between grades in their classifications. If they find that a group of organisms shares a high proportion of primitive characters compared to another group in which the primitive characters are replaced by more "advanced" ones, then the former group is assigned a lower and the latter a higher grade. The appearance of the suite of novel characters heralding the emergence of a new, higher grade is referred to as a *grade shift*.

Cladists point out that several characters cluster lemurs and lorises together with the exclusion of other primates, including tarsiers. The most conspicuous among the shared derived characters are the toilet claw and the tooth comb. The former is a spike-like spur on the second digit of the foot (all other digits having flat nails) that is used for grooming and scratching in a dog-like fashion (Fig. 8.7A). The tooth comb is a set of four or six elongated and compacted teeth growing forward rather than upward from the lower jaw (Fig. 8.7B). It forms a projecting "ladle" with which the animal combs its fur and scoops out soft fruit. Along with several other characters, these two traits therefore unite lemurs and lorises in the strepsirhine clade. The haplorhines are similarly united by a different set of shared derived characters which include a dry nose with reduced whiskers, fused upper lip (Fig. 8.5B), and retina with fovea* but without tapetum lucidum*. The nostrils of Strepsirhini, like those of dogs and many other mammals, are surrounded by an area of moist, hairless skin, the rhinarium (Fig. 8.5A). Haplorhines lack the rhinarium; instead their nostrils are set in dry, hairy skin – they are hair-nosed rather than wet-nosed primates (Fig. 8.5B). The rhinarium and the upper lip of the strepsirhines are split into two halves by a cleft and their upper lip is, in addition, tethered to the gums by the fold of a mucous membrane. Because it is fastened, the lip is restricted in its range of movement and as a consequence, the repertoire of facial expressions is limited. In the haplorhines, the cleft has largely disappeared and the inner tether of the upper lip is greatly reduced; you can convince yourself of this by flipping your lip over in front of a mirror. The haplorhine potential for facial expressions is therefore greatly enhanced, which is an important factor in social interactions. In contrast to the Haplorhini, the strepsirhine retina lacks the fovea but possesses the tapetum lucidum.

All these and several other characters are shared by tarsiers and anthropoids, but are absent in lemurs and lorises. This distribution of the shared derived characters indicates to

---

* The haplorhine retina, the light-sensitive layer, has a yellow spot or *macula* marked by a depression, the *fovea*, which is an area of extremely detailed vision. It lacks however the light-reflecting layer, *tapetum lucidum*, which causes eyes to shine at night when illuminated by a beam of light.

A.

B.

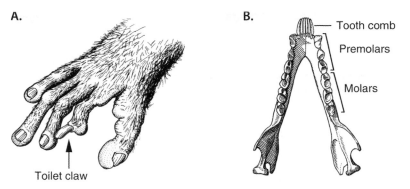

Tooth comb

Premolars

Molars

Toilet claw

Fig. 8.7A,B
Toilet claw (**A**) and tooth comb (**B**) – two shared-derived characters of lemurs and lorises (Strepsirhini). (**A** from Napier & Napier 1985; **B** from Vallois 1955.)

the cladists that in the phylogeny of living primates, the first split occurred between the ancestor of the Strepsirhini and the ancestor of the Haplorhini. The former then split into the lemurs and lorises, whereas the latter diverged into the tarsiers and the ancestor of all the anthropoids in the manner depicted in Figure 8.5B. The classification of the primates must therefore, in the cladists' view, reflect this phylogeny and the principal division of this order must be made between Strepsirhini and Haplorhini rather than between Prosimii and Anthropoidea.

Initially, traditional taxonomists attempted to argue that the first split in fact took place between the ancestor of the group that included lemurs, lorises, and tarsiers, and the ancestor of all the remaining living primates. You can still find this order of splitting presented in obsolete zoology textbooks. The growing evidence, in particular the molecular evidence described in the next section, has, however, rendered this interpretation untenable. Most traditional taxonomists now accept the cladistic phylogeny, but not the classification. They argue that there is a large gulf between lemurs, lorises, and tarsiers on the one hand and the rest of the living primates on the other, and that this fact should be reflected in the division of primates into prosimians and anthropoids. The prosimian group generated by the traditional taxonomists is thus not monophyletic; it is not a clade because its members did not all originate from the same direct common ancestor. Rather, the prosimian group is *paraphyletic*; it combines all the species derived from one ancestor (the lemurs and lorises) with some species (the tarsiers) that are derived from a different ancestor, while placing the remainder of the species descending from the latter ancestor into another group, the anthropoids.

From the cladistic (phylogenetic) point of view, this classification is illogical, but traditional taxonomists do not heed this criticism. Instead, they fall back on the concept of grades to justify the division into prosimians and anthropoids. In essence they say: Look how much the anthropoids differ from the prosimians. They have much larger brains, more closely set eyes, and shorter facial bones that make their faces flatter. They carry their heads in a more erect fashion, their eyes are forwardly directed, and their nasal cavity, as well as the paper-thin scrolls inside it, are much reduced. Their lower jaw is hinged to the braincase above the rows of teeth, and the two halves of the lower jaw, which in prosimians are connected by a joint, are fused in anthropoids, as are the two frontal (forehead) bones. And anthropoids display much more complex sexual and social interactions than prosimians, have a longer period of intrauterine development, as well as longer phases of juvenile and adolescent dependency. All these and several other features constitute an entirely new grade in morphological and behavioral organization. During the transition from tarsiers to anthropoids, a large number of primitive characters linking prosimians to nonprimate mammals were lost to be replaced by novel features which were retained during the evolution of anthropoids all the way to our own species. For traditional taxonomists, the leap from tarsiers to anthropoids is just too great to be ignored in a classification system.

Cladists respond to these arguments by pointing out that the grade concept is nebulous, subjective, and anthropocentric. In other words, it possesses all the characteristics that brought about the downfall of pre-Hennigian taxonomy. How do you distinguish between a gigantic leap that leads to a new grade and a small jump that does not? Where do you draw the line between "lower" and "higher" organizations? How do you resolve a dispute between two taxonomists who have placed the border between two grades in different regions of a phylogenetic tree? Is it not true, particularly in the case of primate classification, that drawing lines between grades is influenced by reference to the morphology and behavior of our own species? And which authority decides when a classification is to be based on monophyletic groups and when on paraphyletic grades? Traditional taxonomists are unable to provide satisfactory answers to these questions. Nevertheless, traditional classifications continue to be used and the grade concept remains popular, especially among anthropologists and paleontologists.

The Deposition of Blood Globules

Of all the mammalian orders, primates are the most extensively studied group, not only at the morphological, but also at the molecular level. Most of the research has, however, concentrated on the human species, while the interest in other primate taxa diminishes with their phylogenetic distance from *Homo sapiens*. As a result, molecular information, although substantial, is unevenly distributed among primate taxa and unsystematic. The only exceptions are two sequencing projects still underway, one by the team led by Morris Goodman and the other by Colm O'hUigin and his coworkers. The two projects have different scopes and goals. Goodman and colleagues are focused on a single genetic system, the globin genes, but comprehensively cover the entire order. They sequenced one locus in 43 primates and a much longer globin region in representative species of the major primate groups. The goal of the project is primarily the reconstruction of the primate phylogeny. O'hUigin and colleagues, on the other hand, aim their effort at expanding the current database to produce sequences of ~50 different genes in representative species of the main primate groups. Their goal is to gather information about the nature of the evolutionary process in primate phylogeny. Here, we describe some of the results obtained by Goodman and his colleagues; data produced by O'hUigin and coworkers are discussed later in this chapter.

Mammals carry multiple *globin* genes, all derived from a single ancestor by repeated duplication (see Chapter Four). Humans have two clusters of globin genes, the α (alpha) cluster on chromosome 16 and the β (beta) cluster on chromosome 11 (Fig. 8.8). The α-globin cluster contains three genes: α1, α2, and ζ2 (zeta 2), as well as four inactive pseudogenes ζ1 (zeta 1), ψα1, ψα2, θ (theta; here the Greek letter ψ, pronounced psi, indicates a pseudogene). The ζ2 gene is expressed in embryonic and fetal red blood cells, the α1 and α2 genes in erythrocytes of an adult. The β-globin cluster is comprised of five genes (ε, γ1, γ2, δ, β) and one pseudogene (η). The ε (epsilon) gene is expressed in embryonic, the γ1 and γ2 (gamma) genes in fetal, and the δ (delta) and β genes in adult red blood cells. Each

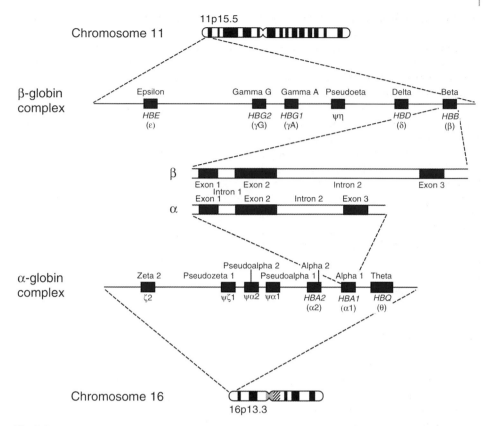

Fig. 8.8
Organization of human α- and β-globin gene complexes and of the α and β genes. Both colloquial (Greek letters) and official (capital Latin letters) gene designations are given. The stripes on the chromosomes are the characteristic banding patterns visible after the application of special stains; designations 11p11.5 and 16p13.3 refer to the bands in which the genes are located.

hemoglobin molecule has two identical chains specified by the α-cluster and two identical chains encoded in genes of the β-cluster. Most of the work carried out by Goodman and colleagues has focused on the β-cluster, which in humans encompasses ~60,000 nucleotide pairs. The researchers elucidated the organization of the β-cluster in representative species of the main primate groups and sequenced the genes from a range of species covering the entire order. Both the organization of the cluster and the sequences are informative in regard to primate phylogeny.

The first duplication of the ancestral gene in early mammals yielded two copies, one of which became a proto-ε gene expressed in embryonic red blood cells and the other a proto-β gene expressed in adult red blood cells (Fig. 8.9). In early placental mammals, a series of additional duplications followed which gave rise to a cluster of five genes, three (ε, γ, ζ) derived from the proto-ε and two (δ, β) from the proto-β genes. This was to be the cluster that the emerging primate order inherited from the ancestor that they shared with some of the other placental orders. Soon after their separation from these orders, the early primates inactivated their η gene, but held on to it as a pseudogene ψη. In lemurs, moreover, the η

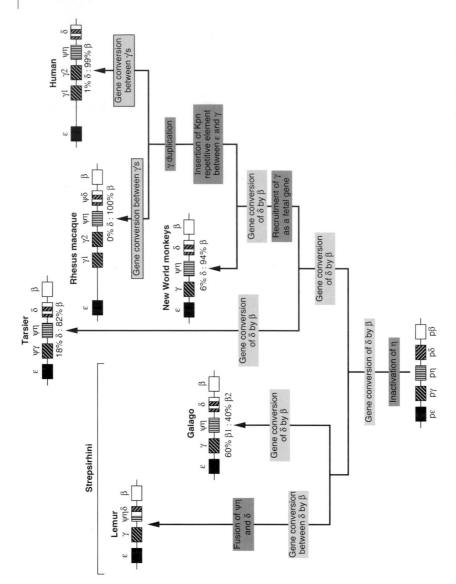

Fig. 8.9

Evolution of the β-globin gene cluster in primates. Genes are identified by Greek letter designations. Designations ρε, ργ, etc indicate postulated ancestral proto-genes; percentages indicate the proportions of the chains in adult blood. (Slightly modified from Koop et al. 1989.)

pseudogene fused with δ and thereby inactivated that gene, too. In the anthropoids, changes in the regulatory region of the γ gene delayed its expression from the embryonic to the fetal stage. A duplication of the γ gene in the stem lineage of the Catarrhini then resulted in the γ1 and γ2 genes now found in all Old World monkeys and apes, including humans.

Sequence comparisons of globin genes at the different loci have revealed that in addition to point mutations, the genes also diverge by another process. To explain, let us consider the δ and β genes. Since they both arose from the proto-β gene, one would expect them to be identical at sites that have not changed since the duplication event and to differ at sites that mutated. Since mutations are random events, one would also expect to find sites with identical and divergent nucleotides intermingled over the entire sequence. Curiously, in some primates one finds instead the δ gene divided into four regions, in two of which the δ and β genes are nearly identical and two in which both identical and divergent sites are intermingled as expected. Goodman and his colleagues explain this finding by invoking a special form of recombination called *gene conversion*.

A double-stranded break that initiates recombination can have two different outcomes. The more familiar one is *reciprocal recombination*, in which the entire segment extending from the breakpoint to the end of the DNA molecule (or to the next breakpoint) is exchanged between the two aligned duplexes. Alternatively, gene conversion can take place in which a short segment of the donor DNA molecule is introduced into the recipient molecule without the former receiving anything in return from the latter (see Chapter Four). Because of the slow rate of nucleotide substitution, the converted stretches can be identified in the sequences as abrupt changes in the degree of sequence similarity millions of

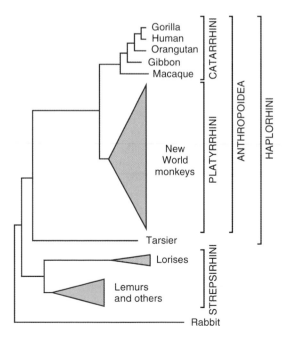

Fig. 8.10
Phylogeny of primates revealed by *HBE1* (ε) gene sequences. The figure is based on a neighbor-joining tree of sequences derived from 34 primate species and an outgroup (rabbit globin sequence) reported by Porter and colleagues (1995).

years after the event. They therefore mark the line of descent extending from the individual in which the gene conversion occurred. Goodman and his colleagues identified several presumptive gene conversions between various primate donor-recipient globin gene pairs and used their distribution among the primate groups to support the phylogeny reconstructed from the sequences of these genes (Fig. 8.9).

The most comprehensive sequence file has been compiled for the region taken up by the ε-globin locus. The region is ~1700 nucleotide pairs long and encompasses both amino acid-specifying stretches (exons) and noncoding stretches between them (introns). Fewer sequences, but still representative of all the major primate groups, are available for a longer segment of ~4700 nucleotide pairs, which includes regions between genes. Maximum parsimony trees based on these sequences (Fig. 8.10) strongly support the division of living primates into the wet-nosed Strepsirhini and dry-nosed Haplorhini; the position of tarsiers as a sister group of the Anthropoidea; the division of the latter into the broad-nosed Platyrrhini and narrow-nosed Catarrhini; and the division of the Catarrhini into two families, the Old World monkeys (Cercopithecidae) and the hominids (Hominidae). The much more limited sequence data set obtained from an assortment of other genes scattered over different chromosomes is in agreement with the phylogeny reconstructed from the ε-globin region sequences. Together, these phylogenies belie the primate classification favored by traditional taxonomists.

The First among the First?

Traditional and molecular taxonomists disagree, however, not only over the position of the tarsiers in the primate phylogeny; they also clash over a much more emotional issue, the position of the human species in the classification system. In the traditional classification, humans are strictly segregated from the apes into their own family of Hominidae that contains a single genus, *Homo*. The apes are divided into the lesser apes, Hylobatidae, encompassing the gibbons and the siamang, and the great apes, Pongidae, embracing the orangutan, the gorilla, and the two chimpanzee species. There are several variants of this scheme, but the principal divisions are retained by all of them. In true Linnean tradition, they all class the human genus *Homo* as "first" among the "first", the primates.

By contrast, both cladistic analyses and molecular phylogenies clearly separate the orangutan from the African great apes (gorilla and chimpanzee), as well as from the human species, but have great difficulty in resolving the order in which the ancestors of the gorillas, chimpanzees, and humans diverged from one another (we return to this problem later). Trees based on the globin sequences and sequences of numerous other genes all show the ancestor of the gibbons and the siamangs separating first from the stem, the orangutan diverging next, and finally the gorilla, the chimpanzee, and the human lineages separating from one another almost simultaneously (Fig. 8.10). Based on molecular phylogeny, Morris Goodman proposes the following classification of the family Hominidae:

Family **Hominidae**
    Subfamily **Homininae**
        Tribe **Hylobatini**
            Subtribe **Hylobatina**
                Symphalangus: siamang
                Hylobates: gibbon

Tribe **Hominini**
    Subtribe **Pongina**, *Pongo*: orangutan
    Subtribe **Hominina**
        Homo (Gorilla): gorilla
        Homo (Homo): human
        Homo (Pan): common chimpanzee, bonobo

In this classification, people are indeed (bloody) apes!

By placing humans into a separate family (or any other higher-ranking category), the traditional taxonomists are once again committing the sin of paraphyly. They break up a clearly monophyletic group comprised of gorillas, chimpanzees, and humans, and then lump gorillas and chimpanzees together with orangutans and segregate the human species from this artificial group. The combination orangutan, gorilla, and chimpanzee is a paraphyletic group because it excludes humans, even though they share a common ancestor with the three apes. In their attempt to justify this grouping, the traditionalists fall back on the same reasoning they use to legitimize uniting the tarsiers with the Strepsirhini into the prosimians – the loss of a large number of primitive characters during the transition from apes to humans.

While you may not care about the position of the tarsier among the primates, the chances are that you have a strong opinion about the position of humans that compels you to side with the traditional taxonomists. "Come on," you might say, "is it not obvious that there is a great gulf between us and the apes? Not only are we "naked" (hairless) and walk erect on two legs (bipedal), more importantly we speak, write novels, compose symphonies, develop relativity and quantum theories, and send probes to Mars; you name it, we do it. Not only is an ape unable to do any of these things, it cannot even be taught to do them!" Taxonomists would be able to add numerous other characters to your list of differences including a larger brain relative to body size, more complex brain, in particular its neocortex, a small face, short canine teeth, several skeletal adaptations to bipedalism, and hands with remarkable manipulative capacity (think of Arthur Rubinstein playing the piano).

To what extent are your argumentations valid? They are based on two suppositions. First, that genetic distance can be used to classify organisms into species, genera, families, and the rest of the taxonomical categories. And second, that individual character states have different weights, that some are more important than others for classification purposes. Specifically, you place a far greater weight on the characters that distinguish humans from the chimpanzee and the gorilla than on those that differentiate the chimpanzee from the gorilla. And you do so because you regard the step taken by the ancestor toward the human species as a much greater *advance* than that taken toward the chimpanzee and the gorilla.

Your argument, however, is highly subjective and brazenly anthropocentric: it is subjective because you would be hard-pressed to specify the yardsticks by which you measure distances and weigh characters at the organismic level if asked to do so, and anthropocentric because you gauge the importance of character states from the standpoint that the human species is the most important element in the universe. Neither of the two suppositions can be justified rationally. In fact, it was precisely the widespread reliance on these suppositions that provoked the cladistic revolt.

Cladists avoid subjectivism and anthropocentrism by abandoning the use of genetic distance for the purpose of defining relationships among organisms altogether. They delineate interorganismic alliances purely on the basis of shared derived characters, whereby one such character is sufficient to establish the existence of a clade (monophyletic group). If more than one shared derived character supports a given clade, so much the better, for this increases the confidence in the alliance of the taxa. Only when there is conflict between characters (some support and others contradict the existence of a particular clade) do cladists resort to quantitation: if the majority of the shared derived characters supports a given relationship, it is considered genuine and the contraindicatory characters are explained away in terms of "false guides", usually as the result of homoplasy. Cladists gauge the relationship between two organisms by the number of bifurcating branch splits separating them on a cladogram. Most closely related are those taxa that are separated by a single split; the more splits separating two taxa, the less related they are considered to be to each other. Gauging is therefore as objective as it can get at the organismic level and concerns about the position of the human species in the system do not influence it.

Like cladistic analysis at the organismic level, analysis at the molecular level steers clear both of subjectivity and anthropocentrism, but it does so by different means. As we learned in Chapter Five, molecular phylogenetic analysis is carried out in two different ways; the molecular data are either converted into genetic distances or – in the character state analysis – they are not. In the former, objectivity is achieved by using a standard, quantifiable distance – either a nucleotide substitution in a nucleic acid or an amino acid replacement in a protein. The unit is the same for all the organisms, be they bacteria, flies, or human beings, for all organisms evolve and undergo substitutions. It can be counted, added up, averaged, compared, and used as a yardstick for measuring distances between closely, as well as distantly related taxa. Nothing of this sort is available at the organismic level, at which there is no standard unit of distance, no common yardstick, and direct comparison between taxa is both ambiguous and difficult, especially between very diverse organisms.

Character state analysis has much in common with cladistic analysis. The phylogenetically informative substitutions used in the former are in essence shared derived characters and the maximum parsimony or maximum likelihood methods of drawing phylogenetic trees correspond to methods of cladogram assembly. Hence most of what we said about the objectivity of the cladistic method at the organismic level also applies to character state analysis methods at the molecular level.

Cladograms and phylogenetic trees, whether produced by distance or character state analyses, are, however, descriptions of genealogical relationships among taxa; they are not, in themselves, depictions of taxonomical classifications. Cladists recognize only one classification category – the clade – so that their classification is nothing more than a hierarchical serialization of clades (see Fig. 7.1). In a cladogram, a higher order clade includes a lower order clade, which itself contains a clade of a lower order still, and so on, whereby the next clade is always separated from the preceding one by a single branch split. In a phylogenetic tree, in contrast to a cladogram, the node at which a single branch splits into two symbolizes an ancestor. In cladograms, the entire bifurcation symbolizes a monophyletic relationship and nothing else, certainly not the existence of a common ancestor at the split point: instead of originating from a common ancestor, taxa A and B that are grouped together by a shared derived character may have originated, for example, one from the other. Theoretically, therefore, a classification can be based on the nodes of a tree, if a mode is found of relating them to the categories of the Linnean system. Commonly, this mode is actually built into the procedure of naming species. When a new species is described, it is placed, if only tentatively, into the Linnean hierarchy of categories. When molecular phylogenetic analyses are carried out, the classically defined categories are imposed artificially on the tree. If there is a clash between the imposed grouping and that indicated by the tree, further analyses are performed to resolve the discrepancies. Otherwise, the imposed ranking is accepted, although in some cases, objections are occasionally raised by some investigators. Dissatisfaction with the imposition of the traditional classification on the molecular phylogenetic trees of the primates, as expressed by Goodman and others, is just one example of the tension between the traditional and cladistic/molecular interpretations of relationships among organisms.

More recently, however, the dissenting voices have grown louder; more and more often the opinion is expressed that the Linnean categories above the species level should either be abolished entirely or at least be redefined on the basis of precisely specified, standardized criteria. One redefinition, proposed by Willi Hennig more than 30 years ago and resurrected recently, is to base all ranks except the species on the divergence time of the taxa. Species would then be defined as before by biological criteria, including reproductive isolation (absence of gene exchange between populations for reasons other than lack of contact). All other categories would be specified by a time scale superimposed on the phylogenetic tree. For example, species that diverged from their common ancestor within a time frame of 2–5 my ago would be assigned to distinct genera; genera that separated from a common ancestor 35–36 my ago would be classified as belonging to different families; and so on. An application of this proposal to the primate phylogeny places humans, chimpanzees, and gorillas into a single genus, as in Goodman's classification; all great apes including humans are placed in a single tribe; Old World monkeys and apes into one family; Catarrhini and Platyrrhini into a single superfamily, and so forth.

There are two main reasons why biologists have not exactly bent over backwards to embrace any such proposal. First, biologists, and not only biologists, tend toward inertia and abhorrence of change once things become established. They apparently feel that inadequate though the present classification system may be, it is not bad enough to justify renaming and reclassifying half of the known taxa. Second, divergence time estimates are, as we shall learn in the next chapter, notoriously unreliable. Renaming taxa every time a new estimate has shifted them into a different time window would be a nightmare. Ultimately, however, something will have to be done to fit the present system into the changing landscape of biological sciences. As the molecular data grow and the divergence estimates improve, it might indeed become feasible to realize Hennig's dream. Reclassification of primates and the reassignment of our own species to its proper place in nature might be a logical starting point for the revision.

Appearance is not Everything

How large really is the difference between *Homo sapiens* on the one hand and chimpanzees or gorillas on the other? "Huge", say the traditional taxonomists, who base their assessment on the appearance of organisms – their morphology and behavior. "Very small", counter molecular biologists, who measure distinctiveness in numbers of nucleotide substitutions or amino acid replacements. Before attempting to reconcile these two very different assessments, let us be a bit more specific about them. Morphological and behavioral differences do not submit themselves easily to quantitation; all we can do here is to refer the reader to the partial list in Table 8.1. Molecular differences, by contrast, can be quantitated and compared quite well.

In the past two-to-three decades, numerous estimates of the molecular difference between humans and chimpanzees (gorillas) have been attempted by a variety of methods as soon as these became available. Among the oldest were estimates based on the electrophoretic or immunological comparison of proteins and comparisons of the reassociation ability of DNA molecules. All three estimates are indirect. The first is based on the change of a protein's mobility in an electric field when the charge of one of its constituent amino acid residues is altered by a mutation. The second relies on a mutational-

Table 8.1
Selected character states distinguishing humans from other primates.

- Hair cover of the body is extremely reduced.
- Hair distribution in males and females is different.
- Skull is balanced upright on the vertebral column (in other primates it meets the column at an angle).
- Humans have chins, other primates do not.
- Human spine is S-shaped, that of other primates is relatively straight.
- The length of the spine is less than 120 percent that of the body.
- Ilium (one of the bones of the pelvis) is broader than high.
- Female breasts are permanently enlarged starting from the onset of sexual maturity.
- If not in the state of sexual arousal, penis is suspended.
- Erect penis is more than 10 cm long.
- Penis contains no bone.
- Sperm density is very low (60 million per milliliter).
- Heel bone has elongated prominence.
- Gestation period is lengthened (280 days).
- The left upper chamber of the heart (atrium) is higher than the right chamber.
- Thumb is elongated.
- Ear breadth is only 50 percent of its length.
- Canines are small and are not differentiated between sexes.
- Canines erupt before premolars.

Based on Groves 1986.

ly-induced loss of the protein's reactivity with a specific antibody. In both instances, a large number of proteins controlled by different loci are compared and from the proportion of altered and unaltered proteins the percentage of identical proteins is calculated. In the third method, solutions of DNA isolated from humans and chimpanzee (gorilla) are heated to break the hydrogen bonds in the double helix and to separate the two complementary strands (a phenomenon called *heat denaturation*). The solutions from the different species are then mixed and slowly cooled to allow all the complementary strands to reestablish the bonds between them and thus reform the duplexes. In addition to *homoduplexes* formed between complementary strands of DNA from the same species, the formation of *heteroduplexes* also occurs, in which a chimpanzee DNA strand anneals with a complementary human strand. The greater the difference in the sequences of the strands from the two species, the fewer hydrogen bonds form between them and the less stable are the resulting heteroduplexes. The stability of the heteroduplexes can be measured by heating the solution again and determining the temperature at which half of them dissociate back into single strands. By comparing the observed with empirically obtained standard values, the percentage of unpaired (mismatched) nucleotides between the two DNAs can be estimated.

These older techniques have largely been superseded by newer ones in which the DNA or protein changes are determined directly by sequencing. The human genome has now been sequenced and there are plans to sequence the chimpanzee genome as well. Thus far, however, only sequences of a few tracts of the chimpanzee genome are available, but they are sufficient to obtain a reasonably accurate estimate of its average difference from the human genome. All the estimates obtained both by the older and the newer techniques have yielded very similar figures: the genomes of these two species differ at approximately 1–2 percent of their nucleotide sites. In one study based on the comparison of 51 human and chimpanzee nuclear genes (62,530 nucleotide pairs) and six intergenic regions, O'hUigin and coworkers found 0.83 percent differences in the protein-specifying parts (1.09 percent for synonymous and 0.75 percent for nonsynonymous sites) and 1.44 percent differences in all sites that did not specify a protein or any other product. Similar values have also been obtained by Feng-Chi Chen and Wen-Hsiung Li. The human and chimpanzee mitochondrial genomes (excluding the control region), which evolve faster than nuclear genomes, differ at 8.46 percent of their sites.

Are these differences large or small? Neither – they are exactly as expected. There is a simple method for calculating how many differences the human and chimpanzee lineages should have accumulated since their divergence from the most recent common ancestor. If substitutions accumulate at a rate µ per site per year, then the expected number of substitutions differentiating the two species is the rate multiplied by the time t that has elapsed since the divergence of the human lineage from the common ancestor plus time t that has elapsed since the divergence of the chimpanzee lineage from this ancestor (i.e., together $2t$). Hence the sequence divergence (genetic distance, see Chapter Eleven) d is given by $d = 2\mu t$. The nucleotide substitution rate in primates appears to be between 1.0×10^{-9} and 1.3×10^{-9} substitutions per site per year and the divergence time of either of the two lineages from the common ancestor is estimated to be 5 my (see next chapter). The expected difference is therefore $d = (2)(1.0$ or $1.3 \times 10^{-9})(5 \times 10^{6})$ or 1.3 percent.

Science writers and even some scientists appear to be flabbergasted when confronted with the sequence divergence between the human and chimpanzee genomes, but they shouldn't be, because the observed value matches the expectation very well. It would indeed have been astounding if the value had turned out to be much greater than 2 percent.

Since the sequence divergence includes substitutions that occurred in both the human and the chimpanzee lineage, the proportion of substitutions that occurred in the human lineage must have been approximately one-half of the total, that is ~0.6 percent. Since a single dose of the human genome contains 3.12 billion nucleotide pairs and since 0.6 percent of 3.12 billion is 15 million, does this mean that 15 million substitutions are responsible for the *sapientine characters* – the morphological, physiological, and behavioral characters that make us distinctly human? By no means, because approximately 97 percent of the human DNA contains no genetic information whatsoever and is apparently also without a function (the so-called *junk DNA*). If substitutions occur with equal probability in the functional and nonfunctional DNA, changes at only 3 percent or 450,000 sites at the most can be expected to influence the sapientine characters. In reality, however, this number is probably much smaller, because not every substitution in a coding sequence has an effect on the phenotype. In the 51 genes that O'hUigin and coworkers analyzed, 22 percent of the substitutions differentiating humans from chimpanzees were synonymous, which means that they did not change the structure and function of the encoded proteins at all. Subtracting these substitutions leaves 351,000 to lead potentially to functional changes, but even this number is a gross overestimate. Analysis of the nonsynonymous differences that have been identified in the coding parts of the 51 genes strongly suggests that few, if any, have an effect on the function of the encoded proteins. If the 51 sequenced genes are representative of the entire functional set, the majority of the 351,000 substitutions probably has nothing to do with the emergence of the sapientine characters in the human lineage. How many of the 351,000 substitutions are actually responsible for the sapientine characters can only be determined when the chimpanzee genome has been sequenced and analyzed.

Several laboratories are currently hunting the genes associated with sapientine characters, but no successes have as yet been reported. At present, three genes are known that differentiate humans from chimpanzees and other primates functionally, but their involvement in the development of any sapientine morphological or behavioral character is not obvious.

The first of the three genes codes for tropoelastin, a precursor protein of elastin, the major component of the elastic fibers in the extracellular matrix of connective tissues in the lungs, in the walls of large blood vessels, and in tough, bone-connecting bands of tissue (ligaments). Crosslinking of elastin fibers leads to the formation of a three-dimensional network with rubberlike properties. In species possessing elastin, the tropoelastin gene consists of 36 exons, but in humans it consists of only 34 exons and in all other Catarrhini 35 exons. All Catarrhini, including humans, lack exon 35, while humans alone also lack exon 34. Although the truncated gene continues to produce protoelastin, the crosslinking capability of the latter is presumably different to that of the protein encoded in the full-length gene. It has been speculated that the truncated protoelastin influences the function of human blood vessel walls.

The second gene is a member of a large family of loci which specify keratins, the proteins of intracellular filaments in cell layers covering body surfaces, such as the epidermis of the skin, and in hair and nails, which are derived from the epidermis. The "soft" keratins of the epidermis and other epithelia are encoded in a different set of genes (*KRT*) than the "hard" keratins of hair (*KRTH*). Both the soft and the hard keratins are of two types, acidic (A) and basic (B) and are organized in A-B pairs in the filaments. In humans, there are nine genes encoding acidic keratins (*KRTHA*) and six genes specifying basic keratins (*KRTHB*). One of the *KRTHA* genes has a stop codon in the middle of its sequence – it is a nonfunctional pseudogene (*KRTHAP1*). This defect has been found in all humans tested, but not in the chimpanzee or gorilla, in which the corresponding gene appears to be functional. The inactivating mutation is estimated to have occurred 240,000 years ago, presumably before the emigration of *H. sapiens* out of Africa (see Chapter Ten). The consequences of this inactivation for human hair growth remain undetermined.

The third gene codes for cytidine monophospho-*N*-acetyl-neuraminic acid (CMP-NeuAc) hydroxylase, an enzyme that increases the rate of a chemical reaction replacing a single hydrogen atom in *N*-acetyl-neuraminic acid (NeuAc) by a hydroxyl group, OH, to produce *N*-glycolyl-neuraminic acid (NeuGc; Fig. 8.11). Both NeuAc and NeuGc, jointly referred to as sialic acids, are sugars which are part of carbohydrate chains attached to proteins and other compounds on the surfaces of cells in a variety of tissues. Sialic acids often form the tips of the linear or branched carbohydrate chains. The CMP-NeuAc hydroxylase gene is functional in the chimpanzee, other nonhuman primates, and other mammals, but in humans it is rendered nonfunctional by a large deletion that expunges a vital part of the gene. As a result humans lack the enzyme and are therefore incapable of converting NeuAc into NeuGc. What adjustments the human species must have made in its physiology to compensate for the loss of this one form of sialic acid is not known. It is believed that when present, NeuGc is involved in interactions between cells. Indeed, cell-surface receptors have been identified that specifically bind the sugar, but the significance of this binding for the physiology of the organism remains unclear. Some investigators speculate that the loss of NeuGc in the human species was associated with protection against an unspecified pathogen. Several cases are known in which parasites use specific sugars of the host to gain entrance into its cells. If in the past a particularly dangerous pathogen evolved a receptor specifically binding NeuGc of the human ancestors, the host may have found it a lesser evil to do without this sugar than have it misappropriated by the parasite.

NeuGc is not the only carbohydrate that the human species treats in a rather discretionary manner. The well-known ABO blood group system is another example of a situation in which the absence of a sugar not only seems to cause no harm, but might even

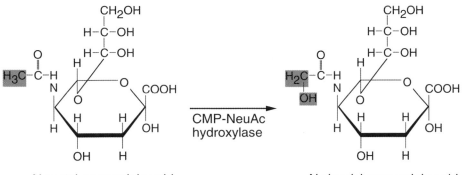

Fig. 8.11
Conversion of *N*-acetyl-neuraminic acid (NeuAc) to *N*-glycolyl-neuraminic acid. CMP-NeuAc hydroxylase, the enzyme that catalyzes this reaction, is inactive in humans but functional in other primates. The atomic groups altered in the reaction are highlighted. CMP-NeuAc, cytidine monophosphate-*N*-acetyl-neuraminic acid.

prove beneficial under certain circumstances. The terminal residue of certain core carbohydrate chains on human red blood cells is either a D-galactose (Gal) or N-acetyl-D-galactosamine (GalNAc; Fig. 8.12). Persons with the Gal-ending carbohydrate chain produce antibodies specific for the GalNAc-ending chain, presumably in response to immunization by certain microorganisms bearing similar chains. Persons with GalNAc-ending chains have, for similar reasons, antibodies in their blood against the Gal-ending chains. Red blood cells exhibiting the Gal-ending chains are therefore clumped together (agglutinated) by the antibodies in the serum of a person with GalNAc-ending chains, and vice versa. The agglutination reactions thus identify the subject as having the A (Gal-ending) or B (GalNAc-ending) blood group type.

The addition of Gal or GalNAc residue to the core is catalyzed by an enzyme encoded in the ABO locus on human chromosome 9. The gene at this locus occurs in three main variants (alleles). One variant specifies the enzyme *N*-galactosyl transferase which catalyses the addition of Gal to the core. The second variant codes for the enzyme D-galactosaminyl transferase which speeds up the addition of GalNAc to the core. These two proteins are very similar in their sequence but differ at a few critical amino acid positions in the functionally active region of the molecule, and these differences determine whether the enzyme adds Gal or GalNAc to the carbohydrate chain. The third allele codes for a nonfunctional protein which cannot add either Gal or GalNAc and the red blood cells are then of the O type – they are agglutinated neither by the Gal- nor by the GalNAc-specific antibodies. The frequency of the O allele is greater than 50 percent in most human populations but can reach 100 percent (in South American Indians, for example). In the latter populations, therefore, the situation resembles that of NeuGc in the entire human species: the O-type individuals do not seem to be handicapped by the absence of both terminal residues on their red blood cells. On the contrary, under certain circumstances the absence might be avantageous for them for similar reasons, as the absence of NeuGc might be beneficial for the entire human species. Of the four human blood groups (A, B, AB, O), B and AB are absent in the chimpanzee, while A, AB, and O are not represented in the gorilla.

Does the possession of different blood groups in humans affect their characters? Some Japanese believe that it does and in a biological attempt at fortune telling have their fate

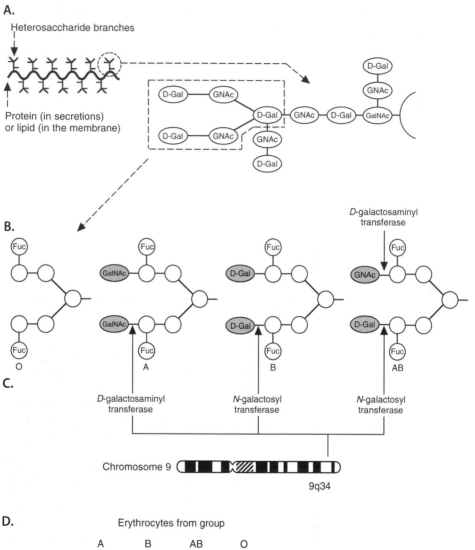

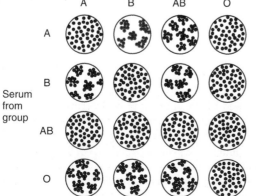

predicted on the basis of their blood group types. For about 100 yen, they obtain a message from a vending machine with four slots labeled A, B, AB, and O, and from these they learn about their personalities and receive advice on how to lead a happy life in harmony with their blood group type. It is an interesting variant of the superstition of persons, including at least one United States president, who consult their horoscopes before making important decisions. Numerous scientific studies have been carried out to discover whether the possession of a particular blood group type has any effect whatsoever. Associations with various traits, ranging from a tendency to suffer from a "hangover" to frequency of defecation, and from sadism to flat feet, have been reported. Most of these reports could not be confirmed in independent studies. The only connection that has endured is the higher frequency of stomach ulcers and cancer in group A subjects, but the statistical significance even of this association has been questioned by some authors.

Once the molecular differences between humans and chimpanzees are known and those responsible for the distinctiveness in appearance are identified, what will they be like? In the absence of any hints from comparisons between other species (in no pair of species have the differences in appearance been explained at the molecular level) one can only guess. From the study of mutations that alter the adult phenotype of humans and other mammals, as well as from the analysis of mutations that alter the development of model organisms, it appears that almost any morphological character is the result of a complex molecular pathway which involves large cohorts of genes. Mutations in many of these genes may lead to an alteration or disappearance of the character even though all the other elements of the pathway remain in place over long periods of time, presumably because they also participate in pathways leading to other characters. In the same way as a character can be lost as a result of a single mutation, it can also be regained if the function of the mutated gene is restored by another mutation. The return-to-function mutation can either take the form of a reversal of the original change or that of a different mutation that compensates for the effect of the first one.

The reappearance of a character believed to have been present in a remote ancestor is known as *atavism* (from Latin *atavus*, ancestor). The resurrection of the character indicates that the complex pathway leading up to it is still intact and that even the gene responsible for the loss of the character is still present in a form that can be reactivated. Extrapolating from these observations it can be speculated that the assembly of the pathway leading to a particular character began long before the actual appearance of this character. The various elements of the pathway in question may gradually have become available from other pathways, but only when the final element appeared did the new pathway become functional. In this manner, a complex character may not just be lost, it may also appear to have been created in a single step, in an extreme case by a single mutation. Small

---

Fig. 8.12A-D
The human ABO blood group system. **A.** The AB blood group antigens reside on complex molecules which consist of a protein or lipid backbone and branched polysaccharide side chains. **B.** The core polysaccharide, with the composition shown, is recognized as blood group O. Addition of N-acetyl-galactosamine (Gal-NAc) or D-galactose (D-Gal) to the terminal sugars of the core creates the blood group antigens A and B, respectively. The addition of both these sugars generates the AB blood group. **C.** The addition of the terminal sugars is catalyzed either by the D-galactosaminyl or the N-galactosyl transferases which are encoded in allelic genes on chromosome 9. **D.** The antigens are detected by antibodies specifically reacting with them. The reaction leads to the clumping (agglutination) of red blood cells (small circles). In the absence of the appropriate antigen, the red cells remain dispersed in the field of vision (large circle). Fuc, fucose; GNAc, N-acetylglucosamine. (**A** and **B** from Klein & Patten 1986; **D** from Klein 1982.)

changes at the molecular level would then appear to have a great impact on the appearance of the organism at the morphological level.

To put some flesh on the bones, consider, as an example, human nakedness. One of the characteristic features accompanying the emergence of mammals from theriodont reptiles was the appearance of hair on the outer body surface. A hair is in essence a long column of keratinized cells growing out from an onion-like nest in a deep pit in the skin, the hair follicle. The stimulus to form a hair follicle ensues from the interaction of two principal skin layers, the outer epidermis, which is of ectodermal origin, and the inner dermis, which is derived from the mesoderm. The long and complex pathway, from the initial interaction of the two tissues to the appearance of a mature hair, involves many genes. They are genes that determine the distribution of the sites in the skin at which the interactions leading to hair follicle formation occurs; genes that code for proteins involved in these interactions; genes that control the cell divisions by which the hair column grows and that determine the column's shape, length, and color; genes responsible for keratinization of the column; and genes that coordinate the entire process.

The reptilian ancestors of mammals lacked hair but possessed scales which, like hair, are produced by the interaction between the epidermal and dermal skin layers and which, again like hair, are formed by a complex developmental pathway. It is highly unlikely that the hair-growth pathway was assembled from scratch through the evolution of brand new elements. It is far more probable that it arose by modification of the scale-growth pathway supplemented by the recruitment of elements from other existing pathways.

There can be little doubt that the ancestors of humankind had bodies which were covered by a coat of hair similar to that found in other primates. The strongest evidence of this is the observation that rare atavistic mutations lead to the recurrence of hair growth in humans to produce a coat of hair resembling that of certain other primates. Since the Middle Ages, some 50 cases of a condition clinically classified as *generalized congenital hypertrichosis* have been recorded (Greek *hyper*, above, over; *thrix*, hair; here "congenital" means "hereditary"). Some of the unfortunate sufferers of this disorder were cruelly paraded in circuses or at exhibitions as "human werewolves", "hair men", "dog men", or "ape men" because most of the surface area of their bodies was covered by dense hair as observed in our ape ancestors. In one of the best studied cases involving 19 individuals of a Mexican family, the origin of the condition could be traced back to a gene on the X chromosome. Presumably a mutation in this gene restored its function and thus initiated the expression of the pathway leading to full hair cover. This pathway must therefore still be intact in every one of us and it takes only a small step to its manifestation.

The difference between possessing a coat of hair or appearing naked is actually far less dramatic than most people think. We are not really "hairless" on most parts of our bodies; in fact, we have almost the same density and distribution of hair follicles in our skin as chimpanzees or gorillas. Most of our body hair is generally much shorter and finer, however, so that it is almost invisible. Regardless of the mechanism responsible for the transformation from clearly visible to almost invisible hair, the cases of hypertrichosis demonstrate that this mechanism can be flipped on or off rather easily at the molecular level. This may also hold true for many other molecular mechanisms responsible for the sapientine characters of our species. The impression of a gap between the small differences at the molecular level to the seemingly large differences at the phenotypic level may arise because the intervening steps between the genotype and the phenotype, between the DNA and the appearance of an organism, remain unidentified. It will be thrilling to find out what these small difference really are, to identify the genes in which they reside, and to see how they are translated into the ability to speak, to walk on two feet, or to write this sentence.

The Question of Questions: Who is our Sister?

In his *Origin of Species* published in 1859, Charles Darwin provided overpowering evidence that extant species are interrelated via extinct ancestors, yet he conspicuously left one species out of his considerations. Instead, he concluded his book on the famous note that "Light will be thrown on the origin of man and his history". The person who first threw light on the origin of man was not Darwin himself but his younger, more combative colleague and friend, Thomas Henry Huxley. In 1859, when he reviewed the *Origin* for the *London Times*, 34-year-old Huxley was a respected zoologist and comparative anatomist who did not hesitate to cross swords with such authorities of the field as Richard Owen. At that time, however, he did not see certain facts that became plain to him only after reading Darwin's book. Almost overnight he became the first apostle of a new creed and a staunch supporter of Darwin and his theory of evolution. Only four years after publication of the *Origin*, Huxley published a book entitled *Evidence as to Man's Place in Nature,* in which he took it upon himself to fulfill the promise of the last sentence in Darwin's *magnum opus*. In a technical, yet highly readable style, *Evidence* addressed the "question of questions for mankind – the problem which underlies all others, and is more deeply interesting than any other", the biological relationship of our species to the rest of the living world and, indeed, to the universe as a whole. Summarizing the evidence from comparative morphology and developmental anatomy, much of it produced by himself, Huxley reached the conclusion that "it is quite certain that the Ape which most nearly approaches man, in the totality of its organization, is either the Chimpanzee or the Gorilla".

And so the news that the bishop's wife wished not to become generally known, was out in the open. On the streets of London, in the salons in Paris, and in the beerhalls in Berlin, the origin of the human species had momentarily become the primary topic of conversation. Extending one's pedigree to apes – which a few years earlier would have been a grave insult – was becoming an acceptable proposition. One did not exactly boast about it, but as an option it was often found to be preferable to certain alternatives. Huxley's rebuke of Bishop Wilberforce at the famous "Oxford Debate" is too well known to be repeated here. Somewhat less familiar is its Parisian version involving Alexandre Dumas père, and so we reprint here the dialogue from R. Hendrickson's *World Literary Anecdotes**:

> "You are a quadroon, Dumas?" a boorish acquaintance asked the French novelist.
> "I am, sir," said Dumas.
> "And your father?"
> "A mulatto, sir."
> "And your grandfather?"
> "A Negro, sir."
> "And may I inquire what your great-grandfather was?"
> Finally Dumas could suffer the fool no longer.
> "An ape, sir," he thundered. "My pedigree commences where yours terminates."

Huxley's "…either the Chimpanzee or the Gorilla" has remained a challenge to evolutionists for almost a century and a half. During this period, comparative morphology and developmental anatomy have been supplemented and upgraded by newer techniques such as cladistic analyses of morphological characters, tests based on antibody reactions with human, chimpanzee, and gorilla proteins, comparisons of chromosome

---

* Facts on File, New York 1990, p. 84

structure and organization, DNA-DNA hybridization, electrophoretic analyses of proteins, and ultimately sequencing of both mitochondrial and nuclear genomes. Yet even after so many years and the application of all these sophisticated methods, some authorities consider the *trichotomy problem* as unresolved. Why trichotomy and why this persisting uncertainty?

The first of these two questions is easy to answer. At issue are the relationships between three species: human (H), chimpanzee (C), and gorilla (G). Theoretically, there are three ways in which the three species can be related to one another genealogically. First, the gorilla lineage diverged early on from the common ancestor of all three species, and the human and chimpanzee lineages diverged from their common ancestor later. Second, the chimpanzee lineage diverged first, and the human and gorilla lineages diverged from each other later. Third, the human lineage diverged first, and the chimpanzee and gorilla lineages split subsequently. And fourth, the three lineages diverged from their common ancestor simultaneously. Using the species abbreviations H, C, and G, these four possibilities can be symbolized as (H,C)G; (H,G)C; (C,G)H; and (H,C,G), respectively, whereby always the species outside the parentheses is the one postulated to have diverged first. The last of the four alternatives is a *trichotomy*: instead of the bifurcations or two-some splits normally expected, it represents a three-some split.

The issue is a *problem* because the various methods, and the same method in the case of sequencing, applied to various parts of the genome have yielded different answers. Each of the four alternatives has been found to be supported by some of the methods and some of the genomic segments. Why is this so? To answer this question, we concentrate on the results obtained by sequencing DNA segments, but – with some modifications – the answers also apply to the results obtained by the other methods.

There are two main reasons why the sequences have provided ambiguous answers. First, the lineages of the three species apparently diverged from one another within a relatively short time interval, probably no longer than 1-2 my. Critical for our consideration is the length of the period after the first lineage split off and before the separation of the second and third lineages. To resolve the trichotomy, during this interval mutations must arise and become fixed in the common ancestor of the second and third lineages. The substitutions that occurred in this common ancestor will then be shared only by the second and third lineages and so differentiate them from the first lineage. Fixation of mutations, however, takes time, and the larger the population, the longer it takes. If the number of substitutions that took place in the ancestor of the second and third lineages is low, the information may not be sufficient to resolve the order of splitting. Many of the *gene trees* then become different from the *species trees*.

The second reason why the trichotomy problem persists is the random assortment of ancestral polymorphisms among the emerging lineages. To explain, let us assume for the sake of argument that the gorilla lineage was the first to diverge from the common ancestor and that the divergence of the human from the chimpanzee lineage followed later (Fig. 8.13) – a case of the (H,C)G phylogeny. Assume further that genes at three loci, *1*, *2*, and *3* have been sequenced in these three species, as well as in an orangutan serving as an outgroup species. Finally, we assume that at each of the three loci, two alleles existed in the common ancestor of the three species, which we symbolize by the nucleotides A, G, C, or T occupying a specific site in the genes' sequences. Each of the three genes may yield a different phylogenetic tree depending on the manner in which the polymorphisms sort themselves out among the species lineages (Fig. 8.13). In the case of gene *1*, the "C" allele was inherited by the gorilla lineage and the "T" allele by the common ancestor of human and chimpanzee, so the gene supports the (H,C)G tree. In the case of gene *2*, by contrast,

SPECIES TREE

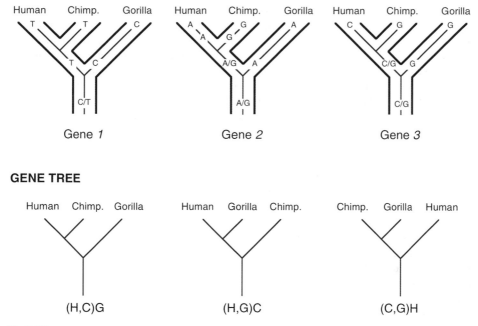

GENE TREE

Fig. 8.13
How polymorphism present in the ancestor of three species can result in three different gene trees depending on the segregration of alleles into the founding populations of these species. Only one of these trees corresponds to the species tree; the other two provide incorrect information about the phylogeny of the species.

the "A" allele was passed on to the gorilla lineage, whereas the common ancestor of human and chimpanzee inherited both alleles which then sorted themselves out, the "A" allele being passed on to the human lineage and the "G" allele to the chimpanzee lineage. This gene therefore favors the (H,G)C tree. Finally, in the case of gene 3, the "G" allele was transmitted to the gorilla lineage and the "C" and "G" alleles were supplied to the human-chimpanzee ancestor; subsequently the "C" allele was inherited by the human lineage, whereas the "G" allele was passed on to the chimpanzee lineage. Thus the gene 3 tree supports a (C,G)H relationship between the three species. Each of the three genes yields a different gene tree, only one of which presumably represents the species tree. But which? Since the fixations of one or the other of the two alleles is achieved by random genetic drift, the identity of the alleles fixed in the three lineages depends purely on chance. In addition to the possibilities depicted in Figure 8.13, there are several others, each of them favoring one of the three possible phylogenetic trees.

When different genes indicate different relationships between the species, how do we decide which of the gene trees represents the true species tree? Since stochastic processes are to be blamed for the incongruence of gene and species trees, we must resort to statistics to arrive at the most probable phylogeny. To explain, let us once again postulate the presence of two alleles, *a* and *b*, at a given locus in the ancestral population before the divergence of the human, chimpanzee, and gorilla lineages (Fig. 8.14). There are two splits in this assumed (H,C)G phylogeny: first the gorilla lineage is separated from the ancestor of the human and chimpanzee lineages and second, the latter two lineages diverge from each other. Theoretically, two principal situations can occur. In each, both alleles "survive" the first split, but then one of them (either *a* or *b*) is lost before the second split, or the two alleles survive the second split and one of them is lost subsequently in each of the three lineages. If the fixations and losses of alleles are completely random, then – assuming the (H,C)G tree – we can expect at least 50 percent of the genes to support this tree, 25 percent to support the (H,G)C and another 25 percent the (C,G)H tree (Fig. 8.14). Hence if we sequence a large number of genes and find that more than 50 percent of them support the (H,C)G tree, we can conclude that this probably represents the true phylogenetic relation-

Fixation occurs before second split

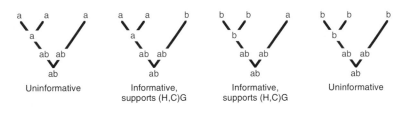

| Uninformative | Informative, supports (H,C)G | Informative, supports (H,C)G | Uninformative |

Fixation occurs after second split — the same allele is fixed in human and chimpanzee lineages

| Uninformative | Informative, supports (H,C)G | Informative, supports (H,C)G | Uninformative |

Fixation occurs after second split — a different allele is fixed in human and chimpanzee lineages

| Informative, supports (H,G)C | Informative, supports (C,G)H | Informative, supports (C,G)H | Informative, supports (H,G)C |

Support for
(H,C)G	4	50%
(H,G)C	2	25%
(C,G)H	2	25%

Fig. 8.14
Why it is necessary to sequence many genes to obtain statistical support for a particular species tree. Explanation in the text.

Table 8.2
The trichotomy problem: Which species is closest to our own? The evidence of multiple genes in three major studies.

Authors	Number of genes	Total length (nucleotide pairs)	Number of genes (percent) supporting phylogeny			
			(H,C)G	(H,G)C	(C,G)H	(H,C,G)
Satta and colleagues (2000)	45	46,855	23 (51)	8 (18)	8 (18)	6 (13)
Chen and Li (2001)	53	24,237	31 (58)	10 (19)	12 (23)	0 (0)
O'hUigin and colleagues (2002)	51	62,530	26 (51)	10 (20)	12 (24)	3 (6)

ship between the three species. Similarly, if more than 50 percent of the genes were to support the (H,G)C or the (C,G)H trees, we would be justified in assuming these to be the correct species trees.*

There is, however, another complication that we must consider. The genes at a given locus may differ among the species at more than one site and the various sites may be found to indicate different phylogenetic relationships – the sites are *incompatible* with one another. One site, for example, may support the (H,C)G phylogeny while another favors the (H,G)C or the (C,G)H phylogeny. How does such a situation arise? Here again, there are two principal possibilities – recombination and homoplasy (convergent evolution). Recombination ordinarily takes place between alleles of the same locus, but if the gene has duplicated tandemly, it can also lead to interlocus exchanges. By reshuffling polymorphic sites within a locus or between loci, recombination may create incompatibilities between them. Independent occurrence of similar substitutions, whether by back mutations or multiple hits, may have a similar effect. Indeed, it is often difficult to decide whether an observed incompatibility is the result of recombination or homoplasy. Both processes, however, occur with a low frequency so that site incompatibilities are a greater problem with more distantly than with closely related sequences.

Sequences of some genes of the mitochondrial genome support the (H,C)G, those of other genes the (H,G)C phylogeny. The sequence of the entire mitochondrial DNA favors the former. Since, however, the mitochondrial genome is inherited in one piece, its testimony is equivalent to that of one nuclear gene and so must be considered as equivocal. The number of determined nuclear genes sequenced in humans, chimpanzee, gorilla, and orangutan or some other outgroup species has been increasing steadily over the last several years. Earlier, several authors analyzed a small number of these genes and either reached an ambiguous conclusion or obtained tentative evidence in favor of the (H,C)G tree. The results of major recent studies are summarized in Table 8.2. In all these studies, the majority (51 to 58 percent) of the loci support the (H,C)G phylogeny, whereas approximately equal numbers of loci (15 to 22 percent) support either the (H,G)C or the (C,G)H phylogenies, and no or very few loci favor the (H,C,G) trichotomy. The combined data lead to three

---

* An underlying assumption of these deductions is that all 14 cases depicted in Figure 8.14 are equally likely to occur. This may often not be the case. If, for example, alleles *a* and *b* differ in their frequencies, some of the cases are more likely to happen than others.

conclusions. First, the chimpanzee rather than the gorilla is the nearest living relative of *Homo sapiens*, and this relationship is strongly favored by statistical analysis. Second, transmission of polymorphism from ancestral to descendant species was common in the evolution of the three species. Its consequence is that a significant proportion of the gene trees is incongruent with the species tree. And third, the splitting of the lineages leading to the three extant species took place in a relatively short time interval.

In cladistic terminology, the human and chimpanzee species form a monophyletic group or a clade – they are *sister species*. After all that we have said thus far, it is perhaps not necessary to point out that the human species did not evolve from the chimpanzee, but rather that the two species merely shared a most recent common ancestor. They are the two different tips of a branched twig. We shall learn in Chapters Nine and Ten that the human lineage, after its divergence from the chimpanzee lineage, bifurcated repeatedly and passed through several species. Presumably, the same is also true of the chimpanzee lineage, although evidence in support of this presumption is not yet available. In the chimpanzee lineage, two species, the products of the most recent bifurcations, have survived – the common and the pygmy chimpanzee. In the human lineage, only one species has survived. One can sometimes read that of the two chimpanzee species, the bonobo is more closely related to humans than the common chimpanzee. This claim is, however, false, as attentive readers of this book will undoubtedly realize. Since the two species diverged from each other after the chimpanzee lineage separated from the human lineage, they are *equally* related to the human species. That the bonobo appears to bear a great resemblance to humans in appearance and behavior (more slender body than common chimpanzees, higher frequency of standing upright, shorter arms compared with legs, smaller skull, smaller teeth, less size difference between sexes, obsession with sex, and face-to-face copulation) must therefore be the result of independent evolution – a good example of how misleading superficial similarities can be.

And Roaches will Inherit the Earth

The identification of the chimpanzee as our sister species concludes our climb of the Tree of Life, which extended over four chapters. Because we scaled the Tree with the aim of reaching a specific end point, our species, it focused on progressively smaller sections of the Tree and ignored vast other parts. In this chapter we zeroed in on primates to identify our position among them and so to answer Thomas H. Huxley's question of questions. We encountered two antipodal views concerning the nature of biological classification, which come to a head most conspicuously in the discussion on primate classification and the assignment of *Homo sapiens* within this scheme. On the one hand we have the view entrenched in a tradition that extends back to Linnaeus, which insists on preserving the original concept of grades as a measure of advance and progress toward a higher level of organization in the time intervals between lineage splittings. On the other hand, the newer, cladistic view bans grades, reveres clades, and does not think highly of the Linnean system. Accordingly, there are two opinions on the place our species occupies in nature. The traditionalists separate *H. sapiens* from all other primates (and from all other living forms) and erect a special category containing our own species exclusively. This assignment is based on subjectively measured changes in appearance which are anthropocentrically assessed as a great leap in the progression toward a higher level of organization. Cladists are not impressed by subjectively determined leaps, however large they might be, and molecular biologists find no evidence of anything unusual taking place in the human lineage at the

DNA level. Cladists, and some molecular biologists at least, therefore group the human species together with apes all the way down to the genus level.

At a more general level of discussion, it must be stated that the separation of humans from animals into a category that segregates it from the rest of the living world is not supported by the biological sciences. The beginnings of every characteristic that might be considered as distinctly human (speech, bipedalism, culture, and so on) are found in animals already. Biologically, humans are animals. The creation of a special category encompassing only humans in the biological classification system must therefore be viewed as a throwback to the days when religion had a strong influence on philosophy and science, and the widespread belief in the immortal soul, which only humans, but no animals, possessed.

The tendency to overestimate our own importance is an age-old misdemeanor. Standing alone on a vast expanse of land, encircled by the horizon and vaulted by the sky's hemisphere, you may easily succumb to a feeling of being the focal point. From here, it is only a small step to believing that you are indeed the center of the universe, and that you are special. Nicolas Copernicus attempted to dispel this illusion half a millennium ago. In biology, the Linnean classification system, with its built-in hierarchies of categories, is very much an open invitation to place yourself on a pedestal as the pinnacle of all living things. All you need to do is work through the inclusive Linnean categories in a stepwise fashion. Here is how you proceed: by virtue of your motility, multicellularity, and inability of your body to synthesize its own food, you belong to the kingdom Metazoa (animals). In your development you transiently possess a flexible, rod-like dorsal structure, the notochord, so you are a member of the phylum Chordata. You have mammary glands, hence you belong to the class Mammalia. Your nails and grasping hands assign you to the order Primates. And your ability to speak, along with the rest of the sapientine characters, puts you in the family Hominidae, of which you are the only living member. If you were to depict each of these categories as one slab resting on top of the preceding one, you would end up with a pedestal and you at the top (Fig. 8.15A). Does that not make you the Lord of Creation?

Charles Darwin provided sufficient evidence to make it clear that this is an illusion, too. For by the same logic by which you proclaim yourself to be the Lord of Creation, cockroaches can do the same (Fig. 8.15B). Here is how *they* might see things from their perspective: "We cockroaches are motile, many-celled, and dependent on other organisms for food, so we belong to the kingdom Metazoa. We are joint-legged and hence members of the phylum Arthropoda. Our bodies are divided into head, thorax, and abdomen, three pairs of legs, and two pairs of membranous wings, all of which assigns us to the class Insecta. We have leathery forewings that cover folded hindwings and so we belong to the order Orthoptera. Finally, our flattened bodies and the hood that extends from our necks over our heads make us members of the family Blattoidea. Hence, we, the roaches, are at the apex of the classification pyramid (Fig. 8.15B). We are the best, the most successful creatures of the living world; we are special products of creation. We have conquered the world and are now living on all continents and islands. We fly in jumbo jets, we sail with ships, and drive in trucks. One day we will board a spacecraft and fly to the moon. We have outsmarted the humans who consider themselves the smartest. They feed us, provide us with shelter, and transport us free of charge. They try to get rid of us, but we make monkeys out of them: every time they come up with a new poison, we quickly invent an antidote. Long after the last human being has disappeared from earth's surface, we will still be around, fully adapted to the new conditions and more successful than ever before."

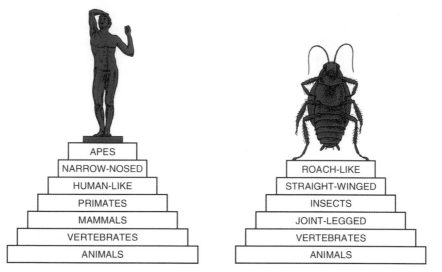

A. THE SUPREMACY OF MAN   **B. THE SUPREMACY OF THE COCKROACH**

APES	ROACH-LIKE
NARROW-NOSED	STRAIGHT-WINGED
HUMAN-LIKE	INSECTS
PRIMATES	JOINT-LEGGED
MAMMALS	VERTEBRATES
VERTEBRATES	ANIMALS
ANIMALS	

Fig. 8.15A,B
The illusion of supremacy in the living world can easily be generated from hierarchical classifications of organisms. Any species can thus be placed at the apex of a pyramid and presented as the Lord of Creation.

This kind of reasoning may look like an attempt to be cute, but in reality it serves to illustrate the important point that a scientific classification of organisms does not provide any basis for elevating one species above all the rest. Different species excel in different ways, cheetahs are the fastest runners, dolphins the fastest swimmers, eagles have the keenest eyesight, moths the most sensitive sense of smell, and we people have the reasoning power that enables us to live in different parts of the world. Our place in nature is, therefore, not on the apex of a pyramid; if anything, we are merely one of the grains in a multitude of sandstone blocks from which pyramids can be assembled in a myriad of different ways. Or better still – and this is the real lesson to be taken home from the four chapters devoted to the Tree of Life – we are an insignificant leaf on an immense bush. We are a leaf that, like all other leaves, sprouted from a bud, unfolded, and will one day be shed to be replaced by new leaves. For like individuals, species are transient and, on a geological time scale, their life span is brief.

Centuries ago, a Chinese poet Li-Tai-Po captured the transience of human existence in evocative verse. One hundred or so years ago, a German, Hans Bethge, translated the verses freely into his native tongue, and in this version they inspired Gustav Mahler to the first movement of *Das Lied von der Erde* ("The Song of the Earth"):

Das Firmament blaut ewig und die Erde
Wird lange fest stehn und aufblühn im Lenz.
Du aber, Mensch, wie lang lebst denn du?
Nicht hundert Jahre darfst du dich ergötzen
An all dem morschen Tande dieser Erde!

The heavens are ever blue and the earth
Will long stand fast and blossom forth in
spring.
But you, O man, how long will you live?
Not one hundred years may you enjoy
yourself

Seht dort hinab! Im Mondschein auf den Gräbern
Hockt eine wild-gespenstische Gestalt –
Ein Aff' ist's! Hört ihr, wie sein Heulen
Hinausgellt in den süssen Duft des Lebens!
Jetzt nehmt den Wein! Jetzt ist es Zeit, Genossen!
Leert eure goldnen Becher zu Grund!
Dunkel ist das Leben, ist der Tod.

With all the rotting trifles of this earth!
Look down there! In the moonlight on the graves
There crouches a wild and ghostly form –
It is an ape! Listen, how its howling
Rings out amidst the sweet scent of life!
Now take up the wine! Now, friends, it is time!
Drain your golden cups to the depths!
Dark is life: dark is death.*

It is our species, the ape, that is howling in the moonlight on the graves amidst the sweet scent of life.

---

* English translation: The DECCA Record Company Limited, London, 1968.

CHAPTER 9

Of Time and the Tree

The Time Scale of Evolution

*U*nfathomable Sea!
whose waves are years,
*Ocean of Time, whose waters
of deep woe
Are brackish with the salt
of human tears!
Thou shoreless flood,
which in thy ebb and flow
Claspest the limits of mortality,
And sick of prey, yet howling
on for more,
Vomitest thy wrecks on its
inhospitable shore;
Treacherous in calm, and terrible
in storm,
Who shall put forth on thee,
Unfathomable Sea?*

Percy Bysshe Shelley:
Time

The Tree Rings of Evolution

Although we have reached the particular leaf that was our destination during the long climb of the Tree of Life, we are not done with it yet. There still remains an important aspect to be considered and one we have ignored until now – time. The form in which genealogy is commonly depicted gives the impression that it deals primarily with spatial relationships. The principal metaphor of genealogy, the tree, is a geometric figure, a composition in space. Genealogists themselves speak of *topologies* when describing their trees, referring to arrangements in space. Yet, in reality, genealogy is first and foremost a relationship in time, a relationship between an individual, a generation, or a species that existed at an earlier time and that which came after it. Even the key genealogical metaphor is intimately yoked with time, for a tree grows with time and its growth *marks* time. In exploring the question of branch arrangement and the tree's topology, we have until now been delving into the spatial aspects of tree growth, but having done this we must turn to the temporal aspect, to the time in which the branching takes place.

Time! Countless books and tractates have been written about it and every philosopher of repute has considered it his duty to vociferate on its nature. To ancient Greek commoners, Time, Chronos, was a god, one among the crowd on Mount Olympus. To Newton it was a current flowing unhindered onward. Shelley imagined it as an amorphous body of water with waves, ebbs, and storms. To Einstein it was the fourth dimension of space, and to surrealists an assembly of sluglike watches crawling over boulders on a shore.

Time and space are intimately connected. We are aware that time passes because of events taking place in our surroundings, because of changes in space. And we use events to *measure* time, especially those that occur repeatedly and with some regularity. In our daily life we gauge time by the rotation of the earth around its axis and by its revolutions around the sun. We call the unit defined by one full rotation day and that defined by one complete revolution solar year. We then divide one day arbitrarily into 24 parts – hours, one hour into 60 minutes, and one minute into 60 seconds. One solar year, which is similarly divided into 12 months, comprises 365 days, five hours, 48 minutes, and 46 seconds.

We can, however, measure time by any other periodic change. In the case of a tree, for example, the passage of time is marked by the tree's growth, manifested externally by the appearance of new branches and internally by the tree (growth) rings. In most trees, branches sprout out too irregularly to function as reliable timekeepers or *chronometers*. Nevertheless, lower branches are generally older than branches higher up the tree because they sprouted earlier. So the succession of branches from the base to the apex represents a rough record of tree growth in time. You could number the branches starting either from the most recent or from the very oldest, and express any time point in the history of the tree simply by counting the branches that grow on it back or forward until you reach the starting point. Although many of the older branches may have withered away and fallen off, a record of their existence still remains – the knots and gnarls on the tree trunk. If you need a more precise measure of growth, you turn to the tree's internal chronometer, the tree rings. They reflect the change of seasons in temperate regions, and so the revolutions of the earth around the sun. Each spring, the cells lying beneath the bark divide and produce large, thin-walled, water-conducting vessels, the earlywood. Later in the growth season, smaller, denser cells with thick walls form, the latewood. The contrast between the dark, dense latewood and the light, low-density earlywood is visible as the border between two successive growth rings. Because the tree rings reflect the same periodicity on which our conventional clocks are based, the time they measure is directly convertible into conventional time units, one growth ring corresponding to one year.

In genealogy, time is measured and expressed in generations. In a genealogical tree or pedigree, generations can be regarded as corresponding to the branches of a real tree. Like branches, generations are also an inaccurate measure of time. Although their turnover is periodic, the overlap between generations and the variation in their length make them a rather unreliable chronometer. Like the old branches of a tree, past generations are not directly accessible, so their existence must be inferred from archival records. And like the branches of a real tree, generations can be counted backward from the most recent one or forward (and backward) starting from any past generation. Corresponding to the tree rings are the historical settings in which past generations existed. The dates of these settings serve to convert time expressed in generations to time expressed in years.

In conventional phylogeny, the passage of time is marked by the comings and goings of species. Here, as in conventional genealogy, only the living species are directly tractable, whereas the existence of extinct species is inferred indirectly from the traces they have left behind, most commonly the fossil record. And here, too, time can be expressed in terms of the position a species occupies relative to the entire sequence of extinct forms or by dating the setting in which the species lived and giving the date in years.

Finally, in *molecular* phylogeny, the flow of time is punctuated by those changes in the structure of DNA and protein molecules that spread through the entire species. The most common types of change are nucleotide *substitutions* in DNA molecules and amino acid *replacements* in proteins. The stepwise accumulation of these changes is akin to the ticking of the *molecular clock*. The principle on which the molecular clock measures time differs from that of conventional phylogeny or genealogy. In conventional phylogeny one can measure the time separating a living species from an extinct species or one extinct species from another but in molecular phylogeny such measurements are generally not possible. Here, not only does one measure the time separating molecules rather than species, the measurement is generally taken between molecules of *extant* species. Direct measurements of time between molecules of extant and extinct species or between molecules of two extinct species are possible only under exceptional circumstances.

The consequences of this difference between conventional and molecular phylogenies can be illustrated by a hypothetical example (Fig. 9.1). Consider extant species X and Y. Their molecules have been sequenced and the differences between them counted. The count (the distance) is a measure of time, the number of molecular ticks registered since the lineages leading to X and Y diverged from each other. The count is the *evolutionary distance* because on phylogenetic trees it is portrayed as the amount of space separating two points (see Chapter Four). In reality, however, the separation is temporal and the count is actually a measure of *time*. To be able to compare sequences of different lengths we normalize the molecular time by expressing it on a per unit basis, usually per nucleotide site or amino acid position. The expression "x-number of substitutions per site" then becomes a common currency, a tick of molecular time, which can be used independently of conventional or any other units. As long as one stays within the framework of molecular time, all temporal relationships among molecules can be expressed in terms of substitutions per site. In this respect, temporal relationships on a molecular phylogenetic tree are again similar to those specified by the succession of branches on a real tree.

What, then, can the X- and Y-derived molecules tell us of the *temps perdue*, of things past? If competently interrogated, they reveal their own past, but very little about the species in which their ancestor was once present. We can assume that they are derived from a single molecule which lacked all the substitutions now differentiating them. This molecule must have been borne by an individual, a member of a species Z ancestral to X and Y. We can envision several scenarios for the evolution from ancestor Z to extant

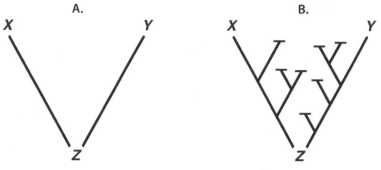

Fig. 9.1A,B
What molecules do not reveal. A comparison of DNA or protein molecules extracted from extant species X and Y does not reveal whether there was a single split (**A**) or multiple splits (**B**) involved in the derivation of these species from their last common ancestor, Z. In **B**, all the lineages terminating with the horizontal bar have become extinct.

species X and Y, but the molecules do not allow us to choose among them. In the simplest case, we could postulate Z to be the direct ancestor of X and Y (Fig. 9.1A). In more complicated scenarios, there could have been any number of intermediate species in the evolution from Z to X and Y (Fig. 9.1B). The ancestor Z could have become a founder of two lineages, one leading to X and the other to Y. In each lineage an unspecified number of successive branch splits may have occurred and resulted in species, all of which are now extinct (Fig. 9.1B). From each split, one of the two species would have produced the next generation of species continuing the line of descent toward the extant form. The two lines would also have propagated the DNA molecules which accumulated the substitutions along the way that now distinguish them from each other.

When probed about all these events, however, the molecules remain silent. They do not reveal whether the scenario was simple or complex, and if the latter, how many intermediate species each of the two lineages passed through. By using the proper method, we can reconstruct the sequences of all the intermediate molecules as they gradually accumulated different substitutions in the two lineages, but we cannot relate them to the species that must have carried them. Molecular phylogeny is what its name implies it to be – a genealogy of molecules; it is not a genealogy of species. At best, the phylogeny of molecules provides a caricature of species phylogeny by implying the existence of some, but not all of the ancestral species. A phylogeny based on sequences of molecules extracted from *n* extant species implies that *n*-1 ancestors (nodes) existed, while in reality the number of ancestors (extinct species) may have greatly exceeded the number of living species in any section of the Tree of Life. Of these "invisible" ancestors molecular phylogeny bears no testimony; the ancestors become "visible" only when paleontologists discover their fossilized remains.

Molecular phylogeny is also impotent in terms of relating molecular to conventional time. Since substitutions that mark molecular time occur independently of the earth's revolutions around the sun, molecular phylogeny lacks the equivalent of tree rings. To convert substitutions into years, molecular biologists must humble themselves and ask their colleagues, geologists, paleontologists, and archeologists, for help. To understand the form in which the help is delivered, we must first review a few basic facts about the fossil record.

A Rock Concert

You may have had a similar experience: sitting at the swimming pool reading a book, you swat a fly which has been annoying you for a while, and it falls to the ground. A minute later, however, in the corner of your eye you see that the fly is still moving. As you bend over to finish it off, you realize that it is dead alright but that a battalion of ants has descended on it from the bushes nearby and is now transporting the carcass to their nest. You are witness to the efficient and rapid disposal of a dead body by predators and scavengers. The quick decomposition of carcasses into their components and ultimately into molecules and atoms is the rule in nature. If it were not, we would be knee-deep in corpses, as one paleontologist quipped.

An escape from this natural recycling process is rare and occurs only under special circumstances: a woolly mammoth that is stuck in the muds of a tundra during a brief thaw and then freezes to death; a saber-toothed tiger that slides into a tar pit; an insect trapped and sealed in a tree resin; and an antelope buried in a dune during a sandstorm. By far the commonest mode of escape, however, is burial at sea, although here, too, the odds are heavily stacked against the organism leaving behind any sign of its former existence. A floating dead body is not likely to be overlooked by scavengers. If it is only made up of soft parts, not one piece of it will remain. If it has hard parts as well, such as shells, bones, teeth, or some form of an outer skeleton, in the best of circumstances, only fragments of these may remain. The body must sink rapidly to the bottom to stand a chance of escaping obliteration. However, even on the sea floor there are scavengers and certainly any number of microbes to decompose the carcass. The indigestible parts may then simply crumble away under the assault of the water itself or the minerals dissolved in it. Their only chance of escape is to be blanketed by deposits before decay sets in or progresses too far. A shower of particles is falling down continuously onto the sea floor. These include sand grains, mud motes, organic debris, as well as floccules of minerals that were dissolved in the water but fell out from the solution because of chemical reactions, a rise in temperature, or evaporation. If sufficiently dense, the blanket protects the body parts from the destructive effects of moving water masses and insulates them from oxygen and from microorganisms dependent on it. The cover may stop the decay, but it does not necessarily protect the remains from alterations.

With time, the crushing weight of the deposited material, the *sediment* (from Latin *sedimentum,* settling), and of the water masses pressing down on the sediment flattens the remains and the deposits themselves undergo compaction, which squeezes out any entrapped water and cements the particles together into a hard *sedimentary rock*. In this manner a sediment of mud or clay (material consisting of particles less than 0.004 mm in diameter) turns into *mudstone*. A deposit of silt, composed of particles ranging in diameter from 0.004 to 0.06 mm, becomes *siltstone*. A layer of sand, which geologists define as rock-derived material comprised of particles larger than 0.06 but smaller than 2 mm in diameter, forms *sandstone*. And gravel, particles larger than 2 mm in diameter, gives rise to a *conglomerate* or *breccia*. A particularly common type of sedimentary rock is *limestone* which is composed principally of crystalline calcium carbonate, $CaCO_3$, the *calcite* (from Latin *calcalx*, lime). Limestone can be derived from particles of calcareous (calcite-containing) rocks, from debris of calcareous organisms (chalk, for example, is derived from deposits of floating microscopic marine plants and animals), or from chemical processes that precipitate crystalline material from a solution. And so selected corpses are given truly royal burials, entombed in sepulchers that surpass pyramids, dolmens, and memorials erected for the corpses of human sovereigns in both size and magnificence. Alas, by the time the tombstone is completed, the corpse is no longer a dead body; it has turned into a *fossil*.

The word derives from Latin *fodere*, which means to dig, and was used originally in reference to any object retrieved from the earth. Currently, however, its usage has become restricted to any material indicator of past life, not only body remains but also footprints, wormholes, and feeding trails – the *trace fossils*. An insect enclosed in amber (solidified tree resin), a mammoth entombed in an ice block, or an antelope mummified in dry, decay-free desert air – these are all fossils.

When you find a fossil "on the rock", how much of the life form that produced it is still present? The answer depends on the type of fossil you have collected. If it is an impression of bark from a prehistoric horsetail tree in a coal block, then all you have is solidified dust of carbon that the tree has turned into. If you chance upon an ammonite* in a limestone quarry, you have essentially recovered the lime that filled the shell of a mollusc whose soft body was devoured by a predator, although a very small quantity of the altered shell material may still be detectable on the surface of the lime filler. If you visit the Petrified Forest in Arizona, what you see scattered around you in the desert are *petrified fossils* in which the original organic material has been replaced, molecule by molecule, by silica (silicon dioxide, SiO_2) in such a way that the original structure of the wood, all the way down to the cellular level, has been preserved. In a paleontological or anthropological museum you will see bones and teeth of human ancestors on display. If they are of a relatively young age, they may not be altered to a great extent, especially the teeth, which are particularly resistant to change. They may even contain some of the organic material, mostly collagen and other proteins, but possibly even DNA fragments. In older bones, however, the organic material is likely to have decayed and been dissolved away. The bones then appear light and crumbly. In bones that are older still, the spaces around the bone crystals have become *permineralized*, impregnated with minerals, mainly calcium and silica, deposited from the solution of water circulating through cracks in the rock. Such bones are hard, compacted, heavy, and strong, and so are better able to resist the weight of the sediment piled up above them.

Most fossils are derived from hard parts of organisms. Fossils of soft-bodied creatures or soft parts of an organism are rare and valued highly by paleontologists. They owe their origin to a rapid burial under unique circumstances. Swift burial protects the corpse from being devoured by scavengers, from decomposition by microorganisms, from rotting, as well as from decay caused by physical and chemical agents such as heat or acid water. A large amount of finely grained sediment must cover the body either while the organism is still alive or shortly after it has died and provide an airtight vault made of noncorrosive material. Mudslides, eruptions spewing out volcanic ashes, and floods are nature's common methods of quick burial.

Most fossils are formed in sediments deposited in the sea. Others may arise in inland bodies of water – lakes, swamps, and rivers. Fossils of terrestrial origin are not very common because conditions on land are generally inconducive to fossilization. Even bones that have not been destroyed by predators and scavengers are likely to crumble rapidly if they remain exposed to the elements – wind, sun, and water. Even when buried, they are dissolved in acidic soils in a relatively short time. Fossils of terrestrial organisms are therefore generally rarer than those of marine forms, and most of those that are found originated from bodies swept into a lake or sea by water currents.

---

*Ammonites are extinct marine molluscs allied to the nautilus and cuttlefish and characterized by a coiled, partitioned shell. The name reflects the resemblance their shell bears to *cornu Ammonis*, the horn of Ammon, that can be viewed on the ancient statues of Jupiter Ammon.

Add to all this the fact that most sedimentary rocks are altered beyond recognition by recycling; that most fossils are inaccessible to paleontologists in the depths of the earth; and that many exposed fossils are destroyed or lost before they are found – and you will appreciate just how capricious and unreliable the fossil record can be.

A Calendar of Earth's History

Ever since its violent birth some 4.6 billion years (by) ago, the earth has not come to rest. It churns beneath the surface like oil heated in a pan. Parts of it quake violently from time to time, and at places it erupts cataclysmically with igneous magmatic discharges. Its perilously thin crust is fissured into plates like the shell of a cracked egg. The plates themselves are never at a standstill, but glide ever so slowly on the sluggish currents of the molten rock underneath them. Along the cracks on the ocean floor, molten rocks well up and out, pushing the plates apart. At its leading edge, one plate slides below the next or two collide with each other like ice blocks piled up on a river when the great thaw comes. Parts of the crust are thus being continuously formed while others are melted down for recycling. Where the plates collide, the crust crumples like the skin of a baked apple and new mountain ranges are thrust skyward. Yet no sooner have the mountains risen than they are assaulted by the combined forces of heat, cold, water, and wind and the rocks begin to crumble. Blocks break up into slabs and boulders, boulders into cobbles, cobbles into pebbles, pebbles into sand, and sand into silt. As if toiled over by a giant carpenter, the mountains are gradually planed down until, after many millions of years, they are reduced to the flatland from which they rose. Later a new uplift may surge again and the cycle will repeat itself.

As the rocks disintegrate into rubble by this *weathering* process, the debris rolls down the slopes and into the valleys, where wind, but especially water, will carry it into lakes and seas. In the sea, the fragments ultimately come to rest as they settle down, piece after piece, particle after particle, forming a *layer* or *stratum* (*bed*) of sediment. Depending on its source, the deposited material can be gravel, sand, or silt. It may be dark due to the organic carbon it contains, white because of a high lime content, red because of the presence of iron, or have some other color. When the source of the material or the conditions change, so does the appearance of the deposited material. A new layer then begins to form which may differ from the preceding one in the size of the particles, in its compactness, color, and other properties so that the transition from one layer to the next is distinctly recognizable. As time passes, layer after layer is deposited, one on top of the other, and the whole stack ultimately hardens into a multilayered sedimentary rock. The various layers of the stack remain hidden from sight unless exposed by weathering, most dramatically when a river slices through it and opens up a chasm.

A spectacular example of bared sedimentary rocks is the Grand Canyon in the northwest corner of Arizona (Fig. 9.2). It cuts through a plateau which was once the site of a range as high as the Rocky Mountains and was located at the edge of the developing North American continent. The range was planed down to sea level by weathering and the plain then became a seashore. As the sea level fluctuated as a result of changes in the planet's climate, parts of the plain became the seabed in periods in which the level rose, and dry land when the ocean retreated. The areas that were under water accumulated sediments, layer upon layer, but when the climate changed and the sea level dropped, the sedimentation was temporarily interrupted, only to be resumed as soon as the ocean returned. These interruptions would later appear as gaps in the rock record. Depending on the extent of the rise

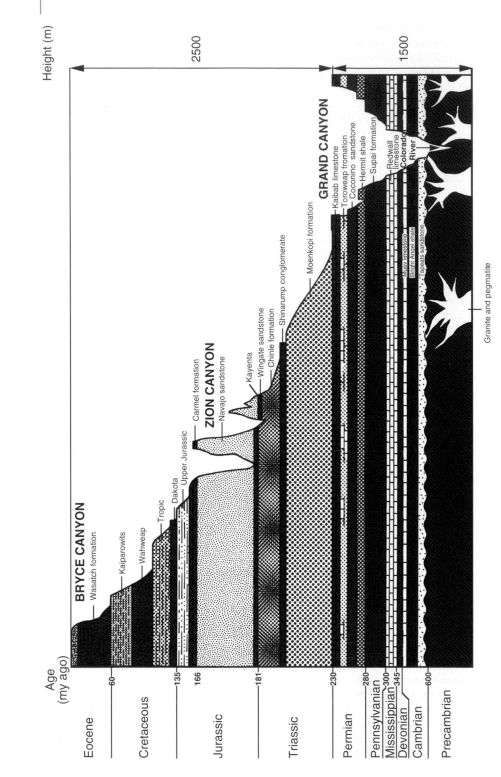

Height (m)

2500

1500

Age
(my ago)

BRYCE CANYON

Wasatch formation

Kaiparowits

Wahweap

Tropic

Dakota

Upper Jurassic

Carmel formation

ZION CANYON

Navajo sandstone

Kayenta

Wingate sandstone

Chinle formation

Shinarump conglomerate

Moenkopi formation

GRAND CANYON

Kaibab limestone

Toroweap fromation

Coconino sandstone

Hermit shale

Supai formation

Redwall limestone

Colorado River

Muav limestone

Bright Angel shale

Tapeats sandstone

Granite and pegmatite

Eocene

60

Cretaceous

135

166

Jurassic

181

Triassic

230

Permian

280

Pennsylvanian

300

Mississippian

345

Devonian

Cambrian

600

Precambrian

in sea level, the ocean reached different altitudes on the rolling plain and submerged different areas of land. The same area could thus find itself first in shallow water and then again as part of a deep ocean floor. In the former period, its deposits were mostly of the sandy type, whereas in the latter silt and mud largely accumulated. In this manner the stacks of sediment grew and when they hardened into rocks, layers of white limestone alternated with layers of gray shale (a sedimentary clay deposited in thin slices which split apart easily) and green to reddish sandstone. When later the plain was raised and the ocean retreated once and for all, the Colorado River cut through the sediments, carving out deep gorges over a period of many millions of years, one of which is now the Grand Canyon.

The walls of the Grand Canyon resemble the bound pages of books in which the leaves are intentionally cut to unequal width to give the edge a rugged appearance. Each layer in the stack revealed in the cross section of the canyon's wall corresponds to one page of a book and like pages, the layers are arranged in chronological order. The last page in the rocky book is the layer that was deposited first on the foundations of the closely shaved ancient mountain range. This page is therefore the oldest in the stack. The second-to-last page was the next to have been deposited and is therefore the oldest but one. Moving up the stack by climbing the canyon's wall, the layers become progressively younger; finally, the layer beneath the topsoil is the youngest, having been deposited just before the ocean's very last retreat. Since the deposition of each layer spanned a certain time interval, the sequence of layers from bottom to top also represents a sequence of time intervals from the first flooding of the plain to the ocean's final retreat. The sequence is a geological chronometer. It is a calendar of Earth's history in which each leaf, each layer, corresponds to a geological "day", a time unit. Theoretically, therefore, we should be able to specify geological time by simply referring to a particular page in the geological calendar, to a specific layer.

But the matter is not that simple. The problem is that the calendar is always local. As long as we limit ourselves to the area of the Grand Canyon, we can use the page numbers (the names of the layers) as a relative time reference in the sense that a higher number specifies an older layer. We can even extend the reference system to the Bryce and Zion Canyons in the vicinity of the Grand Canyon because we can trace some of the layers through the entire region (Fig. 9.2). If, however, we try to compare a gorge in North America with one in East Africa, for example, we seem to come to a dead end. Imagine you are told to look up page number 432 in a book. Which book?, you will ask immediately. Without having any details of the book's author, title, or edition, there is not much you could do with such an instruction. Similarly, unless geologists have some means of establishing exactly which layer in the Grand Canyon corresponds, in terms of age, to a certain layer in the East African Great Rift Valley, they cannot compare these two sites.

Fortunately, geologists of the nineteenth century had found a way of carrying out such comparisons. Until now, we treated the layers in a stack as if they were blank pages bound

---

Fig. 9.2
A geological chronometer exposed in the Colorado Plateau by the Colorado River and other geological changes. Each layer of sediment, from the oldest near the bottom of the Grand Canyon to the youngest at Bryce Canyon, is a leaf in the calendar of the regional history of earth. The layers were deposited over a period of more than 600 my, during which time a warm, shallow sea covered the area, and then receded, only to return again and again. Some 10 my ago, movements of the earth's crust raised the entire region high above sea level and weathering began to expose older and older layers. (Based on Beiser & the Editors of Time-Life Books 1963.)

into a book. The pages were arranged in a fixed, chronological order, but appeared otherwise to bear no date that could be used to identify their position in the sequence once removed from it. But in reality, the pages are not blank. Like pages of a calendar, they bear an imprint that specifies their date. The imprints are the fossils indented in the sediment layers.

Fossils are derived from species that emerged, existed for a while, and then became extinct. They had a beginning and an end, two points that specify a time interval. Further, while some species became extinct, others were already emerging, and as their time came to go, new species took their place. With some luck, a paleontologist might expect to find fossil species whose intervals of existence (life spans) appear in chronological order and cover much of the period of life's history. Further still, since layers of sediment pile up in chronological order and species also come and go in a chronological order, the two orders should be correlated. Older (lower) layers should contain fossils of species that existed earlier than species whose remains are found in younger (higher) layers. By concentrating on fossils of species with a worldwide distribution and relatively short life span (the so-called *index fossils*), a paleontologist should be able to identify layers from the same time interval in different parts of the world by showing that they contain the same type of fossils.

And this is indeed what paleontologists in the last one-and-a-half centuries have done. Often one species is not enough to date a layer, but a group of species (*fossil assemblage*) normally does the job quite well. Using sets of index fossils for the different layers they encounter in sedimentary rocks, paleontologists have been able to measure time in earth's history independently of conventional time expressed in years. They have developed their own time scale divided into a number of intervals, each interval defined by the presence of a particular fossil assemblage in sedimentary rocks. The scale is referred to as *chronostratigraphic* because it is based on the chronological order of strata (layers) in the sedimentary rocks, as revealed by their fossil content. It is a hierarchical scale in that the entire earth's history is divided into three *eons* (Archean, Proterozoic, and Phanerozoic), the Phanerozoic eon is subdivided into three *eras* (Paleozoic, Mesozoic, and Cenozoic), and each of the three eras is subdivided further into *periods*, some of which are partitioned into *epochs* and *ages* (Fig. 9.3).

The *Archean* eon was originally defined as extending from the beginning of earth's existence to the time at which the chemical composition and characteristics of the formed rocks began to change. When, however, it was later discovered that the changes apparently occurred at different times in different places, the definition of the Archean/Proterozoic border was set by convention at 2.5 by ago. It is the only border on the geological scale that is not defined by fossils. The *Proterozoic ("early life") eon* extends from this artificial milepost to the time when fossils of animals and plants first became highly abundant in the rocks, this being taken as the beginning of "visible life", the *Phanerozoic eon*. Since the first period of the Phanerozoic eon is called *Cambrian*, the time interval of earth's history that occurred before it is conveniently referred to as the *Precambrian* era. It encompasses the Archean and the Proterozoic eons.

Precambrian was once thought to be devoid of fossils. It was assumed that life did exist at this time but had left no tracks – it remained hidden. Life's apparent entry onto the stage of visibility was thus taken as the beginning of the Phanerozoic eon. In the meantime this contention has proved to be false. Fossils are not absent in the Precambrian, but they are not abundant either. They are therefore difficult to find and even when found, they often appear very strange in comparison to those of the Phanerozoic era. The oldest and, ironically, the most conventional among them are the *microfossils*, visible only when examined with the aid of a microscope. To study them, a rock must be sliced into millimeter-thin sec-

Eon	Era	Period and system		Epoch and series	Time before present (my)	Appearance of biological forms
Phanerozoic	Cenozoic	Quaternary		Holocene	0.01	
				Pleistocene	1.6	
		Tertiary	Neogene	Pliocene		
					5.3	Earliest hominids
				Miocene		
					23.7	Earliest hominoids
			Paleogene	Oligocene		
					36.6	
				Eocene		
					57.8	Earliest grasses
				Paleocene		Earliest large mammals, extinction of dinosaurs
					66.4	
	Mesozoic	Cretaceous				
					144	Earliest flowering plants
		Jurassic				
					208	Earliest birds and mammals
		Triassic				
					245	Age of dinosaurs begins
	Paleozoic	Upper	Permian			
					286	
			Carboniferous	Pennsylvanian		
					320	Earliest reptiles
				Mississippian		
					360	Earliest winged insects
		Lower	Devonian			
					408	Earliest vascular plants (ferns, mosses) and amphibians
			Silurian			
					438	Earliest land plants and insects
			Ordovician			
					505	Earliest corals
			Cambrian			
					570	Earliest fish
Proterozoic	Precambrian					
					2,500	Earliest colonial algae and soft-bodied invertebrates
Archean						Life appears
					4,600	

Fig. 9.3
The chronostratographic scale for the earth's history. Each of the divisions of this geological time scale was originally delimited by the occurrence of a specific fossil assemblage in sediments deposited in that interval. The scale in years was superimposed on the fossil-defined scale at a later point.

tions, which are then cemented onto a microscope slide and ground so thin that light is transmitted through them. The existence of life in Precambrian is also indicated by *stromatolites* (from Greek *stornynai,* to spread out and *lithos,* stone). They are a special kind of sedimentary rock formed by communities of microorganisms, dominated by photosynthesizing cyanobacteria. Many cyanobacteria give off sticky, gelatinous slime composed mostly of carbohydrates. The slime enables daughter cells to stay together after the division of the parental cell and so to form long filaments that cling together in felt-like

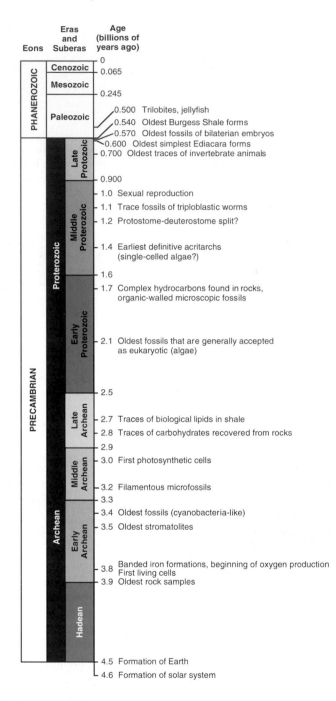

Fig. 9.4
Early evolution of life in the Precambrian.

mats. From time to time a layer of mud accumulates on top of the mat, and when this happens, the cyanobacteria slip out of the gelatinous envelope, crawl to the surface of the mud layer and form a new mat. In this manner the structure grows layer after layer, so that when the hardened rock is cut through, it often has the appearance of a cabbage head.

The three eras of the Phanerozoic – Paleozoic, Mesozoic, and Cenozoic – are demarcated by dramatic changes in the fossil record. Toward the end of an era, the disappearance of species prevails over the emergence of new ones – a widespread *mass extinction* sets in. The beginning of a new era is then marked by a grand scale recovery from the cataclysm. Of the two mass extinctions defining transitions between eras, the one occurring at the end of the Paleozoic (the Permian extinction) was the more extensive, but that at the close of the Mesozoic (the Cretaceous-Tertiary or K-T event) is more widely known because it led to the demise of the dinosaurs. There may have been several reasons for each mass extinction; some were apparently external (the collision of a large celestial body with the earth in the case of the K-T event).

The most spectacular, large-scale expansion of life occurred at the dawning of the Paleozoic era. Although there is no evidence that it was preceded by a mass extinction phase, it happened so rapidly and was so extensive that paleontologists speak of a *Cambrian explosion*. The major events marking the pages of the earth's calendar are summarized in Figures 9.3 and 9.4.

The Clock in the Rock

The chronostratographic scale described in the preceding section marks *relative time*: it specifies the order in which time intervals follow each other, but not their length. How long was the Cambrian period? Was the Ordovician period longer or shorter than the Cambrian? By definition, each of these two periods must have been shorter than the Paleozoic era but were they also shorter than, for example, the Mesozoic era? For a long time, paleontologists had no answers to such questions because the thickness of the sedimentary rocks was not a reliable indicator of time. One could assume that under stable conditions, sedimentation proceeded at a more or less constant rate, but even then the rates could have varied depending on the concentration of particles in the suspension. When the conditions changed, when an especially wet period was followed by a very dry one, for example, the amount of deposit changed correspondingly. The paleontologists' dream of being able to date rocks in units of *absolute time* (i.e., in years) only came true when a clock was found in the rocks in the form of radioactive isotopes.

Atoms of distinct elements differ in the number of protons in their nuclei. Atoms with the same number of protons but different numbers of neutrons are *isotopes* of the same element. The nuclei of some atoms are unstable either because they are too large and hence have too many neutrons relative to protons or vice versa, or because they are in an energetically excited state induced, for example, by the collision with another particle. The unstable nuclei tend toward an energetically stable state by emitting either α particles consisting of two protons and two neutrons (= α *decay*); β particles consisting of high-speed electrons (= β *decay*); or γ rays comprised of high-energy photons (= γ *decay*). Atoms (isotopes) with such nuclei are said to be *radioactive* (*radioisotopes*). The decay may change the number of neutrons or protons (or both) in the nucleus and so transmute one isotope or one element into another. Disintegration of a nucleus is unpredictable, but the time interval during which one-half of the nuclei in a given collection disintegrates is a characteristic of each isotope – its *half-life* (Fig. 9.5; for a mathematical interpretation of the half-

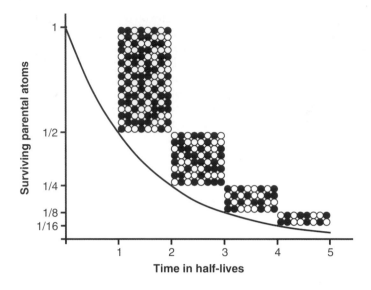

Fig. 9.5
The clock in the rocks: the decay of a radioisotope. Here an element is envisioned to have consisted initially (time 0) of identical radioactive atoms. With time, the atoms of this parental radioisotope (open circles) decay and convert into a new isotope (element; closed circles). The time needed for one-half of the atoms present in the initial collection to disintegrate is the half-life of the radioisotope. Of the remaining parental radioisotopes, one-half of the atoms decay again during the same time interval (two half-lives from time zero); and so on. Thus, during each half-life interval the amount of the parental radioisotope is reduced by 50 percent, compared to the amount present at the end of the preceding half-life interval.

life concept, see Appendix Four). Since half-lives of isotopes are known, they can be used to date rocks that contain them.

The concept of half-life is based on the notion that, in terms of absolute numbers, fewer and fewer atoms decay within consecutive time intervals of the same lengths. If that seems strange to you, think of decay as governed by the same probability laws as coin tossing. Take a case in which you toss 1000 coins all at once. You can expect approximately one-half of them to be heads and the other half tails. Now discard the 500 coins that came up heads (which is equivalent to removing atoms that have already decayed because they cannot decay a second time) and toss the remaining coins again. Again you can expect one-half of them to be heads, but the absolute number will be approximately 250 rather than 500. If you repeat this procedure, in each toss approximately one-half of the coins will always come up heads, but the actual number of coins landing with head facing up will decrease progressively. In terms of probability, the coin-tossing experiment mimics fairly accurately the process of radioactive decay.

Most natural radioactive isotopes are created in the interiors of large stars, the nuclear reactors of the universe. When the reactor runs out of fuel, its core collapses and the star explodes as a supernova, ejecting its matter, including the radioactive elements, into space. When the earth formed from just such material, the radioactive elements were incorporated and from this point on began to decay: the clock in the rock was set. It was then reset repeatedly during the earth's history whenever a particular type of rock formed.

Geologists recognize three basic types of rocks, sedimentary, igneous, and metamorphic. Sedimentary rocks we are already familiar with. Igneous rocks are formed when molten magma cools (here "molten" means liquefied by heat and "magma" is molten rock formed within the earth; when it reaches the surface, it is called *lava*). Metamorphic rocks arise from igneous or sedimentary rocks when these are subjected to high temperatures and pressure. The cooling of molten rocks is accompanied by the crystallization of the minerals and it is during this process that the radioisotopic clock is reset. Crystallization is a selection process in which specific elements are extracted from the magma for their inclusion in regularly repeating arrangements of atoms, the crystal of a mineral. Because crystals normally form from either the radioactive *parental isotope* or the *daughter isotope* to which the former decays but not from both, the daughter isotope later found in the crystal is assumed to have accumulated since the rock formed. The ratio of the parental-to-daughter isotope therefore dates the rock: the more daughter isotope and the less parental isotope the crystal contains, the more time has elapsed since its formation. *Radiometric dating,* the measuring of the age of rocks by means of radioactive isotopes, is made possible by the constancy of the decay rate of each isotope. Neither high temperature, nor pressure, nor chemical reactions have the slightest effect on the decay, because they involve the outer sphere of an atom rather than the nucleus in which the decay takes place. The application of radiometric dating to sedimentary and metamorphic rocks is beset by complications. In the former case, the method normally measures the time since the rock and not the sediment derived from it formed. In the latter case, heat and pressure may drive the daughter isotope out of the rock and thus reset the clock.

Several radiometric methods have been developed, differing in the type of isotope used and the manner of its detection. The choice of isotope determines the age range to which the method is applicable. A golden rule is to use an isotope with a half-life close to the expected age of the rock. If an isotope with a very much shorter half-life were chosen, most of it would have already decayed. By contrast, if an isotope with a much longer half-life were selected, the parental-to-daughter ratio would have changed too little for an accurate measurement. The four most commonly used decay systems are potassium 40 $\rightarrow$ argon 40, rubidium 87 $\rightarrow$ strontium 87, uranium 235 $\rightarrow$ lead 207, and uranium 238 $\rightarrow$ lead 206. (Here the numerals represent the *mass numbers* of the individual isotopes – the number of protons and neutrons in their nuclei; the arrows indicate the transmutation of one element into another by radioactive decay.) The half-lives of the parent isotopes in these four systems are 1.3, 47.0, 713, and 4.5 by, respectively. These and a few other systems have been used to superimpose absolute dates on the scale defined by the succession of sediment layers and the fossils contained in them. The superposition defines the lengths of the individual time intervals of the chronostratographic scale and thus turns it into a *chronometric scale* (Fig. 9.3).

For the study of human evolution, however, the half-lives of the parent isotopes used in these decay systems are far too long to be of much use. An exception is the potassium 40 $\rightarrow$ argon 40 or the *K-Ar method*. Potassium, a common element in the earth's crust and a frequent constituent of many minerals, has three natural isotopes, the highly abundant ^{39}K, the less abundant ^{41}K, and the rare ^{40}K. (Here the superscript is the mass number of each element.) The radioactive ^{40}K isotope decays either by β-emission to ^{40}Ca (about 90 percent of all disintegrations) or by electron capture (a process in which a proton absorbs one of the atom's electrons and so turns into a neutron) to ^{40}Ar (the remaining 10 percent). Although potassium 40 has a long half-life of 1.3 by, sensitive techniques have been developed to detect very small amounts of the daughter isotope ^{40}Ar and these enable dating to within time intervals of 10,000 to 100,000 years ago. Because argon is an inert gas which does not react with other elements, it is generally not incorporated in crystals when these form from a cooling magma. And since the argon formed by the decay of ^{40}K remains trapped within the crystal, the ^{40}Ar/^{40}K ratio can be determined and the age of the rocks under- and overlaying the sediment containing human remains can be calculated from the half-life of potassium 40.

For the most recent time interval (< 40,000 years), a reliable and accurate method of dating organic material, the *radiocarbon* or *carbon 14 technique,* has come into wide use. There are three carbon isotopes, the common ^{12}C, the rare ^{13}C, and the very rare ^{14}C, of which only ^{14}C is radioactive. Carbon 14 is produced in the stratosphere some 16 km above the earth's surface by the constant bombardment of atoms with cosmic rays, mostly high-speed, high-energy protons and α particles from sun flares and other sources in our galaxy. The bombardment produces free neutrons, some of which are absorbed by the nuclei of the nitrogen 14 atoms, the most common atoms in the atmosphere. The absorption leads to the ejection of a proton from the ^{14}N nucleus and thus to the transmutation of ^{14}N to the unstable ^{14}C, which has a half-life of 5715 years. Some of the ^{14}C interacts with oxygen atoms to produce carbon dioxide which, when air currents carry it down to the earth's surface, is used by plants in photosynthesis to produce ^{14}C-containing organic compounds. From plants, carbon 14 is passed on to animals so that all living forms end up possessing some of it. Some of the carbon dioxide is dissolved in oceanic water and then converted by marine organisms into ^{14}C-containing calcium carbonate in their shells. Immediately after its creation, carbon 14 begins to decay. In a living organism, however, the ^{14}C content is continuously being replenished through the incorporation of a new isotope. This comes to a halt when the organism dies and from this moment on, the amount of ^{14}C in any organic remains (bones, teeth, wood, charcoal, shells) gradually decreases. Compared to the present level, the decrease reaches a measurable proportion after about 200 years. After 40,000 years, corresponding to approximately seven half-lives of the isotope or less than one-hundredth of its original activity, its determination becomes unreliable. The radiocarbon method is therefore suitable for dating organic remains that are older than 200 years, but are not much older than 40,000 years.

The application of the radiocarbon method is not entirely unproblematic, however. The two main difficulties are the amount of material necessary for testing and the fluctuation of the ^{14}C/^{12}C ratio in the atmosphere. The former is now largely resolved. In the original method, relatively large amounts of carbon were needed for reliable measurements and these were often not available. The problem was then circumvented by dating not the actual remains, but other organic material found at the site. Since, however, it was not always certain that the material was deposited at the same time as the remains, errors in dating could not be excluded. In its new version, the sensitivity of the technique has been greatly increased and the amount of carbon required has been reduced correspondingly. The second problem is usually avoided by correcting the age estimate by a factor that takes into account the ^{14}C/^{12}C ratio at the time the remains were deposited. The past ratios are estimated from sources such as trees with a long life span. Because corrected and uncorrected estimates may sometimes differ considerably, archeologists distinguish the two by writing the time indicator in capital or in small letters, respectively (B.C. or b.c., i.e., "before common era").

The Clock in the Molecule

After paleontologists found a way of expressing geological time in conventional units, molecular biologists were next in turn to relate molecular time to the geological scale and to express evolutionary rates in years. There is, however, a fundamental difference between a conventional clock and a molecular clock in the manner by which they measure time. In the electronic wristwatch on your arm, the alternating electric current of the battery causes a quartz crystal to expand and contract rhythmically, and the vibrations are then translated into a digital display or into a movement of the hands. The vibrations are so regular that shifts in the position of each hand or changes in the displayed numbers are separated by time intervals of equal length precisely. The molecular clock, by contrast, marks the passage of time in intervals which vary in length randomly about a mean – the clock is stochastic (see also "The Clock Symphony" section in Chapter Four). This is because of the randomness of the two processes, mutations and random genetic drift, which are responsible for the time-marking events, the substitutions. When and where the molecular copyist, the DNA replication machinery, makes a mistake is a matter of pure chance. Similarly, the mutation drift in frequency of an allele toward fixation in a gene depends on the caprices of Lady Luck. However, the mean time interval between two consecutive mutations at a site or between two consecutive fixations is generally unwavering. It may seem highly inappropriate to base a clock on events that fluctuate about a mean, but then the radiometric clock is no different in this regard and it has proved to be of great service to both geologists and paleontologists alike. The description of the molecular clock that follows draws on the arguments presented in the "Clock Symphony" section of Chapter Four and the "Horse Kicks and Monsieur le Baron" section of Chapter Five and the reader may find it helpful to return to these sections before continuing with the present chapter.

The concept of the molecular clock was formulated on the basis of comparisons between protein and DNA sequences from different species. Let us therefore consider two such sequences, A and E, the former derived from an extinct species that was the ancestor of an extant species bearing the E sequence. Both sequences are L sites long and they differ at n_d number of sites by substitutions that occurred in the time interval t separating E from A. To be able to compare this difference with that of other sequence pairs, we express it as proportion p, sometimes referred to as the p *distance*, where $p = n_d/L$. When the indicated division is carried out, it gives the number of substitutions per site per time interval t. If the interval is long, we can expect the p value to be an underestimate of the actual number because at some sites two or more substitutions may have occurred, of which we register only one. We therefore use one of the correction formulae to estimate how many unobserved substitutions may be hidden behind the observed differences, and to obtain the correct proportion p_{pc}.

For the sake of argument, let us now assume that the accumulation of substitutions proceeded at stochastically regular intervals so that the individual intervals, though of different lengths, do not deviate too much from the mean. The p then becomes a gauge of time: the more substitutions found between two sequences, the longer the interval separating them. By using one substitution or some other accepted number of substitutions as a unit, we can then measure time, without reference to conventional time units. In other words, we can measure *relative* time in terms of the number of substitutions and make inferences about the relative age of the species from which the sequences are derived. A tree drawn by any of the distance algorithms provides just this information, since its distance scale is in reality also a relative time scale. Substitutions can, however, only be used to measure time if they accumulate at statistically regular intervals. If they occur erratically, they are of little use as clocks. This issue, whether substitutions occur at regular intervals or not, is at the center of the controversy surrounding the molecular clock and we return to it in the next section.

The number of substitutions per site per year is the *rate of substitutions*, r. The mean per site number of substitutions within a time interval t is then the rate multiplied by the length of the time interval, or rt. The probability of 0, 1, 2, 3, ...k substitutions occurring at a given site during the time interval t is given by the terms of the Poisson distribution which has the general formula

$$P(k;t) = \frac{(rt)^k}{k!} e^{-rt}$$

The probability of no substitutions (k = 0) in time t is therefore e^{-rt}. If the mutations underlying the substitutions are of the neutral type, then the substitution rate r equals the mutation rate μ and the mean rt in the above formula becomes μt (see Chapter Four).

Our assumption at the onset of these considerations is, however, rarely fulfilled: normally we have no access to the ancestor's DNA to be able to sequence it. Instead, we have two sequences, E1 and E2, of two extant species derived from the ancestor sequence A. We have not one line of descent (A→E), but two, one from A to E1 and another from A to E2. Each line had t time to accumulate substitutions at rate r, so together they underwent 2rt substitutions accounting for the sequence divergence d between E1 and E2. By rearranging the equation d = 2rt, we obtain r = d/(2t), which tells us that to measure the rate r, we must know d and the time t.

Obtaining d is not a problem, but determining t is, and we devote the whole of the next section to it. But assuming that we do find a way of obtaining t as well, we can then determine r for different pairs of sequences that diverged from their respective common ancestors at different times, and use these to test the rate constancy. If r is indeed constant, then a plot of different d values against corresponding t values should give a straight line. Once we have established rate constancy, we can then use the rearranged equation t = d/(2r) to determine the divergence time in years of any pair of sequences homologous to E1 and E2. We have calibrated the molecular chronometer and are now ready to clock molecular evolution, provided that r is approximately constant. But is it?

Since the controversy surrounding the molecular clock hypothesis first erupted, theoreticians have come up with a number of reasons for a supposedly inconstant substitution rate, related to the variation in either the mutation rate or the rate of fixations. At least three factors have been proposed that might influence mutation rates: fidelity of DNA replication, generation time effect, and metabolic rate. As to the first of these, the widely held belief is that most mutations arise as uncorrected errors of DNA replication. Because of transient structural shifts (tautomerism, see Chapter Four) in nucleotides of the template strand, the wrong nucleotide is incorporated into the forming strand; if it goes unnoticed by the proofreader and uncorrected by the repair machinery, it ends up as a mutation. Any decrease in the efficiency of proofreading and repair could therefore be mirrored in the magnitude of the mutation rate. Sloppy proofreaders and unconcentrated repair mechanics will overlook misincorporations more often than meticulous workers will do. A good example of such an influence on mutation rates is provided by RNA viruses (i.e., viruses whose genetic information is stored in RNA rather than in DNA) and mitochondrial DNA. The machinery that replicates viral RNA lacks proofreading and repair ability and as a consequence, the mutation rate of the viruses is millionfold higher than that of most other living forms. Inefficient repair is presumably also responsible in part for the ten- to hundredfold higher mutation rate of mtDNA compared to nuclear DNA (tenfold in the coding region, hundredfold in the control region). Smaller differences in the efficiency of proofreading and repair may exist between organisms on different branches of the Tree of Life and possibly also between regions of DNA molecules within the same organism.

The effect of generation time on substitution rates can be explained by comparing mice with humans. The generation time of a mouse is approximately one year, that of a human about 20 years. During their respective generation times, the germ cells in females of both species undergo roughly 24 divisions (see Chapter Four). Assuming that in each replication one DNA segment of a certain length suffers one mutation, then the female mouse accumulates 24 mutations in that segment in one year, whereas a woman accumulates the same number in 20 years. Although the mutation rate per generation will be the same in the two species, the per year rate will be approximately 20 times higher in the mouse than in the human. And since substitution rates are measured in years rather than in generations, mice might be expected to evolve faster than humans.* Generally, differences in substitution rates could be expected between any organisms differing in their generation times. Furthermore, the substitution rate should be higher in males than in females of the same species because the production of mature sperm cells is generally preceded by more cell divisions than the production of mature eggs (see Chapter Four). Correspondingly, there are more opportunities for mutations to occur during DNA replication in males than in females.

The metabolic rate hypothesis rests on the observation that some of the chemical reactions, the sum of which goes under the name *metabolism,* generate highly reactive oxygen molecules with free electrons. When these *oxygen radicals* gain access to DNA, as may happen in mitochondria in particular, they may attack and break DNA strands, remove bases from the nucleotides, or cause other forms of damage. Repair of the damage may then result in mutations. Since some species have higher metabolic rates than others, they may also vary in their mutation rates.

In addition to these three proposed influences on mutation rates, there may be others, some of which may be connected to intrinsic differences between the four nucleotides. Indeed, there are certain patterns in the occurrence of mutations that remain largely unexplained. These patterns are revealed by comparing sequences that can reasonably be assumed to have evolved at a constant rate so that any observed difference can be attributed exclusively to mutation rate effects. Particularly suitable for such comparisons are pseudogenes, which are genes that have lost their function and therefore evolve by random genetic drift alone. If each of the four types of nucleotide had an equal probability of mutating, we would expect 25 percent of all mutations to be from G to some other nucleotide, 25 percent from T to another nucleotide, and the same applying to mutations from A and from C. Similarly, when a nucleotide, G for example, mutates, it should have equal probabilities (1/3) of changing to A, T, or C. When these predictions were tested, however, by comparing mammalian pseudogenes, a significant deviation was observed. For example, G mutated with the highest relative frequency of 33.2 percent, followed by C (31.7 percent), A (19.1 percent), and T (15.9 percent). When G mutated, it changed most frequently to A (20.7 percent), and far less frequently to T (7.2 percent) or C (5.3 percent).

---

*It cannot be expected, however, that the rate difference will be twentyfold. Since the human-mouse comparison involves substitutions in sequences rather than a direct observation of the number of mutations, and since these substitutions have been accumulating since the divergence of the two species from their most recent common ancestor, it must be assumed that for some time after the divergence, the various species in the two lineages had similar generation times. The twentyfold difference in generation time may have arisen later in the evolution of the lineages and hence the difference between the evolutionary rates of human and mouse could be expected to be much less than twentyfold.

Disparities such as these were also observed for A, T, and C, the three other nucleotides. Some of these biases are explained by tautomeric shifts responsible for the preponderance of transitions (A↔G, C↔T) over transversions (all other changes). But the shift does not explain, for example, why G apparently mutates more often than any other nucleotide. Why there should be a difference in the mutability of the four nucleotides is not known, but it implies that different segments of the DNA molecule may vary in their mutation rates because of variability in their nucleotide composition.

In addition to intrinsic effects on mutation rates, neighborhood influences are also involved. If, for example, a C finds itself next to a G in a DNA strand, it can be recognized by an enzyme that adds a methyl ($-CH_3$) group to it after DNA replication. The resulting 5-methylcytosine deaminates (loses an $-NH_2$ group) spontaneously and thus converts to a thymidine. The next-door neighbor, G, thus increases the mutability of C. All these and other effects may ultimately translate into differences in substitution rates as a consequence of variability in mutation rates.

The major factor that might be expected to influence the rate of fixation is selection. If all the mutations were neutral, their fixation would be effected exclusively by random genetic drift and would be equally probable. Nonneutrality imposes a "charge" on mutations, either negative or positive, which makes them subject to negative and positive selection, respectively. By eliminating many of the mutations as they arise, negative selection lowers the substitution rate. In an extreme case, if all mutations were deleterious, the mutation rate could be normal, but the rate of fixations, and thereby also the substitution rate, would drop to zero. In less extreme situations, the more mutations eliminated by negative selection, the fewer candidates remain for fixation. By contrast, the more advantageous mutations arise, the higher the probability that some of them will be fixed and hence the higher the substitution rate.

Selection can, however, be expected to influence substitution rates only in those parts of the genome that have a function, primarily the genic regions. The nonfunctional parts should evolve largely or exclusively by random genetic drift. Since by all the available evidence, 97 percent of the human genome is nonfunctional, fixations of mutations in these parts should be effected by drift alone. This does not necessarily mean, however, that the substitution rates in the various nonfunctional regions will be the same and invariant over long time intervals. They may vary because of differences in mutation rates.

In the functional, protein-specifying parts of the genome, a distinction must be made between mutations that alter the protein sequence and those that do not – between nonsynonymous and synonymous mutations, respectively. The former can be subject either to negative or positive selection. If a particular amino acid residue at a specific position is essential for the function of a protein, a mutation that alters it has no chance of being fixed in the population. The more such residues a protein has, the lower the fixation rate. Evolutionists then speak of the protein being subject to *functional constraints*. Depending on the magnitude of the functional constraints, different proteins can be expected to have different substitution rates. Nevertheless, as long as the function of a protein does not change, the functional constraints remain the same and so should also the substitution rate.

If the function of a protein does change, for example as a result of changes in the environment, and an altered amino acid residue at a particular position increases the efficiency of the protein in carrying out the new function, the probability of fixation of the mutation responsible for the alteration also increases. The ensuing positive selection is therefore commonly thought of as an accelerating factor in sequence evolution. While this could be true in principle, the effects of positive selection on overall fixation rates should be expected to be fairly small to negligible. If the adaptations to a new function involve very few residues, while the changes at other positions are neutral, the effect on the overall rate will hardly be recognized. Furthermore, once the adaptive change has been made, negative selection will replace positive selection to protect the positions from further alterations. The episode of positive selection will therefore be of short duration and hence without a noticeable influence on the long-term rate. In other words, positive selection here and there or now and again is not going to have any large impact on the substitution rate measured in a long DNA segment over an extended time interval.

The effects of selection, however, do not need to be restricted to the nonsynonymous sites of a gene. You will recall that for each of the 20 different amino acids found in proteins there is more than one codon in the DNA. The amino acid proline, for example, can be specified by any of the following four codons: CCT, CCC, CCA, or CCG. Note that the codons have the same nucleotides at their first and second positions, but differ in the nucleotide occupying the third position. It does not seem to matter which nucleotide is present at the third position; as long as the first two nucleotides are CC, the codon will specify proline. Yet, an analysis of the relative frequencies with which the four codons occur in different genes of various organisms does not reveal the 1:1:1:1 ratio one might have expected were the codons truly equivalent. Instead, one of the four codons is usually found to occur more often, while the others are relatively rare. This *codon usage bias* is largely species-specific: some species preferentially use the CCT codon, while others may prefer one of the other three codons. A similar bias also exists in the cases of codons specifying 17 other amino acids.

The reason for the bias apparently lies in the mechanism by which the codon of a gene is translated into an amino acid. The translation takes place via two intermediates, the messenger or mRNA and the transfer or tRNA. The mRNA is a single-stranded copy of the gene with one difference: the T of the DNA is replaced by a U in the mRNA. The tRNA then carries out the actual translation from the mRNA into protein. Each tRNA molecule has two critical functional sites, one with which it binds a free amino acid and another, the *anticodon*, with which it recognizes a specific codon in the mRNA molecule. The cytoplasm of a cell contains more than 30 different kinds of tRNA molecules, which differ for the most part in their functional sites only. Each kind binds only the particular amino acid specified by the codon of the mRNA which it recognizes via its anticodon – the three nucleotides complementary to the codon. Some of the anticodons, however, recognize more than one codon.

The explanation of the codon usage bias is provided by the observation that in each species, the proportions of the various tRNA types present in the cytoplasm correspond to the frequencies of synonymous codons in the genome. This correlation suggests that natural selection is behind the codon usage bias, but whether positive or negative selection is responsible for it is still not known. Selectionists, of course, invoke positive selection. According to them, an organism carrying a mutation that enables it to use the most abundant tRNA type has an advantage over individuals who rely on the less common types, especially in those cases in which large amounts of a protein must be manufactured rapidly. Indeed, the codon usage bias is most pronounced in highly active genes and is much less apparent, or absent altogether, in genes with intermediate or low expression. (Cells would seem to produce disparate quantities of tRNAs in the first place because of variation in the number of copies of the encoding genes. Genes specifying the abundant tRNA types are represented by more copies in the genome than those encoding the less frequent forms.) Neutralists, on the other hand, attribute the codon usage bias to negative selection, which tends to eliminate many of the mutations that would make the codon interact with anticodons of the uncommon type. In other words, they invoke a form of functional constraint operating at the level of codon-anticodon interaction. Whatever the case may be, some form of selection indeed seems to be influencing the nucleotide composition of synonymous sites in a gene.

Taking all these potential influences on the magnitude of the substitution rate into consideration, we certainly cannot expect to find a molecular clock that, like Big Ben, chimes precisely, every hour on the dot, for the whole realm. The clock is neither universal (the same for all organisms), nor global (the same for different DNA regions within an organism). There are indeed differences in the rate of evolution between genic and nongenic regions, between nongenic regions occupying different parts of the chromosome, between different genes, between exons and introns, between different parts of an exon, and between different sites of a codon. Some genes (e.g., hemoglobin α, cytochrome c, albumin, fibrinopeptide) evolve in a clock-like fashion, while others (e.g., superoxide dismutase, glycero-3-phosphate dehydrogenase) apparently do not. Mouse and rat (rodent) genes may evolve somewhat faster than primate genes, and monkey genes may have a higher substitution rate than human genes. Yet, despite all these complexities and notwithstanding headlines such as "The molecular clock is gone …", "Molecular clocks run out of time", or "Molecular clock mirages" in the popular press, the clock is an important complement to dating based on the fossil record. To use it, however, one must take the complicating factors into account and make sure that, at best, they do not interfere with the time estimate, and, at worst, that their effect is minimized. In particular, one must check the constancy of the clock in each individual case and judiciously prune out sites, regions, or genes that show clear signs of deviation from constancy. No, a molecular counterpart of Big Ben there will never be, but a chronometer of sufficient accuracy to fill in gaps in the fossil record and sometimes even repudiate claims based on it, is with us to stay.

Shoemaker's Last

Doubts concerning the constancy of substitution rates are one of two major problems that beset the notion of the molecular clock. The other is the calibration of the clock. The word "calibration" derives from the Arabic *qualib*, shoemaker's last, which refers to the process of making a series of identical pairs of shoes by using one pair as a mold. When all the pairs are finally made, what remains is the model, the shoemaker's last. To calibrate time therefore means to use one kind of clock to specify the time unit of a different clock. To calibrate the *molecular* clock amounts to relating the tick-tock of substitutions to years. The most direct way of calibrating the molecular clock would be to actually measure the interval between two succeeding substitutions, but this is not possible because the time is much longer than the measurer's life span. The clock must therefore be calibrated indirectly. Since we count substitutions by comparing the DNA sequences of two living species, we must find out how much time has elapsed since the two sequences were one. This divergence time corresponds approximately to the time the two species took to evolve from their most recent common ancestor. If we could, therefore, identify the ancestor in the fossil record and determine, by radiometric methods, the age of the rocks in which its remains are deposited, we should be able to calculate the rate of substitutions per site per year. Provided that the rate is constant, we can then use this unit rate to draw a scale in years, apply it to a phylogenetic tree, and thus estimate the divergence times of those species for which no fossil record is available.

Alas, things are not this simple. Complications arise because of uncertainties concerning the fossil record. To give an example of this, let us consider once more two closely relat-

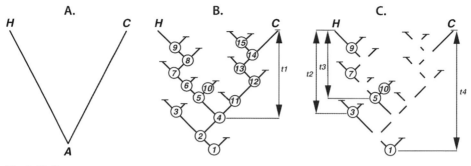

Fig. 9.6A-C
Matching sequences with fossils. **A.** Comparison of DNA (protein) sequences obtained from extant species H and C suggests the existence of a common ancestor A from which the two derive. **B.** The lineages from A to H and from A to C may have entailed many other ancestors (numbered) and splits. **C.** Of these, the fossil record only reveals the existence of some in the $A{\to}H$ lineage and none at all in the $A{\to}C$ lineage (indicated by gaps). The paucity of the fossil record may lead to the erroneous identification of the most recent common ancestor and incorrect estimates of the H-C divergence time ($t2$, $t3$, $t4$) as compared to the actual divergence time ($t1$ in **B**).

ed species, the human (H) and the chimpanzee (C), and their most recent common ancestor (A). Molecular evolutionists depict the divergence as a splitting of A into two lines, one leading to H and the other to C (Fig. 9.6A). The actual divergence, however, must have been much more complex, consisting of several interposed species in each lineage, all now extinct, each species resulting from a separate splitting event (Fig. 9.6B). In an ideal situation (which virtually never occurs), paleontologists would have complete fossils of all extinct species numbered 1 through 15 in Figure 9.6B. In this case, they might be able to reconstruct the complete phylogeny reasonably faithfully on the basis of morphological similarities between the species. In reality, however, paleontologists have only a fraction of the ancestral species, and from each species they possess a mere fragment of its skeleton. In this situation, they can easily be misled by false resemblances and either misidentify some of the fossils or reach wrong conclusions about their interrelationships. If, for example, they knew only about the existence of the species shown in Figure 9.6C, they might conclude that the most recent common ancestor of H and C was No. 1, No. 3, or even No. 5, whereas in reality, as shown in Figure 9.6B, it was No. 4. In this case the practical consequence for the molecular clock is that it would be calibrated not by the time interval $t1$, as it should have been, but by the interval $t2$, $t3$, or $t4$ instead, which are all different from $t1$. The divergence time of H and C sequences would be overestimated ($t2$, $t3$) or underestimated ($t4$), as would all other divergences estimated from the clock calibrated by one of these three time intervals.

However, even if the fossil record is reasonably complete, the identification of the most recent common ancestor is not easy. In general, to identify the ancestor, paleontologists choose characters distinguishing the species H and C and follow them backward in time in the collection of fossils available to them. Based on the characters, they sort the fossils into those presumably representing the H and those representing the C lineage. In each lineage, the fossils are then arranged into a regressive series in such a way that the characters gradually resemble those of the living species less and less, but at the same time resemble the characters of the other lineage more and more. The youngest fossil in which the characters of the two lineages merge is then identified as the most recent common ancestor.

A good example of a character that defines the *H* lineage is the size of the brain, or more accurately the capacity of the cranium, the set of bones that enclose the brain. Living non-human apes have an average cranium capacity of no more than 460 cubic centimeters. The average cranium capacity of modern humans is 1400 cubic centimeters. A fossil cranium is therefore assigned to the *H* lineage (rather than the *C* lineage) when its capacity lies between these two values. Its position in the *H* lineage is then determined according to the specific value of the cranium capacity: the greater its value, the closer to *Homo sapiens* the fossil is placed in the lineage. Similarly, the closer the value is to that of the nonhuman ape, the nearer the fossil is to the most recent common ancestor, provided the other characters indicate that it is related to *H* and *C*. If, however, the fossil record is incomplete in the time interval approaching the stage in which the expected ancestor lived, it is almost impossible to judge whether the fossil is that of the most recent ancestor or some other (i.e., whether it is No. 4 or Nos. 1, 3, or 5 in Figure 9.6C). And since there is no way of knowing whether the record is complete, an element of uncertainty always remains regarding the estimate of the divergence time of any two extant species.

Just *how* unreliable the paleontological identification of an ancestor can be is illustrated by the *Ramapithecus* story. Paleontologists are probably sick and tired of hearing the story repeated over and over again, but since it conveys an important message, it is worth reiterating. The story begins in the 1870s in the Siwalik Hills, a mountain range at the foothills of the Himalayas in north Pakistan and northeast India. There the British paleontologist, Richard Lydekker, found a fragment of an upper jaw with teeth, derived from an extinct ape species. He named the new species *Paleopithecus sivalensis* or Siva's ancient ape, Siva being the god of destruction and regeneration in the Hindu sacred triad, which also includes Brahma, the Creator, and Vishnu, the Preserver. The story is taken up again in 1910-1915, when another British paleontologist, Guy Pilgrim, affiliated to the Geological Survey of India, collected numerous fossils of extinct apes in west Pakistan and north India. Some of these he assigned to a new genus and species which he named *Dryopithecus punjabicus*, the tree ape of Punjab.

After that there was another pause until 1932, when the Yale University at New Haven, Connecticut, sent three geologists to the Siwalik Hills to "probe Gondwana-land mysteries", as announced in the headlines of the local newspaper. In the hills, at a place called Haritalyangar, the youngest member of the expedition, the graduate student George Edward Lewis, collected various fossilized fragments of upper and lower ape jaws. Since the upper and lower jaws were never found together, Lewis concluded that they came from two different, previously undescribed species. Back at Yale, Lewis analyzed the fossils and named them *Ramapithecus brevirostris*, the short-snouted ape of Rama, and *Bramapithecus thorpei*, Brahma's ape of Thorpeus, Rama being a mythical hero and an incarnation of Vishnu. Lewis also pointed out that the name *Paleopithecus* had been used before Pilgrim adopted it to designate – of all things – an extinct reptile, and he suggested therefore that the species should be renamed *Sivapithecus*, the ape of Siva.

Like Lydekker and Pilgrim before him, Lewis toyed with the idea that some of the extinct apes to whom the fossils belonged might be on a direct line of descent to humans. But rebuked on this subject by Ales Hrdlicka, the don of American anthropology, he decided not to make any public statements about it but to mention this possibility in his unpublished doctoral thesis. *Ramapithecus* in particular seemed to Lewis to show an affinity to humans. From the shape of the jaw fragment, Lewis deduced that the ape was rather flat-faced (hence the designation *brevirostris*, short-snouted) and had relatively short canines (on the basis of the size of a root and the depth of the socket, since no canines were preserved in his fossils). Both these characters were pronounced in humans, but not in apes.

The similarity of *Ramapithecus* to humans became a focus of interest 30 years later for two other Yale paleontologists, Elwyn Simons and his younger colleague David Pilbeam. In the intervening years, it had become clear that *Ramapithecus* must have been distributed widely not only over the Indian subcontinent, but also in Africa, Europe, and Turkey. In Africa, however, some of the fossils appearing to be *Ramapithecus* were named by Louis S.B. Leakey (founder of the renowned clan of fossil hunters) as *Kenyapithecus*, the ape of Kenya. Simons and Pilbeam therefore had far more material at their disposal than Lewis 30 years earlier. After analyzing it, they believed they had found additional important resemblances to humans, in particular thickened enamel, the thin layer capping the teeth, and the curve traced by the teeth around the jaw – the *dental arcade*. In apes the arcade is more of a ⋂ shape, whereas in humans it is shaped like this ⌒. Simons and Pilbeam did not have access to a complete *Ramapithecus* dental arch, but they pieced together the fragments Lewis had found in such a way that they seemed to have a human rather than an ape-like shape. The two then crusaded through the paleontological community, rapidly convincing its members that *Ramapithecus* was the direct ancestor of the human lineage. In the Siwalik Hills, no rocks suitable for radiometric dating were available, but the layers in which the *Ramapithecus* fossils were found were assigned to Late Miocene, making them about 15 my old. Some *Ramapithecus* fossils could have been up to 30 my old. The notion of *Ramapithecus* as the most recent common ancestor of humans and living great apes and as existing at least 15 my ago thus became generally accepted. It prevailed for some two decades before the whole story collapsed.

In 1967 two biochemists, Vincent M. Sarich and Allan C. Wilson, published an article in *Science* in which they presented evidence that the most recent common ancestor of *Homo sapiens* and nonhuman apes probably lived around 5 my ago, and certainly no longer than 8 my ago. Paleontologists immediately realized, of course, that if true, this claim would effectively dispose of *Ramapithecus* as a hominid, as a member of a lineage leading from the most recent common ancestor to modern humans. As Sarich later put it: "anything earlier than 8 million years cannot be a hominid no matter what it looks like."

Upon what evidence did Sarich and Wilson base their self-assured (many a paleontologist thought *arrogant*) claim? Paradoxically, on evidence provided by the molecular clock! They compared serum albumins of various living primates by the immunological technique mentioned earlier (quantitative precipitation of the protein with antibodies obtained by the immunization of rabbits against purified albumin). In these tests, the intensity of the reaction mirrors the biochemical difference (the number of substitutions) between the immunizing and the test albumin. It thus indirectly measures the evolutionary divergence of proteins derived from different species. From other tests, Sarich and Wilson concluded that serum albumins of primates apparently evolve at a constant rate. From the fossil record the authors then assessed the divergence between Old World monkeys and apes to have occurred 30 my ago and used this value to calibrate the albumin molecular clock. And so they arrived at the figure of 5 my ago as the divergence time between humans and chimpanzee or gorilla.

Readers interested in getting a taste of the paleontologists'/anthropologists' reaction to the Sarich-Wilson article are referred to page 233 in the book *The Red Ape. Orang-utans & Human Origins* by the anthropologist Jeffrey H. Schwartz.* Yet Sarich and Wilson ulti-

---

* Houghton Mifflin Co, Boston 1987. In this book, on the basis of morphological and paleontological data, Schwartz argues that neither the chimpanzee, nor the gorilla, but the orangutan is the closest living relative of humans, which is a view no longer held by anyone.

mately prevailed, the younger age of the most recent common ancestor became generally accepted, and *Ramapithecus* is no longer one of the candidates for this title.

How did this happen? Was it a "paradigm shift" such as that claimed by the philosopher Thomas Kuhn to be responsible for the replacement of one view by another when younger, more vocal scientists embrace a new paradigm, leaving the dons to carry the earlier view to their graves? Apparently not, because even the original protagonists of the *Ramapithecus* hypothesis were ultimately converted. The change was driven not by a power struggle, but by the weight of evidence. The *Ramapithecus* hypothesis became so fraught with errors that it ultimately simply had to go. The *Dryopithecus punjabicus* turned out not to be *Dryopithecus* at all. The upper jaw of *Ramapithecus* and the lower jaw of *Bramapithecus* were found to belong to the same species. *Ramapithecus* and *Sivapithecus*, most paleontologists now believe, are females and males of the same species; *Ramapithecus*, *Bramapithecus*, *Sivapithecus*, and *Kenyapithecus* are therefore all synonyms. The shape of the *Ramapithecus* dental arcade, which appeared to resemble that of humans, is now believed to be the result of forcibly fitting two unrelated pieces of a jaw together. And the remaining human-like characters of *Ramapithecus* are currently thought to have arisen by convergent evolution.

All this, however, would not have sufficed to convince paleontologists that Sarich and Wilson were in the right had there not been additional evidence, much of it provided by paleontologists themselves and all of it consistent with the younger age of the most recent common ancestor. Numerous new hominid fossils have been found since and none are significantly older than the age indicated by the molecular data. The hominid lineage therefore now seems to merge with the panid (chimpanzee) lineage at around this date. The hominid-panid convergence is, in fact, one of the best documented lineage mergers, and as such is one of the most reliable yardsticks for the calibration of the molecular clock. Even in this case, however, the yardstick is only approximate: the divergence of the hominid-panid lineages might have taken place as early as 4.4 my ago or as late as 8 my ago, and there is no reliable information available at the present time that could be used to improve the precision of the estimate.

The intention behind retelling the *Ramapithecus* story was by no means to malign or denigrate paleontology. Blunders of this kind can and do take place in any discipline, including molecular biology. The purpose of relating it was to drive home the message that paleontology alone is no more reliable than molecular evolution and that only when the two disciplines join forces can there be hope of achieving greater reliability and sharper precision than with each operating independently.

The Thousand and One Clocks

Although the notion of the molecular clock is now almost 40 years old and although the clock has been in use more or less continuously since 1962, its application to measuring time on the grand scale of the Tree of Life began only recently. Not surprisingly, therefore, the application is still suffering from growing-up pains and as a result, the time estimates it has produced must be regarded as tentative. The estimates can be expected to become more reliable, however, as a result of advances on three fronts. First, the number of sequences available for analysis will undoubtedly grow exponentially both in regard to different genomic regions and to different organisms, thus enabling molecular evolutionists to apply stricter criteria to their selection of appropriate chronometers. Second, the methods of sequence data analysis are continuously improving. And third, the Tree of Life, on whose accuracy the estimates rest, is also steadily becoming far more robust.

The standard procedure for generating a molecular time scale for the Tree of Life is to use as many protein sequences as possible. Protein rather than DNA sequences are preferred, even though, in most cases, the former are obtained by translation from the latter. The reason for this preference is historical more than anything else: protein sequences became available earlier than DNA sequences and many methods of sequence analysis were developed for them specifically. To be able to compare newer with older data, it is more convenient to stay with amino acid sequence data. Collections of protein sequences rather than single proteins are used to offset the stochastic variation of the clock. Since each protein evolves at a different rate, there are as many clocks in the collection as there are proteins. Each protein must therefore be calibrated separately, but when this is done, all should give the same divergence time estimate for a given pair of species. In reality, however, they don't because of the clock's stochastic behavior. The individual estimates are therefore averaged and the mean is then taken as the most probable divergence time. The range within which there is 95 percent probability of a value being the actual divergence time is indicated by a bar – the *confidence interval*.

Many of the proteins chosen for the time estimates are enzymes involved in cellular metabolism. This choice was made because the main metabolic pathways and the enzymes that propel them were established early on in the evolution of life and are therefore shared across life's domains. They are thus well suited for comparisons of distant taxa. In the standard procedure, each protein sequence is compared with all other sequences in the set (i.e., pairwise comparisons are made), the differences are counted, and the counts are corrected for multiple hits by reference to an appropriate probability distribution model. Adjustments are then made for possible differences in the replacement rate between different positions in the sequence (again by reference to a suitable probability model), tests are carried out to determine whether the clock may not have accelerated or slowed down in different stages of evolution, and the clock is then calibrated by the identification of divergence times indicated by the fossil record. Since the most comprehensive fossil record available is that of vertebrates, the number of replacements differentiating two vertebrate taxa whose divergence has been estimated from the fossil record is used as a unit of calibration. The divergence times of organisms such as bacteria and archaea, for which suitable fossils are not available, are then obtained by extrapolation. Evolutionary distances obtained for different pairs of vertebrate taxa are plotted against the divergence times of these pairs determined from the fossil record, and a line is then fitted as best can be to the points of the plot (Fig. 9.7). The line is then extended and the deep divergences corresponding to the greater evolutionary distances of the poorly fossilizing taxa are read off the graph.

The most comprehensive divergence time estimates obtained by this procedure thus far are those of Russell E. Doolittle and his colleagues. These researchers retrieved 823 protein (enzyme) sequences from the databases representing all three domains of life, and within each domain the main taxonomical categories. They then analyzed the data by the method described above and calibrated the individual protein clocks – not quite one thousand and one yet! – by reference to six pairs of vertebrate divergences indicated by the fossil record: mammal-mammal, placental mammals-marsupials, mammals-birds/reptiles, amniotes-amphibians, tetrapods-fishes, and jawed vertebrates-jawless vertebrates, which give divergence times of 100, 130, 300, 365, 405, and 450 my ago, respectively. By extrapolation from the plot of these values they then obtained the following divergences: echinoderms-chordates, 590 my ago; deuterostomes-protostomes, 850 my ago; schizocoelomates-pseudocoelomates, 1045 my ago; fungi-animals, 1272 my ago; plants-animals, 1215 my ago; protists-plants-animals-fungi, 1545 my ago; archaea-eukaryotes, 2409 (2261) my ago; bacteria-

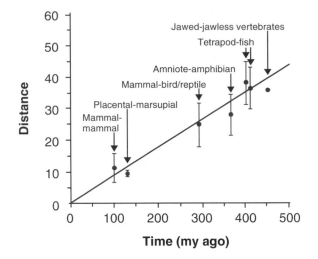

Fig. 9.7

Calibration of the molecular clock. Proteins of groups of vertebrates, for which divergence times are reasonably well indicated by the fossil record, are compared and distances between them calculated. The distances are then plotted against the divergence dates provided by paleontologists. The fact that the straight line fits the points quite well indicates that the proteins have indeed evolved with the regularity of a clock. The increment of distance corresponding to a time unit in years is then used to estimate the divergence time of other organisms for which the fossil record either does not exist, is incomplete, or is less reliable. Vertical bars indicate the likely range within which there is 95 percent probability that a certain value is the actual divergence time (Modified from Feng et al. 1997.)

eukaryotes, 2188 (2080) my ago; archaea-bacteria, 3784 (3125) my ago (the values outside the parentheses are the upper limits, those in the parentheses are the means). The deep divergences were obtained after the exclusion of sequences encoded in genes that may have been transferred horizontally between domains. Separate analysis indicated that most of the horizontal transfers occurred after or at the same time as the origin of eukaryotes, when organelles such as the mitochondria arose from endosymbiotic bacteria.

The oldest divergence – that between archaea and bacteria – is more than one billion years older than the same divergence estimated by Doolittle and colleagues one year earlier from a more limited dataset. The mean estimate of 3.1 by ago is younger than the oldest traces of life on earth (see Fig. 9.4). The oldest microfossils, found by J. William Schopf in the rocks of the Warrawoona Group in northwestern Australia, are 3.4 by old. They have a variety of forms, but most are threadlike and closely resemble living photosynthesizing cyanobacteria.

That the life forms existing at that time were indeed capable of photosynthesis is indicated indirectly by *banded-iron formations* (BIFs). The metal occurs in nature in two forms, ferrous iron (Fe^{2+}) and ferric iron (Fe^{3+}), which differ in the number of electrons in the outer shells of the iron atom, as well as in physicochemical properties. Ferrous iron is present in the deep regions of the earth; it reaches the outer regions through cracks in the ocean floor, dissolved in hot water. When it gets to the surface water layers, it reacts with molecular oxygen, O_2, dissolved in the water, to form insoluble ferric oxide, Fe_2O_3, the mineral hematite and the main constituent of rust. The tiny grains of rust then rain on the ocean floor to form reddish-colored sediments. All the major deposits of iron ore on earth were formed in this manner. Their formation required enormous amounts of oxygen which, geologists believe, could only have been released into the water and into the atmosphere by photosynthesizing organisms. The earliest iron ore deposits have been dated back to 3.8 by ago, which is also the age of the oldest sedimentary rocks, the oldest rocks of any kind being dated to 3.9 by ago (see Fig. 9.4). The deposits appear in great abundance 3.5 by ago and begin to dwindle 2 by ago. Typically, the deposits resemble a multilayered cake, with dark, rusty-colored, iron-rich layers alternating with pale bands poor in iron (hence the name BIF). The banding pattern indicates changing periods of intensive hematite deposition followed by periods of minimal deposition, the reasons for which are not entirely understood. Seasonal variation is one possibility, although periods of population expansion followed by periods of crashes due to oxygen poisoning of the photosynthesizing bacteria is also feasible, along with various other factors. At any rate, the existence of BIFs 3.8 by ago suggests that cellular life was already well established by this time. The considerably younger *mean* molecular estimate of the bacterial-archaeal divergence can be reconciled with the geological/paleontological evidence by postulating that the oxygen was produced by the common ancestors of bacteria and archaea or that the upper value of the molecular estimate (3.8 by ago) rather than the mean is closer to the actual divergence. The latter explanation is supported by the observation that the microfossils resemble cyanobacteria, which presumably arose after the divergence of bacteria and archaea. The molecular clock's estimate of the separation of eukaryotes from archaea (mean of 2.2 by ago and upper limit of 2.4 by ago) is also in reasonable agreement with the paleontological estimate. The oldest fossils that are generally accepted as eukaryotic are coiled, ribbon-like algae from 2.1-by-old rocks in Michigan.

Some of the other molecular estimates contradict those based on the fossil record and for no other group has this had such profound consequences as for the Metazoa. To paleontologists, as you will recall, the origin of the Metazoa seemed so sudden that they began calling it the *Cambrian explosion*. They made the geologically instantaneous appearance of fossils in the sedimentary rocks into a milepost between two stages of the earth's history, the Precambrian and the Phanerozoic. Yet, the fossils did not appear *so* suddenly that, when radiometric dating was later introduced, the beginning of Cambrian could be fixed on the absolute time scale unambiguously. In the end, the date of 543 my ago was chosen arbitrarily. In the Cambrian sediments, the explosion was particularly pronounced in a few sites on earth, the *Lagerstätte* (German for "resting place" or "lodging"), exhibiting a concentration of a great diversity of fossils at one locality. At these sites, the fossils are of exceptionally high quality and even include impressions of soft body parts. The most famous of the Cambrian Lagerstätte are the *Burgess Shale*, a quarry located near the Burgess Mountain in the Rockies of British Columbia, Canada; the *Sirius Passet* on Peary Island in north Greenland; and *Chengjiang* in the Yunnan Province, South China.

At Burgess Shale, at an elevation 2300 meters above sea level, erosion has exposed layers of shale formed by the compaction of beds of clay and mud from the time when this part of the North American continent was a seashore. Here, a few meters underwater, a plain teeming with life extended for some distance from the shore and then dropped vertically, at a cliff's edge, some 70 meters down. From time to time, perhaps triggered by violent storms or by the earth's movements, great masses of the plain's mud slid over the cliff, whisking away all the lagoon's inhabitants and burying them at the bottom. Here, at the foot of the cliff, deprived of light and oxygen, the entombed animals died, and as the mud eventually turned into fine-grained shale, their bodies were transformed into some of the most exquisite fossils of soft-bodied animals ever found. The periodic mud slides produced layer after layer of fossil-rich sediments – a geological herbarium, enabling the paleontologists to leaf through and study fossils as if they were pressed flowers. When the Rocky Mountains formed less than 20 my ago, the sediments were lifted to their present position, where they caught the paleontologists' attention at the beginning of the last century. In the cornucopia of fossils, paleontologists have identified some 170 different animal species. The earliest metazoan fossils of the *Burgess Shale fauna* have been dated to 540 my ago. A "mere" 20 my later the fauna contains representatives of all major modern phyla – and that, on a geological time scale and in the eyes of the paleontologist, is indeed an explosion!

There are several controversial issues surrounding the interpretation of the Burgess Shale fauna, but the two most profound ones are, first, whether the phyla really diverged within 20 my and, second, whether in addition to the phyla that survived into modern times there were also many others that in the meantime have become extinct. As to the first question, we will see shortly that the phyla did not arise within 20 my, and that the actual time it took for the various body plans to diverge still remains unresolved. As regards the second question, like a child who has discovered a treasure-trove of toys in the attic, some paleontologists appear to have been carried away in their interpretation of the Burgess Shale's faunal diversity. Not only did they literally view some of the fossils wrong side-up, but they also discerned novel, extinct body plans in some of the fossils that were later shown to represent surviving phyla. A sober assessment of the fauna confirms that a few of the original phyla may indeed have become extinct, but that on the whole there was no great fanning out of body plans in the Cambrian followed by a random extinction of most of the original phyla so that if the process were repeated again, the outcome would be entirely different.

Back to the first question: how much time did it take for the phyla to diverge? Here, too, the enthusiasm expressed by paleontologists for explosions was dampened somewhat by later developments. Paleontologists themselves provided some of the dampers, but the fire extinguishers were contributed by molecular evolutionists. Paleontologists gradually realized that the transition from an eon of no fossils into one of abundancy was not as abrupt as originally thought. In 1946, Reginald C. Sprigg examined an abandoned mine called Ediacara (presumably an aboriginal word for "foul waters") in the Flinders Ranges north of Adelaide in South Australia, and discovered that the rather coarse-grained sandstone contained numerous impressions of soft-bodied, weird-looking creatures. Some of the impressions were disc-shaped, vaguely resembling jellyfish. Others were elongated, like a worm, while others still resembled fern or palm fronds more than animals of any known kind. In addition, tracks, trails, burrows, and fecal pellets came to light, such as only moving creatures could produce. Later, fossil assemblages related to *Ediacaran fauna* or *biota* were also discovered at several other places, in Russia, Poland, Namibia, United States, and Canada.

What kind of creatures left these impressions and trace fossils? One group of paleontologists believes that they were early metazoans, related to present-day sponges, cnidarians, worms, and arthropods. A second group argues that the creatures were metazoan alright, but were unrelated to any animals alive today. Finally, a third group of paleontologists maintains that the ediacaran organisms were not metazoans at all, but *vendozoans* (a name derived from Vendian, a subdivision of the Proterozoic eon), representatives of a separate kingdom, now extinct. The controversy remains unresolved, but evidence is now accumulating that some of the ediacarans at least were indeed soft-bodied metazoans. The evidence consists of burrows which could only have been left behind by worms; of body impressions very similar to those of simple molluscs; of markings interpreted as having been made by a mollusc radula, a horny, teeth-covered band in the mollusc's mouth; of fossilized embryos; and of microfossils representing tiny marine sponges identifiable by their spikule, minute siliceous or calcareous elements that support the spongal tissues. The exquisitely preserved sponge microfossils and fossilized embryos were found in 570-my-old phosphate-rich deposits comprising the Doushantuo formation in the central Guizhon province of south-central China. In some of the embryos, the clearly recognizable cells are arranged in a pattern characteristic of bilaterally symmetrical animals, indicating that not only sponges, but also bilaterian phyla were already in existence by that time.

The various ediacaran biotas have been dated back to a time interval of 570 to 540 my ago. One of them (discovered in the Mackenzie Mountains in Canada) has been tentatively dated to 600-610 my ago. Taken together, the combined paleontological evidence indicates that several metazoan phyla already existed 570 my ago and included a number of those that are still in existence today. The metazoans therefore had at least 50-100 my to develop their different body plans. In reality, however, they may have had an even longer, a *much* longer time. In a recent report, Adolf Seilacher and his colleagues described burrows found beneath microbial mats which they argue were excavated by triploblastic, worm-like animals. The sediments in which these trace fossils occur are part of the Chorhat Sandstone in the Son valley of central India, and have been dated to 1.1 by ago.

Attempts to date the origin of the Metazoa and the divergences of the animal phyla on the basis of collections of protein sequences have recently been undertaken by several groups of molecular evolutionists. All these studies support the notion that the divergences occurred long before the Cambrian period. There is, however, a considerable discrepancy among the various studies regarding the actual dates of the divergences, presumably due to the selection of sequences, selection of taxa, calibration of the molecular clock, and methods of analysis used. In the most conservative study (carried out, incidentally, by some of the same researchers who in another publication argue that the molecular clock is a "mirage"), the protostome-deuterostome split has been dated to 670 my ago. (Curiously, the authors claim that this value is in agreement with paleontological data, although it is 100 my older than the oldest Ediacaran fossils!) Most of the other studies provide an older estimate of the protostome-deuterostome divergence (Fig. 9.8). The least conservative estimate gives a date of 1.2 by ago, which is close to the trace fossil-based date.

All in all, the studies support the view that the various metazoan body plans were established long before the Cambrian period began. This raises the question why so little fossil evidence has been found for the Precambrian stages in metazoan evolution. The two obvious answers are either that no evidence exists or that it has been sought in the wrong place using the wrong methods. The fact that in the last few years, paleontologists have been steadily pushing back the date of the origin of animal phyla suggests that the trend will continue and that the latter of the two possibilities will turn out to be true. Already, however, there is little justification for retaining the old notion of the Cambrian explosion when referring to the sudden origin of metazoan body plans at the beginning of the Cambrian period. On the other hand, the term is warranted when used in reference to the dramatic increase in the abundance of readily available macrofossils. The reasons for the sudden rise in the quality and quantity of fossils in the sediments remain obscure. The appearance of hard, easy-to-fossilize body parts is probably only one factor among many.

Molecular clock-based estimates of divergence times between some of the branches comprising the metazoan tree are given in Figure 9.9. These estimates generally give older dates than those based on the fossil record. For example, the oldest primate fossils are

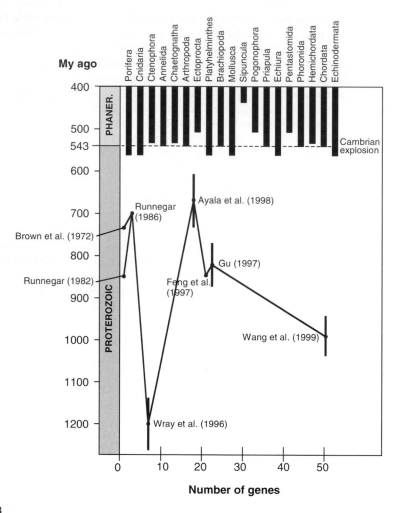

Fig. 9.8
Fossils versus molecules: estimates of the divergence times of various animal phyla. Vertical bars at the top indicate the extent of the fossil record; the points on the curve are estimates based on the indicated number of protein molecules (genes) by different authors (see *Sources*). Vertical bars over points indicate the likely range within which there is 95 percent probability that the value is the actual divergence time. Phaner, Phanerozoic. (Modified from Wang et al. 1999.)

dated to 65 my ago, whereas the sequence-based estimate for the divergence of primates from the tree shrews is 85 my ago. Considering the general tendency of the fossil-based dates to be shifted back with time, the sequence-based divergence times may not be too far off the mark.

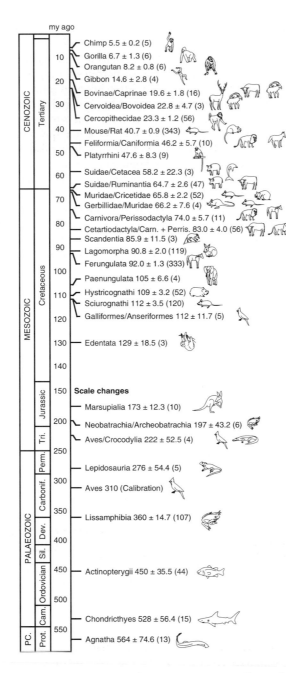

my ago

Fig. 9.9

A molecular time scale for vertebrate evolution. Dates are given in millions of years before present or as divergence times of the groups separated by a slash (/). The ± values indicate the likely range within which there is 95 percent probability that a certain value is the actual divergence time. The figures in parentheses are the number of genes on which the estimates are based. The geological time scale is provided on the left-hand side. Cam, Cambrian; Carbonif, Carboniferous; Dev, Devonian; PC, Precambrian; Perm, Permian; Prot, Proterozoic; Sil, Silurian; Tri, Triassic. (From Kumar & Hedges 1998.)

Time is Relative

This chapter has set out two very different perceptions of time – that of paleontologists and that of molecular evolutionists. Paleontologists perceive time, poetically expressed, as *treacherous in calm and terrible in storm*, as long stretches of eventlessness punctuated by brief episodes of dramatic content. They do not see the sand grains raining down steadily on the ocean floor, or the accidental burials of cadavers – the wrecks vomited upon time's inhospitable shore. They see only the final outcome of a long process, a bed, a layer of deposit with or without fossils. And only when they can distinguish one layer from that below it and that above it, do they know that something dramatic has taken place at the point of transition. Similarly, paleontologists tend to ignore individual variation in the wrecks of time, the fossils, or dismiss it as insignificant, and are always on the look-out for leaps and bounds involving large groups which they can categorize as new species and match their entries with the succession of layers. In this manner, they see fossils come and go, their entrances and exits appearing instantaneous, like those of actors on a stage. Expressed differently: paleontologists do not watch the movie of life's history on a film screen in a darkened theater; they scrutinize it frame after frame in a cutting room. To the cutters, even the scenes that fade in and out appear as sharply distinct, discrete frames. When they then superimpose the absolute time scale on the column of layers, the illusion of intermittency becomes even more pronounced. Greatly impressed by this spasmodicity, some of the paleontologists then make the mistake of extending it to the entire process of evolution.

Molecular evolutionists see things differently. To them, the flow of time appears smooth rather than jerky, gradual rather than spasmodic. To them, time is more like a *shoreless flood, which in its ebb and flow claspest the limits of mortality*. When they compare nucleotide or amino acid sequences, they assume that the discontinuities they observe in the number of substitutions distinguishing two species exist because the intermediate forms have become extinct. If the proteins or DNA molecules of these intermediates could be sequenced, the gradual, smooth increase in the number of differences along a lineage would become apparent. The superposition of an absolute time scale on the molecular scale then reveals, in the opinion of some molecular evolutionists at least, a steady accumulation of substitutions at more or less regular intervals – the ticking of the molecular clock.

Because of this difference in perception, it is not surprising that paleontologists and molecular evolutionists often disagree in their interpretation of evolutionary events, especially in regard to their timing. Yet, *because* the disagreements stem not from differences in facts but from disparate perspectives, there is hope that one day a consensus will be reached and a peace treaty drawn between the two disciplines. On that day, the Tree of Life will be embellished with three time scales – paleontological, molecular, and absolute – all in complete harmony with one another.

Until then, however, things remain relative.

"Time is relative, right, Dr. Einstein?"

"Absolutely!"

CHAPTER 10

The Narrow Road to the Deep North

Hominids and the Origin of *Homo sapiens*

Lucy,
požehnaná mezi ženami,
před třemi milióny let,
když ještě nebyly legendy,
nýbrž jen láskyplné hledání
lupů v kožiše,
čekala na někoho
na břehu jezera,
ale přišla jen smrt,
vhodná pro druh
Australopithecus afarensis,
malá smrt s chůzí člověka
a lebkou opice.

Lucy, vykopaná,
částečně nanebevstoupená
po doplnění sádrou,
zapomenutá říkanka
na-koho-to-slovo-padne,
čeká dál.

Lucy,
Blessed among women,
Three million years ago,
When there were still no legends,
Only the loving search for
Dandruff in the fur,
Was waiting for someone
at the lakeshore,
But only death came,
Appropriate for the species
Australopithecus afarensis,
Small death with human gait
And monkey skull.

Lucy, dug out,
Partially ascended
After being filled up with plaster,
Forgotten nursery rhyme
Eenie-meenie-minie-mo,
Continues to wait.

Miroslav Holub: Hominization. (From Czech/Translated by J.K.)

The Records of Weather-Exposed Skeletons

Matsuo Bashō, the greatest of the Japanese *haiku* poets, undertook three long journeys during his lifetime and described them in the form of travelogs, in which prosaic narrative alternates with *haiku*. The description of the first journey, *The Records of a Weather-Exposed Skeleton*, opens thus:

> Following the example of the ancient priest who is said to have travelled thousands of miles carrying naught for his provisions and attaining the state of sheer ecstasy under the pure beams of the moon, I left my broken house on the River Sumida in the August of the first year of Jyōkyō among the wails of the autumn wind.
> Determined to fall
> A weather-exposed skeleton
> I cannot help the sore wind
> Blowing through my heart.*

In the spring of 1689, Bashō left for the third of his major journeys. The nearly fifty-year-old poet, an old man by the standards of his time, sold his house in Edo (now Tokyo) and headed north, relying entirely on the protection of temples and the hospitality of fellow poets. The Japanese North was an unexplored, mysterious territory at that time and Bashō's journey was thus a venture into the unknown and without return – a symbol of a journey through life toward eternity. Bashō's trip is described in one of the classics of Japanese literature, *The Narrow Road to the Deep North*. The description of the first journey can be interpreted as symbolizing humanity's journey northward from its birthplace in Africa, a journey of which weather-exposed skeletons left a tantalizing record. And so we have borrowed from Bashō's classics not only the title of this chapter that deals with the expansion of the human species into the far north and then throughout the world, but also the title of this section in which we briefly summarize the fossil evidence of humanity's journeys.

If the phylogeny of placental mammals suggested by the multigenic studies (see Chapter Eight, particularly Figure 8.2B) is correct, primates, Dermoptera, and Scandentia shared an immediate common ancestor with glires (rodents and lagomorphs). The ancestor lived in the Laurasia, where the earliest primates also emerged. Fossil remains of these "archaic" primates exemplified by the Plesiadapiformes were found in Europe and North America. They resemble primates so little that some authorities question whether they should be included in this order at all. The archaic primates had elongated faces, small brains, and clawed digits; they lacked the postorbital bar; and their bodies were the size of a mouse or squirrel. They flourished between 65 and 55 my ago, then began to decline, and became extinct at about 35 my ago. They were replaced by primates of a "modern aspect", particularly the Adapiformes and Omomyidae, which began to diversify approximately 56 my ago and which had all the characteristics of living primates mentioned in Chapter Eight. Fossil remains of the "modern" extinct primates, like those of the archaic primates, are restricted to Europe and Asia, much of which was covered by tropical and subtropical forests at that time. Later, when a general trend toward a cooler climate set in and the tropical forests began to recede toward the equator, the primates followed them and have remained allegiant to this habitat, with some notable exceptions, to this day.

Primates reached Africa 35 my ago, at the latest. The earliest primate fossils on African soil were found in the sedimentary deposits of the Fayum depression in Egypt. And it was

---

*Bashō: *The Narrow Road to the Deep North and Other Travel Sketches*, pp. 51.

presumably in Africa that the ape or *hominoid* lineage separated from the rest of the Catarrhini. The African origin of hominoids is indicated by the earliest fossils that can be assigned to this lineage (Fig. 10.1). They are distinguished from other catarrhine fossils by the relative increase in brain size, by the absence of a tail, by a thickening in the enamel of the molar teeth, and by various adaptations in the limb skeleton to a new mode of locomotion. Most of the fossils, which first appear in the Oligocene 25 my ago, are restricted to East Africa. The first hominoids seem to have inhabited tropical forests; they lived largely in trees, were quadrupeds, and fed on hard fruits. The most numerous among them were various species of the genus *Proconsul*, widely distributed in the forests of Uganda and Kenya between 22 and 17 my ago. Others included *Dendropithecus*, *Rangwapithecus*, *Oreopithecus*, and the so-called Lothidok hominoid, which is among the oldest hominoids discovered thus far, dating to 27-24 my ago.

Later in evolution, the hominoids spread out of Africa (if that is where they emerged), first to Europe and then to Asia. This expansion may therefore have been the first journey north for the new group of primates. (Other primates, Catarrhini, Platyrrhini, and Strepsirhini, were by then distributed over several continents and subcontinents.) It may

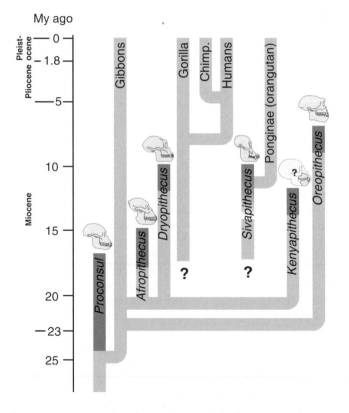

Fig. 10.1
A family tree of hominoid primates (apes). Question marks indicate uncertainty regarding possible connections to other lineages. A geological time scale is added to indicate the presumed divergences and the duration of the lineages. Bars highlighted in color indicate persistence time as documented by the fossil record; the gray bars indicate presumed persistence. Skulls have been reconstructed according to fossil evidence. (Based on Andrews & Stringer 1993.)

have been undertaken by a group of hominoids summarily referred to as *dryopithecines*, some of which apparently abandoned the tropical forests of East Africa for the subtropical, more open temperate forests of North Africa and Europe. The group, distinguished by a set of characters pertaining to tooth morphology (enlarged premolars, molars with low cusps and flattened biting surfaces, robust but low-crowned canines), consists of three subgroups. These are the *afropithecines* (*Afropithecus*), who were confined to Africa some 17 to 15 my ago; *kenyapithecines* (*Kenyapithecus, Sivapithecus, Griphopithecus*, the first two familiar to us from the preceding chapter), whose remains have been found in both Africa and Europe and dated to 15 to 12 my ago; and *dryopithecines proper* (*Dryopithecus*) who inhabited Europe from 12 to 8 my ago (Fig. 10.1). The phylogenetic relationships among the subgroups and those of the entire group to other hominoids are uncertain.

At some point during the reign of the dryopithecines a new group of hominoids emerged which was to produce the most recent common ancestor of the chimpanzee, gorilla, and the human. Regretfully, only a few fragmentary fossils document this event. A set of fossilized jaws and teeth from Macedonia and Greece, gathered in the genus *Graecopithecus* (= *Ouranopithecus*), is thought by some paleoanthropologists to belong to the line of descent leading to the African great apes and humans, and by others to be more closely related to *Sivapithecus* and so part of the line that culminated in the orangutan. A single upper jaw fossil from the Samburu Hills in Kenya, which possesses some of the characters distinguishing the African great apes, might be related to the gorilla or to the chimpanzee. Apart from these, there are no other fossils available that could document the evolution of the chimpanzee and the gorilla (i.e., the lineages that could be called *panid* and *gorillid*, respectively). The absence of fossils is either an indication that the immediate ancestor of these species lived in a region now covered with tropical forest in which the chances of finding fossils are minimal, or that they have simply not yet been found.

The glaring absence of fossils in the chimpanzee/gorilla lineage contrasts with the relative abundance of fossils in the hominid line of descent. Here, not just bone fragments, but whole skulls and even whole skeletons, as well as footprints and countless artifacts, have been retrieved from the earth's rocky archives. Since neither panid nor gorillid fossils have been found at the sites at which hominid bones have been unearthed, the ancestors of the African great apes must have preferred different habitats from those occupied by human ancestors. Or could it be that some of the fossils currently assigned to the hominid lineage are in reality panid or gorillid remains? How do paleoanthropologists decide whether a fossil is hominid rather than panid or gorillid?

The two chief characters that distinguish the hominid lineage are bipedalism and small canines. The switch from quadrupedalism to bipedalism required numerous adaptations, not only of the bones and muscles of the hind legs, the pelvis, and the hip joints, but also other parts of the skeleton, including the skull. Paleoanthropologists therefore do not require leg or pelvic bones to tell whether a creature walked on two or four feet. They can answer that question, for example, by examining the skull (Fig. 10.2). The major adaptation to erect posture and bipedalism in the skull is a positional change of the *foramen magnum*, the large opening through which the spinal cord enters the cranium to connect with the brain (Fig. 10.2). In quadrupedal primates, as well as in many other vertebrates, the opening tends to lie toward the back of the skull, so that it meets the horizontal or sloping vertebral column at an angle. In bipedal hominids, by contrast, the opening is at the base of the skull, which is positioned at the tip of the vertebral column like a ball on a scepter. As for the canines, in primates other than hominids these teeth are longer than the incisors, premolars, and molars, and in some species they are longer in males than in females. Long canines are therefore viewed as an adaptation not only to a particular mode of feeding, but also to the assertion of dominance by displays of aggression or fighting. In hominids a gradual trend toward shortening of the canines is clearly discernible, presumably as an adaptation to a different diet and to an alternative method of asserting dominance. In humans, the canines in both females and males are no longer than the other teeth. (Only in Dracula and his blood-sucking allies, the vampires, do they revert to their ancestral length!)

The position of the foramen magnum, the length of the canines, and a few other criteria enable paleoanthropologists to identify a fossil as having originated from a hominid ancestor. But this is only the first step in the fossil-identification process. The next two steps are to identify the species from which it is derived, and to determine its phylogenetic position among the various species of the hominid lineage. Both steps are enormously difficult and paradoxically become even more so as greater numbers of fossils are found. With a mere handful of hominid fossils, each dated to a different time interval, it seemed relatively unproblematic to agree on their identities, if not always on their names or their relationships. It was not so long ago that textbooks commonly depicted the hominid lineage as a linear progression starting with an *Australopithecus* species and moving on via *Homo habilis* and *H. erectus* to *H. sapiens*. Since that time, however, the number of hominid fossils has greatly increased and concurrently with the increase, the entire picture has become more complex and more confusing. It has become clear that hominid evolution has been anything but a linear progression in which one species neatly transforms into a new one, which then gives rise to a different species again. Instead, hominid evolution has occurred by what evolutionary biologists call *adaptive radiation*, a process in which a founding species fragments into multiple populations, either simultaneously or in rapid succession, each of which then begins to evolve toward a new species by adapting to the environmental niche it has colonized (see Chapter Eleven).

At present, there are probably as many interpretations of hominid evolution as there are researchers studying it. Nevertheless, there are a few aspects of it on which most, if not all, researchers seem to agree. Paradoxically once more, a greater consensus of opinion can be found on the earlier stages of evolution than on the later ones. By general agreement, hominids originated in East and South Africa from an ancestor they shared with the African great apes. The group of hominids closest to this most recent common ancestor are the *australopithecines* (literally, "southern apes"; Fig. 10.3). They have left behind some of the oldest fossils on which evidence of bipedalism and reduced canine size can clearly be recognized. In many other respects, however, they apparently resembled the two living African apes, for example in the size of their brains and in their relatively long snouts (Fig. 10.4). They were of a fairly short stature, their height ranging from 1 to 1.5 meters and their weight varying between 45 and 80 kilograms. Australopithecines existed for about three million years, from 4.4 to 1.3 my ago. The number of species produced by this first major radiation in the hominid lineage is uncertain, but at least ten have been named – seven *gracile* (more lightly built) and three *robust* forms. Whether all australopithecines should be regarded as one genus or as several genera is debatable. The robust forms are now commonly assigned to a separate genus *Paranthropus* (literally "beside man"), whereas the gracile forms are classified into four genera (Fig. 10.3).

At present, it is controversial which fossil derives from the oldest hominid. Until recently, this title was generally bestowed on *Ardipithecus ramidus,* whose bone fragments were found at Aramis, Middle Awash, Ethiopia (Fig. 10.5), at a site believed to be 4.4 my old (in the language of the Afar *ramid* means "root"). The teeth and bones of this species share several morphological characters with the chimpanzee, indicating that it may have been close to the most recent common ancestor of humans and chimpanzees. This inference is further supported by the fact that the fossil remains were found in sediments containing an abundance of fossilized wood, seeds, and monkey bones, suggesting that *A. ramidus* may have been a woodland creature.

In the fall of the year 2000, however, a French team of paleoanthropologists and geologists discovered the fossilized remains (broken femurs, bits of lower jaw, several teeth – 13 fragments altogether) of a creature they believe was a hominid much older than *Ardipithecus ramidus*. They named it *Orrorin tugenensis* because "Orrorin" means "original man" in the dialect of the tribe in the Tugen Hills region of northwestern Kenya, where the remains were found in the lake and river sediments. For the general public they nicknamed their discovery the "Millennium Ancestor" because of the year in which it was

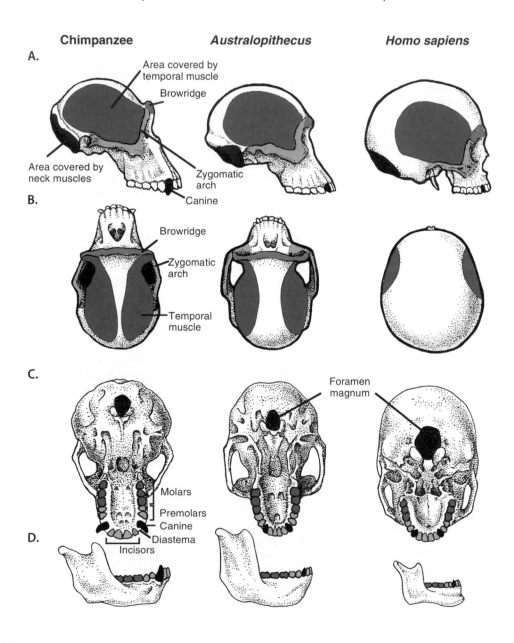

found. The claim that the remains, which are estimated to be 6 my old, are of hominid origin is based primarily on two observations. First, the shape and size of the hipbone (femur) suggested to the French paleoanthropologists that these individuals walked upright, and second, the small, square, thickly enameled molars indicated to them that evolutionarily, it was closer to the genus *Homo* than to any of the known australopithecines. They concluded that *Orrorin tugenensis* and not australopithecines was on the direct line of descent to *Homo*. This interpretation, however, is contested by some paleoanthropologists, who find the argument for bipedalism weak and who point out that enamel thickness of hominid molars is a highly variable character. Rather than being on the direct line to humans, it cannot be excluded that *Orrorin* was actually on the line to the chimpanzee. The issue is still under debate.

The much publicized "Lucy" of the species *Australopithecus afarensis* (literally "southern ape of Afar") was found only 20 kilometers from Aramis, at Hadar in the Afar region (Fig. 10.5). This species, which is estimated to have existed from 3.8 to 3.0 my ago, is morphologically distinct from *Ardipithecus ramidus* in that it shares more characters with later hominids. Numerous other fossils currently classified as *A. afarensis* may belong to more than one species. *A. afarensis* certainly does not appear to have been the only hominid species living during the time interval from 4 to 3 my ago, as indicated by the recent unearthing of a distinctive jaw in Chad and several skull fragments at Lomekwi near Lake Turkana dated to 3.5 my ago. The former fossil has been assigned to a new species, *Australopithecus bahrelghazali*; the latter fragments have been interpreted as representing a new genus and named *Kenyanthropus platyops* in reference to the relatively flat facial contour. In addition to these species, there existed at least two gracile forms, *Australopithecus africanus* in South Africa and *Australopithecus garhi* in Ethiopia, and at least three robust species, *Paranthropus aethiopicus* and *P. boisei* in East Africa, as well as *P. robustus* in South Africa (Fig. 10.3). *A. africanus* remains were originally found in a cave near the town of Taung (and hence came to be known as the "Taung child"), approximately 130 kilometers north of Kimberley (Fig. 10.5). Later, fossils of the same species were also

---

Fig. 10.2A-D
Skulls of a chimpanzee, an *Australopithecus*, and a modern human (*Homo sapiens*): A comparison. **A.** Side view. **B.** View from the top. **C.** The braincase with the upper jaw (maxilla) viewed from the bottom. **D.** The lower jaw (mandible) viewed from the side. **A** and **B.** In the chimpanzee and australopithecines much of the cranium (braincase) is covered by the temporal muscle which moves the lower jaw and is required for chewing (mastication). Later in the hominid lineage, the area covered by the muscle decreases as the entire chewing apparatus loses its power. Similarly, the chimpanzee and some of the australopithecines have powerful neck muscles for holding the head in a horizontal position, but later in evolution, as the skull gradually becomes balanced on top of the spinal column, the area covered by these muscles also decreases. In the chimpanzee and australopithecines the browridges and the zygomatic arch, which support the chewing muscles, are strongly developed, whereas in modern humans the browridges are reduced in size and flattened so that they are no longer visible when viewed from the top; the zygomatic arch is much less pronounced. **C** and **D.** The transition from quadru- to bipedalism was accompanied by repositioning of the foramen magnum, the opening through which the spinal cord connects with the brain, from the back to a more central position on the underside of the cranium. The palate has been tucked under the braincase and the shape of the dental arch has changed from being nearly rectangular in the chimpanzee, through a U-form in the australopithecines, to parabolic in modern humans. Change in diet has led to marked alterations in dentition, in particular the canines. The long, projecting canines of the chimpanzee, separated from the incisors in the upper jaw by a gap (diastema), have gradually become shorter so that in modern humans they project no farther than the other teeth. The reduction in chewing power has been accompanied by a reduction in the size of the lower jaw in modern humans. (Modified from Beneš 1994.)

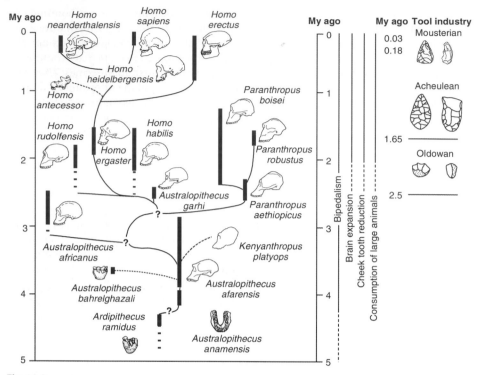

Fig. 10.3
The hominid family tree (one of several possible interpretations). The right-hand side of the figure shows the documented (solid line) and postulated (broken line) emergence and duration of key hominid characters, as well as the duration of three major cultures as identified by characteristic stone tools. The relationships between the individual species (if this is indeed what they were), as shown by the thin connecting lines, is extremely hypothetical and controversial. The vertical bars indicate the documented (solid) or postulated (broken) life spans of the species, symbolized by cartoons depicting their skulls or skull fragments. *Orrorin tugenensis*, whose hominid status is currently disputed, is not included in the diagram. (Modified from Strait et al. 1997 and R.G. Klein, 2000.)

found at other locations in South Africa, as well as in Ethiopia, Kenya, and Tanzania. The fossils have been dated to between 3 and 2.5 my ago. Several characters of the robust australopithecines (e.g., a crest on the top of the skull supporting the heavy jaw muscles, a flat, broad face, massive, long jaw, and large molars) are unique among the hominids.

The relationships among the australopithecine species and their relationship to the genus *Homo* have not been resolved. Almost all possible permutations have been proposed and the scheme shown in Figure 10.3 is only one of many. As with the earliest known hominids, australopithecines, and in particular the skeleton nicknamed "Lucy", continue to draw the avid attention of both experts and public alike. Lucy herself has not only been the subject of countless articles, but was also the inspiration for at least one artist, the Czech poet and scientist Miroslav Holub.

One of the australopithecines is believed to have given rise to the genus *Homo*, presumably in Africa. The genus is defined by the onset of a trend toward a dramatic increase in brain size (Fig. 10.6), a de-emphasis on processing food by intensive chewing, as reflected by changes in the size and distribution of chewing muscles, and the development of cul-

Australopithecus afarensis

Paranthropus africanus
(Taung child)

Paranthropus aethiopicus
(Black skull)

Australopithecus robustus

Paranthropus boisei
(OH5)

Homo habilis

Homo ergaster
(KNM-WT 15,000)

Homo erectus

Homo heidelbergensis
(Steinheim)

Homo heidelbergensis
(Petralona)

Homo neanderthalensis

Homo sapiens
(Cro-Magnon)

Fig. 10.4
The hominid portrait gallery: head reconstructions based on available fossilized remains. The portraits of *A. robustus*, *H. habilis*, *H. erectus*, and *H. neanderthalensis* are by the Czech artist Zdeněk Burian; all others are by another Czech artist, Pavel Dvorský. (From Beneš 1994 and Wolf 1989.)

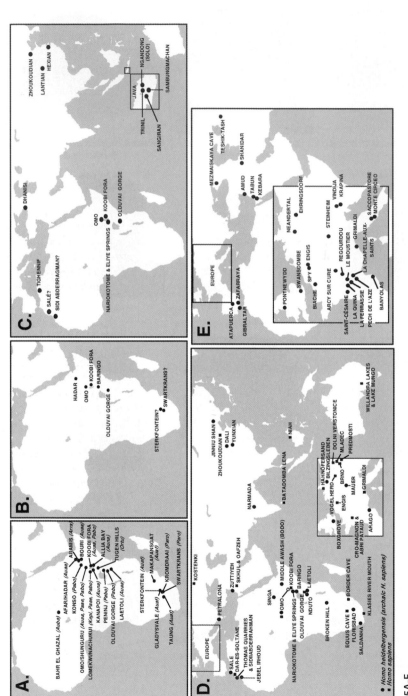

Fig. 10.5A-E
Main sites at which hominid fossils have been found, including those mentioned in the text. A Australopithecines; *Arra: Ardipithecus ramidus; Auaa: Australopithecus afarensis; Auaf: Australopithecus africanus; Auan: Australopithecus anamensis; Auba: Australopithecus bahrelghazali; Auga: Australopithecus garhi; Kepl: Kenyanthropus platyops; Ortu: Orrorin tugenensis; Paae: Paranthropus aethiopicus; Pabo: Paranthropus boisei; Paro: Paranthropus robustus;* B *Homo habilis, H. rudolfensis* and related forms; C *H. erectus* and *H. ergaster;* D *H. heidelbergensis;* E *H. neanderthalensis.* (Based in part on Stringer & McKie 1996.)

| Chimpanzee (400 cc) | Australopithecus afarensis (450 cc) | Australopithecus africanus (500 cc) | Homo habilis (700 cc) | Homo erectus (950 cc) | Homo sapiens (1350 cc) |

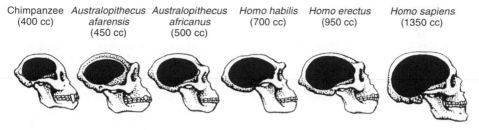

Fig. 10.6
The enlargement of the brain during hominid evolution (a chimpanzee brain is shown for comparison). The mean brain volumes are given in cubic centimeters (cc). Note that as the brain enlarges, the
face becomes smaller as a result of the expanding braincase. (From Beneš 1994.)

ture as manifested in the manufacture of tools. Some of the other characters distinguishing the early *Homo* species from the australopithecines included a more rounded head,
smaller, less projecting face, less massive jaw, larger front teeth but smaller cheek teeth,
open-U-shaped tooth row, shorter arms in relation to legs, more modern hands and hip
bones, and fully modern feet. (Here "modern" implies a closer resemblance to *H. sapiens*.)
The earliest named *Homo* species is *H. habilis* (literally "dexterous human"; Fig. 10.4), but
the name has been controversial right from the time it was introduced in 1964. Some experts felt that the fossil material was not distinct enough to be regarded as indicative of a
new species, but they could not agree on the species the bones did actually represent. One
group of experts took them for advanced australopithecines, while others argued that they
represented an early *H. erectus*. Although additional *habilis*-like fossils have been found,
the controversy has still not been resolved. If anything, the enrichment of the fossil trove
has made the issue even more controversial. This is because, first, the fossils show an unex-
pectedly wide range of variation in morphology and second, the bones often display a mix
of characters, resembling australopithecines in some and later species of *Homo* in others.
The consensus now seems to be that there may be more than one species in this group, but
that the species are difficult to distinguish. In addition to *H. habilis*, the designation
H. rudolfensis has been introduced for some of the early African forms of *Homo*. Needless
to say, the phylogenetic relationships among these forms, as well as their relationship to
australopithecines and later *Homo* species, is highly contentious. Purely for convenience
we refer to the group as *habiline hominids*.

Habilines lived in East and South Africa between 2.5 and 1.65 my ago, and so apparently
coexisted with australopithecines in the same geographical region. Reports of habiline fossils found beyond Africa (Israel and Indonesia) have not been substantiated. Habilines
were between 1.2 and 1.5 meters tall and weighed about 50 kilograms. Representative habilines are the owner of the KNM ER 1470 skull found at Koobi Fora on the eastern shore of
Lake Turkana in northern Kenya ("ER" stands for "eastern Rudolf", Lake Rudolf being the
former name of Lake Turkana), OH24 or "Twiggi", and OH62, the "Dik-dik Hill hominid"
from the Olduvai Gorge in northern Tanzania (Fig. 10.5).

Since habiline remains have been found together with primitive stone artifacts, it is believed that these hominids may have been the first to fashion stone tools (hence their
name) and perhaps even build simple shelters. Habilines may have been the founders of
the *Oldowan toolmaking industry* named after the Olduvai Gorge sites in Tanzania. This
designation refers to a simple method of using one cobblestone as a hammer to chip off
flakes from another cobble, the core (Fig. 10.7A). The sharp edges of the core can then be

A. OLDOWAN INDUSTRY

Chopper   Biface   Hand axe

B. ACHEULEAN INDUSTRY

Hand axe   Cleaver   Flakes

C. MOUSTERIAN INDUSTRY

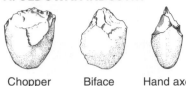

Cordiform   Convex   Levallois   Mousterian
hand axe   side scraper   point   point

Fig. 10.7A-C
Representative stone tools of three early tool industries, Oldowan (**A**), Acheulean (**B**), and Mousterian (**C**), associated with *Homo habilis, H. erectus,* and *H. neanderthalensis,* respectively. The industries are named after the sites at which the tools were found. Levallois is a variant of Mousterian industry named after a suburb of Paris. (**A** and **C** from Campbell 1988; **B** from Oakley 1959.)

used as choppers for bashing bones. The industry, which is limited to Africa, particularly to East Africa, emerged 2.5 my ago and lasted for about 850,000 years.

If there are doubts about the appropriateness of assigning the habilines to the genus *Homo*, there is no dissent regarding such a classification of another group of hominid remains which emerges in the fossil record approximately 2 my ago. The first of these hominids to be identified was originally named *Anthropopithecus erectus*, but later renamed *Pithecanthropus erectus* and finally *Homo erectus*. The first *H. erectus* fossils were discovered not in Africa, but in Southeast Asia, on the island of Java. Only later were fossils presumably belonging to the same species found in Africa and at other sites in Asia. The Javan fossils were originally thought to be of much younger age than those found in Africa, which was consistent with the notion that *H. erectus* originated in Africa and then spread to East Asia, Southeast Asia, and possibly also to Europe. A more recent, still somewhat controversial redating of the Javan fossils suggests, however, that they may be as old as the African *H. erectus* fossils, if not older. Although paleoanthropologists still believe that *H. erectus* originated in Africa, mainly because there is no evidence for the presence of its possible ancestors anywhere else, this belief still needs to be substantiated.

The original *H. erectus*, the Java Man, and forms closely related to him, are distinguished from all earlier hominids by a large brain cavity with an average volume of little less than 1000 cubic centimeters (Fig. 10.6). They differ from the hominids that followed them by their highly prominent, heavy browridges above the eye sockets, a low, sloping forehead, and flatter faces (Fig. 10.4). Much of the rest of the skeleton, however, already resembled that of modern humans. Fossils closely matching these and other characteristics of *H. erectus* have been found not only in Indonesia (Sangiran, Trinil village on the Solo River, Sambunmachan, Ngandong in central Java), but also in East Asia (Lantian, Hexian, Zhoukoudian near Beijing in China, originally named *Sinanthropus pekinensis*, the Peking man), in East Africa (Olduvai Gorge, Nariokotome and Eliye Springs, Koobi Fora at East Turkana, Omo), in North Africa (Tighennif in Algeria), and Eurasia (Dmanisi in the Republic of Georgia; Fig. 10.5). The fossils have been dated to 1.8 my ago (Sangiran in Indonesia, Olduvai Gorge in East Africa), 1.7 my ago (Dmanisi in Georgia), 0.7 my ago (Tighennif in Algeria), and 0.5 my ago (Zhoukoudian in China). The East African fossils resembling the Southeast Asian *H. erectus* differ somewhat from the latter and are therefore considered by some paleoanthropologists to constitute a separate species, *H. ergaster*. They may represent the stock from which the emigrants originated who presumably left Africa approximately 2 my ago.

The notion that *H. erectus* originated in East Africa some 2 my ago and shortly afterwards began to spread to other regions of the Old World – Eurasia, East Asia, and Southeast Asia – is generally accepted. But what happened subsequently is highly controversial. At some point in later evolution, the descendants of *H. erectus/ergaster* gave rise to a species characterized by a large brain cavity (average volume of 1400 cubic centimeters; Fig. 10.6), a gracile skeleton (manifested in the shape of its long bones and the thinness of the skull bone), a short, high-domed skull with the face tucked under the braincase, a prominent chin, and less prominent browridges (Fig. 10.2) – the species to which we all belong, *H. sapiens*. When, where, and how this happened is currently the subject of acrimonious debates.

In the interval between the emergence of *H. erectus* and the appearance of fully modern *H. sapiens*, numerous other hominid forms appeared, some of which (the earlier ones) did show a resemblance to *H. erectus* but nevertheless differed from it, while others (generally the later ones) resembled *H. sapiens* more closely than *H. erectus*, but again differed from both. Some paleoanthropologists lump all these forms together in a single category which they tentatively refer to as the "archaic" *H. sapiens*, and contrast it with the "anatomically modern" *H. sapiens*. Other paleoanthropologists prefer to split the forms into several species, two of which currently seem to be gaining wider recognition – *H. heidelbergensis* and *H. neanderthalensis*.

The former is named after the city of Heidelberg in Germany, in the vicinity of which the first fossil (a lower jaw) was discovered in a quarry near Mauer in 1907. As currently defined, *H. heidelbergensis* encompasses fossils found at many different sites in Europe (in addition to the Mauer jaw dated to 600,000 years ago, there are 350,000-year-old fragments of a skull found at Bilzingsleben, Germany; a 300,000-year-old skull from Petralona village near Thessaloniki, Greece; and others), Africa (East, South, and North), the Near East (Zuttiyeh in Israel), Narmada Valley in central India, and in China (Jinniu Shan, Dali, Yunxian). The fossils, most of them less than 1 my old, vary in appearance, some of them resembling *H. erectus*, others bearing a closer resemblance to *H. sapiens*. The braincase of *H. heidelbergensis* is usually, but not always, smaller than that of modern humans, but the skull is more robust, has heavy browridges, and no chin. Because of the fossils' heterogeneity, some paleoanthropologists argue that *H. heidelbergensis* is in reality an assembly of different species.

African *H. erectus* (*H. ergaster*) and African as well as European *H. heidelbergensis* took the technology of stone tool manufacture one great step forward by introducing the

Acheulean toolmaking tradition, named after the town of St. Acheul, near Amiens in northern France, where many implements of this type have been found. The tradition dominated stone toolmaking from 1.5 my ago to 150,000 years ago and was widespread in Africa, Europe, and parts of Asia. The hallmark of the Acheulean toolkit was a hand axe fashioned by flaking a cobble from opposite sides with a hammer stone and then flattening it into two faces with converging edges that met in a sharp point (Fig. 10.7B). A broad, heavy, hammer-like butt allowed the tool to be held in the hand. Hand axes and similar artifacts were used for cutting meat, woodwork, and digging. Their production required not only considerable manual skills, but also the ability to form a mental image of the desired implement prior to manufacture.

H. neanderthalensis derives its name from the Neander Valley (originally *Neanderthal*, now *Neandertal* in Germany) in the Rhine province near Düsseldorf, where their first fossil remains were found. The valley is named after Joachim Neumann, a seventeenth-century German poet who achieved a modest reputation for composing the texts of church hymns *à la Lobe den Herren* (The Lord be praised), but who has now largely been forgotten, even in his native country. His name, however, lives on in its Greek transliteration (Neumann = New man = Neander) as the designation of a one-time romantic gorge, which Neumann often sought out for inspiration. It was in this part of western Germany, in the Feldhofer Cave, that a fragmentary skeleton was discovered in 1856 by workers quarrying limestone. The skeleton later became known as the Neandertal man. The Neander Valley people resembled modern humans very closely in their anatomy, if not in their behavior. The braincase of the Neandertals was of the same size or even larger than that of *H. sapiens*, but its vault was lower, broader, and longer, almost circular when viewed from behind. The face was very long and narrow, extremely protuberant along the midline, with both the eye sockets and the cheekbones sweeping backwards (Fig. 10.4). The nose was large and the browridges were well-developed in the center but less so at the sides (they gave a double-arched or beetle-browed appearance). The teeth were smaller than those of *H. erectus*, except for the incisors, which are believed to have been used as a clamp for food, tool making, and skin processing. Neandertals were robust, muscular, short (average height between 1.5 and 1.7 meters, average weight of 65 kilograms), and stocky with relatively short forearms and legs.

The Neandertals greatly improved the manufacture of stone tools and expanded the variety, so much so that archeologists have given a special name to their artifacts – the *Mousterian industrial tradition*. It is named after Le Mousterier, a cave site in the Dordogne province in France, where an especially rich collection of artifacts has been discovered. In contrast to earlier traditions in which flakes were chipped from a stone and the remaining block or core was used as a tool, in the Mousterian tradition, the flakes themselves were made into tools by undergoing fine retouching with other stone tools (Fig. 10.7C).

The Neandertals were big-game hunters and gatherers who were capable of adapting to the harsh conditions of Europe's Ice Ages. They buried their dead in simple graves, cared for infirm or disabled relatives, and lived in caves and tents. They appeared on the scene some 300,000 years ago and then, approximately 30,000 years ago, bones positively identifiable as belonging to the *H. neanderthalensis* disappear from the fossil record. Neandertals were distributed over western, central, and southeastern Europe, as well as southwestern Asia (Near East); they apparently did not inhabit Africa or Southeast Asia. Some of the Near East individuals were typical Neandertals, while others may have represented related forms.

In addition to *H. heidelbergensis* and *H. neanderthalensis*, the names of at least three other species have been proposed for regional fossil assemblages that are based on unique combinations of characters. They include *H. rhodesiensis*, a name occasionally used for some of the African fossils that are categorized by others as *H. heidelbergensis*; *H. soloensis*, a name used for some of the Southeast Asian fossils otherwise assigned to *H. erectus*; and *H. antecessor*, the proposed designation of fossils recently excavated at the cave site of Gran Dolina, Sierra de Atapuerca, Spain. *H. antecessor* is believed by its discoverers to be the 780,000-year-old ancestor of Neandertals and modern humans.

A Restless Species

To summarize: the most recent common ancestor of the human and the chimpanzee lived somewhere in sub-Saharan Africa approximately 5-6 my ago, and the earliest members of the hominid lineage, presumably the long-armed, chimp-faced but bipedal australopithecines, remained confined to the African continent. Between 2.5 and 2.0 my ago, presumably one of the australopithecines evolved into the first species of the genus *Homo*. The identity of this species is uncertain, the first uncontested member of the genus being a large-brained meat-eater, *H. ergaster/erectus*. *H. ergaster/erectus* may have been the first hominid to spread out of Africa, although some paleoanthropologists hypothesize that an earlier species could have led the way. The expansion presumably began slightly less than 2 my ago. Its progress is chronicled by the presence of *H. erectus*-like fossils in the Middle East and adjacent regions from 1.7 my ago, in East Asia possibly 1.9 but certainly 1.1 my ago, and in Southeast Asia perhaps 1.8 my ago. In Western Europe, *H. erectus*-like fossils are all younger than 1.0 my. From this distribution, the migration route of *H. erectus* has been reconstructed as leading from East Africa northbound along the coast, across the Suez isthmus and the Sinai Peninsula into the Middle East and the regions around the Black Sea (Fig. 10.8). From there, the path turns eastward, possibly again following the coastline before crossing the Indian subcontinent and splitting into a northbound branch extending into East Asia, primarily China, and a southbound branch entering Southeast Asia and Indonesia. Some paleoanthropologists believe that in the latter region, *H. erectus* used primitive watercraft to cross open waters between islands. Somewhat later, for reasons that are not entirely clear but could be connected with the climate of the Ice Age, another branch of *H. erectus*-like hominids turned from the Middle East westward and entered Europe. During the period of expansion, this species began to change, to differentiate into distinct variants, so that by the time it settled down, for example on Java, it already differed from the populations inhabiting the African continent. The variation was sufficiently extensive to lead to disagreements among paleoanthropologists regarding the identity of *H. erectus* in the different regions.

As stated earlier: what happened afterwards is disputatious. The hominid fossil record of the last 1.5 my has been interpreted in three different ways that are referred to as the candelabra, the multiregional, and the uniregional models (hypotheses) of the origin of modern *H. sapiens* (Fig. 10.9). The *candelabra model* was promulgated by the paleoanthropologist Carleton Coon. He argued that the different populations of *H. erectus* established on the various continents evolved independently of one another toward subspecies of a new species, *H. sapiens*. The subspecies correspond to the four main geographical races – the Caucasoids centered in Europe and Western Asia, the Mongoloids in Central, East, and Southeast Asia, the Australoids of Greater Australia, as well as the Congoids and the Capoids of Africa. The proposition is referred to as the "candelabra" model because its graphic representation resembles a branched candleholder (Fig. 10.9A). It is a bizarre hypothesis in that it postulates an unprecedented mode of species origin. In the standard models of speciation, isolated populations evolve into distinct species which may then

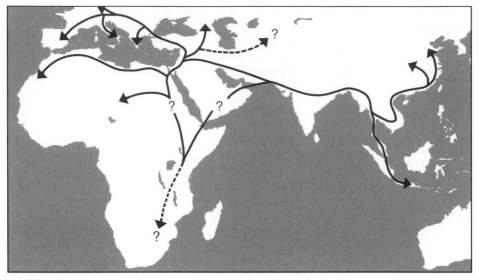

Fig. 10.8
Postulated routes of expansion of *Homo erectus* (and *H. heidelbergensis*) from East Africa. Routes indicated by broken lines are not documented in the fossil record.

fragment into subspecies by further subdivision. The candelabra model turns this process on its head by proposing that instead of diversification into different species, isolated populations evolve toward subspecies of the same new species. To explain the appearance of characters shared by the subspecies and diagnostic of the new species, the candelabra hypothesis must surmise that all the mutations responsible for the appearance of these characters arose independently at least four times and were also fixed independently in each of the four populations, which would make it an extraordinary case of parallel evolution. The hypothesis was severely criticized by the anthropologist Ales Hrdlicka and the geneticist Theodosius Dobzhansky, both of whom considered it racist. The label of racism came to haunt Coon for the rest of his life and if any paleoanthropologists adhere to the hypothesis today, they certainly do not dare to admit it publicly.

The *multiregional hypothesis* espoused in particular by Milford H. Wolpoff and Alan G. Thorne superficially resembles the candelabra model in that it postulates the emergence of

Fig. 10.9A-C
The three main hypotheses of the origin of *Homo sapiens*. **A.** The discredited candelabra model envisioned independent emergence of *H. sapiens* from *H. erectus* in four different regions by parallel evolution. **B.** The multiregional model postulates the evolution of *H. sapiens* from *H. erectus* in four geographical regions by spreading of *H. sapiens*-specific characters from the population of their origin to all other populations (thin lines). At the same time, characters responsible for regional continuity have remained restricted to the population (region) of their origin. **C.** The uniregional (Out of Africa) model assumes that lineages founded by *H. erectus* in the different regions became extinct in all of them except in Africa. The African lineage gave rise to *H. sapiens* which spread into all the regions previously inhabited by *H. erectus* and its descendants relatively recently. Geographical names in boxes are those of sites at which fossil evidence for the presence of *H. erectus* and its descendants has been documented. (**B** and **C** modified from Stringer 1990.)

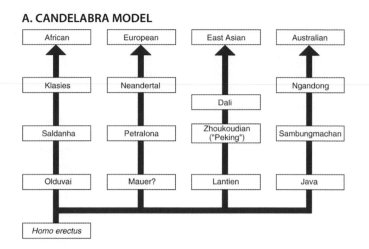

A. CANDELABRA MODEL

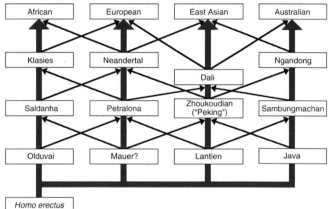

B. MULTIREGIONAL MODEL

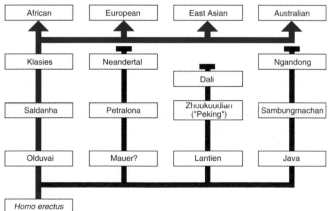

C. UNIREGIONAL MODEL

the same species, *H. sapiens*, from the same ancestor, *H. erectus*, in four different geographical regions (Fig. 10.9B). It differs, however, from Coon's hypothesis in that it invokes continuous gene flow among the populations and so sidesteps the requirement for an inordinate degree of parallel evolution. Any form of the multiregional hypothesis must explain the appearance of two types of characters – those that differentiate the populations in the various regions from one another and those that are shared by the different populations but distinguish *H. sapiens* from *H. erectus*. The Wolpoff-Thorne hypothesis does this by conjecturing that regional differences are the result of mutations that arose locally and then remained confined to the population of each region. These mutations are thus responsible for *regional continuity* – the conservation of characters some of which are claimed by the proponents of the multiregional hypothesis to have persisted in each region over a period of nearly 2 my. The hypothesis asserts further that the shared, *H. sapiens*-specific characters are the result of mutations that arose in one population and then spread by gene flow to all the other regional populations. Because each of these mutations arose only once, its fixation in the different populations does not represent parallel evolution.

According to the proponents of the multiregional hypothesis, both types of character are fixed by selection, one type regionally and the other globally. The regionally-specific characters are fixed by selection tending to adapt the population to specific local conditions. The dissemination and fixation of the *H. sapiens*-specific characters is promoted by selection that tends to improve the adaptation of the global population. In this fashion the global population evolves into a new species while retaining at the same time the various adaptations to local conditions differentiating the forms inhabiting the various regions. In essence, therefore, the multiregionalists invoke an anagenetic (phyletic) mode of evolution in which one species (*H. erectus*) transforms gradually, without splitting, into a single new species (*H. sapiens*).

This is also how the multiregionalists interpret the hominid fossil record of the last 2 my. They strive to arrange the fossils from each region into a linear sequence progressing gradually via various intermediate forms from *H. erectus* to *H. sapiens* (Fig. 10.9B). The younger the fossils, the less similar they are to *H. erectus* and the more they resemble *H. sapiens*, while keeping their regional idiosyncrasies. In Indonesia, one set of the intermediate forms is represented by the Ngandong fossils, while in Europe the intermediates include the Heidelberg (Mauer) and the Neandertal fossils, which other paleoanthropologists assign to separate species, *H. heidelbergensis* and *H. neanderthalensis*, respectively.

The proponents of the *uniregional hypothesis* and at their forefront the paleoanthropologists Chris B. Stringer and Peter Andrews, see things differently. They divide the same sets of fossils in each of the regions into two categories, one related to *H. erectus* and the other to *H. sapiens*. They interpret the former category as illustrative of the radiation of *H. erectus* in the different regions, a radiation that produced different forms, all of which became extinct. They view the latter category as the product of a process in which *H. ergaster* evolved in Africa into *H. sapiens* and then spread, less than 200,000 years ago, into the regions already occupied by the descendants of *H. erectus*, as well as into Australia and the Americas (Fig. 10.9C). The hypothesis therefore posits that there was only one region in which *H. sapiens* originated and that this region was Africa. Because of the latter postulate, it is commonly referred to as the *Out of Africa* model.

The two fundamental differences between the two models pertain to the mode of evolution of modern *H. sapiens* and the time period during which the evolution took place. The multiregional hypothesis posits that *H. sapiens* evolved simultaneously from different populations in several regions, and that during its evolution genes flowed from one population to another. The uniregional hypothesis, on the other hand, postulates that *H. sapiens* evolved from a single population in one region (Africa) and then colonized other regions of the globe. Furthermore, the multiregional hypothesis asserts that the evolution of *H. sapiens* began shortly after the arrival of *H. erectus* in the different regions less than 2 my ago. The uniregional hypothesis purports that *H. sapiens* expanded from Africa less than 200,000 years ago and arrived in the Middle East 100,000 years ago, in East Asia 30,000-40,000 years ago, in Southeast Asia 60,000 years ago, in Australia 50,000 years ago, in Europe 40,000 years ago, and in America 15,000-35,000 years ago (Fig. 10.10). According to one version of the hypothesis, there is no direct relationship between modern *H. sapiens* and any fossils older than 200,000 years in any of the regions except Africa. Specifically, the Ngandong fossils in Indonesia and the Neandertals in Europe are terminal branches ultimately replaced by the spreading modern *H. sapiens*. The various uniregionalists differ, however, in their views of what might have happened in Europe. In particular, the relationships among the forms referred to earlier as *H. ergaster*, *H. antecessor*, *H. heidelbergensis*, *H. neanderthalensis*, and *H. sapiens* are highly controversial. Did *H. ergaster* reach Europe to give rise to *H. antecessor*, who then evolved into *H. neanderthalensis* along one branch and into *H. sapiens* along another, as depicted in Figure 10.3, or were these species interrelated in some other way? How do the various forms of *H. heidelbergensis* (the "archaic" *H. sapiens*) fit into the picture? What exactly did the "replacement" of all these forms, but in particular that of *H. neanderthalensis* by *H. sapiens* amount to? Direct elimination? Economic outcompetition? Another version of the hypothesis admits interbreeding between species and "swamping" of Neandertal genes by the genes of *H. sapiens* as a result. And if Neandertals were not on the direct line of descent to modern *H. sapiens*, how is it possible that they resemble the moderns to such an extent that, as often remarked, if they were dressed in contemporary clothes and walked down the 42nd Avenue in New York City, hardly anybody would turn their heads after them? Would one not have to invoke a con-

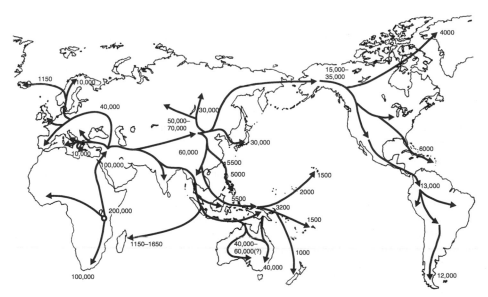

Fig. 10.10
The spreading of *Homo sapiens* out of East Africa as postulated by the uniregional hypothesis. The estimated dates of arrival of modern humans at sites at which their past presence is documented by paleontological and archeological records are indicated. All dates are before present (BP).

siderable degree of parallel evolution to explain the resemblance? Contrary to its presentation at times in the media, the uniregional hypothesis is neither a single hypothesis, nor is it without gaping holes calling for explanations.

Where Numbers Count

Since paleoanthropologists cannot agree on how the fossil data should be interpreted, non-paleontological evidence that could resolve the controversy must be sought instead. To do this, however, we must resort to a somewhat different strategy and to methods other than those used thus far. In all earlier phylogenetic reconstructions, we inferred evolutionary relationships from comparisons between species. Now, however, we have reached a time interval in human evolution from which only one species, *Homo sapiens*, has survived; all other species that once existed in this lineage have become extinct. Since the extinct species left no DNA behind, or at best only highly degraded traces of it, the molecular approach may seem to be failing us in the phase of human evolution in which it is needed most. In reality, however, our lack of DNA from extinct species does not bar us from analyzing the phase in which these species lived. Information about it should be archived in the DNA of *Homo sapiens* itself, specifically in the differences distinguishing the six billion living individuals from one another. To retrieve this information, we must identify and analyze genetic differences among individuals in samples taken from the human population, either directly or indirectly.

The direct way of uncovering genetic differences is through sequencing corresponding DNA segments from many individuals chosen from different geographical regions. If the segment is long enough, it will contain differences among the individuals that have arisen as a result of the sequential accumulation of mutations in the manner described in Chapter Three (see Fig. 3.7). Some of these mutations will be shared by a certain proportion of the individuals, different ones by different individuals. The degree of sharing will reflect the extent to which the DNA sequences are related just as the sharing of substitutions mirrors phylogenetic relationships among species. DNA segments that share more mutations can be assumed to be more closely related genealogically than those sharing fewer mutations. By applying methods used to reconstruct species phylogenies, the different DNA segments of the same species can be arranged into a genealogical tree. Within the tree, groups of DNA segments are interconnected via hypothesized ancestral segments, which in turn are ultimately all derived from a single segment. This "ancestor of all ancestors" is postulated to have possessed none of the mutations that differentiate the living individuals. The tree is not a genealogy or a pedigree of the individuals, however; it does not depict pairs of individuals that begot particular descendants. Rather, it is a genealogy of the DNA segments.

Potentially the tree can provide three types of information about the origin of the human species. First, under the assumption of a molecular clock, the accumulation of mutations can be expected to proceed linearly with time and can therefore be used to estimate the age of the most recent common ancestor of all the different forms (alleles) of the DNA segment now present in the human population. By averaging similar estimates for different DNA segments (loci), one can then calculate when the genetic diversity maintained in the present human population began to accumulate. Second, clustering of the variants into groups according to their similarities can be taken as an indication that the global human population is divided into subpopulations, and the pattern of clustering can be used to infer the evolutionary relationships among the subpopulations. And third, an association of the oldest subpopulation with a particular geographical region can be taken as an indication that the human species originated in that region.

The direct approach to information retrieval is conditioned on the availability of DNA sequences. In the indirect approach, by contrast, it is only necessary to distinguish the products of allelic genes, either proteins or carbohydrates synthesized with the help of proteins. The two most common ways of distinguishing protein or carbohydrate variants encoded in allelic genes are by immunological typing and electrophoresis. In the former, antibodies specifically recognize blood group antigens on the surface of erythrocytes and other cells, while in the latter, proteins are sorted out in an electric field according to their charge (see Chapter Eight). The advantage of the indirect techniques is that a large number of blood or other samples can be tested rapidly and inexpensively. Their disadvantage is that the actual differences between the alleles at the DNA level remain unidentified: the methods detect phenotypic rather than genotypic differences between individuals. The resolution of the indirect methods is therefore lower than that of the sequencing techniques and the information they yield is of a different nature.

The example of the human A and B blood group antigens illustrates this difference. You will recall from Chapter Eight that the A and B antigens are encoded in alleles of a single *ABO* locus and that the alleles specify enzymes with somewhat different specificities for carbohydrates. The enzymes participate in biochemical reactions that add different sugar moieties to a carbohydrate chain on the surfaces of red blood cells. Typing with A- and B-specific antibodies divides humans into four categories, the four blood groups. We can therefore distinguish four phenotypes in the human population, which are encoded in the various combinations of three alleles, *A*, *B*, and *O*, at the *ABO* locus. (In reality, the situation is more complex than just described but the complications need not concern us here.) Obviously, we cannot treat the results of ABO typing in the same fashion as the sequencing data gathered by the direct approach, since the possession of, for example, the *A* allele or the A phenotype by two randomly chosen persons does not necessarily signify that the two are more closely related to each other than either of them might be to a person carrying, say, the *B* alleles. Lacking in such comparisons is the quantitative aspect, which in the case of sequences enabled us to cluster individuals according to the number of mutations they have accumulated in the studied DNA segment.

The DNA segment in which the *ABO* locus resides undoubtedly accumulates mutations in the same manner as other segments and these mutations differentiate individuals and groups of individuals from one another. Unless, however, one of these mutations changes the ability of the encoded enzyme to add a particular sugar residue to the carbohydrate chain, the mutations go unnoticed. Assessed by the indirect method, the *ABO* locus appears to exist in three forms only, the *A*, *B*, and *O* alleles. Is the *A* allele more related to the *B* than to the *O* allele? We cannot tell without sequencing them and even if we could, it would not help us, because there would be little we could do with three alleles in terms of genealogical reconstruction.

Yet, the ABO typing results do contain quantitative information suitable for genealogical reconstruction, even if it is of a different nature than that netted in sequence comparisons. When we align two sequences, we express their similarity as a *proportion* of sites at which they differ in the total number of sites compared. Such a comparison cannot be made among the *A*, *B*, and *O* alleles defined by the immunological method. We can, however, use the results of the immunological typing to compute the proportion of, for example, the *A* allele in the total number of genes at the *ABO* locus in the gene pool and thus obtain the *frequency* of the *A* allele in the population. Instead of individuals, we can compare *populations* and use this information to assess the genealogical relationships between them.

The assumption on which we base this assessment is that populations resembling each other in their gene frequencies are more closely related than populations that differ greatly in the frequencies of these alleles. Why should this be so?

The global human population is not genetically homogeneous and it was even less so in the past. It was, and to some degree still is, divided into subpopulations which often occupied distinct geographical regions and which differed in their genetic make-up. The reason for this is that descendants tend to settle down not too far away from their antecedents and siblings usually remain in close proximity to one another. Consequently, new mutations, if not lost by random genetic drift, achieve the highest frequency in the area of their origin. Communities of individuals are then formed whose members are genetically more similar to one another than they are to members of other neighborhoods. These subpopulations are, however, not isolated from one another. Migration coupled with intermarriages spread the new mutations, once they have achieved an appreciable frequency, from one subpopulation to another. The interplay of the rate at which new mutations arise, the random genetic drift or natural selection to which they are subjected in each subpopulation, the degree of inbreeding, and the rate of migration then determine how similar or different the subpopulations will eventually be to one another. Since migration of individuals is constrained by distances, the frequency of a given allele often decreases gradually from the area of its origin – it forms a *cline*. Genes appear to flow downward from a fountainhead.

Because of the interplay of all these factors and the complex history of populations, unraveling their genealogy is quite a knotty undertaking. To be able to extract the desired information from the population data, it is necessary to try to fit the data into a simplifying, analytically manageable model. Principally, there are two such models that are reminiscent of those depicting species evolution – cladistic and anagenetic. In the cladistic model, populations are assumed to evolve by a splitting process: an ancestral population splits, in the simplest case, into two subpopulations, each of which then splits once again into two more sub-subpopulations, and so on. The essential difference when compared to species cladogenesis is that the bifurcating populations are only partially isolated from one another – they continue to exchange genes. In the simplest situation, the degree of isolation increases proportionally with the number of bifurcating nodes separating any two populations. The genealogy of populations that have evolved in accordance with this model can be represented by a tree drawn on the basis of genetic data reflecting the degree of isolation.

In the second model, reminiscent of anagenetic species evolution, a population expands into new areas in one or in several directions, without splitting. In the absence of splitting, isolated subpopulations cannot arise because all parts of the population are genetically interconnected by gene flow. Nevertheless, the population may differentiate genetically into a series of local groups, for example according to the geographical distances separating them. Groups located close together can be expected to be genetically more similar to one another than groups separated by larger distances simply because in the former there is a greater opportunity for gene exchange than in the latter. This differentiation may reflect the history of the population, for example its expansion into new territories. The advance of the population may then be chronicled in gene frequency clines along the path of progression. This model does not render itself easily to depiction in the form of a tree; a *network* of interconnected groups is a more appropriate description of the populations' history.

This description of gene behavior in populations may at first appear paradoxical: on the one hand, we expect gene frequencies to keep a record of the evolutionary history of populations, yet on the other, we acknowledge random genetic drift to be the main factor influencing gene frequencies. If gene frequencies fluctuate at random from generation to generation, how can they reflect relatedness between populations? The paradox disappears, however, when we deal with large populations in which the change from one generation to the next is small. If the frequency of an allele is 1 percent in a population with an effective size of 10,000 individuals, it takes 204 generations to increase the frequency to 2 percent. For the human species with a generation time of 20 years, this amounts to 4080 years. For alleles that have reached frequencies of 25 percent, 50 percent, 75 percent, and 99 percent, the corresponding numbers for an increase of 1 percent are 244, 308, 456, and 1840 generations, respectively, or 4880, 6160, 9120, and 36,800 years. If a large population therefore splits into two large subpopulations and these begin to drift apart in relative isolation, unbridled by gene flow and natural selection, or if in the anagenetic mode two subpopulations are isolated by a long distance, after 8000 years the gene frequencies in the two subpopulations can be expected to differ on average by 2 percent, if they drift in opposite directions. Hence, if we encounter three subpopulations, X, Y, and Z, two of which, X and Y, are rather close in their gene frequencies, while the third, Z, is far more different, we can assume that X and Y are more closely related to each other than either of them is to Z. We can assume further that Z separated from the ancestral subpopulation of X and Y earlier than X and Y did from each other.

Things are not this simple, however. Complications arise because of the other factors that influence gene frequencies – natural selection, migration, and mutations – and these must be taken into account. Natural selection may accelerate the divergence of subpopulations by promoting differential spreading of alleles. If, for example, the possession of allele $A1$ improves the adaptation of individuals in subpopulation X to the environment it occupies, while allele $A2$ supports the adaptation of individuals in subpopulation Y to the conditions in which they live, then the frequency of $A1$ will increase in X and that of $A2$ will increase in Y to a greater extent than might be expected if they were influenced by drift alone. The effects of natural selection may, however, be restricted to certain mutations and certain genes. Migration, by contrast, will affect all genes. Any exchange of genes between subpopulations works against their divergence by genetic drift and so helps retain the record of their history. By exchanging genes, daughter subpopulations derived from a parental subpopulation tend to remain similar to each other. When they split again or expand into new territories, the similarity in gene frequencies remains highest between the most recent splits and between the geographically closest subpopulations. The older the splits and the more distant the subpopulations are geographically, the lesser the similarity between them. New mutations, too, might be expected to contribute to the historical record of the populations because by spreading from their birthplace, they provide a marker of distance – spatial or otherwise – between subpopulations and so presumably point to their evolutionary relatedness. Only if a sudden drastic reduction in the size of the subpopulation or some other cataclysmic change occurs in the history of a population, may the record in the form of gene frequencies be obscured or even erased altogether. Barring such accidents, however, the pattern of gene frequencies should reflect the history of a population from its origin to its present state reasonably well. To uncover this record, the gene frequencies must be processed, statistically evaluated, and the data converted into a visual representation of the history.

Numbers into Trees

The task we now face is not unlike that of converting sequence differences into a graphic depiction of species phylogeny or of intraspecies genealogy. We can therefore fall back on some of the same concepts and principles we made use of earlier, in particular the concept of distance. As in the case of sequence differences, we can view differences in gene frequencies as a gauge of the distance between two points. In the case of sequences, we obtained the distance by dividing the number of differences between two sequences by the total number of sites compared. This gave a *proportion* of sites that differentiate two sequences. Since gene frequencies are basically also proportions (the ratio of the number of copies of a particular allele to the total number of genes in a pool), the simplest way of expressing the distance between two populations is by subtracting the gene frequency in one population from that in the other. The result is the *genetic distance* as opposed to the *evolutionary distance*, which is based on the number of substitutions differentiating sequences of distinct species.

Since there are two or more alleles at each polymorphic locus, we must subtract the frequencies of corresponding alleles in the two populations and then sum up the individual differences. If we were to do it in this straightforward way, however, the result would always be zero. If, for example, population X contains alleles $A1$ and $A2$ at frequencies $x1$ and $x2$, respectively, while population Y contains the same two alleles at frequencies $y1$ and $y2$, respectively, then the genetic distance between X and Y is $(x1 - y1) + (x2 - y2)$, where $x1 + x2 = 1$ and $y1 + y2 = 1$. After removing the parentheses, we obtain: $x1 + x2 - y1 - y2 = 1 - 1 = 0$. To avoid this problem, we can either take the absolute values of the differences (i.e., ignore the sign of the result) or square the differences. Squaring a number amounts to multiplying it by its own value and since multiplication of two negative numbers gives a positive value, the summation of the differences gives a zero value only when the gene frequencies in the two populations are the same. For example, if $x1 = y1 = 0.3$ and $x2 = y2 = 0.7$, then $|0.3 - 0.3| + |0.7 - 0.7| = 0$, where $|x|$ stands for an absolute value of x. Regardless of whether absolute values or squares of the differences are taken, the maximum value the sum can reach is 2. This happens when no allele is shared between the two populations. For example, if $x1 = 0$, $x2 = 1$, $y1 = 1$, and $y2 = 0$, then $|0 - 1| + |1 - 0| = 1 + 1 = 2$. If we desire the genetic distances to range – like gene frequencies – from 0 to 1, we simply divide them by 2.

The disadvantage of these simple measures of genetic distances is that they do not correlate well with the divergence times of populations. The maximum distance of 2 (or 1), for example, can be obtained with two, three, or any number of alleles. Yet, if each allele is the result of one mutation, it obviously takes a longer time to accumulate three mutations than it does two. To develop a measure of genetic distance that increases linearly with time, it is necessary to find a connection between gene frequency and mutation rate, u, or more accurately the expression $2ut$, which stands for the number of mutations per DNA segment that have accumulated since populations X and Y diverged from the ancestral population Z. The expression represents the mutation rate u multiplied by the time t from Z to X plus the time t from Z to Y.

The most widely used solution to this problem was developed in 1972 by Masatoshi Nei. It is referred to as the *standard genetic distance* or *Nei's distance* and is embodied in the formula

$$D = -ln(I)$$

where D is the standard genetic distance, ln the natural logarithm, and I the genetic identity specified by the formula

$$I = \frac{J_{XY}}{\sqrt{J_X J_Y}}$$

Here, J_{XY} is the probability that by sampling populations X and Y, we choose the same allele from both populations; J_Y is the probability that by sampling population X, we pick up the same allele twice; and J_Y is the probability of selecting the same allele twice from population Y. The D formula is derived by obtaining these probabilities in successive generations after the divergence of X and Y from Z. The derivation is, however, beyond the scope of this book. Instead, we give an example of how the formula is used.

Take a case of a single locus with four alleles at frequencies $x_1 = 0.05, x_2 = 0.05, x_3 = 0.90, x_4 = 0$ in population X and $y_1 = 0.10, y_2 = 0.01, y_3 = 0.79, y_4 = 0.10$ in population Y. We first calculate the J values:

$$J_X = \sum_{j=1}^{4} x_j^2 = (0.05)^2 + (0.05)^2 + (0.90)^2 + 0^2 = 0.8125$$

$$J_Y = \sum_{j=1}^{4} y_j^2 = (0.10)^2 + (0.01)^2 + (0.79)^2 + (0.1)^2 = 0.6442$$

$$J_{XY} = \sum_{j=1}^{4} x_j y_j = (0.05)(0.10) + (0.05)(0.01) + (0.90)(0.79) + (0)(0.1) = 0.7165$$

From these values, we calculate the identity index:

$$I = \frac{J_{XY}}{\sqrt{J_X J_Y}} = \frac{0.7165}{\sqrt{(0.8125)(0.6442)}} = \frac{0.7165}{\sqrt{0.5234}} = \frac{0.7165}{0.7235} = 0.9903$$

And finally, we calculate the genetic distance:

$$D = -ln(I) = -ln(0.9903) = 0.0097$$

I represents the probability of drawing the same allele from populations X and Y. It can be written as $I = e^{-2ut}$ [because $D = -ln(I) = -ln(e^{-2ut}) = 2ut$]. If we take $u = 4 \times 10^{-6}$ per gene per generation, then we have

$$D = 2ut = (2)(4 \times 10^{-6})t = 8 \times 10^{-6}t$$

Hence $t = D/(8 \times 10^{-6}) = (9.7 \times 10^{-3})/(8 \times 10^{-6}) = 1.2 \times 10^3$ generations and we conclude that the populations X and Y diverged 1200 generations ago.

Earlier in this book, we pointed out some of the pitfalls of making phylogenetic inferences based on information from a single DNA segment and emphasized the need to study as many loci as possible. This becomes an absolute imperative when using gene frequencies to reconstruct evolutionary history. To offset the effects of the statistical noise superimposed on the phylogenetic signal, it is vital to gather information on a large number of loci from many different populations and involving samples that consist of hundreds of individuals.

First Trees of a Forest

Blood group and electrophoretic typing may be relatively inexpensive, but it nevertheless involves a considerable effort if the mass-scale imperative is to be respected. Generating gene frequency data on hundreds of genes in thousands of individuals from scores of human populations is clearly not a mission that can be accomplished by a single person, or even by one laboratory. It calls for the participation of a network of research groups strategically distributed throughout the world with easy access to populations in the different corners of the globe. Also, the expense incurred by the typing often has to be justified by aims other than satiation of curiosity about human origin. Blood groups fit the bill nicely.

Although the public is generally aware of only two blood group systems (ABO and Rh), in fact there are dozens, each controlled by a distinct locus. Many of the loci are multiallelic and the alleles occur at appreciable frequencies in most populations. Blood group typing is one of the routine procedures in hospitals of any repute as well as in transfusion stations, for blood transfusions are inconceivable without first matching the donor and the recipient in the main blood group antigens. Similar arguments apply also to the electrophoretic typing of enzymes and other proteins. From the moment clinicians began to appreciate the value of keeping a record of typing results, data on gene frequencies have been accumulating steadily.

By the early 1960s, enough frequency data on blood group genes had been accrued to justify first attempts at using them in phylogenetic reconstruction. Since the approach then was still very tentative and methods to sustain it had still to be developed, it required a person with vision, a good background in mathematics, a broad knowledge of population genetics, and familiarity with the blood group system to make a go of it. Luigi Luca Cavalli-Sforza, then at the University of Pavia, Italy, fitted the role perfectly. Trained in bacterial genetics, Cavalli-Sforza turned to human population genetics after a stint at the University of Cambridge, England, with the founder of modern statistics, Sir Ronald A. Fisher. During 1961 and 1962, Cavalli-Sforza and his colleague from Cambridge, Anthony W. F. Edwards, collated gene frequency data published on five blood group systems (*ABO, Rh, MN, Diego*, and *Duffy*) from 15 human populations inhabiting different continents. To analyze the data, Cavalli-Sforza and Edwards developed a method of calculating genetic distances and using them to draw a tree of the populations. They then arranged the populations into pairs by matching each of them with all the other populations, computed the genetic distances for each pair, combined the frequencies of all 20 alleles at the five loci, and finally drew a tree based on these distances.

It was the first tree of human populations ever produced that was based on gene frequencies and that made use of modern methods of phylogenetic analysis. And like most first attempts, it was not perfect. It grouped human populations roughly according to their geographical distribution (Fig. 10.11) as expected, but paired some populations incorrectly. In particular, it paired African with European populations, excluding Asian populations, a finding which would later be contradicted by most other trees. The imperfections of the first tree can probably be attributed primarily to a combination of two factors: the low number of loci tested and the fact that the genes involved were those coding for blood group antigens exclusively, at least some of which are apparently subject to the distorting influence of natural selection.

Cavalli-Sforza subsequently moved to the University of Stanford, California, where he and his coworkers continued compiling and interpreting genetic data on human populations, as well as producing new data of their own. We mention some of their findings later in this chapter. In the meantime, however, another major figure – Masatoshi Nei, whom we met already on earlier occasions – had entered the field and begun to collate his own database of gene frequencies. Nei trained in mathematical genetics at the University of Kyoto, Japan, but later moved to the United States. Unlike many other theoreticians, he has a broad and deep understanding of general biology which roots his theoretical research firmly in the real rather than the virtual world. He has played a key part in the development of the theoretical framework and the methodological basis of modern population and evolutionary genetics.

Nei and his coworker Arun K. Roychoudhury began to compile human gene frequency data in 1972 and reported intermittently on their progress and on the various inferences they were able to make from them. In 1982, they published a major summary of their analyses in which they presented trees of the various groups of populations, as well as a tree of the global human population. The latter tree (Fig. 10.12) is based chiefly on frequencies of different protein-encoding alleles, as detected by electrophoretic tech-

Fig. 10.11
Not perfect, but historically valuable: one of the first trees of human populations drawn from gene frequency data. The tree was produced by A.W.F. Edwards and L.L. Cavalli-Sforza in 1964 and was based on the frequencies of 20 alleles of five blood group systems. (From Edwards & Cavalli-Sforza 1964.)

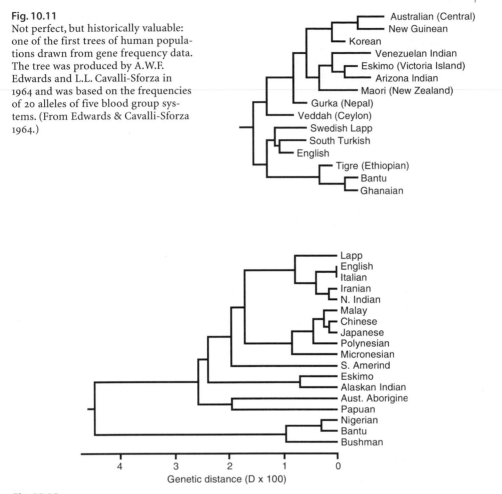

Fig. 10.12
Phylogenetic tree of human populations drawn by M. Nei and A.K. Roychoudhury in 1982 using frequencies of multiple protein-encoding genes determined by electrophoretic methods, as well as blood group genes. The tree-drawing method used was one of the variants based on genetic distance (D), the so-called unweighted pair-group method using arithmetic averages or UPGMA. Note the clear-cut early separation of the three African populations (Nigerian, Bantu, Bushman) from all the other populations included in the study. (From Nei & Roychoudhury 1982.)

niques, although blood group loci are also included in the dataset. The number of loci used varied according to population, but for some of the populations, it was as high as 121. The focus on protein-encoding genes enabled the two researchers to estimate divergence times between populations. For although analysis of proteins by electrophoresis does not yield direct information about amino acid sequence and does not even reveal all the differences between proteins controlled by allelic genes, it does provide an indirect means of estimating the number of amino acid replacements or even the number of nucleotide substitutions involved. From these estimates, divergence times of populations can be calculated.

The analysis enabled Nei and Roychoudhury to reach several important conclusions regarding the evolutionary history of human populations. Of these, we mention here only those directly relevant to the origin of *H. sapiens*. The tree in Figure 10.12 concentrates on three main groups of populations occupying the African, Asian, and European continents. Nei and Roychoudhury refer to these populations as Negroid, Mongoloid, and Caucasoid, respectively. The tree shows two major bifurcations, the deeper between the African and all other populations, the shallower between the Asian and the European populations. From the length of the branches, Nei and Roychoudhury estimate that the first split occurred approximately 110,000 years ago and the second furcation 41,000 years ago. The tree also reveals that the Australian aborigines are most closely related to the Papuans of New Guinea and that populations of Oceania (i.e., Polynesians and Micronesians in the Pacific Ocean) are closely related to some of the Asian populations. On the other hand, the origin of the American Indians appears more complex: their different groups may be related to different Asian populations and South American Indians are genetically quite different from North American populations. Overall, the most straightforward interpretation of the tree is that all existing populations of *H. sapiens* are derived from an ancestral population that existed in Africa. Being the oldest population, it had the greatest length of time to accumulate genetic differences from all other populations and is therefore the most distant from them genetically. The implication is that *H. sapiens* originated in Africa and then spread first to Asia 110,000 years ago and to Europe 41,000 years ago. Some of the Asian populations subsequently colonized Australia, Oceania, and the Americas (Fig. 10.10). This, in essence, is a uniregional interpretation of the origin of *H. sapiens*, as propounded by some paleoanthropologists.

Circus Performance Featuring Eve

In the early 1980s, the field of molecular anthropology was finally ready to extend the molecular approach all the way to the DNA level with mitochondrial DNA as its most easily accessible target. The first study of mtDNA variation in human populations was inducted by Wesley M. Brown. It was still not based on sequencing, but on a technique that approached it – the *restriction fragment length polymorphism (RFLP) analysis* (Fig. 10.13). The technique takes advantage of *restriction enzymes* (*endonucleases*), which are proteins that have been isolated from various bacteria and used to cleave bonds between neighboring nucleotide pairs within a specific sequence motif (see Chapter Five). For example, the enzyme referred to as *Eco* RI because it is isolated from the common gut bacterium <u>E</u>scherichia <u>coli</u> strain <u>RI</u>, recognizes the motif (*restriction site*) GAATTC and cleaves the bond between G and A. In the complementary strand in which the motif reads CTTAAG, the enzyme cleaves the bond between the same two nucleotides so that the broken ends are then staggered symmetrically:

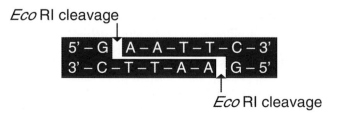

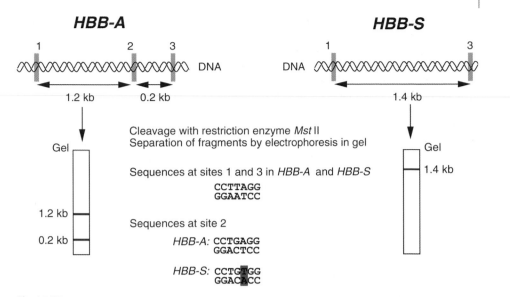

Fig. 10.13
Principle of restriction fragment length polymorphism (RFLP) analysis. *HBB-A* and *HBB-S* are two alleles at the hemoglobin β locus which differ from each other in that the latter lacks one of three sites (No. 2) recognized by the restriction enzyme *Mst II*. As a result, digestion of the DNA with *Mst II* produces two fragments from the *HBB-A* segment (1.2 kb and 0.2 kb in length) and only one fragment (1.4 kb in length) from the *HBB-S* segment. The fragments, separated by electrophoresis in a gel, can be identified by hybridization with an appropriately labeled probe. The enzyme recognizes CCT-TAGG or CCTGGG sequences but not the mutated CCTGTGG sequence. 1 kb = kilobase pair = 1000 nucleotide pairs.

Many other restriction enzymes, however, produce blunt ends. (Note that the motif is a palindrome: it reads the same in both directions, just like the word "Eve" for example. The direction of the two strands is specified by the numbers 5' → 3' and 3' → 5'.) Restriction sites for various enzymes arise by chance in each sequence and are scattered randomly along a DNA molecule. Exposure of a DNA molecule to a particular restriction enzyme therefore cuts the molecule into a specific number of fragments of different lengths. The fragments can then be separated according to their size by electrophoresis in agarose gel. (Agarose is a high-molecular-mass carbohydrate isolated from red algae. When soaked in water, it yields a gelatinous substance with pores that slow the movement of the DNA fragments in the electric field to a different degree depending on their compactness.) The enzymes are highly specific with regard to the restriction sites. When a mutation substitutes one nucleotide pair by another within a site, the enzyme that normally recognizes it fails to do so. As a result, the number and length of the fragments produced by the particular enzyme from the mutated DNA molecule changes in comparison to the DNA that lacks the mutation and the different pattern of fragments between the mutant and nonmutant individuals is recognized as restriction fragment length polymorphism (Fig. 10.13).

Brown isolated mtDNA from 21 persons, divided each isolate into 18 parts, digested each part with a different restriction enzyme, labeled the resulting fragments by attaching radioactive ^{32}P (phosphorus) atoms to their ends, separated the fragments by agarose gel electrophoresis, and visualized them on a film. Digestion with seven of the 18 restriction enzymes did not reveal any differences among the 21 individuals. The differences revealed by the remaining 11 enzymes could all be explained by point mutations in the restriction sites, which accumulated during the time period since the divergence of the individual mtDNA variants from the postulated ancestral sequence that existed before any of the mutations occurred. Since the detected variants differed from the ancestral sequence in 0.18 percent of their sites and since the rate of nucleotide substitutions in human mtDNA is estimated to be 1 percent per one million years, Brown calculated that the mtDNA diversity detectable in the present-day human populations began to accumulate approximately 180,000 years ago. Furthermore, since the DNA samples analyzed by Brown were supplied by three different races, Nei and Roychoudhury could use the data in 1982 to draw a tree in which most of the samples from the Afro-American population diverged earlier from the rest of the samples than most of the Caucasian samples diverged from those of the Mongoloid.

A much more extensive follow up to the Brown study was published in 1987 by Rebecca L. Cann, Mark Stoneking, and Allan C. Wilson. They examined the mtDNA of 147 individuals from five populations – African, Asian, Caucasian, Australian, and New Guinean – using RFLP analysis with 12 restriction enzymes. The authors could thus survey 467 sites in each of the 147 DNAs without sequencing, in total about 11 percent of the 16,569 nucleotide pairs encompassing the human mtDNA molecules. Of the 467 sites, 195 showed variation among the tested persons, revealing the existence of 134 distinct types of mtDNA among 147 tested molecules. A maximum parsimony tree based on these data consisted of two main branches, one bearing only African mtDNA samples, and the other samples from all five populations, including some African individuals. The authors concluded that all the mtDNA diversity they had uncovered stemmed from one woman postulated to have lived about 200,000 years ago, probably in Africa, and that all the populations examined, with the exception of the African population, had multiple origins, implying that each area was colonized repeatedly.

None of these conclusions were new; all had been stated explicitly or were implicit from the reports published earlier by Nei and Roychoudhury. Not even the application of mtDNA to the study of human evolution was unprecedented; in addition to the study by Brown, researchers from several other laboratories had published the results of similar studies prior to 1987. Nevertheless, all earlier studies were squarely ignored by journalists, while the paper by Cann and her colleagues triggered a media frenzy, the likes of which are usually reserved for prominent murder trials. Why this sudden journalistic entrancement?

Publication of the report in a top-publicity, high-fashion journal (*Nature*) and the growing sensationalism of science reporting were probably two major contributory factors. Another was the skillful advertising and the attractive gift-wrapping of the product. Allan C. Wilson, in whose laboratory the study was carried out, was not only a brilliant scientist, but also a gifted salesman, who knew how to choose a "sexy" topic and how to market it. The key slogan in marketing this particular product was contained in a sentence printed in the opening paragraph of the report: "All these mitochondrial DNAs stem from one woman who is postulated to have lived about 200,000 years ago, probably in Africa." * The article was accompanied by an editorial entitled "Out of the garden of Eden" that began with the sentence: "A paper by R. L. Cann, M. Stoneking, and A. C. Wilson on page 31 of this issue reports that Eve was alive, well and probably living in Africa around 200,000 years ago"** And with this opening salvo the circus performance featuring Eve in the leading

---

* *Nature* **325**: 31, 1987.
** *Nature* **325**: 13, 1987.

role took off. Without a moment's hesitation, the major newspapers, magazines, and TV channels snatched up the story and wiggled their hips to its music. Eve was in the headlines, Eve entered the textbooks, Eve became a marketing article in science.

There is nothing wrong in using a catchy metaphor to describe the results of a scientific inquiry, as long as it does not mislead. The Eve metaphor, however, is grossly misleading, for it invokes the false image of *Homo sapiens* stemming from a single "mother of us all", as the headlines proclaimed. "We all share a common mother, our mitochondrial Eve"* one textbook asserted soon after publication of the report by Cann and her coworkers. Metaphors such as this are bound to be taken literally. Indeed, the story was understood by many (some scientists included) as demonstrating the existence of a single woman 200,000 years ago from whom entire humankind is derived in the same sense as you and your siblings are derived from your mother.

This, however, is not what the data indicate. All mutations differentiating mtDNA molecules of persons alive today can theoretically be tracked down to a single ancestral molecule in which the first of these mutations occurred. This molecule was borne by a single woman who, however, was contemporaneous with several thousand other females, just as there is one female among the three billion women alive today who bears the mtDNA molecule that will be ancestral to all mtDNA molecules in the future. Any polymorphism present in the contemporary human population, whether borne by mitochondrial or nuclear DNA molecules, can theoretically be traced back to an ancestral DNA segment in which the first mutation arose. Indeed, if the mtDNA molecules do not recombine and are passed on strictly through the maternal lineage, as is generally believed, variation in the whole molecule can be traced back to a single woman. Tracking the various nuclear DNA molecules back to their common ancestor is more complicated because these molecules do recombine and pass through both female and male lines of descent (see Chapter Three). Nevertheless, it can be achieved using short DNA segments, commonly a single gene, which are not likely to have recombined during the period since their derivation from a common ancestor. The variants now found in the human population in the various nuclear genes therefore all have their single ancestors, male or female, thousands of Adams and Eves, each gene with a different ancestor in a different past generation. There must have been as many Adams and Eves as there are genes – close to 40,000 – each living at a different time and in a different place. Obviously, the Adam and Eve metaphor is not only misleading, but also meaningless, not to mention the fact that it is culturally chauvinistic, since to a good proportion of humankind, these two names mean nothing whatsoever.

The humbug spread by journalists in connection with the paper by Cann and coworkers brought it under closer scientific scrutiny than it might otherwise have undergone. As a result, numerous deficiencies were revealed in the study, so many that one asks oneself how the manuscript could have passed through the journal's refereeing system in the first place. The critics objected that the number of sites tested by the RFLP method was too low to yield reliable genealogical information. They pointed out that the analysis amounted to the examination of a single gene, obviously not enough to exclude the influence of random fluctuation on the evolutionary reconstruction. The method by which the authors rooted the tree (taking the midpoint of the longest branch) was found to be unreliable. Contrary to the authors' claim, the tree they exhibited was not the most parsimonious, and in some of the more parsimonious trees, there was no clear separation of an African population

---

* J. D. Watson, N. H. Hopkins, J. W. Roberts, J. A. Steiz, and A. M. Weiner: *Molecular Biology of the Gene.* Vol. 2: *Specialized Aspects*, 4th ed., The Benjamin/Cummings Publishing Co. Menlo Park, CA 1987, p.1160.

from the rest of the individuals. The "Africans" were disclosed to be Afro-Americans (a fact that the authors concealed in their paper). And finally, the estimate of divergence between the African and non-African populations had a large standard error so that the divergence time could have been as much as 800,000 years.

In the follow-up studies from the same laboratory, the authors attempted to blunt some of the criticism by actually sequencing a short stretch of the mitochondrial genome – approximately 800 nucleotide pairs of the noncoding control region, and by including DNA samples from genuine African populations, as well as chimpanzee DNA, to root the tree. But they found no reason to change any of the conclusions they had reached in their earlier study. They estimated the divergence time of African and non-African populations as 238,000 years ago, which is within the range of 140,000-290,000 years obtained in the previous study. Similar results have also been obtained by other researchers (Table 10.1). An example of a phylogenetic tree based on mtDNA control region sequences that have since been accumulated in many different laboratories is shown in Figure 10.14.

To reduce the error in the divergence time estimates, Satoshi Horai and his colleagues sequenced the entire mitochondrial genome of ~16,500 nucleotide pairs from three humans – one African, one European, and one Asian, as well as four great apes – common and pygmy chimpanzee, gorilla, and orangutan. By assuming that the orangutan lineage diverged from the African great ape lineages 13 my ago, they estimated that the chimpanzee and human lineages diverged from each other 4.9 my ago, and that the split between the African and non-African populations of *H. sapiens* occurred 143,000 ± 18,000 years ago. The 95 percent confidence upper limit of this estimate is 179,000 years, indicating that even if the splitting of the mtDNA lineages now represented in the global human population preceded the population splitting, as it presumably did, there is a high probability that the splitting of the African from the non-African populations did not take place before 179,000 years ago. (The principle of a confidence level is explained in Appendix One.)

Table 10.1
Percent sequence divergence between the two most divergent human mtDNA sequences, the estimated or extrapolated average percent sequence divergence (*d*) between human and chimpanzee mtDNAs, and the time to the most recent common ancestor (TMRCA) obtained from data sets of different authors. For uniformity, the humans and chimpanzee lineages are assumed to have diverged from each other 5 my ago. Since some of the authors used different human-chimpanzee divergence times, the TMRCA values given in their publications may differ from those given in the table.

Investigators and year	mtDNA segment	Percent divergence	d	TMRCA
Cann et al. 1987	Entire molecule	0.57	15.0	190,000*
Kocher and Wilson 1991	ND4-5	0.33	9.6	172,000
Vigilant et al. 1991	Control region	2.90	69.2	210,000
Ruvolo et al. 1993	COII	0.58	10.4	278,000
Chen et al. 1995	Entire molecule	0.29	13.0	112,000
Horai et al. 1995	Entire molecule	1.10	39.0	143,000
	Control region	2.10	70.0	143,000
Watson et al. 1997	Control region	11.00	495.0	111,000
Chen et al. 2000	Entire molecule	0.36	13.0	138,000
Ingman et al. 2000	Entire molecule	0.58	17.0	171,500

*Years.

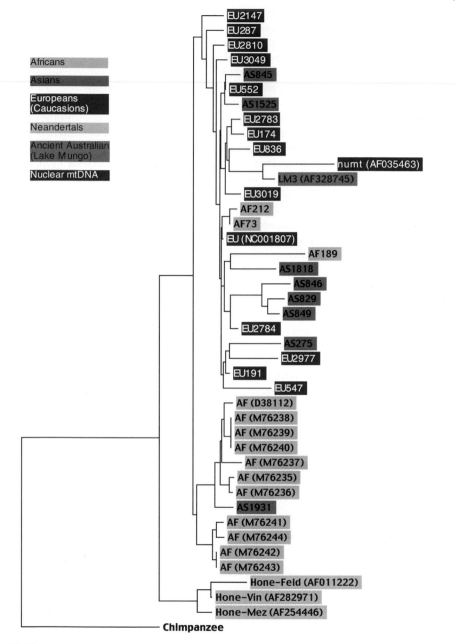

Fig. 10.14

Phylogenetic tree of the human mtDNA hypervariable segment I. The tree was drawn by the neighbor-joining method using sequences chosen from a pool of over 5000 sequences deposited in public databases (the numbers indicate their accession codes). The selection includes extant *H. sapiens* individuals from Africa (AF), Asia (AS), and Europe (EU), three sequences from *H. neanderthalensis* (Hone), one ancient Australian (Lake Mungo) sequence (LM3), one extant *H. sapiens* mtDNA sequence inserted into the nuclear genome (*numt*), and a chimpanzee sequence as an outgroup. (Courtesy of Yoko Satta.)

There are two potential sources of error in making inferences about genealogy from sequences, one concerning the number of variable sites and the other the number of variable segments (loci, genes). The analogy with coin tossing, will help to appreciate this. Imagine two situations. In one, you toss 100 coins and obtain 45 heads and 55 tails, a result that represents 5 percent deviation from the expected 50 heads. You then toss 1000 coins and count 475 heads and 525 tails, this result representing a 2.5 percent departure from the expectation. In general, the more coins you throw, the closer the result will get to the expected value of 50 percent heads. This situation is analogous to sequencing a stretch of 100 or 1000 nucleotides: the longer the segment you sequence, the more exact the estimate of the evolutionary distance between two DNA molecules. Therefore, by sequencing ~16,500 nucleotides of the mtDNA, Horai and his colleagues were able to obtain a far more accurate estimate of the divergence times between molecules derived from different populations than investigators who sequenced only a few hundred nucleotides from each molecule.

In the second situation, you toss one coin many times and discover a slight deviation from the expectation; for example, heads comes up 51 percent rather than 50 percent of the time. Perhaps the coin is simply worn out more on one side than on the other and this gives heads a very slight advantage over tails. Alternatively, a slight bias could have been introduced in the minting process such that all the coins manufactured at one particular mint tend to fall heads up slightly more often than they would otherwise. To clarify this, you test 50 other coins, each many, many times, and if you find the same bias in all of them, you conclude that minting rather than wear and tear is responsible for it. Analogously, if you test only one DNA segment (gene, locus), you do not know whether the result you have obtained is idiosyncratic of the particular segment (or of the particular circumstance, the specific effect of drift, for example) or whether it is a characteristic of the entire population (species). To find out which is the case, you must test a large number of DNA segments. The mtDNA molecule, because it presumably does not recombine, is only one such segment, one unit of behavior (it drifts as a single piece of DNA). Its testing alone, therefore, does not allow a reliable distinction to be made between a random departure from the average and a real difference between populations or species. This is only possible by testing a large number of independently behaving gene segments, which involves scoring loci on different DNA molecules of the nucleus just as Nei and Roychoudhury did for the indirect data sets. In the section that follows, we discuss some of the different types of segments that have been used for this purpose.

The Logbook of the Y Chromosome

The male counterpart of the mtDNA, in terms of its transmission, is the Y chromosome. Like the mtDNA, the bulk of the Y chromosome does not recombine; only short segments at both its ends, the pseudoautosomal regions, regularly exchange parts with the X chromosomes. And as in the case of the mtDNA, this behavior has both advantages and disadvantages for molecular evolutionists. On the positive side, a piece of DNA some 60 million nucleotide pairs long can serve as a logbook in which the accumulating mutations are a record of the chromosome's travels. As we explained in Chapter Three, other chromosomes keep a record of their travels, too, but in their case the logbooks are loose-leaf, and identically numbered pages are perpetually exchanged between them. As a result, the records of different travelers are all mixed up and the sequence of places and dates of the journey are difficult to decipher. The Y chromosome, by contrast, is like a bound logbook that exists in many copies but whose pages cannot be exchanged between different copies.

Each copy therefore contains a full, unscrambled record of the trips undertaken by generations of its owners.

However, unlike the mtDNA, which is passed matrilineally from mother to daughter, the Y chromosome is transmitted patrilineally from father to son. Also unlike mtDNA, the Y-chromosome DNA varies very little. In one of the earliest attempts to determine Y-chromosome variability, Robert L. Dorit and his associates sequenced a stretch of ~730 nucleotide pairs from corresponding segments of 38 human Y chromosomes sampled worldwide, but found no single difference among them. Later, more extensive sequencing and other methods of searching for variability identified several polymorphic sites, but compared to mtDNA and to DNA isolated from nonsex chromosomes (autosomes), the diversity of Y chromosome sequences is greatly reduced. This, too, has its advantages and disadvantages. On one hand the lower variability reduces homoplasy, but on the other, long stretches of DNA have to be analyzed to find a few suitable markers.

Why is the Y-chromosome variability lower than that of mtDNA or of autosomes? The most obvious reason is that in a population consisting of an equal number of females and males, the ratio of autosomes to X chromosomes to Y chromosomes is 4:3:1. Each female carries two autosomes of a particular kind, two X chromosomes, and no Y chromosomes; each male carries two autosomes of a particular kind, one X chromosome, and one Y chromosome. When combined, this gives four autosomes of a particular kind, three X chromosomes, and one Y chromosome. The population size of the Y chromosomes is therefore one-quarter that of autosomes. And since genetic variation is related to effective population size N_e in a manner described in Chapter Eleven, it follows that the smaller the N_e, the lower the variation.

Another reason for the difference in variability is that in some human populations, a few males are reproductively highly successful, while most other males leave no offspring behind. Finally, the absence of recombination over most of the Y-chromosome length may also contribute to the paucity of variability. If an advantageous mutation arises in the nonrecombining part, it spreads through the population by natural selection and sweeps away any Y-chromosome variability that may have existed. This *selective sweep* may explain the exceptionally high number of differences between humans and chimpanzees in the sex-determining gene *SRY*, which resides in the nonrecombining portion of the Y chromosome close to the pseudoautosomal boundary.

Numerous studies of Y chromosome variability have been undertaken by different research teams, most of them designed to resolve relationships within specific groups of populations. Of these, we describe one here that directly addresses the questions we are concerned with in this chapter. A team led by Michael F. Hammer tested 1544 males from 35 populations for markers at eight sites distributed over a region encompassing 29,000 nucleotide pairs of the Y chromosome. At each of these sites, two alleles could be distinguished: at seven of them, the alleles differed in a single nucleotide (e.g., at site number 4064, some Y chromosomes had a G, while others had an A; see Table 10.2); at the eighth site, the YAP difference was more complex. YAP stands for Y *Alu* polymorphic and *Alu* is an abbreviation of a restriction enzyme derived from the bacterium *Arthrobacter luteus*. This enzyme was essential in the identification of a family of repeats now referred to as *Alu elements*. They are believed to be derived from a gene coding for the 7SL RNA, a component of a particle that enables proteins to pass through internal membranes of a cell. From time to time, some of the RNA transcripts of this gene are apparently transcribed back (= *reverse transcription*) into DNA, which is then inserted into the chromosomes at more or less randomly chosen positions. In this manner, approximately one million copies of *Alu* elements have been inserted into the human genome at different times during its evolution.

Table 10.2
Nucleotides at seven polymorphic sites and the presence (+) or absence (–) of the YAP element in ten human Y-chromosome haplotypes, as well as in great ape Y chromosomes. (Based on Hammer et al. 1998.)

Haplotype	Nucleotide (YAP) at site							
	4064	9138	10831	YAP	PN1	PN2	PN3	DYS257
1A	G	C	A	–	C	C	G	G
1B	G	C	G	–	C	C	G	G
1C	G	C	G	–	C	C	G	A
1D	G	C	A	–	C	C	G	A
1E	G	T	G	–	C	C	G	G
2	G	C	A	–	C	C	A	G
3A	A	C	G	+	C	C	G	G
3G	G	C	G	+	C	C	G	G
4	A	C	G	+	C	T	G	G
5	A	C	G	+	T	T	G	G
Great apes	G	C	A	–	C	C	G	G

The YAP is a relatively recent insertion which probably took place in *H. sapiens* after its divergence from *H. erectus*. The YAP element is present in some human Y chromosomes and absent in others.

The alleles at the eight sites were found to occur in ten different combinations or haplotypes in the human populations (Table 10.2). The Y-chromosome haplotypes are numbered consecutively, but since the chronological addition of markers requires the subdivision of some of the haplotypes defined earlier, some of the numbers are differentiated further by letters (e.g., 1A, 1B, 1C, 1D).

Of the ten haplotypes, five were found to be restricted each to a single continent: 1A, 2, and 5 to Africa; 3G to Asia; and 1E to Australia (Oceania). Two haplotypes, 3A and 4, were found to be shared by Africans and Europeans. One haplotype, 1D, was shared by Africans, Europeans, and Asians, but was absent in Australia/Oceania and the New World. And two haplotypes, 1B and 1C, were present in all geographical regions surveyed. Eight of the ten haplotypes were present in Africa, five in Europe, four in Asia, three in Australia/Oceania, and two in the New World.

The relationships between the ten haplotypes can be depicted in the form of a *network* (Fig. 10.15), in which each haplotype occupies one of the nodes (indicated in Figure 10.15 as circles proportional in size to the overall frequencies of each of the haplotypes) and the lines connecting the nodes represent a single-step mutation at one of the eight sites. The network is in reality a form of tree, a dendrogram, which can be obtained by maximum parsimony analysis of the type described in Chapter Five, but in this simple case can also be deduced directly from Table 10.2. The network/tree can be rooted by reference to the nucleotides found at the seven sites in great apes. These nucleotides presumably represent the ancestral states and can be used to deduce the direction of mutations in the individual steps.

There is only one combination of alleles – the haplotype 1A – among the ten found in which all eight sites are in the ancestral states present in the great apes. This combination must therefore be the ancestral haplotype of the ten present in the human populations. Nine of the ten haplotypes are derived from 1A in the manner indicated in Figure

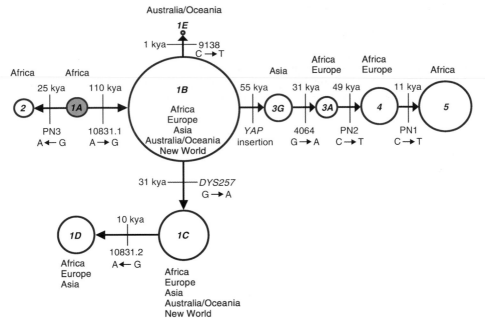

Fig. 10.15

Phylogenetic network of human Y chromosome haplotypes differentiated by nine variable sites. The size of the circle representing a haplotype is proportional to the frequency of that haplotype in the indicated population. The arrows indicate stepwise derivation of the haplotypes from the ancestral haplotype 1A. The bars crossing the arrows indicate individual single nucleotide mutations responsible for the derivation of each haplotype (the sites are identified by code numbers and the nucleotide changes by the short arrows). The estimated age of the mutations is given in thousands of years (ky). (Modified from Hammer et al. 1998.)

10.15 by the arrows and nucleotide changes. The derivation involves one homoplasy: at site 10831, the A nucleotide of the ancestral haplotype was substituted by G to give rise to the 1B haplotype, which then gave rise to 1C by a change at a different site (DYS257). In 1C, however, the site 10831 changed again, 100,000 years later (see below), back from G to A to produce the 1D haplotype. The ancestral 1A haplotype is restricted to Africa and is present at a low frequency in only 5.5 percent of males, as would be expected if Africa were the site at which differentiation into the ten haplotypes began. African populations south of the Sahara desert also contain the highest number of haplotypes (three) restricted to a single continent and the highest number of haplotypes overall (eight). This observation is again to be expected under the assumption of an out-of-Africa migration, since in the oldest populations the Y chromosome had the longest time to accumulate mutations.

Using coalescence analysis, which we describe in the next chapter, Hammer and his coworkers estimated that the differentiation of the nine haplotypes from the ancestral haplotype began approximately 147,000 years ago. This estimate has a standard error of 51,000 years and a 95 percent confidence limit of 68,000 and 258,000 years. The age estimates of the individual mutations that gave rise to the nine haplotypes (Fig. 10.15) range from 110,000 to 1000 years ago. Outside Africa, no haplotype older than 1B has ever been found. The *1B* haplotype presumably originated 110,000 years ago and then became the most frequent and geographically widespread Y-chromosome haplotype. Since *1B* originated in Africa and all haplotypes present in Asia, Europe, Australia/Oceania, and the New World are younger than *1B*, any Y chromosomes that may have existed on these continents at earlier dates must have been replaced by *1B* spreading out of Africa.

This deduction is consistent once again with the supposition that *H. sapiens* originated in sub-Saharan Africa less than 200,000 years ago and then spread throughout the world, replacing earlier forms of the genus *Homo* in the process. At the same time, however, the Y-chromosome data indicate that the population movements may by no means have been unidirectional. The *3G* haplotype, which is restricted to Asia, presumably *arose* in Asia. It then gave rise to haplotype *4*, which in turn gave rise to haplotype *5*. Since haplotype *4* is found in Europe and Africa and haplotype *5* only in Africa, there must have been a movement of males from Asia into Europe and back into Africa. From the estimated age of the mutations that gave rise to the haplotypes *3G*, *4*, and *5* it can be deduced that the reverse migrations occurred between 55,000 and 11,000 years ago.

The X-rated Files

The X chromosome differs from the autosomes in that it is present in males in only one instead of two copies and in female cells, one of the two copies is inactivated; consequently, in both males and females, only one allele is expressed at each locus in each cell. Since, however, the inactivation of the X chromosome is a random process, some cells of a particular female express one allele at a given locus, while others express the alternate form. This situation is therefore different from that in males, in whom the same allele is expressed in different cells of a given individual. In females the two X chromosomes recombine normally, whereas in males, gene exchange takes place only between the short pseudoautosomal segments of the X and Y chromosomes. As mentioned earlier, in a population consisting of an equal number of males and females, there are four autosomes of each kind to three X chromosomes. As a result, the variability of the X chromosome is somewhat lower than that of any of the autosomes.

Several genes on the X chromosome have been studied by different research teams to assess their polymorphism in human populations. The results thus obtained have generally led to similar conclusions as those reached in the Y-chromosome studies. The one exception is the study of the pyruvate dehydrogenase E1-alpha polypeptide 1 or *PDHA1* locus carried out by Eugene E. Harris and Jody Hey. Pyruvate dehydrogenase is a complex of three enzymes (E1, E2, E3) that catalyze the removal of hydrogen atoms from pyruvate, a product of glucose breakdown. The *PDHA1* gene specifies one of two types of subunits constituting the E1 enzyme. Harris and Hey sequenced 4200 nucleotide pairs of the *PDH1* gene, which is a whopping 17,082 nucleotide pairs long. The sequenced part encompassed four of the 11 protein-encoding exons and the intervening noncoding introns. All of the 35 DNA samples tested were from males, each of whom had only one copy of the *PDHA1* gene. By comparing the 35 sequences with one another, Harris and Hey could identify 25 variable sites at which either substitutions of one nucleotide by another or, at one site, a deletion of a nucleotide had occurred. The variations at the 25 sites had taken place in 11 different combinations, the haplotypes *A* through *J* (Table 10.3).

Up to this point, there is nothing unusual about the data in comparison with other sets. The odd result emerged, however, when the haplotypes were shown to fall into two groups, one of which is found only in African populations (haplotypes *C* through *J*) and

Table 10.3
Variable sites in the human PDHA1 gene sequence (in comparison to a chimpanzee sequence) and the frequency of haplotypes defined by these sites in the human populations. (From Harris and Hey 1999.)

Haplotype (allele)	Nucleotide at site 1111111222222233334 4 1 45670233459111244913461 2 5 94790305373236618780080 0 7 44215206936150679596788 9	Number of individuals bearing the haplotype in populations								
		African				Non-African			
		B	S	K	P	F	C	V	M
Chimpanzee	C C G G T T A T G C C G A G A A T A C G G C G C C								
A	– – A C C C – – T G T – – A C – C C – – – – – T –	0	0	0	0	0	2	1	1
B	– – A C C C – – T G T – – A C – C – – – – – – T –	0	0	0	0	5	5	4	0
B1	– – A C C C – – T G T – – A C – C – – – A – – T –	0	0	0	0	1	0	0	0
C	– – – C C C – – T G T – – A C – C – – – – – – T –	1	2	0	2	0	0	0	0
D	– A – – – – – C – – * – T – – – – – T – – T – – –	0	1	0	1	0	0	0	0
E	T A – – – – – C – – – – – – – – – T – – T – – – –	1	0	0	0	0	0	0	0
F	– A – – – – C C – – – – – – – – – T A – – – – – –	0	1	0	0	0	0	0	0
G	– A – – – – – C – – – – – – – G – – T – – – C – T	1	0	0	0	0	0	0	0
H	– A – – – – C C – – * – – – – G – – T – – – C – –	0	2	0	0	0	0	0	0
I	– A – – – – – C – – * A – – – – – – T – A – C – –	1	0	2	0	0	0	0	0
J	– A – – – – – C – – * – – – – – – – T – – – – – –	0	0	1	0	0	0	0	0

The numbers indicating the individual sites should be read vertically from top to bottom. A dash "–" indicates a nucleotide that is identical to that present in the chimpanzee sequence at this site; an asterisk "*" indicates absence of a nucleotide (deletion). Abbreviations of populations: B, South African Bantu speakers; S, Senegalese; K, Khoisan from the Angola/Namibia border; P, Pygmy from the Central African Republic; C, China; V, Vietnam; F, France; M, Mongolia.

the other only in non-African populations (haplotypes A, B, and B1; Table 10.3). Moreover, at one site (number 544), all of the African males had a G and all the non-African individuals an A. In other words, the mutation from G to A (since the chimpanzee has a G at this site, this is presumably the ancestral state) has been fixed in the non-African populations (Table 10.3). On a dendrogram depicting the derivation of the haplotypes by point mutations, the two groups are evident as two main branches which we designate I and II, albeit with one of the African haplotypes (C) appearing on branch I, while the rest of the African haplotypes are on branch II (Fig. 10.16). From the mutation rate of the PDHA1 gene calculated by comparing human and chimpanzee sequences and taking 5 my as the divergence time of these two species, Harris and Hey estimate that the two branches diverged 1.86 my ago. This is also approximately the estimated time at which the most recent common ancestor of all the variants seems to have existed. The ages of the mutations that generated new haplotypes (branch points in Figure 10.16) have been estimated in a similar way. Of these, the mutation at site 544 is of special interest because of its fixation in the non-African populations. Since the estimated age of the 544 mutation is 189,000 years, which the authors argue is older than any of the oldest modern human fossils, the subdivision into African and non-African populations must have preceded the emergence of modern H. sapiens.

These results have been interpreted as supporting the multiregional model of H. sapiens origin and hailed as such in the press. In reality, even if taken at face value and if confirmed

by further study, the results would still be open to alternative interpretations. In the meantime, however, an extension of the study by Ning Yu and Wen-Hsiung Li does not support the
conclusions reached by Harris and Hey in two critical points. By enlarging the size of the
sample, Yu and Li found that both alleles at site 544 as well as all the haplotypes except one
(*D*) are present in both African and non-African populations. This case illustrates how dangerous it is to draw far-reaching conclusions on the basis of results obtained by testing a
small sample taken from a very large population. As far as the multiregional and uniregional
hypotheses are concerned, the *PDHA1* data do not sway the argument one way or the other.

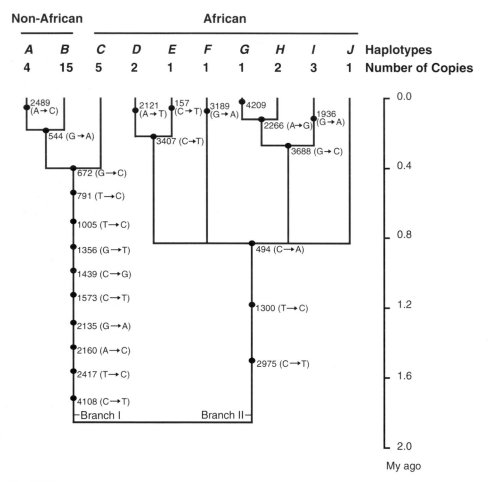

Fig. 10.16
Postulated mode of derivation of haplotypes (alleles) at the pyruvate dehydrogenase E1-alpha
polypeptide 1 (*PDHA1*) locus found in contemporary human populations. Differentiating mutations
are indicated by closed circles on branches and nodes; they are identified by the site number in the
sequence and the presumed nucleotide change (the direction of the change is indicated by arrows).
Where multiple mutations are listed on a branch, the order given is arbitrary; the actual order in
which the mutations occurred is not known. The estimated age of the mutations can be read off from
the time scale on the right-hand side. (Modified from Harris & Hey 1999.)

Opening the A files

Of the three billion nucleotide pairs constituting the haploid genome of a nucleus, approximately 93.3 percent or 2.8 billion pairs make up the DNA of the nonsex chromosomes, the *autosomes*. Of this vast repository of phylogenetic and genealogical information – the autosomal or A files – only a miniscule fraction has been tapped thus far. It has taken some ten years to sequence one human genome equivalent and although the speed of sequencing has accelerated considerably during this time, it is unrealistic to believe that it will ever be possible to sequence the entire genomes of a large number of individuals. Molecular genealogists must therefore select and focus on a small section of the A files only. The choice is not easy. Molecular evolutionists find themselves in the situation of a shopper who faces a large sortiment of goods in a store but has only a small amount of money in his pocket. Not only are there some 40,000 genes to choose from, the genes also take up only about 3 percent of the genome, whereas the remaining 97 percent, taken up by noncoding, intergenic DNA, may be genealogically just as informative, if not more so. Most of the choices are made serendipitously, but tradition also plays a part.

One of the genes with the longest tradition in molecular biology is *HBB*, the gene that codes for the β-subunit of the hemoglobin molecule (see Chapter Eight). Hemoglobin was among the first proteins to be sequenced and to have its three-dimensional structure solved. It was also one of the first genes to be cloned and sequenced and to have its exon-intron organization worked out. Hemoglobin genes figured in an important way in the formulation of the concept of molecular evolution and of the molecular clock. Variants of hemoglobin genes were among the first to be studied molecularly and they provided one of the best examples of the effects of natural selection at the molecular level. And, as described in Chapter Eight, hemoglobin genes have played a critical part in the development of modern ideas regarding the position of *H. sapiens* among the primates.

Not surprisingly, therefore, hemoglobin has also become the obvious choice in studies of genetic variability in human populations. One of the most recent and most extensive studies was carried out in 1997 by Rosalind M. Harding and her coworkers. These researchers sequenced a 3000-nucleotide-pair-long segment of human chromosome 11 collected in Africa, Asia, Europe, and North America, altogether nine different populations. The segment encompassed the entire *HBB* locus (exons and introns), as well as the regions flanking it. By comparing the 349 sequences with one another, it became apparent that the flanking region at one end of the gene contained a *recombination hot spot*, a stretch of about 300 nucleotide pairs in which exchanges between alleles occurred at high frequency. Since recombination complicates genealogical interpretations (see Chapter Three), the authors excluded this stretch from further analysis and concentrated on the remaining 2670 nucleotide pairs of sequence. In this segment they still found 23 apparent recombinants among the 349 sequences. The 18 variable sites, at which one of two different nucleotides could be found, occurred in 29 combinations, of which 13 were recombinant and 16 nonrecombinant haplotypes (Table 10.4). The ancestral states at the 18 sites could be identified by comparison with a chimpanzee sequence, and the sites could be arranged into a network revealing the most likely derivation of the haplotypes from the ancestral haplotype by point mutations (Fig. 10.17). The mutation rate of the *HBB* locus calculated from the comparison of human and chimpanzee sequences could be used to estimate the ages of the 18 mutations (Figures 10.17 and 10.18) and to deduce that the haplotypes began to diverge approximately 800,000 years ago.

Some of the *HBB* haplotypes are distributed globally (e.g., *A1* and *B1* are present in all populations tested), while others are restricted to a single continent or a geographically

Table 10.4
Variable sites in the human *HBB* gene segment in comparison with a chimpanzee sequence.
(Based on Harding et al. 1997.)

Haplotype	Nucleotide at site
	1 1 1 2 2 2 2 2 2 2 2 2
	3 3 5 5 6 9 3 4 4 0 5 5 6 6 7 8 9 9
	5 7 0 3 5 0 5 1 2 0 0 5 3 3 9 7 2 4
	7 9 8 2 0 6 8 6 3 8 3 4 4 6 2 6 4 5
Chimpanzee	A T T T A C C T C T G . . . .
A1	– – – – – – – G – – – G G C A G T G
A2	– – – C – – – G – – – – – – – – – –
A3	– – – – – – – G – – A – – – – – – –
A4	– – – – G – – G – – – – – – – – – –
B1	– – – – – – – – – – – – – – T – – T
B2	C – – – – – – – – – – – – – – – – T
B3	– – – – – – – – – – – – – – – – – T
B4	C – – – – – – – – – – – – – – – C T
B9	– – – – – – – – – – – – – A – – – T
B11	C – – – – – – – – – – – – A C T
C1	– C C – – T G – – C – – – A – – – T
C2	– – C – – T G – – C – – – A – – – T
C3	– – C – – T – – – C – – – A – – – T
C7	– – C – – T G – T C – – A A – – – T
D1	– – C – – T G – T C – C – A – – – T
D2	– – C – – T G – T C – – – A – – – T

The numbers indicating the individual sites should be read vertically from top to bottom. A dash "–" represents a nucleotide identical to the one in the top line; a dot "." indicates that the sequence at this site is not available.

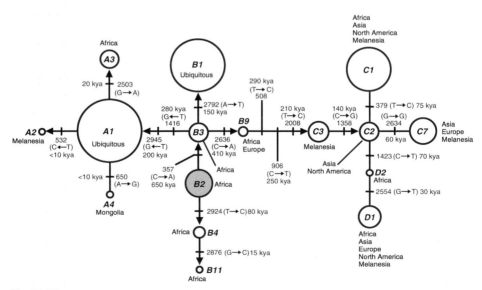

Fig. 10.17
Phylogenetic network of human hemoglobin beta (*HBB*) haplotypes. For explanation of symbols see Figure 10.15. The ancestral haplotype (*B2*) is shaded. (Based on Harding et al. 1997.)

delimited group of populations (e.g., *A3*, *B2*, *B3*, *B4*, and *B11* are present only in African populations, *A4* and *C2* only in Asia, and *C3* only in Melanesia; see Fig. 10.17). The African haplotypes *B2* or *B3* are apparently ancestral to the entire collection. The lineages of *A* and *C* haplotypes are derived from them, while the *D* lineage apparently originated from a *C* haplotype (Fig. 10.17). Populations within a single continent are, however, only poorly differentiated from one another or not at all. Also, the overall genetic di-

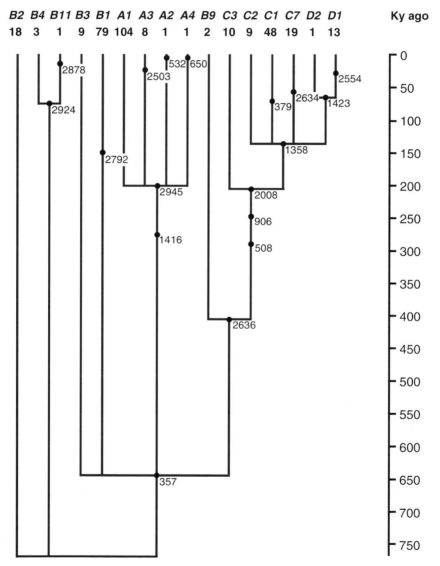

Fig. 10.18
Postulated mode of derivation of hemoglobin beta (*HBB*) haplotypes. Capital letters indicate haplotypes, the numbers below them, the number of individuals found to bear these in the global human population. For explanation of all other symbols, see Figure 10.16. (Modified from Harding et al. 1997.)

versity of the *HBB* region is only marginally higher in African populations than in others. Since the age estimates suggest that the mutations characterizing the Asian haplotypes are more than 200,000 years old, Harding and her colleagues argue that the ancestors of modern humans left Africa and dispersed earlier than the fossil data seem to indicate. Furthermore, because the young, low-frequency haplotypes nevertheless have a world-wide distribution, the researchers also argue that there must have been a substantial, multidirectional global gene flow (migration) in the human populations within the last 100,000 years.

Motifs From the Appassionata

DNA sequences are often compared to sentences written in a four-letter alphabet. Perhaps an even better analogy is with a musical opus composed using a four-tone scale. Indeed, together with his wife, the pianist Midori Ohno, the biologist Susumu Ohno developed a simple procedure for transforming DNA sequences into musical compositions and vice versa. The procedure assigns two consecutive tones in the octave scale to the four nucleotides in ascending order A, G, T, and C:

Ohno's piano arrangement of some of the well-known gene sequences has an eerie, though somewhat tedious sound.

Like a musical composition, DNA sequences have some sections that are unique and others that are repeated in some form or another. In a musical composition, the simplest of the repeats are iterations of one, two, three, or four tones. Corresponding to these in the DNA sequence are runs of one, two, three, or four nucleotides. In Beethoven's piano sonatas, for example, we find sections like this:

or like this

Like this:

or like this:

The first two and the fourth of these examples are from Beethoven's Appassionata, the third is the famous opening passage of the "Moonlight" Sonata. Corresponding to them in the DNA sequence (but not as transcriptions according to the Ohno and Ohno rule) are stretches like this:

AAAAAAAAAAAAAAAA

or like this:

CACACACACACACACA

Like this:

CGGCGGCGGCGGCGGCGGCGG

or like this:

GATAGATAGATAGATAGATAGATA

(In each example, only the sequence of one of the two strands is shown; the sequence of the second strand is, of course, complementary to that of the first.) Each of these repeats is preceded and followed by unique sequences and each is present in multiple copies scattered throughout the human genome. The first of the four repeats shown, for example, makes up approximately 0.3 percent of the genome, the second another 0.5 percent, amounting altogether to 15 million nucleotide pairs. Repeats in which one to four nucleotide pairs (basic units) appear ten to 20 times in each run are called *microsatellites*.

The name has its origin in the observation that when DNA is isolated from a tissue, placed on top of a column of a solution that increases in density from top to bottom, and then spun in a high-speed centrifuge, the bulk of it forms a broad band, while the rest of the DNA fragments gather together in one or several narrow bands as if they were satellites of the main band. Sequencing of such *minisatellite DNA* revealed it to be rich in repeats, much longer and more complex than in microsatellite DNA. The separation of the bulk from the satellite DNA takes place because of a difference in the nucleotide composition (GC content) and hence in density. Microsatellite DNA is not part of the satellite bands; it owes its name purely to its repetitive nature.

An outstanding feature of the microsatellite DNA is its extraordinary variability. It arises when DNA copyists make a mistake either by skipping one repeat unit or by adding an extra unit during DNA replication. The incident is akin to a piano player leaving out a note or playing an extra one in a repeat passage during a performance. This *replication slippage* apparently happens relatively often, in the case of the CA repeat, for example, with a frequency of one in 5000 per locus per generation. Variation in microsatellites and allied repetitive sequences is primarily responsible for the fact that no DNA of any two individuals, not even that of "identical" twins, is ever identical. In this regard, DNA is as reliable as fingerprints in identifying individuals, a fact that has revolutionized forensic medicine. Replication slippage may also be at least one of the mechanisms by which a microsatellite forms initially from an original progenitor sequence.

Since there are thousands of highly variable microsatellite loci in the human genome, they may seem to constitute the ideal material for studying the recent history of the human species. There are, however, certain problems associated with the use of microsatellites in genealogical studies and these have dampened the initial enthusiasm of molecular evolutionists for these genetic markers. One drawback is the lack of a generally accepted model describing the evolution of microsatellite loci. Nobody seems to know how one microsatellite allele changes into another. Since alleles at a particular locus usually differ in the number of repeat units, it is assumed that mutations consist of additions or deletions of single units. Two alleles found to differ by a multiple of repeat units are therefore assumed to have arisen by a stepwise process in which each step entails a single-unit change. It is possible, however, that single-step changes involving multiple units also occur. Since the statistical evaluation of microsatellite variability depends on the nature of the process by which the variability arises, ignorance of the latter places the whole theory of microsatellite variation on somewhat shaky ground. Related to this problem is the question of how to express and measure genetic distances defined by microsatellite variability. Various formulae have been proposed, but the most convenient one appears to be $(2\mu t)^2 = (m_X - m_Y)^2$, where μ is the mutation rate per locus, t the number of generations since populations X and Y diverged from each other, and m_X and m_Y the mean allele sizes in populations X and Y. The distance has a large variance so that many microsatellite loci must be tested to obtain reliable results.

The testing of individual microsatellite loci takes advantage of the fact that each locus is flanked on both sides by unique sequences. Where short stretches of the flanking sequences are known, primers complementary to ~20 nucleotides of the sequence on each side are synthesized and used to amplify the microsatellite between them by the polymerase chain reaction (see Fig. 5.1), and then sequence the amplification product.

Microsatellite variation has been exploited in attempts to obtain dendrograms depicting the presumptive successive splitting of human populations, and in determining the dates of the splits. Figure 10.19 gives an example, taken from a study by Anne M. Bowcock and her colleagues, of such use. Although similar dendrograms have been published by other research groups, in general the trees have proved to be rather unreliable as far as the arrangement of their branches is concerned. This could be due to an insufficiency in the number of loci tested and inadequate statistical data-evaluation methods. Nevertheless, most of the published dendrograms agree in one important point – an early split of non-African from African populations. One advantage of the microsatellite loci is that they can be used to estimate divergence time directly, independently of fossil-based calibration. To date a population split, all that is needed – in addition to allele frequency data – is familiarity with the mutation rate of the loci. Since, however, the rates seem to vary from locus to locus and because reliable estimates are rare, there is still much room for improvement. In one of the most extensive studies of microsatellite variability, the deepest split between human populations has been estimated to have taken place 156,000 years ago.

Tales of Baron von Münchausen...

Up until now we have broached two sources of information about life's ancient history – fossils and molecules of extant organisms. A potential third source are molecules of extinct forms. Life has invented a whole series of molecules which chemists call *organic* and which therefore provide a testimonial to its past existence. Indeed, some of the earliest indications of the appearance of life on earth are not provided by conventional fossils, nor by extant molecules, but by traces of organic molecules in ancient sedimentary rocks. The presence of hydrocarbons (compounds containing only carbon and hydrogen) in particular, but also sterans (compounds related to cholesterol), carbohydrates, and other *molecular fossils* or *biomarkers* has been demonstrated in rocks that are between two and three

Fig. 10.19
An example of a phylogenetic tree depicting genetic relationships among human populations on the basis of information provided by genetic distance calculated from 30 microsatellite loci. The tree was drawn by the neighbor-joining method and the root was placed at the midpoint between the two most distantly related populations (CAR Pygmy and Karitiana). The numbers at the nodes are bootstrap values obtained in 100 resamplings. The segment below the tree shows a unit of distance. CAR, Central African Republic. (From Bowcock et al. 1994.)

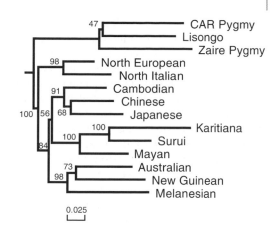

billion years old (Fig. 9.3). Their occurrence in these rocks is therefore taken as evidence for the existence of living forms at these very ancient times. Some of the organic compounds are characteristic of a particular organism and so their presence in the old sediments is used to date the emergence of these taxa. However, by far the most informative molecular fossils should be the proteins and DNA molecules. Molecular evolutionists did not wait for Michael Crichton to tell them that if DNA could be retrieved from organismal remains, a whole new perspective on life's evolution would open up. To be able to consign real instead of implied sequences to the nodes of phylogenetic trees has been the dream of many an evolutionary biologist.

For a while, it almost seemed as if the dream might come true. In 1984, Russell Higuchi and his coworkers successfully isolated DNA from a piece of dried muscle and connective tissue attached to the salt-preserved skin of a quagga that had died 140 years earlier but whose pelt had been stored in a museum in Mainz, Germany. The quagga was a zebra-like mammal characterized by yellowish-brown stripes on the forepart of its body. It once lived on the plains of South Africa but was exterminated through excessive hunting by white settlers in 1878. Taxonomists could not agree whether it was more closely related to zebras or to horses. Higushi and his colleagues were able to retrieve the quagga's mtDNA, sequence a short segment, use the sequence to draw a tree, and thus demonstrate that the species was indeed related to extant zebras and that the quagga-zebra split occurred 3 to 4 my ago.

Almost simultaneously, Svante Pääbo, then a doctoral student at the University of Uppsala, Sweden, succeeded in extracting and sequencing DNA from a 2400-year-old Egyptian mummy. Most of the DNA was broken down into fragments less than 500 nucleotide pairs in length, but some pieces were longer. Pääbo managed to isolate the longer pieces and demonstrated that one of them contained a human *Alu* element together with an unidentified flanking sequence.

These two reports confirmed that DNA, despite its brittleness, can survive the harsh conservation or embalmment procedures to which bodies are subjected in preparation for display in a museum or for storage in a sarcophagus. Although damaged, the DNA was not degraded to such a degree that it could yield no interpretable sequence information. The reports, however, made it clear that there are technical limitations to this kind of analysis, particularly in regard to obtaining sufficient amounts of ancient DNA. Before the DNA could be sequenced, its fragments had to be inserted into an appropriate vector and cloned in susceptible bacteria (see Chapter Five); the cloning procedure, however, requires a certain minimal, nontrivial amount of starting material which most remains of living forms cannot be expected to yield.

Around the time the two reports appeared, however, a technical advance was achieved that sidestepped this limitation: the polymerase chain reaction, PCR, was invented. This method is so sensitive that it can identify a letter writer from the DNA contained in the dried saliva on a licked stamp. In an extreme case, a single DNA molecule such as that present in each chromosome of a sperm cell can be amplified to amounts that are sufficient for cloning and sequencing, and in some instances even for sequencing without cloning. This advance was hoped to give molecular evolutionists the tool they needed to extend the studies of ancient DNA from specimens such as those used by Higuchi and Pääbo to those in which at best only traces of DNA from the extinct organisms could be expected. Before taking a step into the unknown, however, the more cautious molecular evolutionists tested the PCR's potential by applying the technique to DNA samples isolated from sources similar to those used in the first two studies. The application was a clear success and a whole series of reports followed describing DNA sequences of extinct animals such as the woolly mammoth, the cave bear, and the moa, a tall, flightless, ostrich-like bird of New Zealand, hunted to extinction by the Maoris.

Riding on the wave of these successes, the less cautious biologists applied the new technique directly to genuine fossil material, without first checking its limitations. They examined million-year-old bones, fossil-containing rocks, and amber-entombed insects, and reported on their fantastic achievements in fashionable science journals. Each report was introduced with a great fanfare at a news conference and was then followed up by the usual media circus, appearances on talk shows, television and newspaper interviews, and articles in popular science magazines. Among the most spectacular claims was the isolation of DNA from 17-my-old magnolia leaves deposited in clay and from an 80-my-old dinosaur bone.

At the time of this ancient DNA hype, when even some scientists were speculating on Crichton's *Jurassic Park* fantasy becoming reality, other molecular evolutionists, including some of the pioneers in the field, remained skeptical. They chose a much more cautious approach by first attempting to establish the conditions under which DNA preservation could be expected in fossilized materials, as well as defining criteria for authentication of ancient DNA. Svante Pääbo in particular, who in the meantime had been appointed Professor at the University of Munich and later Director of the Max Planck Institute for Evolutionary Anthropology in Leipzig, Germany, initiated a series of experiments in which he and his associates examined the effect of various agents on the DNA. Chemical analyses have revealed that the two most destructive agents are water and oxygen. The molecules of water are triangular in shape, with the oxygen atom located at the apex and the hydrogen atoms at the base of the triangle. The oxygen nucleus draws the electrons away from the hydrogen nuclei and so forms an electronegative region around itself, the two corners of the triangle occupied by the hydrogen nuclei becoming electropositive. The charges of the water molecules may interact with those of the DNA molecule when they arise during spontaneous tautomeric shifts in one of the nucleotides. The interaction may then split the water molecule and a hydrogen atom, the oxygen atom, or the OH group can be added to the base in a reaction chemists refer to as *hydrolysis*. The result is either a modification of the base to a type that normally does not occur in the DNA (as in the *deamination* of the cytosine in Figure 10.20 – the removal of the NH_3 group and the conversion of cytosine into uracil); or the complete removal of the base by breakage of the bond attaching it to the sugar of the DNA chain (as in the *depurination* of the guanine in Figure 10.20).

Each atmospheric molecule of oxygen, O_2, contains two unpaired electrons which are available to form bonds with other substances by accepting electrons from them. When this happens, the electron donor is *oxidized*, whereas the oxygen molecule is *reduced*. The oxygen molecule that has acquired an extra electron becomes negatively charged and so turns into the *superoxide anion*, O_2^-. Superoxide is an example of a *free radical*, a short-lived, unbound but highly reactive group of atoms. (In chemistry, a *radical* is a group of two or more atoms that have at least one unpaired electron. In this sense, molecular oxygen with its two unpaired electrons also qualifies as a radical.) The superoxide anion is highly unstable. At the first opportunity it procures additional electrons from other compounds and in a series of stepwise reactions is converted into two other free radicals, the hydrogen peroxide, H_2O_2, and the hydroxyl radical $OH\cdot$ (the dot symbolizing the unpaired electron). Another free radical, the singlet oxygen, $^1O_2$, is produced by a different chemical reaction or by the action of ionizing radiation on molecular oxygen. (Ionizing radiation is a stream of charged particles that can eject electrons from a neutral atom and so turn it into an electrically charged ion.) Each of the free radicals is, in turn, capable of oxidizing an assortment of compounds by requisitioning their electrons. The oxidation of these compounds changes their properties and so interferes with their normal activities. Some cells exploit these damaging effects of the free radicals in defense against microorganisms, but most cells protect themselves from their effects by mechanisms that restrain, and ultimately evict, the ruffians. If, despite these precautionary measures, the oxygen radicals gain access to the DNA, they may oxidize and so modify the nucleotide bases, cause a loss of bases, break the DNA strands, or crosslink the strands with each other. In a living cell, various mechanisms exist that repair these different types of lesions. But when an organism dies and the repair systems stop functioning, the oxygen radicals, together with a variety of DNA-degrading enzymes unleashed by the cells' demise, rapidly lay waste to the genome. Most of the DNA fragmentation occurs within the first few hours after death. The degradation then continues at a much slower pace effected by various environmental agents, including oxygen radicals generated by the action of ionizing radiation on water. With time, the DNA fragments become progressively shorter until ultimately, nothing is left of the organism's genome.

How long does it take to destroy all the organism's genetic information? Since the answer to this question undoubtedly depends on the circumstances of the fossilization process and on the conditions under which the fossil is preserved, it would be useful to be able to predict, without damaging the archeological material, which piece of it can be expected to yield DNA for sequence analysis. Just such a nondestructive assay was developed by Hendrik N. Poinar and his colleagues in 1996, based on a correlation they found between protein and DNA degradation.

All protein-derived amino acids except one possess a carbon atom to which four different atoms or atomic groups are attached. Picture the four bonds as projecting into space in four different directions (Fig. 10.21A) and then place a mirror at a perpendicular angle to the picture. In the mirror you see a reflection (Fig. 10.21B) which bears the same relationship to the original as your right hand to your left: neither your hands nor the picture and its reflection can be superimposed upon each another. We can designate the picture on the left as the L-form (from Greek *levo*, left) and the reflection on the right as the D-form (from Greek *dextro*, right). It so happens that all the amino acids in proteins are of the L-type. They cannot be converted into their D-forms without breaking bonds and rearranging atomic groups, which never happens as long as the protein is part of a living organism. The conversion or *racemization* of the L- into the D-form does take place, however, in the traces of the dried out proteins in archeological materials. Poinar and his colleagues discovered that the speed of racemization parallels that at which DNA is degraded in the same material. Since DNA is usually found inside the bone (in the cavity once filled with bone marrow and in the inner spaces of the compact part once occupied by bone-forming and bone-destroying cells), whereas protein traces are often present on the surface, measuring the extent of racemization (the ratio of L- to D-type amino acids) is a

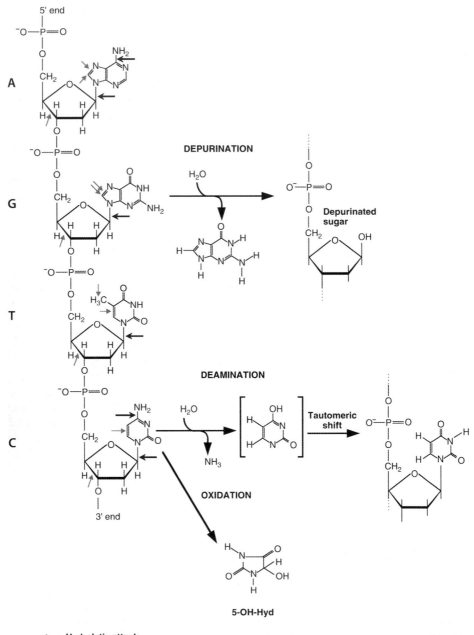

5-OH-Hyd

← Hydrolytic attack
← Oxidative damage

method of pretesting precious archeological material without damaging it. Tests of many archeological specimens have revealed that PCR-amplifiable DNA for sequencing cannot realistically be expected to be found in materials older than 50,000 to 100,000 years, and even then only if the material has been preserved under favorable conditions, such as a cold climate.

Laboratory experiments that tested the resistance of DNA to hydrolysis and oxidation under a variety of conditions, and assayed the state of DNA extracted from fossils of different ages, have cast doubts on the claims of DNA recovery from materials that are many millions of years old. Particularly vocal and courageous among the critics has been the chemist and expert on DNA degradation, Tomas Lindahl. He has argued forcibly that all claims of fossil DNA extractions from organisms that became extinct millions of years ago will ultimately be invalidated. And this is exactly what has been happening in recent years. One after the other, the claims are failing the most critical test of any scientific discovery – reproducibility. In some cases, under rigorously controlled conditions no fossil DNA could be recovered from the same material on which the original reports were based. But in most cases, the purportedly ancient sequences could be shown to be derived from quite modern contaminants.

The longest to resist debunking were reports describing the isolation of fossil DNA from organisms trapped in amber, the stuff so vital to Crichton's book and Steven Spielberg's

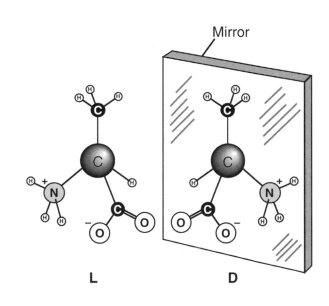

Fig. 10.21
The same and yet different. The two molecules of the amino acid alanine have the same chemical formula, but they differ in the arrangement of atoms and atomic groups bound to the central carbon atom. The molecule on the right (D) is a mirror image of the molecule on the left (L). The L-alanine cannot be converted into the D-alanine without breaking or grossly distorting the bonds. (Modified from Ball 1994.)

Fig. 10.20
Examples of chemical reactions responsible for the spontaneous degradation of DNA molecules with time. The left-hand side depicts four nucleotide residues (A,G, T, C) in a single DNA strand with arrows indicating bonds likely to be disrupted by the reactions (either hydrolysis or oxidation). The right-hand side depicts some of the chemical reactions and their consequences: depurination resulting in the removal of the nitrogenous base (purine) from a nucleotide, and deamination leading to the removal of an amino group (NH_3) from a nitrogenous base (here cytosine) and its conversion via a tautomeric shift into a different base (here uracil). Oxidation of cytosine may also result in 5-hydroxyhydantoin (5-OH-Hyd).

movie *Jurassic Park*. Amber is the fossilized resin of prehistoric trees. Oozing from the trees in its liquid state, the sticky resin trapped insects and parts of plants, and entombed them. As the oils in the tree resin were oxidized, the gummy material solidified and then gradually hardened over a period of several million years. Of the various types of amber, Baltic amber is the commonest. It arose from the resin of pine trees once widely distributed over northern Europe, an area now covered by the Baltic Sea. A recent or fossil resin of various tropical trees in Central America is called copal. The entrapment of organisms was followed by their rapid dehydration (mummification), and since the terpenoids contained in the resin probably prevent the wholesale decay of the organisms, their morphology is often exquisitely preserved right down to the cellular and subcellular level. The resin also prevented the access of water to the entombed bodies and so presumably protected their DNA from hydrolysis. Indeed, when the racemization test was applied to some of the retrieved insects, the D-to-L ratio of the aspartic acid, the standard used in this assay, did not rise above the value at which no DNA preservation is expected. Clearly, if any very old fossil material could be expected to contain ancient DNA, it had to be insect-containing amber.

This expectation was borne out by a flurry of reports describing alleged DNA sequences of 25- to 35-my-old insects (stingless bee, termite, woodgnat), plants, and bacteria retrieved from a series of amber specimens. Topping these fantastic achievements was the purported isolation of DNA from a 120-my-old amber-entombed weevil. Then, however, in 1997, Jeremy J. Austin and his associates published the results of their two-year-long effort to duplicate these results, all of which had ended in failure. These researchers tested more amber-retrieved specimens than all the authors of the previous reports combined, and did so under rigorously controlled conditions. Yet, none of the specimens yielded any trace of fossil DNA. The earlier "successes" were revealed to be the results of experimental artifacts: the weevils in the amber were real, the DNA was not. The failures have not surprised Tomas Lindahl, who has argued all along that amber is permeable to gases, including oxygen, and so all the DNA once contained in it must have long been oxidized and degraded. Thus, it seems that of the science fiction that motivated some researchers in their search for amber-derived ancient DNA, only the fiction remains. All those involved in this fiasco seem to have ended up as losers: the credibility of the authors of the original reports has been damaged, Austin and his colleagues have lost precious time in trying to duplicate what turned out to be artifacts; and the high-fashion science journals have had their reputation undermined by publishing irreproducible results. Only science writers profited by doubling their honorarium: first for reporting about the "successes" and then about the failures. These tales of molecular archeology are in the genre *Les contes d'Hoffmann*, or worse, Baron Münchausen. Tomas Lindahl cannot be blamed for his sarcasm in recently answering the question "But hasn't dinosaur DNA been reported?" thus:

> *Attempts in many laboratories to isolate DNA from dinosaur bones and other antediluvian sources have yielded nothing but modern contaminants. As late as April 2000, one of the major scientific journals contained as a News item a breathless account on the similarity between retrieved dinosaur DNA and bird DNA, starting "They said it couldn't be done. But…" Scientists at University of Alabama sequenced a 130-nucleotide long mitochondrial DNA sequence from dinosaur vertebrae, and found that it was 100% homologous to mitochondrial DNA from turkeys. However, the scientists themselves "remain quite sceptical of our own work" and noted that they had been consuming turkey sandwiches in the laboratory.**

---

* Current Biology 10:R616, 2000.

…and the Tale from the Neandertal

If the dinosaur DNA debacle had any positive effect at all, it was that of raising the crossbar in this scientific high-jump competition. If any further claims of fossil DNA sequences were to be made, they had better meet the new standards which the failed studies helped to define. "Do it right or don't do it at all" became the slogan of the day. The first investigation to meet the higher standards was performed in Pääbo's laboratory and in 1997 resulted in the isolation of DNA from a Neandertal individual found in the Feldhofer Cave in the Neander Valley.

Since the Neandertals inhabited Europe and western Asia from 300,000 to 30,000 years ago, the bones of the younger specimens might be expected to contain traces of their owner's DNA. It was therefore only logical to begin the search for fossil human DNA using Neandertal bones. This expectation was buttressed by the finding that the ratio of D-to-L amino acids in the bone from the Feldhofer Cave was within the range that allowed for DNA persistence. To isolate the DNA, Matthias Krings and his associates cut out a small block from the Neandertal bone with a drill saw, ground it to powder in liquid nitrogen, and treated the powder with a mixture of phenol and chloroform, a standard procedure for DNA extraction. They then used short stretches of nucleotides (primers) complementary to parts of the first highly variable section of the human mtDNA control region (the *hypervariable region I* or HVRI) to amplify DNA fragments about 100 nucleotide pairs in length from the extract by PCR. The products of this amplification were cloned and sequenced. Three independent extractions were carried out, two at the University of Munich, Germany (extracts A and B), and one by collaborators at Pennsylvania State University, University Park in the United States (extract C). Material from each extraction was amplified by more than one PCR and from each amplification, multiple clones were sequenced. Figure 10.22 explains why it was necessary to obtain multiple sequences. Since the bone was that of a single individual, theoretically it should have yielded only one type of sequence. In reality, however, it yielded 17 different sequences in a total of 49. Where did these different sequences come from?

Inspection of Figure 10.22 reveals the existence of two quite distinct groups of sequences. One group of seven sequences is nearly identical with one of the previously published contemporary *H. sapiens* sequences. These seven sequences have clearly been obtained from modern humans, presumably from one or several of the persons who handled the bone prior to investigation. Because it is so extremely sensitive, the PCR method picks up traces of DNA left on the bone by fingerprints or deposits from air droplets after sneezing or even from normal breathing. Krings and his associates went to great lengths not to contaminate the specimen with their own DNA, but of course, generations of paleoanthropologists who had examined the bone before them had exercised no such restraints. One or more of them had apparently left their DNA on the bones for Krings to amplify.

The remaining 42 sequences are classed together and are separated from the seven sequences of the first group by the sharing of nucleotides at eight diagnostic sites: 14, 21, 25, 35, 17, 19, 53, and 55 (Fig. 10.22). The sharing suggests that all the sequences in this group – let us call it the N group – came from a single source. Furthermore, their similarity to the *H. sapiens* sequences is high enough to indicate that the source was indeed a member of the genus *Homo*. But was the source *H. neanderthalensis* or just another *H. sapiens* specimen that differed both from the source of the reference sequence and from the seven sequences in the first group, which we will call the S group? To answer this question, we must compare the N-group sequences to all the *H. sapiens* available, sequences of which there are now over 5500 in the databases. The comparison reveals that at sites 35, 53, and 55, the

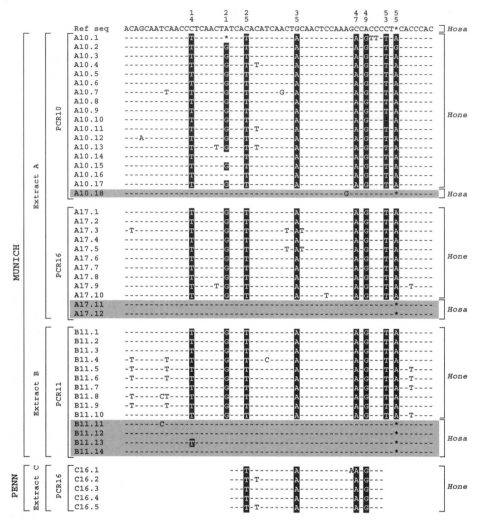

Fig. 10.22

Mitochondrial DNA sequences obtained from Neandertal bones. Three DNA extractions (A, B, and C) were performed on a bone specimen found in the Feldhofer Cave of the Neander Valley near Düsseldorf, Germany, two at the University of Munich and one at the Pennsylvania State University. The extracts were then amplified by the polymerase chain reaction with primers specific for part of the hypervariable region I (here numbered 1-61) and the amplification products sequenced. Two basic types of sequences were obtained: sequences representing variants of modern *H. sapiens* (*Hosa*, highlighted) mtDNA, presumably contaminants, and those quite different from modern human mtDNA, presumably derived from *H. neanderthalensis* (*Hone*). Substitutions at sites 14, 21, 25, 35, 47, 49, 53, and 55 are shared by all the Neandertal sequences (highlighted) but are nearly absent in the modern human sequences. Dashes (-) indicate identity with the top reference sequence; differences are shown as letters. (Based on Krings et al. 1997.)

nucleotides found in the N-group are absent in all of the S-group sequences, but that those present at sites 35 and 53 are also present in some of the nonhuman ape sequences. At the remaining sites (numbers 14, 21, 25, 47, 49), the nucleotides of the N-group sequences are present in some of the S-group sequences but not in others. What is important, however, is that the *combination* of the nucleotides found at these eight sites has not been found in any contemporary *H. sapiens* sequence. This observation alone indicates that the N-group sequences are derived from an individual who belonged to the genus *Homo* whose mtDNA lineage is not represented in the sample of sequences derived from *H. sapiens* and deposited in the databases. Krings and his colleagues concluded that the N-group sequences belonged to an individual of the species *H. neanderthalensis*. According to them, the possibility that the N-group combination does derive from *H. sapiens*, but has simply not yet been recorded is rendered unlikely by the fact that the number of differences between the N- and S-group sequences is three times higher than that between the two most different sequences of the S-group. Similarly, the number of differences between the N-group sequences and those of the nonhuman primates is about one-half that found between the *H. sapiens* and the nonhuman primate sequences. Taken together, Krings and his associates argue, these observations are consistent with the notion that the N-group sequences are indeed derived from *H. neanderthalensis*.

But if this is the case, why are there 18 different sequences in the N group (ignoring the partial sequences obtained from the C-extract) when they are all derived from the same individual? One possibility is that this individual possessed different mtDNA molecules, a situation that geneticists call *heteroplasmy*. While it is true that mtDNA heteroplasmy is well documented in certain organisms, in humans its existence is controversial. It has, however, never been observed to occur among contemporary individuals to an extent comparable to the heterogeneity seen in the N group. Heteroplasmy is therefore a highly unlikely explanation of the N-group sequence heterogeneity. (Note also that there is much less heterogeneity among the seven S-group sequences than among the N-group sequences in Figure 10.22). A more likely explanation is that the heterogeneity is an artifact arising from the antiquity of the DNA. Even under optimal conditions and in the hands of an experienced experimenter, sequences obtained from high-quality DNA cannot be trusted absolutely. Errors can arise at two different levels. First, the enzyme (DNA polymerase) that replicates the DNA strands during PCR amplifications is not infallible and occasionally incorporates the wrong nucleotides that are not present in the original template. Depending on the stage of amplification in which the mistake occurs, nearly all or just a small fraction of the amplified molecules may carry the erroneous nucleotide at the particular site. Presumably, when it replicates a partially degraded and modified template, the DNA polymerase makes more mistakes than it would do normally. Second, the sequencer, the machine that analyses the DNA molecule nucleotide after nucleotide and "reads" the sequence, also makes mistakes. If, for whatever reason, the separation of neighboring nucleotides is not clear-cut, the sequencer may interpret the ambiguous signal incorrectly and make wrong "calls". Both sources of error can normally be controlled by carrying out independent PCR amplifications of the same DNA preparation or, better still, of independent preparations, and by sequencing several clones from each amplification. As is evident from Figure 10.22, despite the fact that they took all the necessary precautions, Krings and coworkers obtained a much higher error rate than normal, presumably because of the changes undergone by the ancient DNA.

Normally, when different sequences are obtained where a single sequence is expected, the one that occurs at the highest frequency is taken to be correct. Among the 36 long sequences of the N group, there are 16 identical sequences that differ from the *H. sapiens* sequence at the eight diagnostic sites. These can thus be assumed to represent the authentic *H. neanderthalensis* sequence. The additional differences found in the remaining 20 sequences must have arisen either as replication or as sequencing errors. The presence of the same error in different clones of the same PCR suggests that the DNA polymerase made a mistake in an early stage of amplification. It is more difficult to explain the same error in independent PCR amplifications or even in different DNA extracts. One possibility is that a certain modification of a nucleotide may predilect it to be mistakenly taken for a different nucleotide by the DNA polymerase. If this were the case, one might be tempted to argue that the substitutions at the eight diagnostic sites are in reality artifacts that have arisen in a similar manner.

To establish firmly that the sequence was of Neandertal origin and not derived from contaminant *H. sapiens* DNA degraded and modified by external factors, two additional controls were necessary: a sequence derived from another *H. neanderthalensis* and a sequence of ancient *H. sapiens* mtDNA, preferably of approximately the same age as the putative Neandertal DNA. The former was to serve as a control of reproducibility; the latter was necessary to demonstrate that the phylogenetic record in the DNA molecules has not been seriously distorted by the decay. In collaboration with other researchers, Krings and his colleagues applied the amino acid racemization test to several other fossil bones of *H. neanderthalensis*, *H. sapiens*, and some animals, but could amplify DNA only from those that were of very recent origin. They attributed their success with the original Neandertal specimen to the circumstances under which these bones were preserved. The age of the Feldhofer Cave specimen is not known, but presumably it is over 30,000 years old. At that time, the Feldhofer Cave must have been a part of a tundra-like environment and the cold temperatures would have slowed down the DNA decay in the remains. The other bones tested were mostly from southern Europe and hence from an environment lacking in factors that could have decelerated the DNA degradation process in a similar way. Later, however, mtDNA sequences of presumably Neandertal origin were obtained both by Krings and others from bones found at southern locations. There must therefore have been other reasons for the initial failure than those suggested by Krings. At any rate, mtDNA sequences not only from additional Neandertal specimens but also from ancient *H. sapiens* bones were indeed obtained; of these later, however.

Another possibility to be considered was that the sequence obtained by Krings and his colleagues was not from a mitochondrion, but from nuclear DNA of *H. sapiens*. Numerous instances have been documented in a variety of organisms, including humans, in which fragments of mtDNA are integrated into chromosomes in the nucleus. The inserted nuclear mitochondrial DNA evolves independently of the genuine mtDNA and one could be misled into believing that its sequence, if taken to be of mitochondrial origin, is actually derived from a different species. Krings and his colleagues tested this possibility by using primers that precisely matched the putative *H. neanderthalensis* but not the *H. sapiens* sequence under conditions that allowed less than one copy of the Neandertal sequence per *H. sapiens* genome to be amplified. Since these primers failed to amplify products from DNA samples of African, European, Asian, Australian, and Oceanian *H. sapiens* origin, Krings and his colleagues argued that the sequence is apparently absent in contemporary human populations.

Starting from the first ~100 nucleotides, Krings and colleagues extended the sequence to cover the entire HVRI of the control region (379 nucleotide pairs). Altogether they obtained 719 nucleotide pairs of mtDNA sequence from one Neandertal individual. They achieved this extension by moving the priming sites progressively further forward, thereby producing a series of partially overlapping clones. The overlap of sites diagnostic for the putative Neandertal sequence assured contiguity of the sequence and at the same time served as a control against an artifact called "jumping" PCR. When two or more different templates to which PCR primers can anneal are present in the reaction mixture (in this case DNA from both the Neandertal and the contaminant *H. sapiens*), the DNA polymerase sometimes starts to replicate one template but then switches ("jumps") to the other. The result is a hybrid amplification product which, if the sequences of the templates are not known, can mistakenly be taken for a novel sequence. Indeed, some of the previously reported "fossil" DNA sequences were apparently modern sequences that had been scrambled by jumping PCR. The use of overlapping clones protected Krings and colleagues from falling victim to this ruse.

In a follow-up study published in 1999, Krings and colleagues extended the analysis to additional 340 nucleotide pairs of the control region, specifically to the *hypervariable region II or HVRII*. Combining the control region segments I and II, but excluding for methodological reasons sites such as those with insertions yielded 663 sites suitable for comparative analysis (Table 10.5). As in the case of the comparisons based on the initial short fragment (Fig. 10.22), those involving most of the control region reveal that the Neandertal mtDNA differs from the various *H. sapiens* sequences on average at three times as many sites as the latter differ among themselves. By taking 4-5 my as the divergence time between the human species and the chimpanzees, Krings and colleagues calculate the substitution rate of the control region to be 0.94×10^{-7} substitutions per site per year and from this figure they estimate the age of the most recent common ancestor of *H. neanderthalensis* and *H. sapiens* mtDNA as 465,000 years. The confidence limits of this estimate are 317,000 and 741,000 years. By the same measure, the various control region sequences now present in the human population had a most recent common ancestor 163,000 years ago with confidence limits of 111,000 and 260,000 years. The fourfold difference in the age of the MRCA of the Neandertal and modern human mtDNAs on the one hand, and the MRCA of all the modern human mtDNAs on the other indicates to the authors that the Neandertal mtDNA gene pool evolved for a substantial time period as an entity distinct from the mtDNA gene pool of modern humans. Cast in the mold of the uniregional hypothesis, Krings and colleagues argue that the *H. neanderthalensis* and *H. sapiens* lineages split from each other 465,000 years ago in Africa, but that the migration of *H. sapiens* out of Africa began only 163,000 years ago.

From all these observations, the authors draw four principal conclusions. First, the sequence they obtained from the Neandertal upper arm bone is a genuine Neandertal

Table 10.5
Number of differences between *H. sapiens, H. neanderthalensis*, as well as common and pygmy chimpanzee mtDNA control region sequences. (From Krings et al. 1999.)

Species compared	Number of differences	
	Mean ± SD	Range
H. sapiens vs. *H. neanderthalensis*	35.3 ± 2.3	29 – 43
H. sapiens vs. *H. sapiens*	10.9 ± 5.1	1 – 35
Chimpanzee vs. *H. neanderthalensis*	94.1 ± 5.7	84 – 103
Chimpanzee vs. *H. sapiens*	93.4 ± 7.1	78 – 113
Chimpanzee vs. *chimpanzee*	54.8 ± 24	1 – 81

SD, standard deviation. Comparisons involving chimpanzees were only based on part of the control region.

mtDNA sequence. Second, the *H. sapiens* and *H. neanderthalensis* ancestral gene pools have evolved separately from each other for a sufficiently long period of time to justify their differentiation into two distinct species. Third, Neandertals have not contributed mtDNA to the *H. sapiens* gene pool, implying that little or no interbreeding occurred between these two species. Fourth, by using the Neandertal sequence to root the group of the *H. sapiens* sequences the African lineages are the first to diverge on the phylogenetic tree; the data therefore support the notion that *H. sapiens* arose recently in Africa as a distinct species that, after a period of coexistence, replaced the Neandertals in Europe and Asia.

Support for the authenticity of the Neandertal mtDNA sequence has been provided by Igor V. Ovchinnikov and his collaborators. These researchers succeeded in obtaining a second Neandertal control region I sequence from the fossilized ribs of a Neandertal infant* estimated by the radiocarbon method to have died ~29,000 years ago. The infant's remains were found in the Mezmaiskaya Cave in northern Caucasus on the border between Europe and Asia. The researchers used the preservation of collagen-like debris as an indicator that DNA might still be present in the bone; they attribute the preservation to the specific microenvironment in the limestone cavern. The 345 nucleotide pairs of the sequence share 19 substitutions with the Feldhofer Cave sequence, relative to the reference DNA (Fig. 10.23). The Mezmaiskaya and Feldhofer Cave sequences differ from each other at 3.48 percent of their sites.

In the year 2000, Krings and his associates described a third sequence retrieved from a Neandertal bone, this time a bone found in the Vindija Cave, Croatia, and dated to 42,000 years ago. The nucleotide diversity of the three Neandertal sequences is 3.73 percent, which is similar to that of *H. sapiens* (3.43 percent), but lower than that of chimpanzees (14.82 percent) and gorillas (18.57 percent). The implications of these findings will be discussed in the next chapter. On a phylogenetic tree, all three Neandertal sequences group together, separately from selected *H. sapiens* sequences.

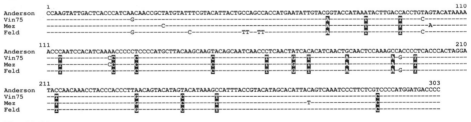

Fig. 10.23
Comparison of three Neandertal mtDNA (hypervariable region I) sequences with the modern human sequence (Anderson). The Neandertal mtDNA was extracted from bones found in the Feldhofer, Mezmaiskaya, and Vindija caves (Germany, Russia, Croatia, respectively). Identity with the top sequence is indicated by a dash. Neandertal-specific substitutions are highlighted. (The sequences are from Anderson et al. 1981; Krings et al. 1997; Ovchinnikov et al. 2000, and Krings et al. 2000.)

---

* Not all paleoanthropologists agree that the bones are those of a Neandertal infant. John Hawks and Milford H. Wolpoff, for example, point out certain discrepancies in the dating of the layer in which the bones were found and the difficulties in distinguishing Neandertal from modern human bones. They argue that the bone from which Ovchinnikov and his colleagues extracted the mtDNA were in fact those of a modern human infant.

The independent recoveries of three putative Neandertal sequences fulfill the first of the two requirements for their authentication. A step toward fulfillment of the second requirement was taken by Gregory J. Adcock and his coworkers in 2001. These researchers obtained ten different HVRI mtDNA sequences from hominid bones ranging in age from 2000 to 62,000 BP. The bones were excavated at two sites in southeastern Australia: Lake Mungo in western New South Wales (four specimens) and Kow Swamp in western Victoria (six specimens). Judging from their appearance, all the remains belonged to the anatomically modern *H. sapiens*, but they fell into two morphologically distinguishable groups – gracile (the Lake Mungo specimens) and robust (the Kow Swamp specimens). The origin of the groups is controversial, although archeologists tend to regard them as being derived from two different sources by past immigrations. In addition to the ten sequences from prehistoric specimens, Adcock and coworkers also obtained samples of HVRI sequences from living indigenous Australians (Aborigines).

On the maximum likelihood tree presented by the authors in their report, nine of the ten sequences are intermingled with sequences of the living indigenous Australians in a cluster clearly separated from the three Neandertal sequences. There is nothing unusual about them and they thus serve as controls indicating that their different degrees of fossilization did not change the mtDNA sequences so dramatically for them to take up unexpected positions on the tree. The tenth sequence, on the other hand, placed itself outside the cluster of contemporary and extinct *H. sapiens* sequences, a position to which Adcock and colleagues attribute great significance. The sequence, Lake Mungo 3 or LM3, was derived from a gracile individual who expired 62,000 ± 6000 years ago. On the tree presented by the authors it groups with the mtDNA sequence inserted into the nuclear genome (the *numt*) in a position which suggests that the two sequences diverged from the rest of the *H. sapiens* sequences after the divergence of *H. sapiens* from *H. neanderthalensis* but before the divergence of all the other modern human sequences (extant and extinct) from one another. According to the authors, the timing of divergence "…implies that the deepest known mtDNA lineage from an anatomically modern human occurred in Australia…" and the data "…present a serious challenge to interpretation of contemporary human mtDNA variation as supporting the recent 'out of Africa' model. A separate mtDNA lineage in an individual whose morphology is within the contemporary range and who lived in Australia would imply both that anatomically modern humans were among those that were replaced and that part of the replacement occurred in Australia."

For those who do not read scientific reports, the journalists conveyed the message in no uncertain terms as a clear home run for the multiregionalists in their match against the uniregionalists. In reality, however, the LM3 sequence does not support any such conclusions. The position of the sequence on the phylogenetic tree depends on the choice of extant *H. sapiens* sequences included in the sample and on the tree drawing method used. In many trees, both the LM3 and the *numt* sequences position themselves *within* the cluster of contemporary human sequences and, as the authors themselves admit, "trees in which the LM3/Insert lineage branched before the MRCA of contemporary human sequences were not significantly more likely than trees in which this lineage diverged after the MRCA of contemporary human sequences." In other words, you can take the tree that you find most appealing and the authors obviously preferred the one showing the LM3/*numt* sequences to be outside the cluster of extant human sequences. However, even if one were to accept this placement as genuine, it would still not provide evidence for the multiregional and against the uniregional hypotheses. It could be interpreted alternatively as evidence that mtDNA lineages existed in prehistoric *H. sapiens* that no longer persist in contemporary humans. The extinct lineages may have been replaced by the ancestors of the currently existing lineages which spread through the human species either by random genetic drift or by selection acting on one or several of the genes borne by the nonrecombining mtDNA molecules.

Of course, the same reservations must also apply to the interpretation of the Neandertal sequences. Their position outside the cluster of modern human sequences is largely reproducible and independent both of sampling and the method used. Nevertheless, it by no means provides evidence that Neandertals did not contribute genes to the *H. sapiens* gene pool and hence that they were "replaced" by modern humans. The apparent absence of *H. neanderthalensis* mtDNA variants in the *H. sapiens* gene pool could reflect the possible extinction of Neandertal lineages by drift or selection following the initial mixing of the two gene pools. Moreover, in the collection of contemporary human sequences, there are pairs that differ from each other to a greater extent than certain contemporary human sequences differ from the Neandertal sequences: the minimum number of substitutions between Neandertal and contemporary human mtDNAs is 13, whereas the maximum number of substitutions in comparisons between mtDNAs of living humans is 22. This observation is, of course, inconsistent with the conclusion that *H. neanderthalensis* and *H. sapiens* were two distinct, noninterbreeding species, because if they were, any overlap in mtDNA variation between them should have been removed by evolution subsequent to species divergence. Although the inconsistency can be explained by postulating multiple changes that obscure the phylogenetic signal at some nucleotide sites, it can be argued – as the proponents of the multiregional hypothesis indeed have done – that this explanation is an ad hoc postulate introduced to avert the downfall of the uniregional hypothesis. In all fairness, therefore, of the conclusions reached by Krings and coworkers and echoed by Ovchinnikov and his associates, only one is warranted: the sequences are most likely of Neandertal origin. The rest are overinterpretations reflecting a subjective bias toward the uniregional model. Viewed objectively, neither the Neandertal nor the ancient Australian sequences resolve the controversy regarding the origin of modern *H. sapiens*. Indeed, it is doubtful that mtDNA studies ever will. Despite the fanfare that accompanied the publication of the sequences of Neandertal and other fossil mtDNA sequences, the actual contribution of these sequences to the resolution of scientific questions has thus far been minimal.

Outside the Realm of Possibility

Although most of the molecular evidence discussed in this chapter favors the uniregional hypothesis, none of it definitely refutes the multiregional hypothesis. Since the latter has not yet been formulated in genetically precise, objectively assessable terms, its evaluation is very difficult. While the uniregional hypothesis unambiguously states that *H. sapiens* originated from a population living in Africa which spread from the continent between 130,000 and 200,000 years ago to replace all previous forms of *Homo* throughout the world, the multiregional hypothesis is more nebulous about its specific claims. Essential to the hypothesis is the postulate of migration and the consequent gene flow between populations, but the proponents of the hypothesis have not committed themselves in regard to the extent and direction of the flow. Similarly, the hypothesis invokes natural selection to explain retention of some characters and spreading of others without specifying the conditions under which the expected effects might be achieved. In this manner, the hypothesis has proved highly resilient to any attempts at refutation, absorbing new findings and adjusting to them after each confrontation.

To curb this evasiveness, Naoyuki Takahata and coworkers have tested the ranges of some of the genetic parameters within which the multiregional model could be a viable hypothesis and have compared the expected values with actual observations reported by various groups of geneticists. Before describing the results of this study published in 2001, let us first remind ourselves that according to the multiregional hypothesis, 1 to 2 my ago one hominid species (*H. erectus*) spread from a single region (Africa) and established founding populations in several other regions (Eurasia, East Asia, Southeast Asia) of the globe. Their subsequent expansion from these founding centers brought the populations in contact with one another and allowed them to exchange individuals and genes (Fig. 10.9B). We must therefore differentiate between events occurring in *regional populations* and those affecting the *global population*, the entire assembly of all regional populations. In the subsequent 1 to 2 my, the regional populations differentiated from one another in certain characters but simultaneously they also changed uniformly by acquiring the same novel characters. Hence we distinguish two types of characters: regionally-specific (i.e., those that distinguish populations of different regions) and globally-specific (i.e., those shared by all the populations but distinguishing the evolving global population from that of the original *H. erectus*).

Corresponding to these two types of characters are two types of mutations. But since mutations are the result of random events, mutations leading to both regionally- and globally-specific characters can occur in any of the regional populations. Yet, the former must remain restricted to one region, whereas the latter must spread to all other regions. Two processes must therefore govern the fate of the mutations, natural selection and gene flow. Regionally-specific mutations presumably impart selective advantage on their bearers in one region, but selective disadvantage in all others. The selective advantage leads to their rapid fixation in the one region by positive selection; the selective disadvantage to their elimination in all other regions by negative selection. (It cannot be assumed that a mutation is advantageous in one region and neutral in all others. If this were the case, the mutation would spread to other regions by gene flow and the character it controls would not be recognized as regionally-specific.) Globally-specific mutations impart selective advantage on their bearers regardless of the region in which they occurred. Migration of individuals between populations then spreads the mutations to other regions so that they ultimately become fixed in the entire global population.

Regionally-specific mutations are responsible for regional continuity of characters. Proponents of the uniregional hypothesis either deny the existence of regional continuity or explain it by independent (convergent), environment-driven evolution of similar characters in different species of *Homo*. In addition to the regionally- and globally-specific mutations and characters, there should also exist a third type – neutral mutations and characters. Their evolution is not influenced by natural selection, neither positive nor negative, and their fixation is effected by random genetic drift. Each of them presumably arises in one region and then spreads by gene flow to others.

A record of the origin, nature, and fate of these three types of mutation should be preserved in the sequences, frequencies, and geographical distribution of the chromosomal segments in which they occurred. Let us consider the three types one by one. For simplicity, we restrict ourselves to three main regions, Africa, Europe, and Asia, assume that the populations in these regions consist of N_1, N_2 and N_3 breeding individuals, and focus particularly on the time interval T_d extending from the founding of the regional populations 1 to 2 my ago to the emergence of anatomically modern *H. sapiens* ~130,000 years ago (Fig. 10.24). Three characteristics of these mutations interest us particularly: when they occur, where they occur, and to what extent they contribute to the diversity of the current human population.

To estimate the time of origin of the mutations, we compare all the variants of a given genomic sequence found in the human population with one another, determine their ge-

A. GLOBALLY ADVANTAGEOUS

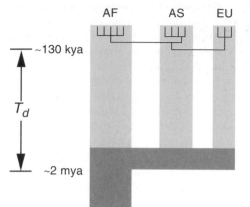

B. REGIONALLY ADVANTAGEOUS

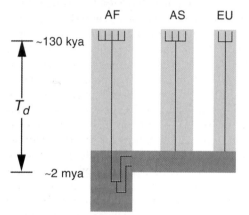

C. NEUTRAL

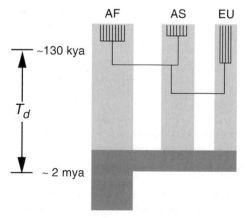

Fig. 10.24A-C
Genealogies of globally advantageous (**A**), regionally advantageous (**B**), and neutral (**C**) mutations expected under the multiregional model of modern human origin. The gene pools of the three regions (AF, Africa; AS, Asia; EU, Europe) are indicated by light shading; the common pool in existence before the separation into region pools is indicated by dark shading. Gene genealogies are shown by colored lines. (From Takahata et al. 2001.)

nealogical relationships by drawing a tree based on the sequences, identify the most recent common ancestor (MRCA) of the variants on the tree, and then calculate how long ago (TMRCA) the ancestor lived. For the calculation we compare the number of nucleotide substitutions accumulated in corresponding segments during the 5 my since the divergence of the human and chimpanzee lineages with the number of substitutions differentiating the two most divergent segments in the human population.

The place (region) in which the MRCA originated – the PMRCA – is identified by using the parsimony principle to infer the ancestral nucleotide at each variable site in the collection of sequences. If the site in the human sequence is occupied by the same nucleotide as that found in the chimpanzee, it is assumed to have remained unchanged since the divergence of the two species: it is surmised to represent the ancestral state. The human sequence that is by this criterion closest to the chimpanzee sequence is then presumed to be the MRCA and the population in which it is found to be its birthplace. If the MRCA sequence occurs in more than one region, the region in which its frequency is highest is taken to be its place of origin.

We thus have two parameters, TMRCA and PMCRA, with which to assess the behavior of the globally-advantageous, regionally-advantageous, and neutral mutations. The globally-advantageous mutations can arise at any time after the founding populations have been established in the different geographical regions. However, most of the mutations that distinguish the anatomically modern *H. sapiens* can be expected to have arisen since the time this form began to emerge ~130,000 years ago, that is after the T_d interval (Fig. 10.24). Advantageous mutations that might have occurred during the T_d period would have been fixed rapidly and so would have replaced any alleles that existed prior to their emergence at the particular locus. No record of diversity reaching deep into the T_d period is therefore expected. Any diversity found in the current population should be of recent origin and represent either recent mutations heading for extinction because of their inferiority relative to the advantageous allele, or a new advantageous allele heading for fixation and replacement of an earlier advantageous allele. In any case, the TMRCA is expected to be short. The PMCRA should not show a preference for any region because the mutations can arise in any one of them. Since the globally-advantageous mutations spread from their birthplace to all other regions, they can be expected to be subject to a considerable gene flow. Since, however, there may be little or no variation within and between regional populations at the locus at which the globally advantageous mutations occur, the entire population can be expected to be genetically homogeneous.

Regionally-advantageous mutations arise independently in each local population. Assuming that regional differentiation begins as soon as the founding populations are established, distinct allelic lineages arise in each region at or prior to the subdivision (Fig. 10.24B). Whenever new mutations, more advantageous than the preceding ones, appear in the individual lineages, they quickly replace the earlier alleles. The genealogy of these genes therefore consists of a single allelic lineage in each region. Only in the current population might alleles be present that are derived from recent mutations, not yet eliminated. The TMRCA of the regionally-advantageous mutations is therefore very long, extending over the entire length of the T_d interval and even beyond it. The PMRCA is in the region in which the ancestral population existed before it split, hence presumably Africa. And since there is no exchange of these mutations between regions, the entire population can be expected to be genetically heterogeneous.

The behavior of neutral mutations depends heavily on the number of breeding individuals (N) in the founding population and on the migration rate between the subpopulations. When the number of breeding individuals is approximately the same in the different subpopulations ($N_1 = N_2 = N_3$) and the migration rate (m) is high, then it can be shown, first, that the average expected TMRCA is approximately equal to $4N_T$, where N_T is the number of breeding individuals in the total population (i.e., $N_1 + N_2 + N_3$), and second, that the PMRCA is evenly distributed among the three founding populations. When the number of migrants per generation is limited, TMRCA is longer than T_d and the most recent common ancestor of all the neutral alleles present in the current population is found in Africa. Only when one or more migrants are exchanged between subpopulations every generation, does the MRCA tend to occur in the T_d time interval. For the MRCA to be found exclusively in Africa, the number of breeding individuals in the African population must be far greater than the number in the combined European and Asian populations.

To quantify the predictions concerning the expected behavior of globally-advantageous, regionally-advantageous, and neutral mutations under the assumptions of the multiregional hypothesis, Takahata and his colleagues simulated the events in the subdivided populations using a computer program designed specifically for this purpose. Armed with the precise predictions, they confronted the model with the available information about the genetic diversity of current human populations (Table 10.6). After disqualifying some of the data sets for various reasons of a technical nature, they were left with ten sets of sequence segments – those described on the preceding pages, as well as a few others. The TMRCA of the different autosomal systems ranged from 0.71 to 1.36 my; that of the X-chromosome systems from 0.41 to 1.59 my; while the TMRCA of the Y-chromosome and the mtDNA data sets was 190,000 and 200,000 years, respectively. The differences between the different sets are as expected when the representation of the sets in the gene pool and other factors are taken into account. For example, the TMRCA of the mtDNA data set is expected to be approximately one-fourth that of an autosomal system because mtDNA is maternally inherited (hence its effective population size, N_f, is one-half that of the autosomal segments) and because mtDNA is haploid (each cell or individual can have only one version of a genomic segment, while the diploid autosomal systems can have two). Therefore, if the expected TMRCA of a neutral gene is 800,000 years (see next chapter), that of an mtDNA segment is ~200,000 years. The observed TMRCA of all ten studied genomic sequence sets lies within the T_d time period. The sample may therefore represent a set of neutral genes. This observation neither refutes nor supports the multiregional hypothesis, unless we specify the total number of breeding individuals in the whole population.

Of the ten genomic sets, nine place the MRCA in Africa and one in Asia (Table 10.6). (Since the latter placement is based on a single genealogically informative site, it carries little weight.) This observation cannot be explained easily by the multiregional hypothesis unless it is assumed that the African founding population was much larger than the founding populations in Europe and Asia. The requirement for a ten-times-larger breeding size in the African founding population compared to the non-African population implies that the genetic contribution of the latter to the current population is minor if any.

It does not improve matters to argue, as some paleoanthropologists are prone to do, that the principal genetic parameters entering the calculations of population size and migration rates are not reliably estimated. It is true that, for example, the evolutionary rates of the genomic segments are difficult to estimate precisely, but even if the current estimates were off the mark by an order of magnitude, which is unlikely, the genetic data would still be at odds with the multiregional hypothesis. Thus, although the number of studied genomic segments is still too small to refute the multiregional hypothesis formally, the data available thus far are sufficient to seriously undermine its basic tenets.

Table 10.6
DNA sequence polymorphism of current human populations, as well as the TMRCA and PMRCA values deduced from it. (From Takahata et al. 2001.)

Sequence segment	Length (nucleotide pairs)	Sample size	Number of polymorphic sites	Number of haplotypes	PMRCA	TMRCA (my)
X-chromosome						
Il2rg	1147	10	0	1	—	0
Plp	772	10	2	3	Af (As)	1.28
Hprt	2676	10	4	4	(Af, As, Eu)	0.53
Gk	1899	10	1	2	As	0.41
Ids	1909	10	0	1	—	0
Pdha1	1657	10	5	3	As (Af, Eu)	1.05
Pdha1	4067	35	24	13	Af	1.59
dys44	1537	10	8	7	Af	1.35
dys44	7622	860	33	64	—	—
Zfx	1215	335	10	10	Af	0.93
Xq13.3	10163	69	32	20	Af	0.56
Xq22	87000	24	102	24	—	—
Autosomes						
Ace	23840	22	74	18	—	1.11
Mx1	565	708	9	10	—	—
Lpl	9692	142	44	76	(Af, Eu)	0.91
β-globin	2998	329	21	15	Af (As, Eu)	1.36
Mc1r	954	242	6	7	Af (As)	0.71
Mitochondrial DNA						
Control region	610	189	201	135	Af	0.20
Y-chromosome						
YAP	2638	16	5	5	Af	0.19

Af, Africa; As, Asia; Eu, Europe; my, million years; PMRCA, place of birth of the most recent common ancestor; TMRCA, time to birth of the most recent common ancestor.

By contrast, the uniregional hypothesis has no difficulty in accommodating the available data. Like the multiregional hypothesis, it easily accounts for the variation in length of the TMRCA by assuming that it is the consequence of ancestral polymorphisms persisting for different lengths of time in the ancestral populations (species) from which *H. sapiens* arose. The concentration of the PMRCA into one region is, in fact, exactly what the uniregional ('Out of Africa') hypothesis predicts. Exceptions may occur because the sorting of ancestral lineages into descendant regional populations is random or because of the gene flow between populations. According to the uniregional hypothesis, the extent of local genetic variation of the current human populations should be low, and this is indeed what the data show. However, the small founding population size in Europe and Asia suggested by the data does not mean that only a few individuals existed at that time. The implication of the genetic data is that the latest exodus from Africa led to the colonization of various areas of Eurasia, but that this process was terminated 30,000 years ago.

Groping Our Way in the Clouds

We began this chapter with the metaphor of Bashō's three journeys to the north and drew parallels between them and the three major migrations out of Africa that presumably took place during the hominoid history. The first out-of-Africa expansion was undertaken by a group of tree apes, dryopithecines, more than 15 my ago. During the second exodus, at least one hominid – H. erectus – spread from East Africa between 1 and 2 my ago. The third exodus occurred between 130,000 and 200,000 years ago and ensured the present-day global distribution of H. sapiens. Of the three, the first two migrations are uncontested but the third has split both classical and molecular paleoanthropologists into two camps: the multiregionalists who deny its occurrence, and the uniregionalists who swear by it. We then devoted most of the chapter to the discussion of evidence for and against the third exodus.

Evidence based on weathered skeletons does not seem to be sufficient to settle the issue of the origin of H. sapiens. If two groups of paleoanthropologists can view the same set of bones and still reach diametrically opposite conclusions, there is little hope of determining whether H. sapiens evolved separately from H. erectus in different regions over the last 1 to 2 my or whether it originated from a single population that began spreading out of Africa and replacing all earlier hominid forms 130,000 to 200,000 years ago.

So far, molecules have been unable to resolve the issue either, but there is hope that they will. Like traditional paleoanthropologists, molecular geneticists remain split into two factions, one favoring the multiregional and the other the uniregional hypothesis. It is, however, probably fair to say that there are more molecular and population geneticists in the latter than in the former faction. Nevertheless, it is also fair to say that the molecular evidence currently available is not entirely convincing. Although it favors the uniregional hypothesis, it does not entirely refute some form of the multiregional alternative. The early divergence of Africans on phylogenetic trees of present-day populations, the assignment of the most recent common ancestors of genetic variants present in current populations at the different loci to Africa, the high genetic diversity of the African populations – all these observations and more are most parsimoniously explained by the recent origin of H. sapiens from an African population, but alternative explanations are also possible. Similarly, the three putative Neandertal mtDNA control region sequences do suggest an early divergence of H. neanderthalensis from H. sapiens, but with reservations, and the sequences certainly do not prove that the two species did not interbreed.

On the other hand, it is somewhat naïve to believe that a band of adventurous hominids marched out of sub-Saharan Africa and did not stop until they reached the Near East. And after this new locality became crowded by population expansion that another group trotted out similarly, only stopping when it reached the present-day Beijing region, where a new population was established; and so on. It is far more likely that the process of expansion was continuous and much more complex, with the expanding populations remaining in contact via back-and-forth migrations. To depict the results of the process in the form of a tree may indeed, as some population geneticists argue, be improper. The seeming discontinuities of genetic distances suggestive of population cladogenesis may be an illusion created by comparisons of extremes in a continuum.

The manner in which we described the two hypotheses of human origin may have left the impression that each is a monolith of ideas. In reality, however, each – but especially the uniregional hypothesis – exists in several versions differing in various degrees of detail and propounded by different scientists. We have deliberately omitted any description of the differences because at present there is little to go by in their assessment. The critical issue at this time is to decide between the principles on which the two hypotheses are based.

Whether the multiregional or the uniregional hypothesis prevails in the end, the description will undoubtedly and by necessity turn out to be a mere caricature of the real events, which must have been so complex that they currently defy comprehension. The nature of the speciation process by which *H. sapiens* arose is the subject of the next chapter. We end this chapter just as we began it: with a quotation from Bashō, this time from the description of his second major journey. He called these travel sketches *The Record of a Travel-worn Satchel*:

> We passed through many a dangerous place... the road always winding and climbing, so that we often felt as if we were groping our way in the clouds. I abandoned my horse and staggered on my legs, for I was dizzy with the height and unable to maintain my mental balance from fear. The servant, on the other hand, mounted the horse, and seemed to give not even the slightest thought to the danger. He often nodded in a doze and seemed about to fall headlong over the precipice. Every time I saw him drop his head, I was terrified out of my wits. Upon second thoughts, however, it occurred to me that every one of us was like this servant, wading through the ever-changing reefs of this world in stormy weather, totally blind to the hidden dangers, and that the Buddha surveying us from on high, would surely feel the same misgivings about our fortune as I did about the servant.

Scientists, like Bashō's servant, are groping their way in the clouds, unaware of the precipices flanking the narrow road that they have taken.

CHAPTER 11

Through the Neck
of a Bottle

The Genesis
and the Genetic Nature
of *Homo sapiens*

O
BOUTEILLE
PLEINE
TOUTE
DE MISTERES,
D'UNE OREILLE
JE T'ESCOUTE;
NE DIFFERES,
ET LE MOT PROFERES
AUQUEL PEND MON CUEUR.
EN LA TANT DIVINE LICQUEUR,
QUI EST DEDANS TES FLANS RECLOSE,
BACHUS, QUE FUT D'INDE VAINCQUEUR,
TIENT TOUTE VERITÉ ENCLOSE.
VIN TANT DIVIN, LOING DE TOY EST FORCLOSE
TOUTE MENSONGE ET TOUTE TROMPERYE;
EN JOYE SOIT L'AME DE NOÉ CLOSE,
LEQUEL DE TOY NOUS FEIST LA TEMPERYE.
SONNE LE BEAU MOT, JE T'EN PRYE,
QUI ME DOIBT OSTER DE MISERES.
AINSI NE SE PERDE UNE GOUTTE
DE TOY, SOIT BLANCHE, OU
SOIT VERMEILLE,
O BOUTEILLE,
PLEINE TOUTE
DE MISTERES.*

François Rabelais: *Pantagruel*

The Oracle of the Holy Bottle

In the concluding chapters of Gargantua and Pantagruel, the ribald classic of fifteenth century French literature, Pantagruel, the son of the giant Gargantua, travels to the Island of the Sacred Bottle to consult the Oracle of the Holy Bottle on whether he should or should not marry. As he enters the Temple, Panurge, his merry travelling companion, is ordered to kiss the Bottle's rim and to sing the bacchic hymn – this chapter's epigraph. The symbolism of the song will hardly be lost on a reader familiar with Gargantua's bibulous tendencies.

Nearing the end of our journey, we too will need to consult an Oracle in our next and final chapter. In preparation for the ritual, we also have to take to the bottle, but not as an inebriate. Before conferring with the oracle on the future of our species, we want to scrutinize the molecular archives for any evidence pertaining to the mode by which the species arose. In the long narrative of the preceding chapters we followed the metamorphoses of species and groups of species along lines of descent that ultimately led us to *H. sapiens*. During this odyssey, one question must have persistently troubled the reader's mind: how does a species arise? Earlier (see Chapter Four) we concluded that at the molecular level, essential to the metamorphosis is splitting of the gene pool and the cessation of gene exchange between the daughter pools. In the absence of gene flow, the pools begin to diverge from each other until at some point, even when the opportunity arises, some of the altered genes begin to thwart any exchange with the sister pool. This explanation is, however, too general to satisfy the reader's curiosity. The aim of the present chapter is therefore to provide more details about *speciation* in general and the formation of *H. sapiens* in particular.

Regretfully, our understanding of the speciation process is still limited. Speciation has been the subject of much speculation (perhaps too much for the good of the subject), but of few empirical inquiries, especially at the molecular level. Consequently, it is a topic rife with hypotheses, and so with controversies, but wanting in facts. The two major controversies are the size of the founding population and the circumstances leading to the cessation of gene flow between sister pools. To emerge as a new species, a group of individuals – the founding population – must somehow disengage itself from the parental population and begin an existence of its own. How large does the founding population have to be?

This question is by no means trivial, because the size of the population influences the nature of the genetic processes taking place within it. There are two extreme views concerning the size of the founding population. According to the *founder effect hypothesis*, a species arises from a pair or from very few individuals that have become isolated from the parental population. In the most familiar scenario, the founders are castaways, stranded on an isolated island, but alternatively they may be the survivors of a drastic decline in the size of the parental population. In the latter case, a graphic depiction of the change in pop-

---

* *O Bottle / Full of mystery, / With a single ear / I hark to thee. / Do not delay, / But that one word say / For which with all my heart I long. / Since in that liquor so divine / That your crystal flanks contain / Bacchus, India's conqueror strong, / Holds all truth, for truth's in wine. / And in wine no deceit or wrong / Can live, no fraud and no prevarication, / May Noah's soul in delights dwell safe and long, / Who taught us use in moderation / Of our cups. Be kind to me, / Let the fair word be said. / From misery set me free / Then no drop, white or red / Shall perish. There shall be no waste of thee. / O Bottle full of mystery, / With a single ear / I hark to thee. / Do not delay.*
Francois Rabelais: *The Histories of Gargantua and Pantagruel.* Translated and with an Introduction by J.M. Cohen, p. 703. Penguin Books, Baltimore, MD 1963.)

ulation size takes the shape of a bottle, with the bottle's neck representing the speciation phase (Fig. 11.1A). The founding period, in this case, is therefore commonly referred to as the *bottleneck phase*. Later, to achieve a sustainable size, the population must increase by expansion from the few founders. When this expansion phase is included in the depiction, the graphic presentation assumes the shape of an hourglass (Fig. 11.1B), the sand-filled vessel formerly used to measure time. The founder effect hypothesis posits that the bottleneck phase is an essential part of the speciation process. Although random genetic drift, natural selection, and inbreeding (the mating of closely related individuals) affect both small and large populations, in the former they lead to a very rapid loss of some alleles and fixation of others. Proponents of the founder effect hypothesis argue that the net effect of all changes in the small founding population is a *genetic revolution* involving the refashioning of the gene pool, which increases the likelihood that a new species will arise.

The rival hypothesis interprets the transition from one species to another as a gradual process without any radical reduction in population size (Fig. 11.1C). Rather than postulating sudden genetic upheavals, it conjectures the accumulation of small changes in the gene pool which differentiate and ultimately isolate it from all other pools. A large population has a good chance of supplying the mutations necessary for the adaptation of the rising species to the new environment that it colonizes. The founder effect hypothesis may be generally better known than the gradual transition hypothesis, but there is in reality very little evidence to support it and even theoretically, it stands on shaky ground.

The circumstances leading to the cessation of the gene flow during speciation are at the heart of two controversial issues, one concerning spatial and the other temporal relationships between the diverging pools. The classical interpretation of spatial relationships, which goes back to Darwin, asserts that to diverge, the sister pools must become separated by a barrier that more often than not takes the form of long geographical distance (e.g., birds stranded on a remote island; Fig. 11.2). Since the migrants colonize lands other (Greek *allo-*) than their original fatherland (Greek *patra*), this model of speciation is called *allopatric* (Fig. 11.3A). The isolation may, however, also occur at much shorter distances, for

A. BOTTLENECK   B. HOURGLASS   C. CYLINDER

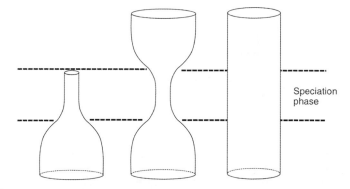

Fig. 11.1A-C
The three icons of speciation. The neck of a bottle (**A**) or the constriction of an hourglass (**B**) epitomize the founder effect hypothesis; this postulates that a drastic reduction in the size of the founding population is a *sine qua non* for the emergence of a new species. A cylinder (**C**) symbolizes a group of hypotheses in which a dramatic reduction in population size is not a condition for the origin of a new species.

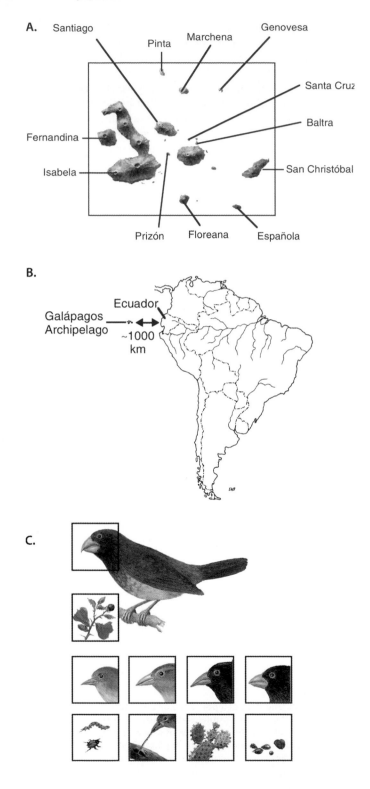

A. Santiago Pinta Marchena Genovesa Santa Cruz Baltra Fernandina Isabela San Christóbal Prizón Floreana Española

B. Ecuador Galápagos Archipelago ~1000 km

C.

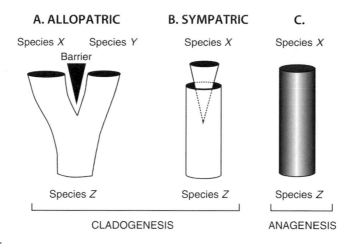

A. ALLOPATRIC   **B. SYMPATRIC**   **C.**

CLADOGENESIS   ANAGENESIS

Fig. 11.3A-C
Different modes of species formation. New species can emerge when a population comprising the ancestral species is split into two (**A** and **B**). In another case new species may arise by the accumulation of
genetic changes within a population without splitting (**C**). The splitting may involve geographical separation of the ancestral population into two populations or more which then occupy different regions
(**A**); alternatively it can occur within the ancestral population in the same geographical region (**B**).

example in a lake in which the water level has dropped to expose a ridge that divides the
original stretch of water into two separate lakes (Fig. 11.4). An alternative to allopatry is
sympatric speciation, in which populations living in the same (Greek *sym-*) place diverge
to form distinct species (Fig. 11.3B).

It is not difficult to imagine how, in the case of allopatry, isolated gene pools accumulate
different mutations, some of which eventually prevent exchange between the populations and
so turn them into distinct species. But it is far less obvious how a new species can arise within an area already occupied by an ancestral species. To propose that the area is fragmented
into distinct environmental niches and that new species arise by adaptive pigeonholing, each
nestling down into a different nook, is not a solution. This could merely be regarded as a case
of scaled-down allopatry with the niches representing separate territories. Under true sympatry, a new species has to emerge from its predecessor without spatial isolation. How could
this happen? A likely explanation is suggested by the fact that individuals constituting a species
differ from one another and that differences in appearance play an important part in choice
of mate. *Sexual selection* can therefore provide an isolation mechanism separating one group
of individuals within a species from the rest and so lead to the emergence of a new species.
Several good examples of this mode of species formation have been documented and the

Fig. 11.2A-C
A classical case of gene pool isolation as a result of geographical distance: The Galápagos Archipelago
(**A**) in the middle of the Pacific Ocean lies at a distance of almost 1000 km from the nearest land – the
coast of Ecuador in South America (**B**). The individual islands of the Archipelago arose as volcanoes
soaring up from the ocean floor and were subsequently colonized by plants and animals that reached
them from the mainland. Among the most prominent colonists of the Archipelago are banting-like
birds known as Darwin's finches (**C**). These birds originated from a founding flock of a single ancestral species which then diverged into several species distinguished by their adaptation (reflected particularly in the shape of their beaks) to different food resources. (**A** and **C** adapted from Beneš 1994.)

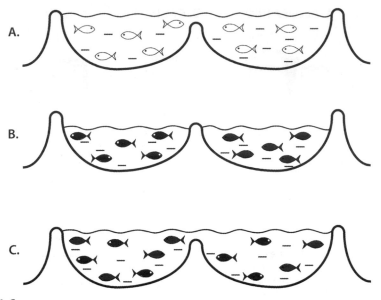

Fig. 11.4A-C
A hypothetical example of allopatric speciation. **A.** A lake is inhabited by a single species of fish. **B.** The water level drops and a barrier emerges which divides the single lake into two. In the two lakes the fish populations accumulate different mutations and so diverge from each other both morphologically and behaviorally. **C.** Much later the water level rises again and the two populations intermingle. They are now so different, however, that they are no longer able to interbreed: they have become distinct species.

existence of sympatric speciation must therefore no longer be doubted.

The issue of a temporal relationship between diverging gene pools is polarized by the cladogenetic and anagenetic hypotheses. The *cladogenetic hypothesis* envisions speciation in the form of population (gene pool) splitting (Fig. 11.3A,B). It is commonly assumed that splitting occurs as bifurcation, a two-pronged branching, but a split into more than two branches is sometimes also admitted. The *anagenetic (phyletic) hypothesis* assumes a linear transformation of one species into another without population splitting (Fig. 11.3C). Here the mere accumulation of mutations along a single lineage differentiates its early stage from its late phases to such an extent that the stages can be regarded as representing different species. Species do evolve in the interval between succeeding splits; this is uncontested, but are these intervals long enough to produce distinct, temporally rather than spatially isolated species of a single lineage? And is it not academic to argue that the temporally isolated populations are distinct species when this contention cannot normally be tested by examining their ability to interbreed?

How long a single species can exist before splitting is uncertain and probably highly variable (see Chapter Twelve). However, in some cases an apparently unbranching lineage is known to have persisted for a very long time indeed and to have changed morphologically to such an extent that distinct species designations for the end points of the interval are certainly warranted. For example, the dodo, the flightless bird that inhabited the island of Mauritius in the Indian Ocean before it became extinct, evolved from flying ancestors related to the pigeon without any evidence of lineage splitting. It is reasonable to assume that the flying ancestor and the flightless dodo would not have been able to interbreed had they ever met and this assumption is certainly no more academic than that made by pale-ontologists every time they assign a fossil to a new species. We have to conclude, therefore, that both cladogenetic and anagenetic modes of speciation exist.

All these different alternatives for each of the two controversial issues have been evoked by researchers contemplating hominid evolution and specifically the origin of *H. sapiens*. Thus, the multiregionalists argue that *H. erectus* originated by a founder effect during a bottleneck phase some 2 my ago. Some uniregionalists, too, explain the emergence of *H. sapiens* via a bottleneck, which they date to a much later time, however. Although the mode of hominid speciation is generally assumed to have been allopatric, elements of sympatry figure in some of the evolutionary scenarios. Finally, while cladogenesis in the formation of hominid species is now the preferred hypothesis, not so long ago hominid evolution was generally interpreted in terms of anagenesis. Even today, the transformation of *H. erectus* into *H. sapiens*, as envisioned by the proponents of the multiregional hypoth-esis, remains a purely anagenetic interpretation.

Queen Victoria's Fishes ...

There are several popular misconceptions about speciation. We mentioned one of them al-ready: that new species are founded by a single pair accidentally sequestered from its parental population. Two other misconceptions are that speciation does not occur or can-not be observed in our times and that it is so slow a process, it cannot be studied directly. In reality, speciation is going on all around us and at a speed fast enough to be discerned. There are numerous groups of species that are a nightmare for taxonomists for the very reason that they are in the midst of speciation and hence difficult to sort out. Some of these groups have been studied extensively and have attained certain popularity not only among the experts, but also beyond their circle. Here we portray two of them to acquaint the read-er with some of the general principles underlying the speciation process.

With the first example we remain in the cradle of humanity, East Africa, but switch our attention to a different vertebrate altogether, the cichlid fish familiar to every aquarist. Cichlids are perch-like fishes that are widely distributed across the whole of sub-Saharan Africa but are especially species-rich in the three great lakes – Victoria, Tanganyika, and Malawi. Nobody knows exactly how many species of cichlid fishes live in these lakes but the number probably exceeds 1000. In Lake Victoria alone about 200 species have been identi-fied, but the actual number could be twice as high, although sadly many of the species are rapidly becoming extinct as a result of human intervention. The great lake, second largest in the world (after Lake Superior), was named after Queen Victoria by the English explorer John Hanning Speke, the first European to reach it. (How boring: the explorers could choose from such a wide range of exotic words in naming the parts of the world that they had dis-covered for the West, yet they opted for such dull alternatives as Lake Edward, Lake Albert, New York or New Jersey!) Zoologist Peter H. Greenwood, who described many of the Lake Victoria species, established that their overwhelming majority occurred nowhere else in Africa, nor in the rest of the world, not even in the rivers around Lake Victoria. Yet, mtDNA analyses by Axel Meyer and his coworkers, as well as by other investigators, have provided evidence that many of the East African species are closely related to one another, forming a group called *haplochromines*. The name refers to the striking variability of these fishes in their color patterns (Greek *chroma*, color), a feature for which they are so highly prized by aquarists. (The name "cichlids" derives from Greek *kichle*, a designation for a group of wrasses, to which the cichlids are related and which are also quite colorful.)

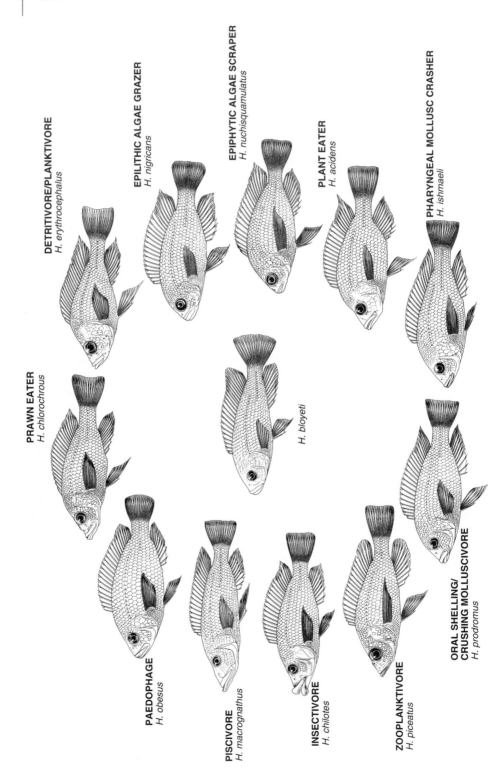

DETRITIVORE/PLANKTIVORE
H. erythrocephalus

EPILITHIC ALGAE GRAZER
H. nigricans

EPIPHYTIC ALGAE SCRAPER
H. nuchisquamulatus

PLANT EATER
H. acidens

PHARYNGEAL MOLLUSC CRASHER
H. ishmaeli

PRAWN EATER
H. chlorochrous

H. bloyeti

ORAL SHELLING/ CRUSHING MOLLUSCIVORE
H. prodromus

PAEDOPHAGE
H. obesus

PISCIVORE
H. macrognathus

INSECTIVORE
H. chilotes

ZOOPLANKTIVORE
H. piceatus

Sandra Nagl and her coworkers used mtDNA sequences to identify a group of hap-lochromines in the East African rivers that is the closest relative of the Lake Victoria fishes. Presumably it was this group that provided the founding stocks of the two currently known lineages encompassing most of the Lake Victoria haplochromine species. Morphologically and behaviorally the Lake Victoria species are quite distinct from one another (Fig. 11.5). The morphological differences are mainly in coloration, body size and shape, as well as shape of the mouth and head. The behavioral differences relate to feeding and breeding habits. Both the morphological and behavioral characters have evolved as adaptations to the various environmental niches that the different species have colonized. As far as could be ascertained, the species normally do not interbreed in the lake, even though they have plenty of opportunity to do so. Experiments by Akie Sato and others have demonstrated, however, that if given no other choice or if fertilized artificially, many of the species do produce fertile hybrids.

In striking contrast to the morphological and behavioral differences, the species are indistinguishable molecularly; this applies to any of the markers that have been tested thus far, including the control region of mtDNA. As a group they are differentiated from the group of river cichlids that apparently produced them, but even this difference is minor. All these observations indicate that the species are quite young. Just how young is suggested by geological studies in which sediments from different parts of the lake were probed and dated. The analysis, carried out by a team led by Thomas C. Johnson, indicates that the vast lake had dried up completely about 18,000 years ago and remained that way until 12,400 years ago. Because the species that now inhabit Lake Victoria do not occur anywhere else, they must have arisen locally. And since the present, relatively shallow lake began to form only 12,400 years ago, the more than 200 species inhabiting the lake only had this short period of time to develop from the two or very few founding species that presumably still live in the rivers nearby. The river species show none of the morphological and behavioral diversity displayed by the lake species. The diversity must therefore have arisen within the short time period of ~12,000 years.

The haplochromines that are restricted to some of the smaller satellite water reservoirs near Lake Victoria may have had even less time to evolve into distinct species. In the few hundred years since the formation of the satellite lakes, these fishes were able to acquire a distinct appearance and behavior. If this interval seems too short to you, just think of the many breeds of dogs! They are, of course, not distinct species and the selection that produced them was not natural, but the breeds probably would have speciated if, in addition to selection for morphological differences, they had also been subjected to sexual selection. Yet, some of the breeds are no more than a few hundred years old.

These observations indicate that the numerous cichlid species of Lake Victoria arose from perhaps only two river species that entered the forming lake ~12,000 years ago. The rapid speciation was driven by adaptation to different environmental niches, in particular to alternative food sources. The different cichlid groups have each specialized to eating or-

---

Fig. 11.5
An example of adaptive radiation. Cichlid (haplochromine) fishes of Lake Victoria are adapted to different environmental niches within the lake. The adaptations relate especially to different modes of subsistence and affect the morphology of the head and the mouth in particular. The different forms apparently do not interbreed in nature (although some of them do in captivity) and are therefore regarded as different species of the genus *Haplochromis* (*H.*). Some taxonomists even assign the species to different genera. All the species are believed to have arisen from a form similar to that shown in the center of the circle, which is found in the rivers around Lake Victoria. The species divergence may have occurred quite rapidly, perhaps within the last 10,000 years. The origin of many new species from a single ancestral species by adaptation to different ecological niches is referred to as adaptive radiation.

ganic particles (detritus) and floating, tiny organisms (plankton); scraping algae from rock surfaces and plants; eating plants; eating snails by crushing their shells with their jaws; crushing snails with a second set of jaws in their throats; gathering drifting small animals (zooplankton); and eating insects, other fish, fish embryos, or prawns (Fig. 11.5). Any morphological modification that improved their efficiency in procuring the particular foodstuff was strongly favored by natural selection and in this manner each group adapted itself to the specific conditions of its particular niche. The different groups (the emergent species) diverged from one another accordingly in a process of *adaptive radiation*.

Simultaneously, the members of each group developed a partiality for their own kind and found ways of recognizing individuals of the opposite sex from the same group. Any character that promoted this partiality and recognition was strongly favored by sexual selection. The ability to distinguish themselves from other groups by color patterns (species) and the ability to perceive the color differences may have been particularly important in the differentiation process, as preliminary observations indicate. The fixation of mutations responsible for the species-recognition mechanisms has barred the different species from interbreeding and so isolated them reproductively from one another. Some of the species may have accumulated mutations that prevent the development of hybrids or make the hybrids infertile should the recognition barrier be breached. Most of the species, however, have not yet reached this stage in the evolution of reproductive isolation and are still able to interbreed when faced with no other choice. In this sense, the speciation of the Lake Victoria cichlids is progressing under our very eyes.

Random sampling of the cichlids' nuclear genes reveals the presence of polymorphisms that are shared not only by the different species in the lake, but also by the Lake Victoria and the river species. In other words, if there are two or more alleles present at a given locus, they are found in the different species in the lake and in the rivers. This situation would not be expected if each species arose via an extreme bottleneck, as postulated by the founder effect hypothesis. If each species were founded by two or very few individuals, virtually all the preexisting neutral polymorphisms would have been lost by random genetic drift. Calculations of the kind we describe later in this chapter indicate that each founding population must have consisted of thousands of breeding individuals.

The short time interval in which the speciation of Lake Victoria haplochromines has taken place suggests that the numerous species emerged almost simultaneously, as if the founding populations radiated into their different adaptive niches in a starlike fashion. In the future, many of the species will very likely become extinct, most of them with human "help", and the polymorphisms, now widespread in these species, will become fixed in the few that survive. It can be expected that the neutral polymorphisms at the different loci will sort themselves out randomly through these fixations. When this happens, the surviving species will appear to have arisen by bifurcation.

… and the Birds of the Tortoise Islands

Our second example of speciation takes us to the Galápagos Archipelago in the Pacific Ocean, the islands of giant tortoises (which is what the Spanish name means), iguanas, boobies, flightless cormorants, tropical penguins, and *Darwin's finches*. Charles Darwin visited the Archipelago briefly in 1835, as part of an around-the-world journey on board *H.M.S. Beagle*, and noticed among the many interesting creatures that he had not seen anywhere else on his trip, a group of small, inconspicuous, finch-like birds. He collected some of them and brought them back to England for identification by expert ornithologists. The birds were revealed to be new species of a group that became popularly known as Darwin's finches. Since then, the group has been studied by several generations of ornithologists and evolutionists for it presented an interesting puzzle: how did the birds get to the Archipelago, which is located at a distance of almost 1000 kilometers from the nearest continent (South America) with not a speck of land in-between (Fig. 11.2)? Are the birds, which resemble one another, really related or are they the result of multiple colonization of the islands by different founders? Where did the founders come from and how many were there? Some 800 kilometers northeast of the Galápagos Archipelago is the minute Cocos Island, the habitat of a single species of finch-like bird; is this Cocos finch related to the Galápagos finches?

The experts could not agree on any answers to these questions. They could not even agree on how Darwin's finches should be classified. After much nomenclatorial wrangling, the finches are now segregated from related birds into a single tribe, Geospizini, which is divided into two main groups, seven genera, and 15 species, based on their morphology, behavior, and ecology (Fig. 11.6). The two groups are the ground and the tree finches. The *ground finches* are so named because they feed on seeds they collect on the ground on the arid islands of the archipelago. Of all the Darwin's finches, they are the most finch-like in appearance with a medium-length, pointed, conical beak, short neck, short legs, and medium-length wings and tail. They are classified into a single genus, *Geospiza*, that comprises six species which differ mainly in body size and the size of their beaks. The five *tree finch* species spend most of their time in foliage and vegetation and only occasionally forage on the ground. They are classified into two genera, *Cactospiza* and *Camarhynchus*, both of which are insect eaters. Of the remaining four species, the Cocos finch, *Pinaroloxias inornata*, feeds predominantly on insects, both on the ground and in trees; the vegetarian finch, *Platyspiza crassirostris*, lives in trees but feeds on fruit; and the two species of *Certhidea*, the warbler-like finches, resemble warblers more than finches because of their small, slender beaks and the manner in which they catch insects.

A recent study of mtDNA isolated from Darwin's finches and from a selection of birds currently living in Central and South America has finally provided answers to the questions posed above. Akie Sato and her coworkers have demonstrated that the species of Darwin's finch form a single group, distinct from other living birds. Presumably, therefore, they are all derived from a single ancestral species that reached the Galápagos Islands less than 2.3 my ago. Of all the living birds studied, the one most closely related to Darwin's finches is the grassquit genus *Tiaris*, and specifically the dull-colored grassquit, *Tiaris obscura*. This species is widely distributed over Central and South America, at the edge of humid forests, in scrub, and in open woodlands. There it feeds on seeds, and builds dome-shaped nests with a side entrance, just as Darwin's finches do to this very day. The dull-colored grassquit itself was probably not the ancestor of Darwin's finches, but the finches and the grassquit *shared* a most recent common ancestor 2.3 my ago. The ancestor presumably no longer exists, for 2.3 my since their divergence is long enough for both the continental and the island-based lineages to evolve into new species. At the Galápagos Islands the ancestral species gave rise to the 15 extant and to an unknown number of extinct species. We can therefore assume that on the continent, the other lineage issuing from the ancestor radiated in a similar way to give rise to a group of finches, one of which, the dull-colored grassquit, now appears to be the closest living relative of the Darwin's finches.

Geologists insist that the Galápagos Islands, which began to form 10-20 my ago, were never connected with the mainland. Moreover, they assert that the distance between the islands and the mainland was never significantly shorter than it is now. Hence, if the islands grew as volcanoes, lifting up from the ocean floor in the middle of a vast expanse of water, we can safely assume that the ancestor reached the shores of the Galápagos Islands from

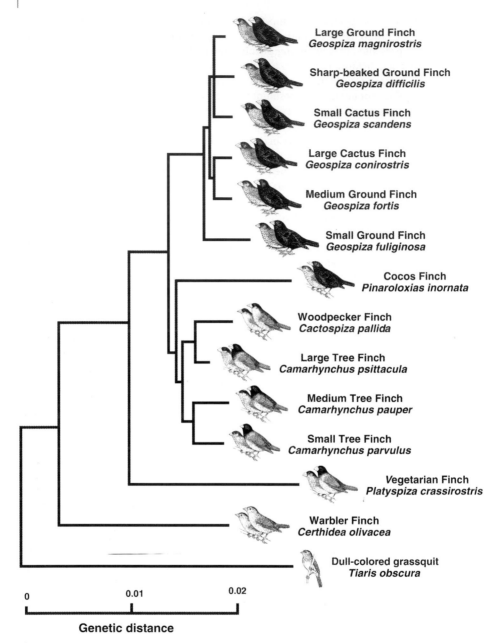

Fig. 11.6
Another example of adaptive radiation. The 13 species of Darwin's finches depicted here have been shown to be closely related to one another by the analysis of their mitochondrial DNA. Their closest relative is the dull-colored grassquit of South America. At some time in the past a flock of grassquit-like birds flying from South or Central America reached the Archipelago and their descendants diverged into new species by adapting to the different ecological niches on the islands (see Figure 11.2). The phylogenetic tree shown here is based on the sequence comparison of the mtDNA control region and the cytochrome *b* gene. The differences were converted into genetic distances and these were used to draw the tree by the neighbor-joining method. (Based on data obtained by Sato et al. 1999 and 2001.)

Central/South America rather than the other way around and this involved a very long, non-stop flight. If ever there were a case in favor of the founder effect hypothesis, surely it would be Darwin's finches. Surely, not more than a couple of birds or even a single inseminated female could reasonably be expected to make such an arduous journey, right? Well, it was apparently neither a single bird, nor even a breeding couple that founded the population, but a whole flock of birds consisting minimally of 40 breeding individuals. This is the conclusion that Vladimir Vincek and coworkers reached after comparing the amount of shared polymorphism at certain nuclear loci of the different species of Darwin's finches.

From the study carried out by Sato and her coworkers, it can be conjectured that the ancestor of Darwin's finches was a finch-like, seed-eating bird. Upon reaching the Galápagos Archipelago, its founding population must have begun to expand and spread to the various islands of the Archipelago. As long as the seed supply sufficed to feed them all, the birds would not have changed their seed-eating habit. But as the population grew and the competition for seeds increased, some of them probably started experimenting with alternative food resources and in this process began adapting to the different environmental niches on the islands. One group became specialists in catching and consuming insects in flight and adapted to this mode of feeding by developing lighter bodies and slimmer beaks. Ultimately the birds of this group changed in their morphology to such an extent that they resembled warblers rather than finches. Other groups began to specialize in sucking out nectar from cactus flowers, munching the soft pulp of cactus fruit, eating young leaves and blossoms, ferreting out insects from cracks with cactus spines or twigs, cracking the eggs of other birds, puncturing the veins of large birds and drinking drops of blood, or twisting out ticks from the skin of reptiles (Fig. 11.2). Here again, any spontaneous change in the body, but in particular in beak morphology, that facilitated the chosen mode of food procurement spread rapidly through the group by natural selection. The groups thus differentiated from one another, intragroup matings came to be preferred over intergroup ones, and the original species gradually diverged into a set of new species.

Just how fast the morphology of the birds can change and how rapidly adaptations can evolve has been revealed by long-term observations carried out on the islands by Peter B. and B. Rosemary Grant. These two researchers have made the ecological and evolutionary study of Darwin's finches the focus of their scientific careers and have contributed more to the understanding of this group's evolution than any other investigator to date. In one particular study, they concentrated on the medium ground finch, *Geospiza fortis*, which has a relatively small beak suited to eating small, soft fruits (Fig. 11.6). The Grants noticed that the beak size varied among the individuals of this species, that at least some of the variability was inherited, and that the average beak size of the population changed depending on the type of fruit available. In a period of severe drought lasting for several years, when the only fruit still available in limited amounts was large and thick-walled, many of the birds died, but those that did survive had a 4 percent larger mean beak size than earlier generations. When the rains returned to sustain the growth of plants with small, soft fruits, the mean beak size of the *G. fortis* finches decreased by about 2.5 percent.

Underlying the heritable changes in the mean beak size must have been changes in gene frequencies in the population. And since change in gene frequencies over generations constitutes evolution, the Grants were actually witnessing the evolution of Darwin's finches in a character pivotal for species divergence. In this case, evolution proceeded first in one direction and then, when the environmental conditions changed, retraced its steps over the next generations. One can, however, well imagine a situation in which the environmental change persists and the beak size continues to increase until it is optimally adapted to consumption of the new type of food. Had the rains not returned in time and the small fruit-producing plants all become extinct, the trend toward a modified beak size and shape and the accompanying change in behavior might ultimately have led to the emergence of a new species.

Molecular studies by Sato and others indicate that the groups of Darwin's finches (the ground, tree, and other finches) are clearly distinct from one another, but that the species within each group are indistinguishable in the genes tested. As in the case of the cichlid fishes, the species of each group share polymorphisms and lack species-specific substitutions in the segments of mtDNA for which they have been examined. Moreover, for a non-expert the species are difficult to tell apart, so much so that it has led to a saying among ornithologists that only God and Peter Grant know which one is which. No matter – the finches themselves apparently have no problem in recognizing who is who because hybrids are rare and occur only under special circumstances such as low population density. These observations indicate that the ground (and the tree) finch species are quite young and that speciation in their case has not yet been completed. They also serve to emphasize that it would be wrong to view speciation as an instantaneous transformation, completed in a few generations – as the founder effect hypothesis seems to be suggesting. Not only cichlid fishes and Darwin's finches, but all the other taxonomically exasperating and confounding groups on this planet are presumably dragging their speciation processes over many generations.

A Just So Story

These two examples, drawn from two different vertebrate classes, imply that the essential features of the speciation process are common to diverse taxonomical groups and can therefore be expected to apply to hominid speciation as well. The fossil record indicates that frequently in the past, two or more hominid forms overlapped in time, if not in space (see Fig. 10.3). Even as recently as 30,000 years ago at least two forms, *H. neanderthalensis* and *H. sapiens,* were contemporaneous in parts of Europe and Asia.

Presumably, the various forms coexisting in the different stages of hominid evolution were adapted to distinct ecological niches and, like the cichlids and the Darwin's finches, were the products of adaptive radiations. However, in contrast to the cichlid fishes and Darwin's finches, in which most of the products of adaptive radiation and the ecological niches are still in existence, of the hominid radiations only one species, adapted to many distinct niches, has survived. The interpretation of the causes and effects of adaptive radiation always involves some elements of guesswork, even in cases available for direct study. As regards hominid adaptive radiations, indications for which are provided only by a few fossil remains and some very indirect and controversial reconstructions of past ecosystems, the hypotheses set up by scientists and some nonscientists as well often come close to telling Kiplingesque Just So Stories. The narrative that follows must therefore be taken not with a grain, but with a lump of salt. Moreover, since the molecular archives have so far contributed very little to these interpretations, we limit the speculations to a mere outline of some of the possible causes and effects of key hominid adaptations. Inspired by the examples of cichlid fishes and Darwin's finches, we seek clues to hominid radiation primarily in adaptations to the different food resources and in sexual selection.

For most of the Cenozoic Era, which started 66.4 my ago and therefore encompassed a large part of primate evolution, there was a general trend toward a cooler and drier climate. This intensified in the Middle Miocene, beginning 15 my ago, when it ushered in a period of major ecological changes all over the world, including equatorial Africa. Up until then, this region had been covered with a largely unbroken belt of dense tropical rainforest, but now the increasingly cooler and drier climate effected a fragmentation of the forest. In the open spaces that formed, the forest was gradually replaced by less dense woodland and grassland punctuated by tree groups – the *savanna*.*

The climatic and ecological changes presented a major challenge to the apes living in the rainforest, among them the common ancestor of the gorilla, chimpanzee, and human. The ancestor may have been the size of a chimpanzee, but it spent more time in the trees than its counterpart, eating stems, leaves, and fruits. The contraction of the forest put the ancestor under extreme pressure, since all of sudden the habitat was insufficient to sustain the entire population. Some individuals were forced to explore alternative food resources and different habitats. The best alternative to the forest's interior was the forest's fringe, but to live there demanded changes in lifestyle; specifically the ancestor had to adjust to a reduced dependence on trees, to an increased terrestrial activity, and to broadening the menu to include nourishment available on the ground.

In the ancestral population, some individuals were undoubtedly better outfitted genetically to deal with the challenges of the new environment than others. As in every population, this one must also have possessed considerable genetic diversity that was responsible for morphological, physiological, and behavioral differences among the individuals. Some of these phenotypic variations may have been a hindrance to the adaptation of their bearers in some characters in the old habitat and more consonant with the new one. Individuals possessing these characters can be expected to have shown a greater interest in exploring the new environment than individuals lacking them. Since the characters were inherited, the progeny of these individuals must have displayed the same tendency. Moreover, any new constellations of genes that appeared in the progeny and improved their chances of survival in the new environment would have been subject to strong positive selection. The process may not have differed significantly from that observed by the Grants in the ground finches on the Galápagos Islands. Over longer spans of generations, any new mutations that increased the chances of success in the ecosystem at the fringe of the forest were thus preferentially passed on to the following generations. In this manner, the accumulation of minor changes gradually led to the emergence of major adaptations. The morphological and behavioral differences between individuals preferring to inhabit outskirts of the forest and those content to remain in the forest's interior would have separated the two populations reproductively. Not only opportunity, but also a preference for mates expressing the adaptive characters would have ultimately led to the segregation of the original population into different, reproductively isolated species. As in the case of the cichlid fishes and Darwin's finches, there was probably not just one, but several species that fanned out from the ancestral ape population.

In the subsequent course of events, as the climate deteriorated even further, much of the rainforest retreated into the mountains that continued to provide humid, though cooler air. The population that avoided the forest fringes moved with the retreating forest into the mountains. There it underwent selection that favored individuals with a larger body size who were better adapted to the colder temperatures, as well as individuals with a taste for

* "Savanna" is a word of Caribbean Indian origin, but it is now commonly used in reference to "a plain characterized by coarse grasses and scattered tree growth, especially on the margins of the tropics where the rainfall is seasonal, as in East Africa". (*Random House Webster's College Dictionary*.) The imprecision of the term is often rectified by qualifiers such as "treeless savanna", "grassland savanna", or "wooded savanna".

the type of trees and succulents that thrive in montane forests. The large-sized apes found the treetops a dangerous place to move about in and consequently they, too, began to prefer a terrestrial over an arboreal mode of life. The outcome of this evolutionary course was the present-day gorilla.

The final phase of the split that separated the gorilla ancestors from the common ancestor of the chimpanzees and the hominids presumably occurred ~8 my ago. No fossils documenting the split have been discovered, but the molecular record discussed in Chapter Eight reveals that it did indeed occur at the indicated time. Since no remains of the chimpanzee-hominid common ancestor are available, the ancestor has no name. We also have no way of knowing whether only one or several species of the chimpanzee-hominid lineage lived at the forest fringes. Since, however, the fringes were undoubtedly fragmented and extended over a large area and since some of the fragments were isolated from one another by distance, if not by other barriers that hindered free movement, the latter alternative seems more likely. Nevertheless, if multiple species were present, only one of them became the ancestor of the chimpanzee and the hominids.

The chimpanzee-hominid lineage persisted for less than 3 my. During that time, between 8 and 5 my ago, the global deterioration of the climate took a dramatic turn for the worse by entering a period of pronounced climatic oscillations in which phases of extreme aridity alternated with somewhat more humid times. Nowhere are these oscillations better documented than in the sediments at the bottom of the Mediterranean Sea. Although it has been known for some time that the Mediterranean Basin dried out during the *Messinian event* (named after the sedimentary rocks near Messina, Sicily), more recent studies indicate that it actually dried out and refilled several times, perhaps as many as 15, during an interval lasting 2.5 my and extending from 7.3 to 4.9 my ago.

Another, but related, sign of the climatic downturn was the widespread formation of glaciers in different parts of the world. A *glacier* is a large mass of ice that moves ("flows") slowly down a slope of high-mountain valleys (a *valley g.*) or spreads outward on a land surface in polar regions (a *continental g.* or an *ice cap*). Glaciers form when the amount of snow fallen during the winter exceeds that which melts and evaporates in the summer. The lower layers of snow are compacted into ice and the whole mass moves under the pressure of its own weight and by the pull of gravity. In the period from 8 to 5 my ago, glaciers began to form at high-altitude sites in the Northern Hemisphere and over the islands of western Antarctica. As increasing amounts of water became locked up in the forming glaciers, the sea level dropped worldwide by 40 meters during this period. Concurrently with these global changes, dramatic regional events were also taking place, such as the uplift of the Himalayas in the final phase of the collision between the Indian subcontinent and Asia 5 my ago. This was also the time when the western branch of the Great Rift Valley formed in East Africa.

The Great Rift Valley is a crack in the earth's crust so huge that it is visible from as far away as the moon. It begins near the Zambezi River in Mozambique and extends northward through East Africa, all the way to Turkey, Lebanon, and Syria in southwestern Asia (Fig. 11.7A). Its origin lies in the plate tectonics, in the movements and interactions (collisions, spreading apart, sliding alongside one another) of approximately 30 rock floes, the *plates*, composing the earth's crust and floating on the magma below them. A local upwelling of magma in the area of the crack is believed to have been responsible for melting the rocks and thinning the crust in this region. At the same time, the interaction of the plate bearing the African continent and part of Southwest Asia with the neighboring plates resulted in tension at various points that was released in a series of fractures and fissures in the area of the weakened crust. Where two of these fractures or *faults* ran parallel to

Fig. 11.7A,B
The system of Great Rift Valley in East Africa (**A**) and a diagram depicting their presumed emergence when two parallel faults formed through plate movements and the block between them sank (**B**). Note the location of Lake Victoria on the plateau between the eastern and western branches of the valley.

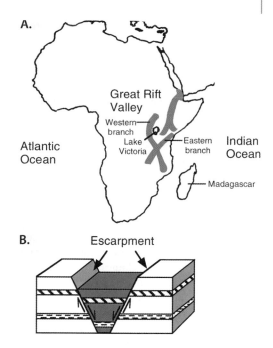

each other, the crust block between them sank, forming a deep valley, a trench-like *graben*, bounded by steep escarpments (Fig. 11.7B). In East Africa, the system of grabens splits into two, the eastern and western branches of the Great Rift Valley. The former runs along the eastern border of Zaire, the latter through central parts of Tanzania, Kenya, and Ethiopia. (Lake Victoria is located on an uplifted plateau between the two branches.) The Great Rift Valley began to form 30 to 40 my ago, but in East Africa most of the eastern branch formed some 10 my ago, whereas most of the western branch emerged, amidst great volcanic fireworks, only about 5 my ago.

This, then, was the physical setting for the chimpanzee-hominid split at the end of the Miocene and the beginning of the Pliocene Epoch approximately 5 my ago. The immediate cause of the split remains a mystery, but the possibilities considered include the formation of the western branch of the Great Rift Valley and the expansion of the Saharan savanna. The former hypothesis, sometimes referred to as the "East Side Story" posits that the western branch of the Rift created a physical barrier, splitting the common ancestor population into two and isolating these from each other. The population west of the chasm evolved, according to this interpretation, into the chimpanzees, whereas the eastside population gave rise to the hominids. The main weakness of this explanation is that the distribution of the chimpanzee straddles the western branch.

The alternative hypothesis pictures changes in the Sahara as having a decisive influence on the conditions in East Africa. In the Oligocene and Miocene, the Sahara, irrigated by frequent rains, was covered by dense forests which were inhabited by several species of primates. Toward the end of the Miocene, however, the growth of the ice sheets in Antarctica cooled the water current that flows from Antarctica northwest along the coast of West Africa. The chilled waters evaporated less rapidly than the warmer ones and provided less precipitation for the western part of the Sahara. Concurrently, the desiccation of the

Mediterranean Basin also led to a substantial decrease in rainfall on the northern parts and the entire Sahara began to lose its forests, which came to be replaced by the savanna. According to this hypothesis, a wedge of the savanna extending from the Sahara then split the population of chimpanzee-hominid ancestors.

It is, however, not really necessary to postulate the existence of a physical barrier to explain the chimpanzee-hominid split. Instead, one can envision that as the East African savanna expanded as a result of the climatic changes, a new niche opened up for those individuals in the ancestral population who, by chance, carried mutations supporting their ability to exploit the new food resources offered by the grassland. As such, the split could have taken the form of sorting out individuals possessing these mutations from those lacking them in the manner described above for the common ancestor of the gorilla, chimpanzee, and the hominids. After extending over many generations, the process may have thus led to the emergence of a population better adapted to life in the savanna than the original population.

Of primary importance among the various adaptations was undoubtedly bipedalism. The advantages attributed to walking erect on the hind legs include having a better view of the surroundings by holding the head high above the tall grass; greater (faster and more enduring) mobility on the ground compared to quadrupedal primate movements; liberating the hands for tasks such as carrying food to a more protected storage place or for wielding sticks and throwing stones to fend off competitors or attackers; and increasing the effectiveness of cooling the body by reducing the surface area exposed directly to sunrays. Upright walking allowed the emerging hominids to spot a potential meal from a long distance, reach it faster than their competitors, chase away any aggressors, and bring back parts of the meal to the relative safety of a tree group. Bipedalism, which entailed redesigning many parts of the skeleton and of the muscle systems, could not have evolved in a single step. It must have been the result of a gradual, stepwise accumulation of many small changes over numerous generations. This must have also been the case with other adaptations to life in the savanna, such as changes in dentition enabling the hominids to consume food that other primates could not process.

The emergence of the first hominids coincided with the transition from one geological epoch to another, from Miocene to Pliocene. Like all such transitions, this one, too, was marked by a conspicuous change in the fossil record, namely by the large-scale extinction of some species (many invertebrates in the oceans and mammals on land) and by the entrance of new species on the scene. The extinctions bear witness to drastic changes in the environmental conditions, the hallmark of which is the onset of climatic fluctuations in the Pliocene Epoch extending from 5.3 to 1.8 my ago. While the general trend toward a cooler and drier climate continued, colder periods began to alternate with warm phases. Moreover, the amplitude of these fluctuations increased as the cold periods gradually grew colder, the warmer periods became hotter, and rainy intervals were wetter.

It was also the time when seasonality became pronounced in many parts of the world. The causes of these fluctuations (as well as of the general climatic trends) remain uncertain, although the wobble of the earth as it rotates on its own axis and the variations in the orbit of the earth around the sun are generally believed to be the chief culprits. In Africa, the Pliocene conditions changed the landscape strikingly. The continent at that time looked as if armies of loggers had rolled over it. From north to south, from east to west, in most regions the dense forests had all but disappeared, to be replaced by sparse woodland and open grassland.

Many species could not tolerate these wide swings in temperature and humidity and they bowed out from the stage. Others were able to adapt to the new conditions and evolved in the process into new species. The hominids belonged to the latter category. They underwent a series of adaptations to adjust to the increasingly dry climate of their habitats, particularly in regard to the new types of food that the vast expanses of grassland had to offer. Their menus became quite variegated, as might be expected from an omnivore, with meat becoming a major staple. The hominids adapted to meat consumption by evolving the ability to mince large chunks into tiny pieces by chewing and so making it ready for absorption in the intestines. Big, flat grinding teeth (premolars and molars) topped by thick layers of hard enamel, and strong mastication muscles relocated in front of the skull where they are most effective were the major anatomical modifications underlying these adaptations.

For the hominids, the land available to them was vast, the opportunities unlimited, and the conditions varied. It is hardly surprising that their expansion was accompanied by differentiation into groups specifically adapted to the regional conditions. Ultimately, some of these groups evolved into distinct, coexisting species adapted to different niches: the hominids underwent their first round of adaptive radiation. The major differentiation was into two lineages, the gracile and the robust australopithecines, but within each group, a more subtle differentiation into related species took place. The long distances between groups undoubtedly contributed to their relative isolation and so to speciation. The anatomical differences between the gracile and robust australopithecines were presumably the result of natural selection adapting the two groups to the consumption of different types of food. The gracile australopithecines subsisted on food that required less chewing than that which the robust australopithecines became specialized in handling.

Like the emergence of the hominids, the appearance of the genus *Homo* was closely associated with a change of geological epochs, this time from Pliocene to Pleistocene. Although the beginning of the Pleistocene Epoch is now dated to 1.8 my ago, the preparatory phase for the first deep freeze heralding its onset started earlier, about the same time to which the first *Homo* fossils date. Pleistocene, known colloquially as the Ice Age, is characterized by the alternation of glacials and interglacials. The former are cold climatic episodes of glacier advance in northern latitudes; the latter are warmer episodes of ice retreat. There were at least 11 or 12 glacial episodes, the last one ending 10,000 years ago, a point of time taken by convention as the end of the Pleistocene. (In reality, however, there is no reason to believe that the interglacial in which we now live will not be followed by another glacial.) In more southerly latitudes, each glaciation episode expressed itself in the onset of extreme aridity caused by a sharp decrease of rainfall, which in turn was the consequence of large amounts of water being locked up in the glaciers.

The severe droughts of the glacials presented an extreme challenge to the inhabitants of the savanna, a test that many mammalian species, including most of the hominids, failed to pass. The hominids that were successful evolved into entirely new forms, the first species of the genus *Homo*. The population that gave rise to these species can be envisioned as consisting of individuals with mutations that enabled them to cope with the challenges of the glaciation episodes. The most important among these were mutations responsible for an increase in brain size relative to the overall body size, specifically the volume of the brain's outer layer, the gray matter of the *cerebral cortex*. The increased size of the cerebral cortex led to an unprecedented increase in *intelligence* – the ability to remember, to use what one has learned to solve problems, to adapt to new situations, and to understand and manipulate one's environment. Increased mental capacities, in turn, resulted in the appearance of *culture*, the system of learned behavior passed from one generation to the next. The first indication of the appearance of culture is the presence of stone tools in association with the fossil remains of *Homo*.

A good dose of intelligence seems to have been vitally important for the survival of the early *Homo* species as the conditions grew increasingly harsher with the approaching Pleistocene. Intelligence was the species' only protection against predators and its only advantage over competitors. Without it, *Homo* was both defenseless and helpless. Although it could run faster than any other primate, it could not outrun a lion or a leopard, not to mention a cheetah. It had no prominent canines to scare off a would-be attacker, no claws to scratch an opponent if attacked, no protective body cover to shield it from a strike. And its senses were no match for those of most other mammals living in the savanna. At night it was almost blind; its sense of smell was pitiful and its hearing ability not much better. The only positive factor in its fight for survival was *Homo*'s intelligence. To rely on its brain and through it on tools, on communication among members of its clan, on learning and remembering, on passing knowledge accrued from one generation on to the next, this was *Homo*'s only chance of coming to terms with the precarious situation in which it found itself in the Pleistocene Epoch.

By ruthless and relentless natural selection, *Homo*'s brainpower grew and grew, especially when the environmental conditions deteriorated even further. Driven by the new challenges, selection led to the extinction of the australopithecines and the replacement of the early *Homo* species, of which there may have been several, by a new set of species typified by *H. erectus*. It was also increased mental capacity that prompted *H. erectus* to move out of Africa in search of new food resources, and enabled it to meet the challenges it encountered in the new environments. But by far the greatest leap occurred with the emergence of *H. sapiens*, and enabled the development of all the aspects of human culture as we know them. What finally triggered this by far greatest increase in brain size remains unidentified. It took place in the middle of the Pleistocene and no unique environmental change can be pinpointed as its decisive cause.

We see, then, that just as in the case of cichlid fishes and Darwin's finches, in hominid evolution the emergence of new species was also associated with adaptation to new food resources. There are further parallels between the speciation processes in fishes and birds compared to hominids, all of which suggest that it is probably fair to describe hominid evolution as a series of adaptive radiations, each of which produced several different species.

The question we now must turn to concerns the size of the populations that founded these species. Did the species emerge from pairs or from very few individuals, as the proponents of the founder effect hypothesis would like us to believe, from a few dozen individuals as the Darwin's finches apparently did, or from thousands of founders, like the different cichlid species?

N and *N_e*

A *population*, let us remind ourselves, is a group of individuals of one species that inhabits the same area and is usually isolated to some degree from other, similar groups. The area can range in size from a fairly small locality to that of the entire globe and must be specified for each population. A *founder population* is the group of individuals from which a new population, and in our case a new species, arises. *Population size* is the number of individuals in a population at a given time. The actual number of individuals in a population, obtained by counting heads, is the *census population size, N*. The statistical study of populations, and the human population in particular – its size, density, distribution, and vital statistics – is the subject of *demography* (from Greek *demos*, people, *graphein*, to write). The study of human populations before recorded history is *paleodemography*.

In historical times, population size and associated parameters were assessed by *census-es* (from Latin *censere*, to assess), by counting heads. Censuses were often ordained when the head of state became suspicious that his tax collectors were cheating him, which they invariably were, or when he wanted to know whether he was in a position to assemble an army large enough to attack a neighboring state. Ancient Babylonians, Chinese, Hebrews, Egyptians, Greeks, and Romans all conducted censuses from time to time and from the records that have survived, we can estimate the size of the human population in the last 4000 years reasonably well (Fig. 11.8). For prehistoric times, however, our estimates must rely on three main sources: the study of archeological remains, comparisons with present-day hunter-gatherers, and the construction of paleodemographic models. The most direct of these three sources is, of course, the first one. From the density of occupied sites and from the number of individual remains at each site, archeologists can make an educated guess as to how many people lived in the different periods. Such estimates take us back to about 10,000 years BP.* As we push further and further back, the more difficult it becomes to estimate population size, because fossilized bones become less frequent – not necessar-

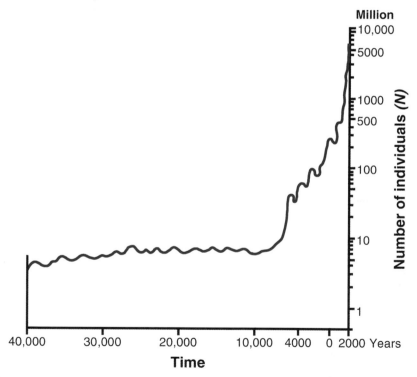

Fig. 11.8
The estimated census size (*N*) of the human population in the last 42,000 years. Note that the scale of the *y*-axis is logarithmic.

---

* BP, an abbreviation for *before present*, is taken by convention to indicate a time before the year 1950 of our era. Hence to convert a date from BP to B.C. (before common era) it is necessary to subtract 1950 years.

ily because fewer individuals lived earlier but because fossilization and preservation of bones depends on a string of lucky coincidences. By the time we reach 1 my BP, fossilized human remains become so scarce that every major find makes newspaper headlines. The more direct counts are then supplemented by much less direct estimates, including those made by computer simulations with model populations.

At 40,000 years ago, *H. sapiens* achieved its global distribution and subsisted by hunting wild animals and gathering wild plants. By taking into account the size of the territory a present-day hunter-gatherer (of whom only a few still exist and all supplement their diet by some form of farming) needs to survive, paleodemographers have come to the conclusion that during the entire Pleistocene, there could not have been more than a few million hominids at any one time.

The situation changed dramatically with the introduction of agriculture some 10,000 years ago. Cultivation of plants and breeding of domesticated animals increased the amount of readily available food supplies and so allowed the population to grow. At first, when the early farmers still had to rely on stone tools, even if the quality of these had been improved by grinding in the Neolithic or the New Stone Age, growth was modest, and the size probably did not exceed the ten million mark. Only when bronze, the alloy of copper and tin, was invented and the farmers began to till the soil with metal tools, when irrigation, fertilization, and other improvements were introduced and the productivity of farming increased correspondingly, did the population grow substantially to reach the 100 million mark. After that the population still continued to grow, but more slowly. Although further improvements in farming practices continued to be made, they were relatively minor and 100 million became the approximate size of the human population which the available resources could support.

Only in the early eighteenth century A.D. did it come to another explosive increase in the human population size. At that time, and in the period that followed and continues right up to the present, significant improvements in crop-growing practices, advances in livestock breeding, the invention of new farm equipment, and the introduction of fertilizers, herbicides, and chemicals designed to control crop diseases improved the productivity of farming tremendously. In response to the growing food resources resulting from this *agricultural revolution*, the human population skyrocketed. One day in October 1999, it reached the six billion mark and there is still no sign that the trend might be stopped in the near future.

A head count, *N*, is vital for a sovereign who needs to know how much cannon fodder he has available, but for the continuation of a genealogical lineage a different count is far more relevant – the *effective population size*, *N*ₑ. Earlier we defined *N*ₑ as the number of individuals in an idealized population that would give rise to the same variance of gene frequency or the same rate of inbreeding as that observed in the actual population under consideration. (Here, *variance* is a statistical measure of the dispersion of values about a mean; *inbreeding* is the mating of individuals more closely related than average pairs in the population.) We also explained that for all practical purposes, *N*ₑ represents the number of breeding individuals in a population. Now, however, we must take a closer look at this all-important population parameter.

Why is the *N*ₑ so critical? In evolution, founded on an unfaltering turnover of generations, what counts is that which is transmitted from one generation to the next. An individual that does not contribute genes to the next generation is like a withered twig on a healthy branch: it has no influence on the future growth of the tree. In a population consisting of 100 individuals of which only 50 breed, the latter alone can potentially influence the fate of the lineage. Here, then, *N* equals 100, but *N*ₑ is 50 and it is this number that is evolutionarily informative.

The concept of effective population size has another salient aspect, however, according to which not even the number of breeding individuals is vitally significant. Consider the 100 individuals again. If they are all genetically identical, it makes no difference whether all, half, or just two of them contribute to the next generation. The gene pool of the following generation will be the same regardless of whether two genes are provided by each of the 100 individuals, four genes by each of 50 individuals, or 100 genes by each of two individuals. A very different situation will arise if all 100 individuals are genetically distinct. Then it *will* matter whether the second generation genes are provided by 100, 50, or just two individuals. If they are contributed by 100 individuals, there is a chance that all the different genes present in one generation will again be represented in the next generation. But if the pool of the next generation is filled with genes from only 50 or merely two individuals, there is no possibility that the composition of the new pool will be the same as that of the previous generation. Some of the different genes will be lost during the turnover of generations. And since it is the genetic composition of a pool (population) that decides its evolutionary fate, it is far more important to consider the composition rather than the number of breeding individuals when discussing the evolution of a population and the origin of species.

There is, of course, a certain relationship between the number of breeding individuals and the genetic composition of a population, but discrepancies between the two are common. A good example of such discordance is provided by the current human population. As mentioned earlier, the census population size of the global population is six billion individuals. The effective population size estimated from the genetic composition of the present-day human population by methods discussed in the next section is 10,000 individuals. The exact number of breeding humans currently living on this planet is not known, but it is clearly well over 10,000. The discordance is even more striking when we compare the population of Africa with that of the entire world. The N_e of the African population is very similar to that of the global population. Although the number of breeding individuals in the African population is certainly smaller than that of the entire world, the effective population size of the two is the same because their genetic composition is very similar.

From this example, it should be apparent that N_e is not to be viewed in terms of a head count, but as a more abstract value, specifically as a measure of the genetic composition of a population. We should keep this in mind every time we are tempted to convert N_e into the number of individuals, breeding or total. Although there are formulae for such a conversion, each taking a different potential cause of discrepancy between N and N_e into account, they are of little use in specific cases because the causes are either not known or difficult to quantitate.

Under these circumstances, would it not be better to replace the term "effective population size" by some other designation related to population composition rather than to a head count? Theoretical population geneticists apparently do not think so, perhaps because they like to reason in terms of abstract populations that are not beset by the imperfections of real populations. They call these imagined populations *ideal* and it was in the terms of just such an ideal population that Sewall Wright defined the effective population size in 1931. What the definition states is this: the real population has a certain genetic composition which is the result of a complicated interplay of many different factors, most of which are intangible. To deal with all these complications is very messy, so let us retreat instead into a Platonic world of ideal forms, here ideal populations. In the ideal world we can account for the genetic composition of the real population by postulating the existence of a population consisting of N_e individuals, all of which have an equal opportunity to mate with any other individual of the opposite sex and all of which obey a few simple rules of civilized behavior. From the ideal world we can then make certain testable inferences about the real world. All this is fine, as long as we do not mix up our worlds.

The Praise of Diversity

If not Chapter Ten, then the preceding section should have made it clear that the genetic composition of a population is our most important source of information about the demographic history of the human or, for that matter, of any other species. But what do we understand by "genetic composition of a population", or as others prefer to say "genetic variation", "genetic variability", "genetic diversity", and "genetic polymorphism"? What do all these more or less synonymous terms mean?

The concept is most easily grasped by first imagining a population *without* genetic variation. This situation does not occur in nature, but it is helpful to conjecture and take it as the starting point in population history, a kind of B.C./A.D. divide. In a population lacking genetic variation, the individuals are identical in all their genes, and indeed in their entire genomes. The closest one can get to such a *tabula rasa*, a clean slate, would be the moment of inception of a species under the founder effect scenario. Although the two founding individuals of a species are by no means genetically identical (no two individuals ever are), many of the differences are likely to disappear in the following generation, when genetic drift in the small population fixes one or the other versions of each gene. This *tabula rasa*-like situation does not last very long, however, since new mutations appear in every generation of the emerging species and begin to differentiate the various individuals. Although most of the mutations will be lost by genetic drift and some will become fixed, the proportion of the mutated genes that have not, as yet, been lost or fixed will grow in the pool until an equilibrium is established in which the loss of mutations is balanced out by the influx of new ones. The stability of the equilibrium, the level at which it is established, the proportion of loci or genomic segments at which variants occur, the frequency of the variants, the number of variable sites at a locus, all these and other characteristics will depend on the properties of the population. They will be influenced by the size of the population, by the changes in population size, by its structure, gene flow, and other features. The sum of the differences among the individuals of a population is what geneticists refer to as *genetic variation*. By studying genetic variation in a population, geneticists strive to reconstruct its history.

To assess genetic variation and to extract the desired information from it, population geneticists use statistical methods specifically designed for this purpose. Some of these methods simply measure the proportion of polymorphic loci in the population, others the number of alleles, the relative frequencies of genes within the population, the total frequency of heterozygotes across multiple loci (the expected heterozygosity), or the number of segregating sites. Some of the methods are applicable to results that distinguish alleles at loci but do not provide information about the nature of the difference between them (as, for example, in the case of genes specifying electrophoretically distinguishable proteins), while others are suitable for the analysis of data that not only distinguish alleles but also specify the altered nucleotides in the DNA sequence. Most, however, are applicable to both types of data. In what follows, we describe one of the most informative methods used in the analysis of DNA sequences sampled from a population; later we mention others.

The method measures the *average nucleotide diversity*, π (pi), of a population. It takes into account the number of alleles at polymorphic loci, the frequencies of the individual alleles, and the number of substitutions differentiating the alleles. An ideal dataset to which it could be applied would be one obtained by selecting ~50 loci scattered randomly over the genome and choosing ~100 individuals at random from a population, and then sequencing all the genes from all the individuals. The real datasets available fall short of the ideal one, but to explain the principle of calculating π, it is simpler to use the latter. There are, in fact, two ways of calculating π, only one of which requires the knowledge of gene frequencies. To illustrate them, let us assume that at the first of the 50 loci, at locus *A*, we found three different sequences among the total of 100, *A1*, *A2*, and *A3*, represented by 50, 25, and 25 sequences, respectively, so that the frequencies of these three alleles are $x_1 = 50/100 = 0.50$, $x_2 = 25/100 = 0.25$, and $x_3 = 25/100 = 0.25$, respectively.

By comparing each allele with the other two, we find that *A1* differs from *A2* at three sites, *A1* from *A3* at five sites, and *A2* from *A3* at two sites out of the 100 compared. By taking $p = n_d/L$ (where p is the proportion of nucleotide differences per site, n_d is the number of nucleotide differences between two alleles, and L the number of sites compared), we use the formula $d = -\frac{3}{4} \ln(1 - \frac{4}{3}p)$ to estimate $d_{A1A2} = 3/100 = 0.03$, $d_{A1A3} = 5/100 = 0.05$, and $d_{A2A3} = 2/100 = 0.02$. These are the estimated numbers of nucleotide substitutions between two alleles. To obtain nucleotide diversities among all sequences, we must multiply the distances by the frequencies of the alleles. Specifically, $\pi_{A1A2} = (2)(d_{A1A2})(x_1)(x_2) = (2)(0.03)(0.5)(0.25) = 0.0075$; $\pi_{A1A3} = (2)(d_{A1A3})(x_1)(x_3) = (2)(0.05)(0.5)(0.25) = 0.0125$; $\pi_{A2A3} = (2)(d_{A2A3})(x_2)(x_3) = (2)(0.02)(0.25)(0.25) = 0.0025$. The nucleotide diversity of locus *A* is then given by $\pi_A = \pi_{A1A2} + \pi_{A1A3} + \pi_{A2A3} = 0.0075 + 0.0125 + 0.0025 = 0.0225$. Note that in the above calculation $(d_{A1A1})(x_1)^2 = 0$, which corresponds to the case in which we happen to choose the same allele, *A1*, twice. The same is true for the *A2* and *A3* alleles. Mathematicians generalize this instruction for calculating π from gene frequencies by using a short formula

$$\pi = \sum_{ij}^{q} d_{ij} x_i x_j$$

where x_i and x_j are the frequencies of the *i*-th and *j*-th alleles, respectively (general designations of any two alleles), d_{ij} is the number of differences (substitutions) per site between the *i*-th and *j*-th alleles, q is the total number of different alleles, and $\sum$ is the summation symbol instructing us to carry out the multiplication "frequency times difference" for all the combinations of the different alleles. Strictly speaking, however, this formula is only appropriate for a case in which the entire population has been tested and the exact frequencies of the alleles are known. Obviously, for large populations, we cannot examine every single gene and must restrict ourselves to testing a *sample* of size *n* from the population and then estimating the frequencies in the entire population from the result. This estimate is associated with a potential bias, a certain probability that it deviates from the real value, but this can be rectified statistically by the use of an appropriate correction factor.

The second way of calculating the average nucleotide diversity is by taking all 100 sequences at each of the 50 loci, comparing each of them with all the remaining 99 sequences in a pairwise manner, and counting the number of differences in each pair. Generally, if there are *n* sequences, the total number of pairwise comparisons is $n(n-1)/2$. In our case, there are $(100)(99)/2 = 4950$ comparisons to be made. Since we are comparing sequences rather than different alleles, many or most of the comparisons will show no difference because they are between identical alleles. The π is then given by the sum of the differences obtained in all the combinations divided by the total number of comparisons made, that is by $n(n-1)/2$:

$$\pi = \frac{2}{n(n-1)} \sum_{i<j}^{n} d_{ij}$$

For multiple loci, we compute π for each locus by using either of the two formulae given above, summing up the values thus obtained, and dividing the sum by the total number of loci tested. In our example of 50 loci, similar calculations are therefore carried out not only for the A locus, but also for the remaining 49 loci (B, C, etc) to obtain the π_B, π_C, ... values. The average nucleotide diversity of the population over the 50 loci is then given by ($\pi_A + \pi_B + \pi_C + ...$)/50.

In a study published in 1991, Wen-Hsiung Li and Lori A. Sadler estimated the average nucleotide diversity of the human population to range from 0.0001 to 0.0011, depending on the nature of the sites compared – whether they are coding or noncoding, synonymous or nonsynonymous, degenerate or nondegenerate, and where in the genome they are located. These values are generally considerably lower than those found in our closest living relative, the chimpanzee. Henrik Kaessmann and colleagues, for example, compared in 1999 the same noncoding regions of human and chimpanzee nuclear genomes and found the π value of noncoding human regions to be 0.00037 or 0.037 percent, whereas that of the corresponding chimpanzee regions was 0.0013 or 0.13 percent – more than three times higher. A similar difference has also been reported for the mitochondrial genomes of these two species. This difference exists despite the great reduction in the census population size of the chimpanzee in recent years and the expansion of the human population. The estimated current census population size of the common chimpanzee is less than 200,000 individuals, compared to six billion humans. These observations have important implications for the interpretation of the paleodemography of our species. Before we turn to them, however, we must introduce another pivotal concept and method of studying genetic variation, the concept of coalescence.

Virtual Piranhas

Compared to the protein level, the DNA level provides a resolution of genetic variation one order of magnitude higher, which most of the older methods of statistical analysis are not well equipped to handle. New methods were therefore called upon to deal with the increase in information content and to take not only the existence of different alleles into account but also the quantitative differences between them as revealed by DNA sequences. The average nucleotide diversity is one of the new generation of methods, and analysis of the *coalescence process* is another.

The theory of the coalescent was conceived by two mathematicians, G.A. Watterson in 1975 and J.F.C. Kingman in 1980. It was then developed and its high math given a more user-friendly appearance by Fumio Tajima, as well as several other population geneticists. The theory represents a reversal in thinking about the genetic processes in a population, a reversal initiated earlier by Motoo Kimura and Tomoko Ohta. While the classical population theory follows processes, such as the fixation of mutations, in forward drive, from the past to the present, the theory of the coalescent puts the gear into reverse and describes events backward in time. To grasp the concept, think of a pool of genes in a recent generation, G_0, and their relationships to genes of past generations G_1, G_2, G_3, etc (see also Chapter Four). Each gene in the G_0 pool is a copy of a gene that existed in the G_1 pool, all the G_1 genes are copies of G_2 genes, which are copies of G_3 genes, and so forth. By depicting the genes by dots or circles, we can indicate their origin from genes of preceding generations by connecting descendancy lines in the same manner by which we depict genealogical relationships of individuals in a pedigree or species in a phylogenetic tree (Fig. 11.9). As we trace the descendancy lines backward in time in this *gene genealogy*, we obtain *gene lineages*, in which the nodes (the dots) represent real objects (genes), whereas the connecting lines are mere mental aids symbolizing genealogical relationships. Several points and observations need emphasizing about the gene genealogy, some of them obvious, others perhaps less so.

First, each gene of a given generation has an ancestor ("parent") in the preceding generation, but not every gene has a descendant in the succeeding generation (Fig. 11.9). This asymmetry is the consequence of the randomness (stochasticity) in the process generating the next generation pool from a pool of the preceding generation. Although each gene has the same probability of having its copy drawn into the new pool, some genes are picked by chance and others are not picked at all to participate in the process. The lineage of the former is thus continued, whereas that of the latter is discontinued, it becomes extinct. Even in a large pool, some genes will always end up without an issue and their lineages will be extinguished.

The second and the most relevant feature of gene genealogy is the fact that while some genes do not pass any copies of themselves into the succeeding generation, others may contribute more than one copy. This fact, too, is the consequence of stochasticity governing the drawing of genes. In a random process, such as the repeated casting of a die, a particular number may not come up at all, it may come up once, or it may come up more than once. In a large pool of a given generation it is therefore likely that there are some genes which are multiple copies of the same gene in the preceding generation. In a graphic depiction of this situation, the descendancy lines of these genes come together or *coalesce* in the same gene of the preceding generation. Because some genes contribute more than one copy of themselves to the new pool, the pool size may remain constant from one generation to the next, even though some other genes do not contribute any copies at all. This process is familiar to us: in Chapter Four, we followed the time arrow forward; here we follow it backward. In the forward direction it represents *divergence*, in the backward direction *coalescence*. Schematically:

Divergence   Coalescence

The third observation follows on from the first two. If there are some genes in each generation that do not contribute any copies to the next generation, others that contribute one copy, and others still that contribute more than one copy, and if the fate of a given gene in the face of these three possibilities depends exclusively on chance, then viewed backward in time, in any generation (say G_0) there will be lineages that coalesce in the preceding generation (G_1) and others that do not. Of the latter, some coalesce in G_2 and those that do not may coalesce in G_3, G_4, G_5 etc (Fig. 11.9). Because some of the lineages that coalesce in more distant generations are themselves the products of previous coalescences, the number of lineages gradually decreases as we follow them back in time. Ultimately, only two lineages remain and when these coalesce, we have reached the *most recent common ancestor* (MRCA) of all the G_0 genes.

A. COALESCENCE OF TWO *G0* GENES

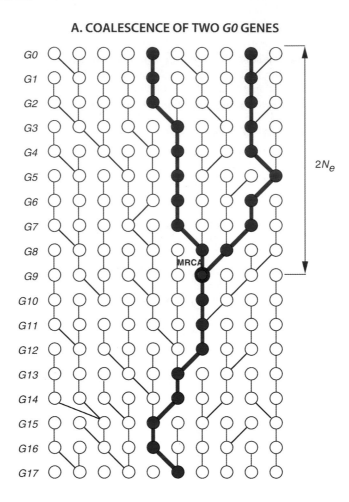

Fig. 11.9A
Population genetics in reverse: the concept of coalescence. **A.** Any two genes selected at random from a current gene pool (*G0*) have a most recent common ancestor (MRCA) in one of the past generations (here *G9* for the highlighted genes) and many other ancestors in the preceding generations. On average, the MRCA occurs $2N_e$ generations ago, where N_e is the effective population size.

Figure 11.9 illustrates some of the points we were making earlier (in Chapter Ten) in connection with the hullabaloo surrounding the Eve (Adam) stories: we stated that the MRCA is a gene (or at most, a mitochondrial genome), not an individual; that genes or genomic segments at different loci have different MRCAs in different generations; and that the MRCA of any set of genes is part of a large gene pool. It also demonstrates another point: the MRCA of genes in one generation (for example our own) may not be the same as that of an earlier generation (of our parents). Since the composition of the gene pool changes from one generation to the next, the MRCAs of genes in different generations can be expected to be different genes in different past generations.

It is perhaps unnecessary to point out that the whole concept of gene genealogy is purely theoretical ("virtual" in modern phraseology), not only because we are always dealing

B. COALESCENCE OF ALL *G0* GENES

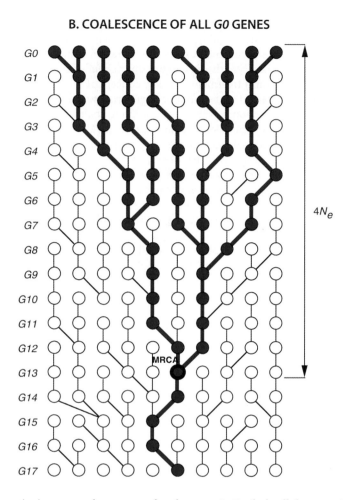

Fig. 11.9B
Population genetics in reverse: the concept of coalescence. **B.** Similarly, all the genes in the *G0* pool have an MRCA in one of the past generations (here *G13*), generally on average $4N_e$ generations ago. Each circle represents a gene; connecting lines indicate descendancy; and highlighted genes and lines are those whose descendancy is being followed backward in time. Note that each gene has one antecedent, but not every gene has descendants. For the sake of simplicity, genes are shown to have two descendants at the most; in reality they can have more than two descendants.

with a *sample* of genes rather than a complete set constituting a gene pool, but chiefly because real gene genealogy cannot be reconstructed. If a gene does not change, we have no way of knowing which gene in the preceding generation was its ancestor and we therefore cannot trace gene lineages and lineage convergences. When a gene mutates, we can distinguish it from other genes and so trace *allelic genealogies* (as we have done in Chapter Ten; see Figures 10.16 and 10.18), but we still cannot distinguish genes within an allelic lineage. All we can do is to create imaginary gene and allelic genealogies and analyze them either mathematically or by computer simulation. We can obtain genealogies for different population histories and find out how they influence genetic variation. In this manner we can determine which genealogy best explains the observed genetic variation of a population and then infer the most likely demographic history of that population. Specifically, we can answer questions regarding the size of the founding population, the time that has elapsed since its foundation, and the occurrence, extent, and duration of bottlenecks.

Although parts of the coalescent theory can be comprehended intuitively, its precise expression requires mathematical treatment. The key concept of the theory is the *mean coalescence time*, the time expressed as the number of generations it takes on average for lineages of two genes selected randomly from a population to coalesce, or the time it takes for lineages of all the genes of a population to coalesce into the MRCA. Intuitively, the coalescence time might be expected to depend on the effective population size because in a small population coalescence of lineages will be more rapid than in a large population. In a small population the probability of inbreeding is higher than in a large population, because many individuals do not have any other choice but to mate with close relatives. And since, by definition, close relatives share recent ancestors, the gene coalescences are much shallower than in large populations in which matings between individuals sharing recent ancestors are much more rare.

In a modified version of the metaphor attributed to Joseph Felsenstein, the behavior of genes in a pool can be compared to that of hungry, cannibalistic piranhas in an overcrowded water tank. In the absence of any other food, the piranhas try to eat each other, and whenever they succeed, two fish become one – they "coalesce". At the beginning, when their density is high and the possibilities for escaping null, it is easy for the fish to succeed in catching one of their neighbors and so many "coalescences" occur within a short time. After a while, however, as their numbers decline, it takes progressively longer for one fish to hunt down another, and the "coalescences" not only become rarer, but are also separated by longer intervals of time. When finally only two fish are left in the aquarium, it takes a very long time for the lucky one to make its catch and so achieve the ultimate "coalescence". The attentive reader has undoubtedly realized that the metaphor is flawed: it lacks the element of replenishment. As we have emphasized more than once already, there need not be any reduction in population size to achieve a coalescence of all genes into the MRCA as one moves backward in time. The genes will coalesce into a single MRCA even when the N_e has remained the same over generations, from the most recent to the more distant ones. Apart from this, however, the metaphor illustrates the coalescence process quite fittingly, albeit only figuratively. To describe it accurately, it is necessary to resort to mathematics. It is not necessary, however, to go through the entire mathematical argument to get the flavor of it; a statement of the principle will suffice.

Consider the case of two genes selected randomly from a population consisting of N diploid individuals and hence containing $2N$ genes at a given locus (Fig. 11.9A). If the individuals mate at random, the effective population size is N. How many generations do we have to travel back in time from G_0 to find the MRCA of these genes? Theoretically, the ancestor could have existed in G_1, G_2, G_3 or in any of the more distant generations, but the probabilities that it did are not the same for the individual generations. Assuming that the number of genes stays the same in all generations and that the genes are subject only to genetic drift, each of the $2N$ genes in G_1 has an equal probability of having been the common ancestor of the two chosen G_0 genes. The probability that coalescence occurred in G_1 is therefore $1/(2N)$ and the probability that it did not occur in G_1 is $1-1/(2N)$ (see Chapter Four which describes the probabilities for the reverse process, gene fixation). If coalescence occurs in G_2, it means that it did not occur in G_1. So, the probability of coalescence in G_2 is the product of two values – the probability that the two genes did not have a common ancestor in G_1 [which is $1-1/(2N)$] and the probability that they did in G_2 [which is $1/(2N)$ as in G_1]. Similarly, the probability that coalescence occurs in G_3 is the product of three values: the probability of noncoalescence in G_1 [which is $1-1/(2N)$], the probability of noncoalescence in G_2 [which is also $1-1/(2N)$], and the probability of coalescence in G_3 [which is $1/(2N)$]. Continuing in this manner and designating the probabilities of coalescence in G_1, G_2, G_3, etc as P_1, P_2, P_3, etc, respectively, we obtain:

$$P_1 = \frac{1}{2N}$$

$$P_2 = \left(\frac{1}{2N}\right)\left(1 - \frac{1}{2N}\right)^1$$

$$P_3 = \left(\frac{1}{2N}\right)\left(1 - \frac{1}{2N}\right)\left(1 - \frac{1}{2N}\right) = \left(\frac{1}{2N}\right)\left(1 - \frac{1}{2N}\right)^2$$

$$P_4 = \left(\frac{1}{2N}\right)\left(1 - \frac{1}{2N}\right)\left(1 - \frac{1}{2N}\right)\left(1 - \frac{1}{2N}\right) = \left(\frac{1}{2N}\right)\left(1 - \frac{1}{2N}\right)^3$$

$$\vdots$$

$$P_t = \left(\frac{1}{2N}\right)\left(1 - \frac{1}{2N}\right)^{t-1}$$

where t is the number of generations. In actual calculations, P_t is approximated by a formula based on the exponential function (e^x; see Appendix Four):

$$P_t = \frac{1}{2N} e^{-t/(2N)}$$

in which t stands for a real number ranging from 0 to ∞ (infinity).

In reality, however, rather than determining whether coalescence occurred in G_1, G_2, G_3, or any later generation, we want to know in *which* generation the coalescence can be expected to occur on average. This latter quantity is the mean of all possible values in the individual generations. But how can we calculate this mean? Once again, we help ourselves with the example of casting a die. Roll the die a number of times and record the outcome, the value of the face that comes up with each throw. Since there are six faces and each has an equal probability of being thrown, the probability of any particular face coming up is 1/6. The *mean outcome* of many throws is then obtained by multiplying each individual outcome by its probability and then adding up the products. The expected mean outcome is:

$$E(x) = 1\left(\frac{1}{6}\right) + 2\left(\frac{1}{6}\right) + 3\left(\frac{1}{6}\right) + 4\left(\frac{1}{6}\right) + 5\left(\frac{1}{6}\right) + 6\left(\frac{1}{6}\right) = 3.5$$

The more throws we make, the closer the real mean outcome will be to the expected mean outcome of 3.5.

Applying this analogy to the coalescence process, we obtain the *expected mean coalescence time*, $E(t)$, in a similar manner

$$E(t) = 1\left(\frac{1}{2N}\right) + 2\left(\frac{1}{2N}\right)\left(1 - \frac{1}{2N}\right) + 3\left(\frac{1}{2N}\right)\left(1 - \frac{1}{2N}\right)^2 + \dots + t\left(\frac{1}{2N}\right)\left(1 - \frac{1}{2N}\right)^{t-1}$$

(Here we have to assume that the "die", the coalescence process, has an infinite number of faces.) This cumbersome expression can be simplified by using the sigma notation for summing up:

$$E(t) = \frac{1}{2N} \sum_{t=1}^{\infty} t \left(1 - \frac{1}{2N}\right)^{t-1}$$

It can be shown, by performing algebraic acrobatics, that the formula reduces to $2N$: the expected mean coalescence time of two randomly selected genes is $2N$ generations (Fig. 11.9A). It is a *mean*, which is to say that the actual value obtained for any two genes may deviate from the expectation and that very many gene pairs must be tested for the real mean to approach the expected mean. Since the expected value of t square, $E(t^2)$, is $(2N)^2$, the variance of the coalescence time, defined by $E(t^2) - [E(t)]^2$, becomes $2N(2N - 1)$. In other words, the standard deviation (the square root of the variance; see Appendix One) is almost as large as the mean.

In the second phase, we extend the analysis of the coalescence process from two to all the genes in the *Go* pool to determine their expected mean coalescence time (Fig. 11.9B). To this end, we first arrange the genes into all possible pairs, define the probability that a given pair will coalesce, and then deduce the expected mean coalescence time of the entire group. The n genes of a pool can be arranged into $n(n-1)/2$ different pairs. If the expected mean coalescence time of a single pair is $2N$, if each pair has an equal probability of being the first to coalesce into its ancestor, and if there are $n(n-2)/2$ gene pairs, then the expected time to the first coalescence is

$$E(t_1) = \frac{2N}{n(n-1)/2} = \frac{4N}{n(n-1)}$$

After the first coalescence, of the original n gene lineages there remain $(n-1)$ lineages (Figure 11.9B). The mean expected time to the second coalescence is therefore:

$$E(t_2) = \frac{4N}{(n-1)(n-2)}$$

[Here, n of the first formula has been replaced by $(n-1)$ and $(n-1)$ has been replaced by $(n-2)$.] After the second coalescence, $(n-2)$ gene lineages are still on hold and the mean coalescence time to the third coalescence is therefore:

$$E(t_3) = \frac{4N}{(n-2)(n-3)}$$

If we proceed in this manner, we ultimately end up with two gene lineages, for which the mean coalescence time is $2N$ (Fig. 11.10). The expected mean coalescence time of all the *Go* genes is then the sum of the individual coalescence times:

$E(T) = E(t_1) + E(t_2) + E(t_3) + \ldots\ldots + E(t_{n-1})$

or

$$E(T) = 4N\left(\frac{1}{n(n-1)}\right) + 4N\left(\frac{1}{(n-1)(n-2)}\right) + 4N\left(\frac{1}{(n-2)(n-3)}\right) + \ldots + 4N\left(\frac{1}{(2)(1)}\right)$$

This long expression can again be simplified by more algebraic acrobatics which we skip once again to reach the final result:

$$E(T) = 4N\left(1 - \frac{1}{n}\right)$$

If n is very large, which it normally is, then $1/n$ is close to zero, and the formula simplifies further to $E(T) = 4N$. We conclude therefore that all the genes of a given generation are derived from a single ancestral gene which existed $4N$ generations ago, whereby our assumption N is equal to the effective population size N_e (Fig. 11.10).

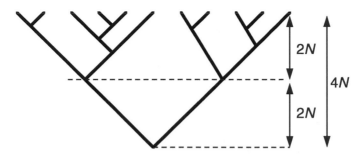

Fig. 11.10
The piranha effect: in terms of the number of generations, it takes the same time for all the genes of the *Go* generation to coalesce into two ancestors ($2N$ where $N = N_e$) as it does for these two genes to coalesce into their most recent common ancestor. The mean coalescence time of all the *Go* genes is therefore $4N$ generations. In this diagram only the coalescing lineages leading to the MRCA are depicted; all other lineages and genes have been ignored.

Several points regarding the coalescence theory need to be highlighted. The first one, which cannot be overemphasized, is that the $2N_e$ and $4N_e$ expectations are statistical averages; they are means toward which individual observations tend. Individual estimates can deviate considerably from the expectation and the deviations are greater, the smaller the N_e, for when dealing with a small gene pool, the stochastic error can be very large. In a computer simulation of the coalescence process that we carried out using a ridiculously small population (five individuals, ten genes), in 20 runs all the *Go* genes coalesced in 11, 40, 17, 13, 20, 8, 35, 44, 8, 13, 13, 28, 28, 24, 20, 12, 19, 11, 18, and 9 generations. Although the variation is large, the mean of these 20 coalescence times is 19.6, which is reasonably close to the expected coalescence of $4N_e = (4)(5) = 20$ generations. Obviously, most of the individual observations, if considered in isolation, would have grossly misled us, but the mean of the observations does not deviate significantly from the expectation. Here, the take-home lesson is: to estimate the coalescence time reliably, use genes at multiple loci.

The second point to emphasize is that the formulae for the expected mean coalescence time have been derived under certain simplifying assumptions which in reality need not always be met. The chief assumption is that the studied genes are neutral and hence that their behavior is determined exclusively by random genetic drift. Some other simplifying assumptions of the classical coalescence theory are that the population size remains the same from generation to generation; that at most, only one coalescence occurs in each generation; that the generations are nonoverlapping; and that a gene contributes maximally two copies of itself to the next generation. Taking all these variables into account would complicate the theory but, as computer simulations show, would not change the theory's main conclusions significantly.

The third point worth highlighting is that it takes the same mean time for all the *Go* genes to coalesce into two ancestors as it does for these two ancestors to coalesce into a single most recent common ancestor of all the genes (Fig. 11.10). If this result seems counterintuitive, recall the piranhas eating each other. A less exotic explanation is that in the backward flow of generations, the number of ancestral lineages of the sampled genes decreases while the number of genes in each generation remains the same. The probability of coalescence therefore decreases as well and is at its lowest when only two ancestral lineages remain. This observation has important implications for the choice of genes in coalescence studies.

The Remarkable Theta

Up until now, we did not ask whether the genes changed at all during the entire coalescence process, or whether the sequences of the *Go* genes are all the same and identical with the sequence of the MRCA $4N_e$ generations back in time. Since $4N_e$ generations can be a long time, mutations are to be expected during this period and these we must now take into account. First we consider the case of two randomly selected *Go* genes and their coalescence into a common ancestor $2N_e$ generations ago. How many differences in terms of nucleotide changes between the two genes should we expect? Since their divergence from the common ancestor, the two lineages accumulated mutations at the mutation rate μ per nucleotide site per generation. Because mutations are rare events, the mutations that accumulated in one lineage are with a high probability different from those accrued by the second lineage. The expected number of mutations accumulated in each of the two lineages is therefore given by the product of two values, the time elapsed between *Go* and the generation of the common ancestor (which is $2N_e$), and the rate at which mutations arise per site per generation (which is μ), hence $2N_e\mu$. Since there are two lineages, the total expected number of differences between the two *Go* genes is $4N_e\mu$. This expression specifies the mean number of nucleotide differences per site between two sequences taken randomly from a population which is, by definition, the average nucleotide diversity π. We can therefore write $\pi = 4N_e\mu$. This expression, which is also commonly referred to as θ (theta), figures in an important way in a variety of population processes and is one of the key measures of genetic variation.

By rearranging the equation $\pi = 4N_e\mu$ we obtain $N_e = \pi/(4\mu)$, which means that if we know π, the average nucleotide diversity, and μ, the mutation rate, we can calculate N_e, the effective population size for the period of $2N_e$ generations. The equation thus achieves a remarkable union of theory and practice – an elegant mathematical theory deduced by a combination of imaginary manipulations of virtual entities and down-to-earth practice in the sweatshop called the laboratory. The media rave every time a stretch of DNA some few thousand years old is sequenced, but the much more spectacular accomplishment of mathematically relating present-day genetic variation to past population sizes has escaped the attention of journalists.

Armed with this formula, we can now estimate the size of the human population in the last $2N_e$ generations. Rather than using the single-locus π value and so laying the estimate bare to the possibility of a large statistical error, we avail ourselves of a mean difference obtained from pairwise comparisons of genes at 50 different loci – the average nucleotide diversity. The best estimates available provide the π value of 0.0008 synonymous nucleotide changes per site (nonsynonymous changes are less suited for the present purpose because they are influenced by negative selection, the strength of which may vary from locus to locus). The estimated mutation rate of human genes is $\mu = 2 \times 10^{-8}$ per synonymous site per generation. And so the estimated effective size of the human population is $N_e = (0.0008)(2 \times 10^{-8}) = 10^4$ or 10,000 individuals. The time period to which this estimate applies spans the last $2N_e$ or $(2)(10,000) = 20,000$ generations. For the generation time of 20 years, this amounts to $(20)(20,000)$ or 400,000 years. If *H. sapiens* really did diverge from *H. neanderthalensis* 400,000 to 600,000 years ago, as the mtDNA sequences and one specific interpretation of the fossil record suggest, this estimate covers most of our species' existence.

A similar estimate is obtained independently from the analysis of mtDNA sequences. Since mtDNA is presumably inherited exclusively via the female lineage and each female is thought to have multiple but identical nonrecombining mtDNA molecules in each of her cells, the mitochondrial genome behaves as if it were a single gene in a haploid condition. Hence, if there are N_f females in each generation, each carrying a single version (haplotype) of the mtDNA genome, there are not $2N_e$ genes in each generation, as in the case of nuclear genes, but N_f mitochondrial genomes. Therefore the coalescence time of two alleles is not $2N_e$, as with the nuclear genes, but N_f. The time available for two mtDNA lineages to accumulate nucleotide changes since their divergence is therefore $2N_f$ generations, and the number of accrued differences per site is $\pi_f = 2N_f\mu$ where μ is the mutation rate of mtDNA, which differs from that of the nuclear genome. The effective female population size is then given by $N_f = \pi_f/(2\mu)$. The sequence databases contain 53 complete human mtDNA sequences and a large number of partial sequences from different parts of the mitochondrial genome. The single-locus π value calculated from 128 control region sequences is 0.0146 and the mutation rate for this region is estimated to be $(20)(7 \times 10^{-8})$ per site per generation (one generation is again assumed to be 20 years). From these two values, the N_f has been estimated to approximately 5000 females. Assuming an equal number of females and males, the estimated effective population size $N_e = 2N_f$ is 10,000 individuals, which is in agreement with the estimate based on the nuclear genes. It is comforting that two independent tests provide nearly identical results.

Identical or Different, That is the Question

There is another method of estimating the effective population size, which does not require knowledge of DNA or protein sequences. Its principle is this: consider again a gene pool at generation Go, from which two genes have been picked at random and compared. Their identity or difference ultimately depends on their sequence, but it can be established without sequencing. Let us designate as F the probability that the two chosen genes are the same and as H the probabiliy that they are different, whereby $F + H = 1$. Assume that the two genes had a common ancestor t generations ago and hence that they had $2t$ generations to accumulate differences. What is the probability that no mutations occurred during the $2t$ interval and hence that the genes have remained identical?

For a mutation rate u per gene per generation, the expected average number of mutations differentiating the two genes is $2tu$. From the mutation rate of 2×10^{-8} per site per generation, which comes to 3×10^{-6} per gene per generation (assuming the average length of a gene coding for 150 amino acid residues), we can therefore expect $(3 \times 10^{-6})(2t)$ or $(6t) \times 10^{-6}$ mutations to occur in the lineages leading to the two genes. This value, however, is a mean, while in any specific case the number of mutations can be 0, 1, 2, 3, etc, where each of these possibilities has a certain probability given by the successive terms of the Poisson distribution (see Chapter Five). The probability of no mutations occurring in one generation is $(1 - u)^2$ for the two genes. So, the probability that the two randomly chosen Go genes are identical (neither of them has suffered a mutation since their derivation from a common ancestor t generations ago) is $(1 - u)^{2t}$. Since u is small, $(1 - u)^{2t}$ is approximated by the exponential function, e^{-2tu}. (For an explanation of the exponential function see Appendix Four.)

In the next step, we consider the probability that the two identical genes had a most recent common ancestor t generations ago. From earlier discussions, we know that the probability is given by

$$P_t = \frac{1}{2N_e} e^{-t(2N_e)}$$

The probability that the lineages leading to the two genes coalesced t generations ago and have remained identical from that time to the Go generation is then the product of two probabilities, P_t and e^{-2tu}. The product is the probability that, provided the two genes coalesced t generations ago, no mutation occurred during this time interval. For the t-th generation, it is

$$\frac{1}{2N_e}e^{-(1/(2N_e)+2u)t}$$

When the products of the individual generations, from 0 to t (infinity), are added up, the formula simplifies to:

$$F = \frac{1}{1+4N_eu}$$

Because $F + H = 1$, $H = 1 - F$, where H is the nonidentity or *heterozygosity* of the two genes. From this relationship it follows that

$$H = \frac{4N_eu}{1+4N_eu}$$

By rearranging the formula, we obtain

$$N_e = \frac{H}{4u(1-H)}$$

Proteins sampled from human populations and encoded by alleles at 121 loci were tested by electrophoresis and the homo- or heterozygosity of the individuals was determined in different laboratories. From the results of these screenings, the average heterozygosity of $H = 0.143$ was obtained. Taking the mutation rate of $u = 3 \times 10^{-6}$ per gene per generation for these loci, the effective population size is

$$N_e = \frac{0.143}{(4)(3 \times 10^{-6})(1 - 0.143)} = 13{,}900$$

Thus three independent estimates of the effective human population size based on different sets of data and obtained by different methods reach a very similar value of approximately 10,000 breeding individuals over a period extending from about 400,000 to 10,000 years ago. Because the effective population size is usually lower than the census size, the actual number of people living in this period must have been more than 10,000. Approximately 10,000 years ago, the human population began to grow, first slowly and then at an accelerated rate. This increase, however, has so far left no detectable mark on the genetic variation within the population, specifically on the sequences of the nuclear genes. The pace at which changes accumulate in these genes is too slow to reflect a population expansion over the last 500 (i.e., 10,000/20) generations.

A Camel Through the Eye of a Needle?

If 10,000 years is not long enough to change the genetic composition of the global human population, is it not possible that during the 400,000 years of human existence there might have been occasions when the population size dropped dramatically, perhaps even down to a few individuals? To examine this possibility, let us consider the extreme case of a reduction in population size to a single pair of breeding individuals – an ultimate bottleneck of Biblical proportions. Let us designate this hypothetical moment in the demographic history of the human population *G0* and count the generations leading from it forward (Fig. 11.11). What might be the consequences of this bottleneck for genetic variation of the population?

For purely didactic reasons, it is helpful to think of genetic variation as having two dimensions. The horizontal dimension, which has been our main focus of interest up until now, is given by the array of different loci, each with a small number of alleles (but at least two; i.e., *A1, A2; B1, B2; C1, C2*; etc). The vertical dimension is apparent at a few loci, each with great many alleles (i.e., *M1, M2, M3, M4, …Mn*). An example *par excellence* of these lo-

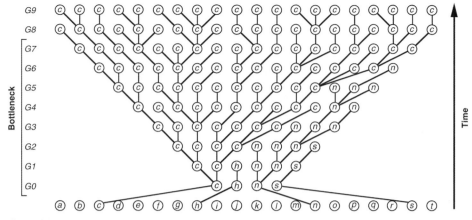

Fig. 11.11
The effect of a bottleneck on a highly polymorphic locus such as those of the *Mhc*. Here the gene pool prior to the bottleneck (prior to *G0*) contained 20 different alleles (*a* through *t*) at an *Mhc* locus. Because of the dramatic reduction in the size of the pool, only four of these (*c*, *h*, *n*, and *s*) were passed on to the *G0* generation. Although in subsequent generations (*G1* through *G8*) the pool grew until it attained its pre-bottleneck size, random genetic drift eliminated three of the four alleles so that ultimately only one (*c*) remained. From this point on, polymorphism could only arise by slow accumulation of new mutations.

ci is provided by the *major histocompatibility complex, Mhc,* which we will describe shortly. The effects of the bottleneck are much more severe on the vertical than on the horizontal genetic variation. Loci such as those of the *Mhc* are therefore particularly suited for ascertaining whether a bottleneck did occur and if so, to what extent.

The differential effects of a bottleneck on the two dimensions of genetic variation are easily grasped by considering the extreme situation of a population crash leaving only two individuals. Each individual possesses all the loci, each in two editions. Therefore, if there were only two different alleles at each of the variable loci, and if at least one of the two *G0* individuals or the two individuals together were heterozygous (had two different alleles) at all of these loci, theoretically there might not be any loss of genetic variation caused by the bottleneck. In reality, however, a substantial loss *will* occur for two reasons. First, the two *G0* individuals may be homozygous (carry identical alleles) at some of the loci. And second, at many of the initially heterozygous loci, random genetic drift can result in fixation of one of the two alleles and hence in homozygosity.

The effect of the extreme bottleneck on the vertical dimension of genetic variation is prompt and dramatic. If there are hundreds of alleles at one locus in a population, as indeed there are at some of the human *Mhc* loci, only four of them at the most will pass through the bottleneck, two from each individual (Fig. 11.11); all the other alleles will be lost. The extensive polymorphism of the *Mhc* loci is therefore like the Biblical camel that has no possibility of passing though the eye of a needle. Furthermore, of the lucky four alleles, all but one might subsequently be lost by genetic drift.

Such are the expectations under the extreme version of the founder effect hypothesis. Even natural selection would be virtually powerless in the face of the strong effects of genetic drift in the small founding population and the incipient species would be expected to begin its tenure with a genome resembling a clean slate in terms of variation. Most of the genetic variation of the new species would have to be generated *de novo* in the post-speciation phase. Even if the bottleneck were less dramatic, its principal effects on the species would only be quantitatively, not qualitatively different from those expected in the extreme situation: the information inscribed on the slate would be partially rather than fully erased.

The extent of erasure would depend on two main factors – the length of time, t_b, of the bottleneck's duration and the size, N_b, of the bottleneck's population. These two parameters are interdependent: if the N_b is very small and the t_b very short, or if the N_b is moderately large and the t_b long, the bottleneck will reduce genetic variation considerably. It is therefore convenient to evaluate their effect together in the form of an N_b/t_b ratio. Calculations indicate that genetic variation of a population diminishes drastically once the N_b/t_b ratio drops below 10, regardless by what means. Thus, for example, ten generations of a bottleneck with an N_b of 100 individuals would have the same effect as 100 generations of a bottleneck with 1000 individuals.

The probability, P, that two neutral genes pass through a bottleneck is that of no coalescence during the bottleneck period. It is given approximately by the exponential function

$$P = e^{-t_b/(2\,N_b)}$$

(The concept of function is explained in Appendix Three; for an introduction to exponential function, see Appendix Four.) For a probability of 0.99 (near certainty), we have $P = e^{-t_b/(2N_b)}$ which gives $t_b = 0.02N_b$. For $t_b = 10$ generations, for example, $N_b = 500$ breeding individuals. In other words, for two neutral alleles to pass with high probability through a bottleneck, the effective size of the bottleneck population would have to be larger than 500 individuals. For more than two neutral alleles to pass, a correspondingly larger bottleneck population is necessary.

Genetic variation of a population is also influenced by the mode of population growth in the post-bottleneck phase. The mode determines the length of time it takes for a population to reach its pre-bottleneck size. To give an example, let us ask how long it would take a population reduced to a single pair to reach its original effective size of 10,000 individuals. Obviously, if the pair had only two children and this pattern were repeated in subsequent generations, the population size would never increase, it would remain constant. A number remains constant if multiplied by 1. To increase a number, it must be multiplied by $1 + r$. For a population to increase, each generation has to produce $N_o(1 + r)$ next-generation individuals, where N_o is the number of breeding individuals in the preceding generation and r is the *growth rate* of the population. For example, if $r = 0.5$ and we start with $N_o = 2$, then in successive generations we obtain $2(1 + 0.5) = 3$; $3(1 + 0.5) = 4.5$; $4.5(1 + 0.5) = 6.7$; $6.7(1 + 0.5) = 10.1$ individuals; and so on. At this 50 percent growth rate (the mean number of offspring per couple being 3), the population would reach 10,000 individuals in 22 generations or 440 years, assuming a generation time of 20 years.* If the start-up population size was ten individuals, the population would reach $N_e = 10,000$ in 18 generations or 360 years. A growth rate of 50 percent is, however, unrealistically high. The growth of the human population in the last few hundred years has been 0.05 percent and in the period extending up to 10,000 years ago it was approximately 0.02 percent. At the latter rate, expansion from two to 10,000 individuals would take 431 generations or 8620 years.

Both the horizontal and the vertical forms of genetic variation are repositories of demographic information. The latter, however, has certain aspects that are lacking in the for-

---

* This figure is obtained by taking $2(1 + 0.5)^t = 10,000$, which can be rewritten as $t(ln1.5) = ln\ 5000$ or $t = ln\ 5000/ln\ 1.5 = 21$.

mer and that are particularly useful in paleodemographic reconstructions. For this reason, we digress at this point from this chapter's main theme to cover some of these aspects. In no other genetic system are the special features more pronounced than in the *Mhc*.

A Major in the Army of Compatibles

A feature of the *Mhc* that continues to command researchers' attention is its involvement in tissue or histocompatibility. A century ago, researchers studying cancer in mice discovered that a tumor which arose in one mouse would carry on growing when transplanted to another, but only if the latter shared the same alleles at a set of *histocompatibility loci* as the donor. Nothing more was known about these hypothetical *H* loci except that they were numerous, that identity at all of the loci was necessary for a successful transplantation, and that their effect was not restricted to tumor cells: the fate of grafts involving normal cells, tissues, and organs was governed by them in a similar manner. A single-locus difference between the donor and the recipient sufficed to initiate destruction (*rejection*) of the transplant. Subsequent research revealed that one group of the *H* loci stood out from the rest in the efficiency with which graft rejection was accomplished. And so the loci came to be divided into minor *H* loci scattered over the entire genome and major *H* loci, usually huddled together in a single chromosomal region. A couple of Nobel Prizes later, the mechanism of the *Mhc*-mediated rejection process came to be understood in principle, if not in detail. Some of the *Mhc* loci were revealed to be so highly polymorphic that among unrelated individuals, hardly any two persons could be found that were matched in all of them. The loci control proteins which are inserted into the membranes of different cell types and displayed on the cell surfaces. Following transplantation, the recipient's immune system recognizes the *Mhc* proteins of the graft as foreign and mounts an attack on the grafted tissue in a manner resembling an immune assault on virus-infected cells.

Organ transplantation is an unnatural act which is practiced by surgeons and researchers only, not by Mother Nature. Researchers therefore puzzled for a long time over the natural function of the *Mhc* loci. When it finally became known yet another couple of Nobel Prizes later, it amounted to a revelation. The Mhc molecules, the powerful stimulators of immune response, turned out themselves to be one of three critically important initiators of adaptive immunity. The latter is what a layperson commonly understands as "immunity" – the refractoriness to an infection by the same agent that an individual had successfully overcome on a previous encounter. The other two sets are the immunoglobulins, better known as antibodies, and T-cell receptors. The two are intimately associated with a fraction of white blood cells called lymphocytes, the T-cell receptors with T lymphocytes (so named because they mature in the thymus), the immunoglobulins with B lymphocytes (which in mammals mature in the bone marrow).

At the time of their synthesis, the Mhc molecules pick up fragments of other proteins (peptides) and display them on the cell surfaces. In an uninfected individual, the peptides displayed are those of the individual's own (self) protein; in an infected person, some of the Mhc molecules display peptides derived from the pathogen (Fig. 11.12). The T lymphocytes are constantly scanning the Mhc molecules on the surfaces of other cells. Mhc molecules with self-peptides bound to them leave the lymphocytes cold, but molecules associated with foreign (nonself) peptides engage the T-cell receptors, and thus activate the lymphocytes. Depending on their genetic program, some of the activated T lymphocytes then differentiate into killer cells which attack the infected cells, while others secrete factors which help the B lymphocytes to develop into antibody-producing cells.

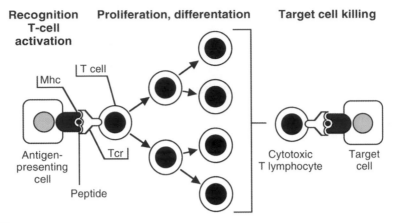

Fig. 11.12
Initiation of adaptive immune response by viral peptides bound by Mhc molecules. Proteins in virus-infected antigen-presenting cells are degraded into peptides which bind to Mhc molecules and are displayed on the cell surface. The Mhc-peptide complexes are recognized by T-cell receptors (Tcr) on thymus-derived lymphocytes which are thereby activated. The stimulated T cells divide repeatedly and differentiate into cytotoxic T lymphocytes capable of killing virus-infected target cells.

A graft from a genetically disparate donor can be rejected for three reasons. First, the recipient's T lymphocytes recognize the donor's Mhc molecules on the transplanted cells as foreign. Second, the recipient's T lymphocytes regard the peptides derived from the donor's broken-down Mhc molecules and bound to the recipient's Mhc molecules as pathogen-derived. And third, the T lymphocytes take peptides for pathogen-derived if they are from any broken-down donor protein which, because of a mutation, differs from the recipient's protein. In each of these three situations, the T lymphocytes initiate an immune attack on the grafted cells. In the third situation, the proteins from which the peptides are derived are encoded in the *minor histocompatibility loci*. The latter are unrelated to the *Mhc* and most of them have otherwise nothing to do with immunity. They become involved in rejection only because they happen to differ between the donor and the recipient.

It is the peptide-binding part of the Mhc molecule (Fig. 11.13) that is highly variable among individuals, the variability enabling different molecules to bind distinct sets of peptides. The amino acid positions involved in the binding of the peptide are known to evolve under the influence of natural selection which favors certain residues over others, namely those capable of interacting with certain amino acid residues of the peptides. The effect of natural selection is manifested in three special features of the Mhc proteins and of the genes that specify them. First, the number of alleles at specific loci is very high. At the human *Mhc*, the Human Leukocyte Antigen or *HLA* complex, at loci *HLA-A, -B, -C, -DRB1*, *-DQA1, -DQB1, -DPA1*, and *-DPB1*, there are 195, 399, 94, 257, 20, 45, 19, and 92 alleles, respectively, but new alleles are still being described. Second, the proteins specified by distinct alleles differ at a number of positions, in some cases as many as 50 or more (Fig. 11.14). By contrast, most proteins controlled by alleles at non-*Mhc* loci differ at a single position, or at a few positions at most. The majority of the differences between products of allelic *Mhc* genes are concentrated in the peptide-binding region, at positions occupied by residues that contact the residues of the peptide. Third, alleles at certain *Mhc* loci can be handed down from one species to another over many species along a phylogenetic lineage. At the *HLA* loci, some of these *allelic lineages* were founded in the common ancestor of apes and

CLASS I **CLASS II**

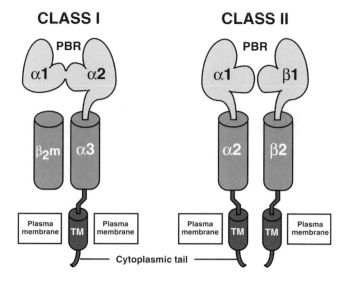

Fig. 11.13
Two classes of Mhc molecules. Class I molecules consist of a single membrane-anchored α-chain associated with the β_2-microglobulin (β_2m) chain which is encoded in a gene outside the *Mhc*. Class II molecules consist of two membrane-anchored polypeptide chains, α and β. The class I and class II chains contain distinct domains, the external α1, α2, α3, β1, and β2 domains, the transmembrane region (TM), and the cytoplasmic tail. The α1 and α2 domains of the class I chain form the peptide binding region (PBR). In the class II molecules, the PBR is formed by the α1 and β1 domains.

Old World monkeys 20-30 my ago and then passed through numerous, now extinct species all the way to the extant species – human, chimpanzee, gorilla and others. Each species through which the lineages passed added a few changes during its period of existence and in this manner the distinct lineages diverged from each other more and more until over 50 differences arose between some of them. The consequence of this *trans-species mode of evolution* is that certain *Mhc* alleles of one species (e.g., human) are more closely related to certain alleles of another species (e.g., chimpanzee) than to some other alleles of the same (i.e., human) species (Fig. 11.15). Because of a widespread trans-species polymorphism, phylogenetic trees of *Mhc* genes can be incongruent with the phylogenetic trees of the species: they reflect the genealogy of the genes and not of the species.

The allelic lineages persist because natural selection prevents the fixation of individual mutations in any of the species. Each mutation spreads through the genes of the allelic lineage in which it arose but not to the genes of another lineage. In each individual and at each locus, natural selection appears to favor the presence of two alleles rather than one, presumably because heterozygosity broadens the spectrum of peptides available for binding by the Mhc molecules of a given individual. If each molecule can bind 1000 different peptides, then the two distinct molecules can bind more than 1000 different peptides (however, generally not as many as 2000 distinct peptides because there is a certain overlap in the binding spectra of the two Mhc proteins). This preference for the persistence of two or more alleles in a population is referred to as *balancing selection*.

In population genetics studies, the *Mhc* has two major advantages over most other genetic systems. First, the large number of alleles at the *Mhc* loci renders the system highly sensitive to short-term changes in effective population size and uniquely suited for the de-

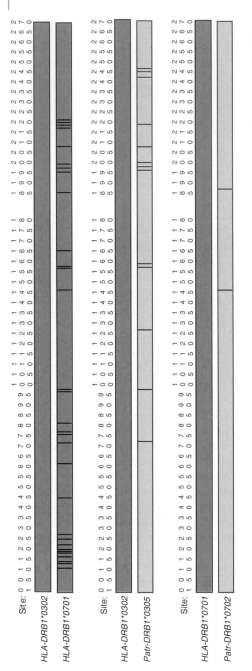

Fig. 11.14

Comparison of parts of different human *Mhc* class II alleles (*HLA-DRB1*0302* and *HLA-DRB1*0701*) with each other and with genes at the corresponding locus of the chimpanzee (*Patr-DRB1*0305* and *Patr-DRB1*0702*). In each pair the upper bar represents a sequence with which the lower sequence is compared. Differences are indicated at the relevant sites by vertical lines. The human sequence is clearly more similar to the corresponding chimpanzee sequence than to the allelic human sequences. (From Klein et al. 1993.)

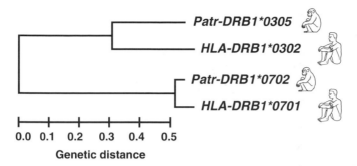

Fig. 11.15
The trans-species character of *Mhc* polymorphism. Phylogenetic tree based on the sequences depicted diagrammatically in Figure 11.14. The human gene is in this case more closely related to the corresponding chimpanzee gene than to its human allele. (From Klein et al. 1993.)

tection of bottlenecks. And second, the long-term persistence of *Mhc* allelic lineages makes the system opportune for estimating the effective population size at the time of and prior to speciation. A few examples of how the *Mhc* and other genetic systems have been used in the study of human paleodemography follow.

Couture For a Pool

Intuitively, you would expect such a dramatic event as a sharp decline in the size of a population, followed by the rapid expansion to its original size to leave an imprint on the population that can be revealed by genetic analysis. The episode should have two major consequences. First, the analysis should reveal an across-the-board reduction of genetic variation in the entire genome, but particularly at the highly polymorphic loci such as the *Mhc*. And second, if the expansion time t_a of the population after the bottleneck is much shorter than the coalescence time of the genes under study, their genealogy will appear to be abruptly truncated. Their relationship will not be tree-shaped (Fig. 11.16A), as it would without a bottleneck, but star-shaped (Fig. 11.16C) because the passage through the bottleneck erases most of the preexisting genetic variation. As new variation begins to accumulate in the post-bottleneck phase, individual allelic lineages will evolve separately for a period of t_a generations and the number of nucleotide differences between any two genes of a pair, d_{ij}, will be more or less the same. Plotted graphically, the frequency distribution of the nucleotide differences will not be a rugged curve (Fig. 11.16B), as it would in the case of no bottleneck, but will display a prominent peak at $d_{ij} = 2t_a\mu$ (Fig. 11.16D) – the expected nucleotide difference per site. The "tails" of the frequency distribution in Figure 11.16D are a reflection of the stochasticity governing the accumulation of mutations in the population.

There are two principal methods of analyzing the imprint of history on genetic variation in a population, one based on modeling and the other on mathematical inference. The essence of the modeling method is to specify a set of parameters determining the genetic behavior of a population, write a computer program simulating the fundamental processes in the population, and then instruct the computer to proceed with the simulation of the population's history. The parameters include the size of the gene pool, mutation rate, and selection coefficient. The population processes concern the mode of drawing genes from

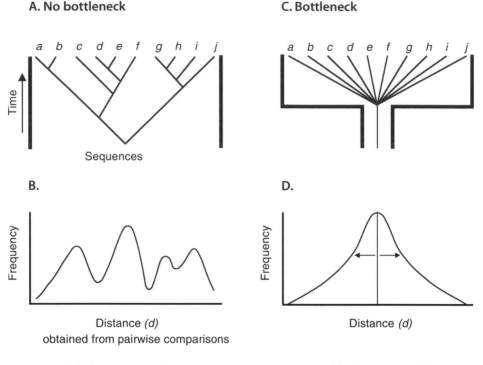

A. No bottleneck

a b c d e f g h i j

Time

Sequences

B.

Frequency

Distance (d)
obtained from pairwise comparisons

Variance: very large

C. Bottleneck

a b c d e f g h i j

D.

Frequency

Distance (d)

Variance: small

Fig. 11.16A-D
Genetic signature of different population histories inscribed in genealogies. **A.** In the absence of a bottleneck, the genealogy of genes constituting the current gene pool is deep and tree-like in its topology. **B.** When pairwise comparisons of the genes are made and genetic distances are plotted against frequencies, a ragged curve with multiple peaks is the result. **C.** After a recent bottleneck, the gene genealogy is shallow and all the lines of descent appear to be radiating from a single point in time. **D.** The plot of genetic distances obtained by pairwise comparisons yields a curve with a single, prominent, narrow peak corresponding to the mean distance, d.

the preceding to the succeeding generation pool, the manner in which generations succeed one another (overlapping or nonoverlapping), and the length of the time period (specified in the number of generations) simulated. After several runs of the program, the parameters are changed and another series of runs is initiated. The results of the individual runs and each series of runs are evaluated statistically and conclusions about the likelihood of the particular results are drawn.

By applying this method to the data available on *HLA* polymorphism, in 1990 Jan Klein and coworkers concluded that since the inception of the human species, the size of the human population has probably never dropped significantly below 10,000 individuals (Fig. 11.17). This estimate was in fact the first to feature the magic number of 10,000 individuals for the human effective population size. Moreover, at the time of the mitochondrial Eve rage, it provided a strong argument against the occurrence of a severe bottleneck in the history of the human population. The same results were also obtained later by other investigators using more sophisticated methods of simulation.

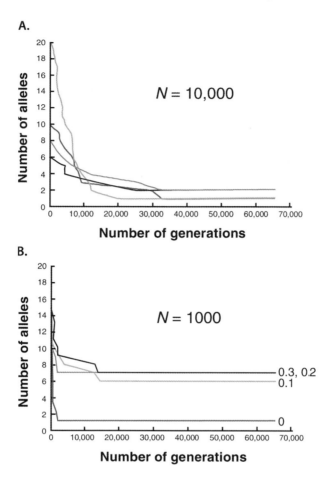

Fig. 11.17A,B
Computer simulation of the loss of *Mhc* alleles over 65,000 generations in a population consisting of 10,000 (**A**) or 1000 (**B**) individuals. **A.** The conditions of the simulation were these: alleles were initially present at equal frequencies; individuals mated at random; there was no loss or gain of alleles through mutations or migration; homozygotes had the same probability of mating as heterozygotes; and the size of the population was the same throughout the experiment. The initial number of alleles was 20, 10, 8, and 6 in the different experiments. Under these conditions and in the absence of selection, loss of alleles occurs by random genetic drift alone. **B.** Here the conditions are the same as in **A** except that in three of the four runs heterozygotes had a 0.3, 0.2, or 0.1 times higher probability of mating than homozygotes. Even though the initial population size was 1000 individuals, most of the 20 alleles present in the founding population were rapidly lost in spite of strong selection favoring heterozygotes. Fewer than one-half of the input alleles could persist in the population for the duration of the experiment. This simulation mimics the effect a bottleneck can be expected to have on the persistence of alleles under selection. In the fourth control run without selection, all 20 alleles except one were rapidly lost from the population. (From Klein et al. 1990.)

A bottleneck in a computer model is simulated by the size reduction of a gene pool followed, after a specified number of generations, by expansion to the original size. By varying the N_b and t_b parameters, it is then possible to study the number of alleles that overcome the bottleneck phase and the conditions under which this happens. The simulations reveal that to pass 40 out of a total of 50 *HLA-DRB1* alleles into the post-bottleneck population with a probability of 95 percent, a bottleneck lasting ten generations and immediately followed by expansion of the population back to its mean size of 10,000 breeding individuals cannot involve less than 400 to 500 individuals (Fig. 11.18, Table 11.1). By taking into account the time required for the growth of the population to 5000 individuals in the expansion phase, and by assuming a growth rate of about one percent per generation (which is about 50 times higher than the postulated average growth rate of the human population during the Pleistocene), the estimate of the minimum bottleneck population size is more than 2000 individuals. Natural selection has very little effect under these conditions. All these estimates are based on the consideration of *HLA* alleles at only one of the highly polymorphic loci; by taking alleles at the other *HLA* loci into account the estimate

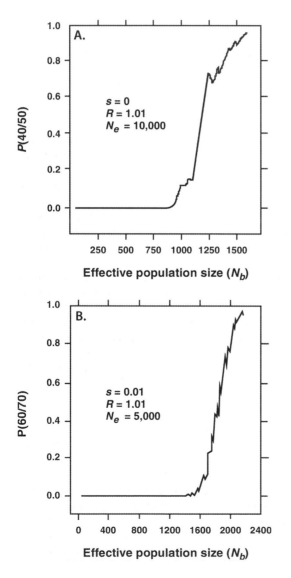

Fig. 11.18A,B
The probability that 40 out of 50 [$P(40/50)$] or 60 out of 70 [$P(60/70)$] *HLA* alleles pass though a bottleneck lasting ten generations and its size is as indicated (N_b). After the bottleneck the population is allowed to grow at a rate of one percent ($R = 1.01$) until the effective population size of $N_e = 10,000$ individuals is reached. Two situations are considered: **A.** The alleles are assumed to be neutral and hence subjected to random genetic drift only ($s = 0$). **B.** The alleles are subjected to balancing selection with a selection coefficient of $s = 0.01$. The graphs represent averages of 300 computer simulations. To pass the indicated number of alleles through the bottleneck with a probability of 95 percent requires minimum population sizes of $N_b = 1540$-1600 (**A**) and $N_b = 2140$-2180 individuals (**B**). Reducing the probability level changes the requirement for the minimum population size very little. Similarly, under these conditions selection has very little effect. (From Ayala et al. 1994.)

Table 11.1
Computer simulation-based estimate of the minimum population (N_b) required to pass 40 out of 50 or 60 out of 70 HLA alleles through a bottleneck with 95 percent probability. (From Ayala et al. 1994.)

s	R	N_e	N_b to pass alleles through bottleneck	
			40/50	60/70
0	—	—	270–300	458–490
0.01	—	—	292–302	454–462
0	1.1	10,000	460–460	
0	1.1	5,000		750–790
0	1.05	5,000	980–1010	
0	1.01	10,000	1540–1560	
0	1.01	5,000		2120–2180
0.01	1.01	5,000		2140–2180

The bottleneck was allowed to last for ten generations ($t_b = 10$), after which time the population was allowed to grow at a rate R until it attained the population size N_e. The alleles were assumed to be either neutral ($s = 0$) or subjected to balancing selection with a selection coefficient of $s = 0.01$. The values obtained are based on 300 computer simulations. In the first two rows the time required for a population to grow to its mean size was ignored.

increases correspondingly. We thus come to the conclusion that if short-term bottlenecks did occur in the history of the human population, the N_b could not have been smaller than several thousand breeding individuals.

Mathematical modeling of HLA polymorphism is complicated by the nonneutrality of the Mhc loci. The complication stems from the necessity of incorporating an additional parameter – selection intensity (coefficient, s) – into the equations describing the behavior of the genes in a population. Fortunately, when considering severe bottlenecks, this parameter can be set aside because it can be shown that if the product of bottleneck population size and selection intensity – an important determinant of population behavior during the bottleneck phase – is smaller than one ($N_b s < 1$), the fate of genes is determined largely by random genetic drift. Hence, if s is about 0.02 and N_b drops down to 50 individuals, the Mhc genes behave as if they were neutral because $N_b s = (0.02)(50) = 1$. And since the probability of passing neutral alleles through a bottleneck of 50 individuals lasting ten generations or longer becomes less than 90 percent, the effective human population must have been larger than 50 individuals at all times. In reality, the population must have been much larger than that because not two, but many Mhc alleles had to be passed through the bottleneck, and because of the large number of generations needed to restore the original population size after the bottleneck. In what follows, we consider the conditions necessary for the passage of multiple Mhc alleles through successive generations.

Let us assume that n different alleles have been found at a given HLA locus in the present-day human population. A long way back in the past there must have been one gene from which all these alleles originated, their most recent common ancestor. Between these two points in time there must have been numerous intermediate stages of progressive coalescence to fewer and fewer alleles, until the last two alleles coalesced into the common ancestor. In other words, at different times there were different numbers of allelic lineages leading to the extant alleles. Let us designate these numbers generally as k. To draw conclusions about the population size, we need to know the actual value of k at a chosen time in the past. To this end, we draw a tree depicting the descent of the extant alleles and then place a time scale alongside it. The number of intersects on a line drawn from the chosen point on the scale perpendicular to the lines of descent then gives us k. Drawing the tree is straightforward; we simply apply one of the methods discussed in Chapter Five. The problem is obtaining the time scale.

When we find two similar genes at the same locus in two distinct species, we assume that the genes diverged before the divergence of the species (Fig. 11.19). To determine how

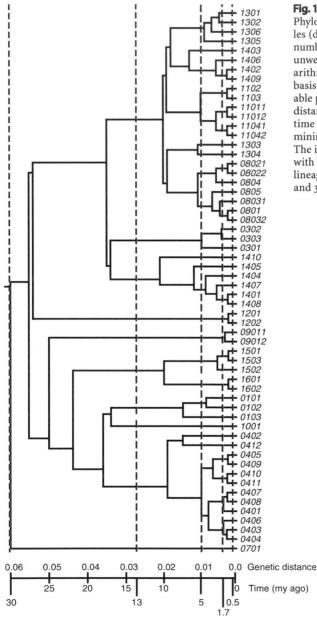

Fig. 11.19
Phylogenetic tree of *HLA-DRB1* alleles (designated by four- or five-digit numbers). The tree was drawn by the unweighted pair-group method using arithmetic averages (UPGMA) on the basis of sequences of the most variable part of the gene (exon 2). The distance scale is complemented by a time scale obtained by the minimum-minimum method (see Fig. 11.20). The intercepts of the broken lines with the tree branches indicate gene lineages that existed 0.5, 1.7, 5.0, 13.0, and 30 my ago.

1301
1302
1306
1305
1403
1406
1402
1409
1102
1103
11011
11012
11041
11042
1303
1304
08021
08022
0804
0805
08031
0801
08032
0302
0303
0301
1410
1405
1404
1407
1401
1408
1201
1202
09011
09012
1501
1503
1502
1601
1602
0101
0102
0103
1001
0402
0412
0405
0409
0410
0411
0407
0408
0401
0406
0403
0404
0701

| 0.06 | 0.05 | 0.04 | 0.03 | 0.02 | 0.01 | 0.0 | Genetic distance |

| | 25 | 20 | 15 | 10 | | 0 | Time (my ago) |

30   13   5   0.5

1.7

long before the species divergence the genes split, we divide their divergence time into two intervals, t_s corresponding to the species divergence time and t_g corresponding to the time from the gene divergence to the species divergence. The coalescence time of the two genes is $t_s + t_g$, but we know only t_s and not t_g. We have no direct way of determining t_g, but we can dodge the problem by choosing a pair of genes whose t_g is close to zero, in other words genes that diverged at nearly the same time as the two species. The coalescence time of these genes is approximately equal to t_s, the species divergence time. By employing this trick, we can calibrate the entire time scale, assuming that the *Mhc* genes accumulate mutations with the regularity of a clock.

How do we find the pair of genes that diverged at the same time as the species? To explain, let us consider three species, human (H), chimpanzee (C), and macaque (M). From each of these species we sequence sets of genes at the same *Mhc* locus, compare each gene of one species with all the genes of the second and third species, and compute the genetic distance for each comparison. In each interspecies comparison (i.e., H-C, H-M, and C-M) we identify the smallest distance and plot it against the species divergence time (Fig. 11.20). We then draw lines through zero and through each of the three points, choose the line passing through the lowest of the three points, and calculate its slope. This lowest point represents the gene pair that diverged closest to the species divergence time, and the slope of the line passing through it is the substitution rate of this set of *Mhc* genes. It is a *minimum-minimum rate* determined by taking two minimal values based on the pairwise comparison of genes. The *minimum-minimum method* was developed and applied to human population size estimates by Yoko Satta and coworkers in 1991.

Once the time scale of the tree is calibrated, the k values can be read off it for any time interval. Figure 11.19 shows an application of the minimum-minimum method to 56 alleles at the *HLA-DRB1* locus. It reveals that 0.5 my ago, all 56 alleles were already in existence. Approximately 1.7 my ago they coalesced into 37 ancestral genes, 5 my ago into 25 genes, 13

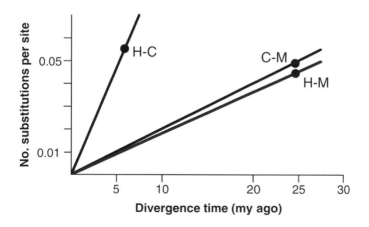

Fig. 11.20
The minimum-minimum method used to calibrate the distance scale in Figure 11.19 and so convert it into a time scale. Sequences of *HLA-DRB1* genes from the human (H), chimpanzee (C), and macaque (M) were compared in all possible pairwise combinations (not shown) and in each combination (H-C, H-M, and C-M) the one showing the smallest distance was identified and plotted here as the number of substitutions per site against the divergence time of the three pairs of species. The slope of the line passing through the lowest of the three points (H-M) was calculated and the value used to calibrate the time scale. (Based on Satta et al. 1991.)

my ago into nine genes, and finally, 30 my ago, into a single ancestral gene. The probability that n alleles originated from k ancestral alleles t generations ago is given by the formula

$$P_{nk}(t') = \sum_{m=k}^{n} \frac{(2m-1)(-1)^{m-k} k_{(m-1)} n_{(m)}}{k!(m-k)! n_{[m]}} e^{-m(m-1)t'/2}$$

where $n_{[m]} = n(n-1)(n-2)\dots(n-m+1)$; $n_{(m)} = n(n+1)(n+2)\dots(n+m-1)$; m is a variable that changes from k to n; $t' = t/(2N_e f_s)$, and f_s is the factor for converting neutral genealogy to the genealogy of alleles evolving under balancing selection. The factor was derived and the structure, as well as the time scale of allelic genealogy, were described by Naoyuki Takahata in 1990.

By taking different values of N_e, we can calculate the t' values from the above formula, and for each t' value we can compute the $P_{nk}(t')$ probability. The N_e that gives the highest probability of k ancestral genes in existence t generations ago giving rise to n alleles now found in the human population is then assumed to have been the effective population size t generations ago. For $k = n$ which, according to Figure 11.20, has been the case for the last 400,000 years, $P_{nk}(t')$ simplifies to

$$P_{nn}(t') = e^{-n(n-1)t'/2}$$

In this case $P_{nn}(t')$ decreases monotonically with increasing t' and N_e cannot be estimated from the curve. It can, however, be estimated from $k = n - 1$ and can be regarded as the minimum effective population size. The estimated N_e values for the last 2 my are between 10,000 and 100,000 individuals, which is once again consistent with all previous estimates.

Necks of Different Size

The various data sets and the different methods of analysis described on the preceding pages all indicate that the $N_e = 10{,}000$ of the human population is rather atypical for a primate species, although apparently not for many other mammals. In 1984, Masatoshi Nei and Dan Graur compiled all the available data on genetic variation in a variety of organisms and from this information computed the effective population sizes of the different species. An N_e of 10,000 was in fact the size they obtained for most of the species. Hence viewed from this very broad perspective, the human effective population size is close to the average; for a member of the primate species, however, it is on the lower side.

If we assume that the N_e of the human-chimpanzee MRCA was similar to that of the chimpanzee, then, formally, the tenfold reduction in population size from 100,000 to 10,000 in the hominid lineage represents a bottleneck. But it is not a bottleneck of the type that the proponents of the founder effect hypothesis claim is necessary to give birth to a new species! Taking into account the duration of the phase of reduced size and the fact that many other species have an effective population size of 10,000, the population of the human species is either a neck without a bottle or bottle without a neck. If a metaphorical designation is necessary for it, we would prefer the "crane's neck". However, setting semantic arguments aside, it is more important to consider when the reduction in size took place and why.

As for the first question, we note that if the reduction occurred a long time ago, the level of *HLA* polymorphism would have been lowered. If, on the other hand, it occurred recently, the diversity at the silent sites of non-*HLA* genomic segments would have been elevated in the extant human population. Population genetic analysis of variation at both the *HLA* loci and non-*HLA* genomic segments suggests that the reduction phase occurred 1-2 my ago and coincided roughly with the emergence of *Homo erectus*.

The second question is more difficult; here we can only provide speculative answers. The explanation we prefer centers around the organization of the population in the *erectus* and post-*erectus* phase. Although no direct information is available on the structure of the *H. erectus* population, we note that this species was the first of all the hominids to migrate out of Africa and to achieve a worldwide distribution. We do not know exactly how it spread, but it seems unlikely – as was already stated earlier – that a single regiment marched out of its base and left stable camps at various sites along the way which then evolved in complete isolation. It is far more likely that there was a continuous flow of emigrant groups out of Africa which remained on the move for most of the time. The mobility was associated with frequent extinctions of some groups and their replacement by others. Combined with the relatively small size of the groups, the instability may have resulted in the loss of genetic variation and so in the reduction of the overall effective population size. The emerging and expanding *Homo sapiens* population continued to undergo this birth-and-death process of local groups and the situation did not change until about 10,000 years ago with the advent of farming and the beginning of a fundamentally different life style. The change, however, occurred too recently to have any noticeable effect on the level of genetic variation in the expanding population.

Real bottlenecks in the history of the human population have been postulated by the proponents of both the multiregional and the uniregional hypotheses. In both cases, these purported bottlenecks of the founder effect type are surmised to have reduced the population to a very small, but otherwise unspecified number of individuals and then to have been followed by a rapid expansion to the present effective population size. The multiregionalists associate this bottleneck with the transition from *H. erectus* to the archaic *H. sapiens* 2 my ago, whereas the uniregionalists connect it either with the split of an ancestral species into *H. neanderthalensis* and *H. sapiens* some 600,000 years ago, or with the migration of *H. sapiens* out of Africa >100,000 years ago. The sole reason for this postulate by both the multiregionalists and uniregionalists seems to be the unfounded belief that speciation cannot take place without a bottleneck. For some uniregionalists, the postulate is based on their mistaken interpretation of the coalescence process (the false Eve concept).

In our analysis of the genetic information available, we find no evidence for a substantial founder-effect-type bottleneck in the history of the human species. Neither at the time of its emergence, nor at any time succeeding it, was the effective population size of the species reduced to just a few individuals. Had this happened, the event would have left traces in the character of the genetic variation, but no such echoes of past founder effects could be ascertained. Just as with cichlids, Darwin's finches, and myriads of other species, hominids, along with their sole surviving representative, appear to have emerged not through a genetic revolution, but through a peaceful transition involving fairly large founding populations.

Be Fruitful and Multiply

Earlier in this chapter we used the examples of Lake Victoria cichlid fishes and Darwin's finches of the Galápagos Islands to point out two fundamental processes of speciation – adaptation to one of the environment's most critical aspects, the food resource, and the complex of interactions associated with mate choice. Adaptation to an environmental niche is achieved by natural selection; mate choice operates via sexual selection. It can be surmised that both natural selection and sexual selection played an important part in hominid speciation.

What the human-chimpanzee ancestor ate, we cannot be sure of. We may not be too far off from the mark, however, by assuming that its feeding habits were similar to those of a chimpanzee. In the wild, the chimpanzee's diet consists of an assortment of fruits, leaves, blossoms, and buds, supplemented by ants, termites, birds' eggs, fish, and on occasion, monkey and pig meat. The first hominids, forced by climatic changes to come to terms with an environment that sustained progressively fewer trees, probably modified their feeding habits by adding grass seeds and other specialities to their menu that the expanding savanna had to offer. In the hominid lineage, the trend toward feasting on hardy vegetable stuff may have culminated in the robust australopithecines with their powerful chewing apparatuses. At the same time, however, a new direction in hominid evolution became progressively more apparent: some of the forms began to use their increasing brain capacity to plan the manufacture of primitive stone tools and use these to broaden their cuisine by including more meat on the menu. Although chimpanzees like to eat meat, they are not really adapted to its regular digestion. Their teeth do not enable them to chew it thoroughly and their gastrointestinal tract only partially digests it so that much of the meat passes through the system undigested. (The chimpanzees seem to be aware of this and often resort to "recycling" by re-eating the undigested pieces that passed through their guts.) Adaptation in their dentition and presumably also in the physiology of their digestive tracts must at some stage have equipped the hominids for meat consumption. This trend led to the appearance of the first hominids representing the genus *Homo* and culminated ultimately in active hunting with the help of an improved stone-tool kit. Hunting and gathering became the predominant mode of subsistence for *H. erectus*, as well as for all the other forms that followed.

Since these various forms all obtained their sustenance from the same food resources, they had to compete for them. Presumably, those forms that were more efficient in tapping the resources than the others had the upper hand. Greater efficiency and success in hunting and gathering were intrinsically associated with sophistication, advanced intelligence, and a larger brain. Ultimately, the more intelligent species outcompeted those less so, until only one was left, our own. According to this scenario, new hominid species emerged either by adapting to different niches with distinct food resources or by becoming more successful in the exploitation of the same resources than their competitors. The adaptations must have been accompanied by changes in morphology, physiology, and behavior, and these again by sexual selection, delineation of populations, and ultimately speciation.

The hunting and gathering mode of subsistence served the hominids well for most of the last 2 my of their existence. As long as their population densities remained relatively low and there was enough game to hunt, they had no reason to search for any radically new modes. After the last glaciation, however, the population densities, which had been rising slowly but steadily for some time, reached a point at which hunting and gathering could no longer feed all the people in certain regions. It was then that pressure developed to explore new ways of obtaining food. Planting crops purposefully to increase the efficiency and yield of gathering, and the domestication of certain animals to assure a durable supply of animal products was an obvious solution to the problem. The scheme was conceived independently in different geographical regions at different times and these regions then became the centers from which the new mode of food production – *agriculture* – spread. The centers arose in areas in which the conditions for the development of agriculture – a climate suitable for farming and the presence of wild plant and animal species appropriate for domestication – were optimal.

It was probably not by accident that one of the first, perhaps *the* first of these centers, arose in southwest Asia, in an area through which all the out-of-Africa emigrants must have passed on their way to more distant corners of the globe. Traffic through this area must have been heavy, the resources for hunting and gathering limited, and the pressure on finding alternative ways of sustenance high. It was here, in the region referred to as the *Fertile Crescent**, that conditions prevailed for combating the rising food shortage. The Fertile Crescent can be divided into two separate regions – Levant and Mesopotamia. Levant includes present-day Jordan, Israel, Lebanon, and Syria; Mesopotamia embraces present-day Iraq. In the Levant, a strip of coastal plain is followed in the easterly direction by a range of highlands and then by a deep valley which is part of the Great Rift running through the Dead Sea, the Red Sea and into East Africa. Beyond the valley is another region of highlands which slope down toward the Syrian Desert. In former times, the highlands were forested and the plains covered by steppe. Mesopotamia in the east encompasses a valley between the Tigris and Euphrates rivers, most of it covered by deposits of silt that have been left by floods over millennia. The climate of the region is *Mediterranean,* characterized by two seasons, a long, dry, hot summer, and a short, rainy, cooler winter. The climate, adequate water sources, fertile soil for growing crops, steppes for grazing animals, and the presence of wild plants and animals responsive to domestication – all these factors undoubtedly contributed to the development of agriculture in this region – plant domestication at around 8500 B.C. and animal breeding some 500 years later.

Other centers of crop and animal domestication emerged in similarly amiable regions in other parts of the world: in China 7500 B.C.; in Mesoamerica (central and southern Mexico together with adjacent areas of Central America) 3500 B.C.; in the Andes of South America and possibly in the Amazon Basin 3500 B.C.; and in the Eastern United States 2500 B.C. In these five regions (including southwest Asia), agriculture arose independently. In an additional four regions, a completely independent origin is less certain: Sahel (the southern flank of the Sahara desert that stretches across six countries from Senegal to Chad) 5000 B.C.; tropical West Africa 3000 B.C.; Ethiopia at an unknown date; and New Guinea, possibly as long as 7000 years ago. The assortment of plants and animals domesticated in the different centers varied according to the range of wild species locally available: wheat, pea, olive, sheep, and goat in southwest Asia; rice, millet, pig, and silkworm in China; corn, beans, squash, and turkey in Mesoamerica; potato, manioc, llama, and guinea pig in the Andes and Amazonia; sunflower and goosefoot in Eastern United States; sorghum, African rice, and guinea fowls in Sahel; African yams and oil palm in tropical West Africa; coffee and teff in Ethiopia; sugarcane and banana in New Guinea.

From these primary and secondary centers, agriculture spread in one of two ways, either by adoption from neighboring farmers, or by the replacement of local hunters and gatherers by invading agriculturists. The former way was apparently how agriculture reached, for example, the Atlantic coast of Europe, whereas the latter seems to have been the way of establishment of agriculture in most of the rest of western Europe 6000 to 3500 B.C. In areas in which agriculture was introduced secondarily, the local farmers often domesticated additional plants and animals. Examples include the domestication of poppy and oat in Western Europe; sesame, eggplant, and hamped cattle in Indus Valley at about 7000 B.C., as well as sycamore fig, chufa, donkey, and cat in Egypt some 6000 B.C.

Agriculture was such an efficient form of food production that as soon as it was developed or introduced, the farmers were able to harvest more food than required to feed their families. As a result, for the first time in the history of humankind, a division split the pop-

---

* Here crescent refers to the delineation of the area by a highland range that curves around the Syrian Desert like a half-moon.

ulation into those that produced food and supplied the excess to others, and those that lived on food produced by the farmers. The nonproducers were free to devote themselves to other activities – manufacturing ceramics, making jewelry, fighting battles and wars, administering, ruling, doing nothing at all, or – worst of all – complicating the life of other people by turning into bureaucrats.

Overproduction of food encouraged the people to settle and establish communities characterized by complex social interactions. Ultimately, it led to the various facets of what historians refer to as *civilization*. Above all, however, overproduction has led to a phenomenal, unprecedented population growth with all of its dire consequences.

In the mythology of some of the descendants of those who may have been the first to switch from hunting and gathering to full-time farming, God orders them: *Be fruitful and multiply; fill the earth and subdue it.* No other Biblical imperative has been so eagerly embraced by humankind. There are 6 billion of us today and our numbers are growing at a rate of nearly 100 million per year. One hundred million! That is more than one-third the population of the United States, more than the entire population of Germany, twice the population of France, and more than ten times the population of Sweden! It is a number that should stagger us and bring us to our senses. But the warnings and visions of poets go unheeded:

> *Have you noticed meanwhile the population explosion*
> *Of man on earth, the torrents of new-born babies, the*
> *bursting schools? Astonishing. It saps man's dignity.*
> *We used to be individuals, not populations.*
> *Perhaps we are now preparing for the great slaughter.*
> *No reason to be alarmed; stone-dead is dead;*
> *Breeding like rabbits we hasten to meet the day.* *

Breeding like rabbits? Perhaps we should now say *Rabbits breeding like humans,* for it is humans, not rabbits, who have steamrolled the planet.

---

* Robinson Jeffers: *Birth and Death*. From *The Beginning* & *the End and Other Poems*. Random House, New York.

CHAPTER 12

Who Are We?
Where Are We Going?

The Present Condition
and the Future of Our Species

Cassandra: . . . they were so blind to their own demise.

Aeschylus: *The Oresteia.*

Homo destructus

In the creation myths with which we opened this book, humans are portrayed as if they were a race of aliens that have arrived out of nowhere to be planted in the natural world, just as proud home-owners place figurines of gnomes in their gardens. Like the gnomes, human beings in myths and in many religions are an artifact, completely out of context with the rest of creation. The human-nature schism is especially pronounced in Judeo-Christian mythology. Adam and Eve have no genealogical relationship whatsoever to any of the creatures in the Garden of Eden, and upon their expulsion are commanded to subjugate the natural world. The first couple and their descendants interpreted the command as a license to enslave, and even eradicate, species as they deemed necessary. Prehistorical and historical records document how willingly they obeyed the command. The records show that whenever humans appeared in a region, a variety of species, especially the large ones, began to disappear. The giant kangaroo, ostrich-like flightless birds, and big reptiles in Australia and New Guinea 40,000 to 30,000 years ago; the wooly mammoth and the wooly rhinoceros in Eurasia 20,000 to 10,000 years ago; elephants, horses, and giant sloths in North America 14,000 years ago; on Mauritius the dodo; on Madagascar the giant lemurs; and on Hawaii the big flightless geese: they all bowed out as humans entered the stage. It is hard to believe that in all these instances, humans were mere bystanders, spectators to the extinction.

But that was just the beginning of the onslaught. The real assault on nature came with the advent of agriculture, when vast expanses of forested land were turned into fields, pastures, and meadows, and with the explosive expansion of the human population exacting ever more space for habitation, manufacture, transport, and leisure. As the development of human culture accelerated, so did the extent of the harm inflicted by humans on the environment. Today the damage is so great that *Homo destructus* would be a more appropriate name for our species.

And the assault continues on all fronts. Although a small fraction of our population has become alarmed and a somewhat greater percentage is at least aware of the problem, the concern is often directed at environments other than one's own. The "environmentally conscious" Europeans in particular like to decry the destruction of rainforests in the Tropics or point their finger at the wastefulness of North Americans, at the same time passionately resisting any environmental measures that would infringe on their convenience or comfort. Even small measures with hardly any economical impact at all, such as introducing a speed limit on German *Autobahns*, are apparently impossible to carry out politically. And try to tell Europeans to stay away from the Alps, which they are rapidly degrading by turning them into a vast amusement park!

Evolutionary biology puts our existence into an entirely different perspective and offers a new philosophy and ethics in regard to our relationship to the rest of the living world. Rather than presenting our species as the designated ruler of Nature's realm, empowered by a higher authority to usurp territories and pass death sentences on other species, it reveals our species to be the product of more than three billion years of evolution that has made us kin to all the other species with which we share this planet. At the DNA level, a mere 1.7 percent of nucleotide differences at silent sites distinguish us from the chimpanzee, 2.0 percent from the gorilla, 4.0 percent from the orangutan, 4.9 percent from the gibbon, 7.7 percent from the macaque, an Old World monkey, 13.1 percent from a New World monkey, and 27.7 percent from a lemur. Continuing in this manner, we can similarly document our kinship to nonprimate mammals, other vertebrates, and through other metazoans and other eukaryotes, all the way to the bacteria in our guts and to the viruses in the "bowels" of the bacteria.

Evolutionary biology reveals that in this community of living things our species does not enjoy any special status relative to the other members. It provides no grounds for elevating one species over all others or for assigning a greater evolutionary importance to any particular species. Although biologists like to speak of "higher" and "lower" or "advanced" and "primitive" species, such terms must not be understood as implying evolutionary inequality among organisms. The first two words simply refer to the fact that some branches split off from the trunk of the phylogenic tree earlier than others and are therefore at a lower position in the standard rendering of the tree. But just as the leaves on the lower branches of a real tree are equally important as those on the higher branches, so all extant species at the tips of the branches in the Tree of Life are coequal in terms of their evolutionary status. It is only by convention that phylogenetic trees are usually drawn with their roots down and crowns up. They can just as well be drawn "upside down", in which case "lower" becomes "higher" and vice versa. Only a person unfamiliar with the evolutionary history of organisms might use "lower" and "higher" in the sense of less and more important species. Anybody tempted to resort to such usage should remember that in reality, the "lower" forms have been in existence for a longer period of time than the "higher" ones and so, if anything, could be considered as more successful evolutionarily.

Similar arguments apply to the other pair of words: here again "primitive" does not imply inferiority and "advanced" superiority. At most, the word "primitive" can be used to indicate that a form lacks certain characters present in an "advanced" form and that in this sense the former is simpler than the latter. There is, however, no evolutionary basis for arguing that complex forms are worthier than simple ones. The continuing coexistence of simple and complex forms signifies that both are on a par evolutionarily. These arguments apply to all characters, including intelligence. Humans, in their unlimited sense of self-importance and arrogance, like to consider themselves as "special" because of their intelligence. In reality, however, intelligence is not restricted to humans and it is not a gift endowed on humans by a supernatural power. Evolutionary biology documents a stepwise growth of intelligence in different branches of the Tree of Life by the same processes responsible for the emergence and evolution of all other characters. Whether the higher dose of intelligence, which humans unquestionably do possess when compared to all other creatures, is such a positive feature in the long term remains to be seen. Thus far humans have used as well as misused it. The uses have propelled *H. sapiens* to the creation of art, science, and technology; the misuse has resulted in the perversion and abuse of these same three aspects of human culture. If we continue with misuse, in the end, it might be our intelligence that will bring on the downfall of our species.

Kinships among species and parity of species are two aspects of the changed perspective proffered by evolutionary biology. Another is the realization that all species, including *H. sapiens*, together with their environments form an integrated system, the *biosphere*.* One way to view the system is as a web or a network in which the knots are the species and the interconnecting cords represent interactions among the species. In this view, the disappearance of a species amounts to a tear in the network which affects other species through local suspension of the interactions, the extent of the influence depending on the size of

---

* The term was apparently first used in 1875 by the geologist Eduard Suess in his treatise on the Alps. The concept of "life's envelope" was then developed in the second decade of the twentieth century by Vladimir Vernadsky. *Biosphere* refers to the totality of all organisms on earth. It is, however, often used as a synonym for *ecosphere* which is that part of the earth inhabited by living organisms. It includes the upper layers of the earth's crust (*lithosphere*), the aqueous envelope of the earth (*hydrosphere*), and the mass of air surrounding the earth (*atmosphere*).

the tear. Small tears are repaired by relatively minor readjustments of interactions, where-as large ones may have a more profound effect on the existence of other species, but also on the overall environment. Biologists are by no means united in their interpretation of the web's vulnerability. At one extreme is the almost mystical view of the entire biosphere constituting a highly organized superorganism in which species, playing the part of cells in a body, are highly integrated and their disappearance is tantamount to bodily injury. At the other extreme is the view that the web is nothing more than a collection of relatively independently existing species, and that it is fairly resilient to intervention in the sense that the loss of one species has little influence on the well-being of the rest of the community. Regardless of one's point of view, there is no denying the fact that the human species is harming the biosphere irreparably.

The Four Horsemen of the Apocalypse

In the first century A.D., an exile on the Aegean Island of Patmos stirred people by the image of God opening four of the seven seals on the scroll in His right hand and thus unleashing the Four Horsemen of the Apocalypse on humankind, the four harbingers of doomsday (Fig. 12.1). Now, 2000 years later, those willing to sense it can feel the ground trembling under the gallop of approaching equine hoofs. And those willing to see can recognize four silhouettes looming up on the horizon, four men on horseback, the bearers of bad tidings.

The first rider, saddled on a white steed, could be called "Habitat Destruction" through population growth, the unrestrained, cancerous sprawl of our species over the globe. There may be other globally-distributed species with census population sizes that go into billions of individuals, but none usurps as much territory for its exclusive use as humans. Not only because of their relatively large size, but also because of their culture, do humans expropriate large amounts of space which they use for their dwellings, their food production, the manufacture of cultural artifacts, transportation, communication, travel, and for recreation. Other species may claim large territories for themselves, too, but they share these areas with untold numbers of other species. They are part of natural ecological systems or *ecosystems* in which different species interact with one another, depend on each other, and also interact with the nonliving part of their environment. Not so the human species. It creates artificial habitats in which it high-handedly tolerates a few species only and excludes most others from gaining entrance. Humans divide species into two categories: desirable (i.e., directly or indirectly serving their needs) and undesirable (pests, weeds). They sanction the former and persecute the latter. In the developed countries (Europe, United States, and Japan, in particular), not a single patch of the original, undisturbed natural habitat remains in the form in which it may have existed before the introduction of agriculture. All land has been turned into an artificial, "humanized" habitat, large parts of which are plastered with concrete and asphalt; it has become a desert, but with incomparably less species diversity than any natural desert. Much of the rest has been turned into fields, meadows, pastures, and timberland, all with greatly depauperated species diversity. Nature has been supplanted by parks operated by humans who have their own ideas of what they should look like and who should inhabit them. The rest of the world follows suit. If the trend continues, there will soon be no undisturbed natural habitats left anywhere on this planet. Orwell springs to mind with a futuristic scenario in which even the parks are ultimately replaced by fully artificial entertainment areas: Disneyland with mechanized maquettes faithfully imitating trees, grass, and animals.

Fig. 12.1
Albrecht Dürer: *The Four Horsemen of the Apocalypse*. 1498. (From Waetzoldt 1938.)

In the second rider, mounted on a red horse, we recognize "Mass Extinction" of life forms and depauperation of *biodiversity*.* In the course of evolution, species are born, persist for a while, and then exit the scene, either because they are unable to adapt to changing environmental conditions, because of a loss of habitat, because of the appear-

---

* Biological diversity or biodiversity of an environment is indicated by the numbers of different species inhabiting it.

ance of new predators, or because they turn into new species. Extinction is therefore part of their normal life cycle. Ordinarily, the extinction of species is spread out evenly along the arrow of time and is balanced out by the emergence of new species, so that the overall biodiversity remains more or less the same. From time to time, however, the extinction of species increases dramatically to assume global proportions and a period of reduced bio-diversity is inaugurated. The extent of the reduction may vary, but when it becomes partic-ularly drastic, paleontologists speak of *mass extinctions*. Paleontologists have identified at least half a dozen mass extinctions in the history of life and have used them to partition the geological time scale into major periods (see Chapter Nine). Particularly severe mass extinctions occurred at the end of the Permian period 245 my ago and at the close of the Cretaceous period 65 my ago. The Permian extinction is estimated to have wiped out 90 percent of all species existing at that time and to have ushered in the Age of the Reptiles. The popularly better-known Cretaceous extinction led to the demise of the dinosaurs, ple-siosaurs, ichthyosaurs, and ammonites and marked the beginning of the Age of Mammals. The causes of these and other past extinctions are largely unknown, but probably multiple. In some of the mass extinctions (in particular the Cretaceous), the impact of a heavenly body on earth seems to have been a contributing factor. Each of the past mass extinctions was followed by a recovery phase in which a large number of new species arose and the pre-extinction level of biodiversity was gradually restored, albeit only after many million years.

Current analyses indicate that we are now at the beginning of a new mass extinction which, if it continues at its present rate, may prove to be the most severe of them all. And this time we can point the finger at the cause unambiguously and unequivocally – our-selves. Evidence of the great dying is all-pervading. The elderly among us will seek many of the species they remember from their younger years in vain, even in places in which they were once found in abundance. By far the most dramatic signs of impoverishment of bio-diversity are apparent in the tropical rain forests. It is the rain forest which harbors most of the animal and plant diversity on this planet and so its loss is a major factor in the current mass extinction. It is often pointed out that a single tree in the rain forest is inhabited by more insect species than are found in the whole of Europe and that in a small patch of rain forest there is a greater diversity of trees than in the entire area of North America. Yet, in the few minutes it took you to read this paragraph more than 100 acres of rain forest have been destroyed by human activity. And with each acre of forest gone there is this much less living space for a multitude of species and consequently a corresponding drop in global biodiversity. It has been estimated that since the appearance of *Homo sapiens*, the rate of extinction of other species has increased by 1000 percent!

The name of the third dark rider is "Pollution". He rides a horse as black as *lutum*, the mud and dirt from which his name is derived, for "pollution" stands for all the various ways of making the environment dirty and harmful to its inhabitants. In this regard, *Homo sapiens* must be singled out as the dirtiest of all species on this planet. Other species, in-cluding those that live in large, crowded colonies (e.g., bees or ants) keep their environ-ment meticulously clean and free of their own wastes. Only humans poison water with sewage, pesticides, fertilizers, and spills; foul air with fuel exhausts, smoke, and industrial chemicals; and contaminate soil with mercury, lead, acid rain*, and radioactive materials. Other species produce biodegradable wastes which are broken down into recyclable com-

---

* Sulfur dioxide and nitrogen oxides produced by burning coal, gasoline, and oil react with water vapor in the air to give rise to sulfuric and nitric acids. The resulting acid rain kills fish and other aquatic life when it enters the water system and destroys trees when it pours down on the forest soil.

ponents by various organisms. Only humans produce and release substances into the environment that no organism would ever touch and if it did, would be poisoned by them. Only when they realized that they were also poisoning themselves did humans begin to express concern about pollution. And even then the concern was limited largely to protecting themselves, and this in certain parts of the globe only. In other parts, fouling and poisoning of the environment continues unabated. In the history of *Homo sapiens,* the simple equation has always been: more people = more waste = more damage to the environment. The extent to which the biosphere is being poisoned by pollution is increasing so dramatically that cynics among the "wise men" suggest that when the time comes, humankind should simply leave all the garbage behind and move on to another planet!

The fourth horseman, riding a pale mare, may answer to the name "Climatic Change". Here again, vacillation of climate is common in earth's history. Geologists can document numerous dramatic changes in climatic conditions by showing, for example, that tropical forests once thrived in regions which are now covered by kilometer-thick layers of ice. It is thus to be expected that environmental conditions will change again in the future owing to the geological processes responsible for the evolution of our planet. Such changes, however, are slow and gradual; they extend over thousands and millions of years, giving the biosphere sufficient time to adjust to them. They are accommodated into the normal turnover of species, some of the forms adapting to them and evolving into new species in the process, and others – those incapable of adapting – bowing out. Now, however, humans appear to be changing the global climate by their activities so fast that organisms cannot be expected to adapt to the new conditions. Organic evolution is simply too slow to cope with this particular environmental crisis in any way other than extinction.

Two phenomena are usually cited as major factors influencing climate change – damage to the ozone layer and the greenhouse effect. Ozone is a form of oxygen consisting of three (O_3) instead of two (O_2) atoms of molecular oxygen. It arises in the upper layers of the atmosphere by the action of high-energy sunrays which split the molecular oxygen into its atoms; these then combine with O_2 to form O_3. The ozone layer serves as a shield that stops most of the sun's ultraviolet rays from reaching the earth's surface by absorbing them. Although the thickness of the ozone layer fluctuates from year to year, during the year, and from place to place, an abnormal thinning of the layer over the Antarctic has been observed since 1980. Depletion of the ozone layer, dubbed the *ozone hole*, is greatest during the winter when the region sinks into darkness and lowest during the summer when sunrays regenerate some of the lost ozone. There is much less shrinkage of the ozone layer over the arctic region and less still over the rest of the globe. Ozone depletion in the upper atmosphere has been linked to the release of industrially produced chlorofluorocarbons (CFCs), as well as other chemicals into the air. CFCs are chemical compounds otherwise absent in nature that are used as propellants in sprays and coolants in air conditioners and refrigerators. Upon reaching the upper atmosphere, the released and otherwise unreactive CFCs are broken down into their constituents by ultraviolet radiation and the chlorine atoms react with ozone molecules, converting them into ordinary oxygen. Because of the climatic conditions existing in the upper atmosphere (air movement, temperature), CFCs are more concentrated over the Antarctic, where their degradation then generates the ozone hole. The consequence of ozone depletion is the bombardment of the earth's surface with high doses of ultraviolet radiation leading to DNA damage and mutations and so to the development of cancer and other pathological conditions.

The state of the ozone layer influences climatic (i.e., weather) conditions only indirectly, for example in connection with the *greenhouse effect*. The latter owes its name to the principle on which greenhouses operate: sunrays penetrate the glass to raise the temperature inside, but the heat then becomes trapped under the glass. In the earth's atmosphere, an increased level of carbon dioxide may act like the glass of the greenhouse: it allows the sunrays to warm the earth's surface but prevents the heat from escaping into space. In recent years, the level of carbon dioxide in the atmosphere has been rising steadily from burning fossil fuels (oil, gasoline, and coal) and from burning expanses of the tropical rain forests. It is estimated that about 5.5 billion tons of carbon dioxide are currently released into the atmosphere every year. Some of the gas is absorbed by the oceans and by the forests to be converted into oxygen during photosynthesis, but the rest accumulates in the atmosphere and may indeed be responsible for some of the recent changes in the global climate. If the current trend in the accumulation of carbon dioxide continues, it is predicted that in the next few decades the average temperature may rise by $3°$-$9°C$ worldwide. Should this happen, much of the polar ice can be expected to melt, the ocean levels would then rise, and droughts, floods, and severe storms such as hurricanes would increase both in frequency and intensity. The wide-ranging disruption of ecosystems accompanied by mass extinction could be the consequence. Since the greenhouse effect warms the earth's surface but cools the upper atmosphere, and lower temperatures favor ozone depletion, it also provides conditions for the enlargement of the ozone hole.

The four horsemen ride in concert: the actions of one are linked to those of the other three with the result that their effects amplify one another. Pollution leads to the extermination of species and causes ecosystems to collapse. Reduction of biodiversity translates into serious damage to the biosphere. Pollution also alters climate, and climatic changes affect the ecosystem and so the biosphere. Changes in biosphere, in turn, influence local and global climate...the effects are all part of one monstrous, vicious circle. Some of the consequences of the concerted assault on the biosphere are predictable, others are unforeseeable, and it is not clear which we should fear more.

Après nous le déluge?

The assault has an impact on the entire biosphere, which includes humans. But humans, incurable anthropocentrists that they are, worry chiefly about consequences for themselves and much less about the broader context of their behavior, although the fate of the biosphere will clearly also determine the fate of our species. What can be expected to happen to the biosphere if the four horsemen are allowed to ride on, if the global human sprawl, the mass extinctions, the devastation of the environment, and the meddling with the global climate system continue at the present rate or even accelerate in the future?

Even when they take the anthropocentric stand and ignore concerns about the biosphere as a whole, ecologists can still find plenty of reasons for sounding the alarm. The most indisputable consequence of the assault, albeit not one that gives a businessperson any sleepless nights, is the loss of knowledge. Like Gauguin, whose musings were the starting point of this narrative, most people are curious about the origin of our species. The preceding chapters should have convinced you that the search for our origin does not begin at the point at which the path leading to *Homo sapiens* separated from that terminating in the chimpanzee; rather, the search has taken us all the way down to the roots of the Tree of Life. The chapters should also have conveyed the message that the study of fossils is one way of learning about our origins, the other being the study of contemporary species, in particular the molecular archives of each species. Viewed from this perspective, it should be apparent that the loss of any species amounts to perdition of an extremely valuable cache of documents. The conflagration of tropical forests, combined with all the other practices that result in the destruction of ecosystems and the mass disappearance of species, are comparable to setting national archives on fire. Worse than that, national archives only contain a record of the past few centuries, whereas the molecular archives borne by each species are priceless chronicles relating more than three billion years of evolution. An arsonist who sets the national archives on fire would be judged as criminal; by allowing the conflagration of species diversity, we too will be decreed larcenists by future generations.

The businessperson who is not swayed by ethical arguments may be receptive to more pragmatic reasoning, based on the realization that the species diversity represents unmined reserves for food production, medicine, and industry. Our food resources are based primarily on species selected for domestication some 10,000 years ago, at a time when the needs of the human population were different from those of the present or the future populations. The well-fed citizen of a developed country may not feel the need to explore the utility of new crop species because those we utilize at present serve us well, but the inhabitant of an underdeveloped country in a different climatic zone may see things differently. Also, if the global climate changes in the future, as it indubitably will, the existing domesticated species of plants and animals may no longer serve humanity adequately. It will then be necessary to modify them or replace them altogether. Biologists estimate that among the 235,000 species of flowering plants still in existence, there may be some 30,000 which have edible parts and hence could potentially provide a source of nutrition. Among these are some that, if domesticated, would surpass, in some regions at least, those currently in use. The numbers are probably lower for animal species, but still substantial. All these untapped resources are being lost by the despoliation of the physical environment and the extinction of species that follows on its heels. For now we may be able to get by with the few domesticated species in wide use, but when the time comes and the need arises for new species to fall back on, they may be long gone.

Ecologists also point out that some 40 percent of all prescription drugs dispensed by pharmacies are derived from living organisms and that the design of most of the synthetic pharmaceuticals was based on the knowledge of their natural relatives. Aspirin is a classical example of the latter category. Salicylic acid, its basic constituent, was originally discovered in meadowsweet, *Filipendula ulmaria*, and was then combined with acetic acid to produce acetylsalicylic acid, a more effective painkiller. Inestimable numbers of potential pharmaceuticals exist in the known species, as well as in those still waiting to be identified, as continuing discoveries of new drugs extracted from living organisms attest.

A wealth of dyes, textiles, fabrics, and other substances suitable for industrial exploitation can be expected to exist in nature, only to be lost in the cataclysm that humans are bringing upon the biosphere. This mass extinction will also wipe out much of the cache of genes for which the burgeoning biotechnological industry would otherwise have undoubtedly found many applications – genes for resistance to diseases, for example.

All these are valid arguments. We fear, however, that they carry little weight with those merely interested in accumulation of wealth and personal gain. It is like telling Madame Pompadour to curb her wastefulness because among those dying of hunger in Parisian streets there might be another Voltaire or Lavoisier. Madame obviously did not care in the slightest about the fate of the people in the street and even less about the future. But she should have, for if she had not had the good fortune to die in time, she would have found herself on the street and under the guillotine. It is the task of the ecologists to convince the top managers who have no thought for the environment that the street and the guillotine are nearer than they think. If their wastefulness, extravagancy, and greed continues, bringing with it devastation of the environment, a total collapse of the biosphere will be the result. Humankind will then find itself up to its eyes in its own filth. The accumulating poisons will strike back with a vengeance. Ills, diseases, and genetic defects will become widespread as a result of the action of noxious substances in water systems, soil, and atmosphere. The East Coast of the United States, including most of Florida, and the west coast of Europe, including the whole of the Netherlands, will disappear under water. Agriculture will collapse, either because fields will remain flooded for months on end or will be scorched by protracted periods of drought...

There will, of course, always be skeptics who remain unconvinced even in the face of these arguments. Some of them will adopt the *après nous le déluge* (after us the deluge) attitude of the mistress of Louis XV. Others will attempt to equivocate or muddle the issue by citing arguments such as: What is more important, saving an obscure owl species or a person's job? Or: Who can blame a Peruvian farmer for clearing a patch of rainforest to feed his family? (The deceitfulness of this argument is that the logger's job will be lost anyway when the forest is gone and that it is not just the owl that the logging endangers but a whole unique ecosystem. Similarly, in a generation or two, the families of the farmer's descendants will go hungry when the fertile, but thin topsoil produced by the rainforest is exhausted and eroded away. In both situations, alternative and more permanent solutions to the predicaments of the loggers and the farmers have to be sought.) Others still like to lock horns with environmentalists on the issue of global warming for which, they insist, the evidence is unconvincing and therefore provides no reason to be alarmed. (In reality, the rise in global temperature and in the content of carbon dioxide in the atmosphere are undeniable facts, while the causal link between these two observations is very likely. The point is, however, that if we wait until the causality is established with certainty, it will be too late for an implementation of measures that might reverse the trend.) Finally, there are those who subscribe to the Biblical ethics of human dominance over nature. They insist that humans have a God-given right to manipulate the biosphere, to harm it if necessary, if that serves their needs and purpose. All the challenges ensuing from the destruction, including climatic changes, will be met, or so they believe, by modern science and technology. (How wrong this ethic is, we pointed out earlier. As for the notion that science and technology will fix all the rising problems, it suffices to remind ourselves how powerless our culture is in the face of truly difficult challenges. For over a century now, scientists have been trying to cope with the problem of cancer – largely unsuccessfully. Even comparatively straightforward virus infections, including that of the human immunodeficiency virus, are still raging, and infections once thought to be under control are again becoming a major health-threat because their causative agents have evolved to become drug-resistant. Similarly, most of the so-called "earth systems engineering and management" or "geoengineering" efforts intended to prevent geological and climatic catastrophes have led to disastrous consequences instead. The impotency of technology in the face of an earthquake, a hurricane, a flood or a drought is all too obvious to viewers of television news reels.)

The human threat to the biosphere is the first of the two major issues *H. sapiens* has to resolve, and resolve fast, if it is to have a future. The first step toward resolution is to acknowledge that we do indeed have one of the gravest problems on our hands since history began and to realize that its solution will require substantial changes in our thinking and in our lifestyles. Turning now to the second issue, we have to return to the question of human diversity and deal with its broader implications.

Patriotism, Nationalism, Racism

The species, as the only biologically definable category, provides an asymmetrically drawn dividing line in biological classification. Most of the other categories (genus, family, order, etc) are positioned above the species level, while only a few are in the sub-species level. The latter, which include variety, subspecies, and race, are poorly defined and ambiguous. Any deviation from the *holotype*, the specimen on which the description of a new species is based, is referred to as *variety*, even when the deviation is in a single morphological character. A *subspecies* is a population or a group of phenotypically similar populations inhabiting a geographically defined region and differing from other populations of the same species in diagnostic characters. *Race* is used by taxonomists either as a synonym of subspecies or as a designation of a local population within a subspecies. Different variants, subspecies, or races of the same species are either known or expected to interbreed if given the opportunity.

Varieties, subspecies, and races have been described for many species, including primates. In the gorilla, for example, three subspecies (races) are commonly recognized, the western and eastern lowland, and the mountain gorillas. The western lowland gorilla (*Gorilla gorilla gorilla**) is restricted to the forests of western Africa, specifically Nigeria south of the Congo River, and is distinguished by its brown or gray coloration and by the extension of the whitish "saddle" down to rump and thighs. The eastern lowland gorilla (*G. gorilla graueri*) inhabits the lowlands of eastern Zaire, is black in color, and its white saddle is restricted to the short hair on the back. The mountain gorilla (*G. gorilla beringei*) lives in the upland regions of Rwanda, the Virunga Mountains of Zaire, and the mountain forests of Uganda at altitudes up to almost 4000 meters. It resembles the eastern lowland gorilla, but has longer hair, especially on the arms. The three subspecies also differ in other characters, for example in the size of the jaws and teeth, as well as the shape of the face and the body.

All this is biological reality which raises few emotions. Taxonomists may disagree on the number, delineation, naming, indeed on the very existence of the subdivisions in a particular species, but other than that they find nothing objectionable about the notion of species consisting of populations between which gene flow has been reduced, because of the geographical distance between them, for example. When it comes to *Homo sapiens*, however, the feathers fly. Biologically, *H. sapiens* is a species like any other and as such it might also be expected to be differentiated into subspecies, especially since its global distribution creates opportunities for adaptation to different climatic conditions and so for morphological divergence. But the existence of human races is a highly charged issue and the reason for this is a streak in human nature which we refer to as intolerance.

The roots of the streak are probably biological. Primates, like many other animals, are territorial creatures, each individual or group of individuals claiming a certain area for its activities and defending it against potential competitors of the same species. In the human species, territoriality has acquired a cultural aspect in addition to its initial biological function (monopolization of resources and mates). Humans organize themselves into communities which differ in their origin, traditions and customs, language, religion, behavior, social structure, and political views. The communities often occupy politically delineated regions from which inhabitants of other communities tend to be excluded. Members of a community are expected to show devotion and loyalty to the group and this

---

* In the binomial scientific nomenclature the first name indicates the genus and the second name the species. The subspecies of the same species are distinguished by adding a third name, even if in some cases it is the same as that of the genus and of the species.

allegiance is often fostered by a variety of means ranging from education, through indoctrination, to outright brainwashing. The most widespread fragmentation of the contemporary human society is into *nations* inhabiting a defined territory and possessing a central government. The word derives from Latin *nasci*, to be born, which reflects the origin of many nations from individuals united by genealogical ties – families expanding into clans, tribes, and tribal conglomerates. The territory occupied by a nation, the *country*, is often transformed into an abstract symbol, *patria*, the land of the fathers, which then becomes the subject of reverence referred to as *patriotism*. Children are taught at school to love their country, soldiers are asked to die for their country if necessary, politicians have the interests of their country continually on their lips, and journalists center their reporting on the politics, economy, culture, science, and sports of their own country. Patriotism assumes a range of forms, but implicit (and sometimes explicit) in them is the belief that one's own nation is better than all others. From this belief it is only a small step to the conviction that the people of other nations are inferior, less worthy than people of one's own country, and to xenophobia, fear of, and hostility toward foreigners. A particularly obnoxious form of patriotism is referred to as *nationalism* (the exaltation of one nation over all others) and *chauvinism* (undue partiality to a subject, like the blind, bellicose devotion to Napoleon Bonaparte by the French soldier Nicholas Chauvin).

Originally, no distinction was made between the terms "nation" and "race"; they were used interchangeably (and sometimes still are), but later the former came to designate a country under one government and the latter a group of people of common descent.* For some, however, both terms came to be associated with the notion of superiority of one nation or one race over all other nations or races. The former is usually associated with cultural and social superiority, whereas the latter implies biological (and usually also cultural) preeminence. The belief in the supremacy of one race over others is called *racism*, although some left-wing activists extend the usage of this word to the belief that races exist as biological entities. The two interpretations must, however, be separated from each other, for if human races do exist, it certainly does not automatically indicate that one of them is preeminent. It is in the sense of alleged superiority of one race that the term "racism" is used in this text.

What, if Anything, is a Race?

Is the human population subdivided into races? The answer depends on one's understanding of this term and people speaking of race should therefore first specify what exactly the word means to them. The original division of the human species into races was based on visible, conspicuous differences among individuals. European travelers to Africa noted that people on this continent tended to have dark brown or black skin, kinky black hair, brown eyes, a broad nose, and thick lips. Similarly, explorers returning from Central and East Asia reported that the inhabitants of this part of the world had yellowish skin, coarse, straight hair, and a fold of skin extending from the eyelids across the inner corner of the

---

* The term *nation* is probably the older of the two, as it is derived from Latin and was imported into English via Old French. Its first documented use in English is in the verse romance *Kyng Alisander* composed by an unknown author before 1300 and describing the life and deeds of Alexander the Great. The term race, in the sense "individuals of a common descent", is of uncertain origin; it has been traced back to Italian *razza*, used in 1580, which became French rasse, and from there was taken up in English. The word "race" in the sense of a competition is of independent origin and is derived from Old Icelandic.

Fig. 12.2A-C
Differences in eye morphology among Europeans, Asians, and Blacks. **A.** European female. **B.** Asian male. **C.** African male (all three approximately 20 years old). In **B**, note in particular the fold of skin extending downward from the surface of the upper eyelid to the root of the nose and covering the inner corner of the eye – the *epicanthic fold*. (From Williams & Warwick 1980.)

Fig. 12.3A-C
Differences in head morphology of an African (**A**), Asian (Japanese, **B**), and European (Italian, **C**) as depicted by Peter Paul Rubens (*Studies of Black Person's Head;* Brussels, Musées Royaux des Beaux-Arts), Kitagawa Utamaro (*Deeply Concealed Love;* Tokyo National Museum), and Sandro Botticelli (*Portrait of a Youth;* National Gallery Washington).

eye (a feature called the *epicanthic fold*; Fig. 12.2). These characteristics contrasted with those of the Europeans who tended to have fair skin, light-colored hair, either straight or wavy, blue eyes, a narrow nose, and fairly thin lips (Fig. 12.3). These three types came to be called the Negroid, Mongoloid, and Caucasoid* races, but we shall refer to them as Africans, Asians, and Europeans, respectively. Later, the list of races was extended to include peoples of other continents, subcontinents, and island groups, specifically American Indians, Australians, Indians, and Oceanians (Melanesians, Micronesians, Polynesians). Closer study, however, revealed a considerable variation within the major races in the

---

* The designation "Caucasian" can be traced back to the founder of physical anthropology, Johann Friedrich Blumenbach, who first used it in his doctoral thesis *De generis humani varietate nativa liber* (Of natural human varieties), published in Göttingen, Germany in 1775. In choosing the name, Blumenbach was motivated by the belief that some of the most paradigmatic representatives of the European race lived in the region around the Caucasus Mountains.

characters used initially for their delineation. Thus, for example, 36 shades of skin pigmentation have been described, ranging from jet black to almost white. Using these, as well as other characters, additional races – up to several hundred – were described and anthropologists were neither able to agree on the most suitable classification system, nor on how detailed the subdivision should be.

As the number of races grew, so did the skepticism. The large number of races, the existence of intermediates, and the hierarchical grouping of clusters of individuals suggested to some anthropologists that the characters on which the classification was based varied continually and hence that any partitioning into discrete groups was arbitrary and without biological significance. It appeared that any two groups delineated by one or more physical characters could be linked up by a series of intermediates providing a smooth, uninterrupted transition from one group to the other. The continuous line seemed to stretch from family units all the way to the global population. Moreover, the classification based on one visible character often did not match that based on other characters. From all these observations, some anthropologists concluded that it was an exercise in futility to classify humans into races and that it would be best to abandon the whole concept of human race. This conclusion also appeared substantiated by the misuse and abuse of the concept in some countries for the purpose of discrimination, oppression, and even genocide.

The proposal to scrap the concept of race altogether is currently only one extreme in a range of views. It is certainly not shared by all anthropologists and is by no means the majority opinion of the public at large. It appears to be a conclusion reached more on the basis of political and philosophical creeds than on scientific arguments. Correspondingly, anthropologists who do hold this opinion often attempt to shout down their opponents rather than convince them by presentation of facts. Their favored method of argumentation is to label anybody who disagrees with them as racist. The public, however, seems unimpressed by their rhetoric. It refuses to believe that the differences they see are a mere figment of their imagination. A lay person can tell with a high degree of accuracy where individuals come from just by glimpsing their features, as the following quote attests to:

> "White man! Lookim this-feller line three-feller man. This-feller number-one he belong Buka Island, na 'nother-feller number-two he belong Makira Island, na this-feller number-three he belong Sikaiana Island. Yu no savvy? Yu no enough lookim straight? I think, eye-belong-yu he bugger-up finish?"
>
> No, damn it, my eyes-belong-me were not ruined beyond repair. It was my first visit to the Solomon islands in the Southwest Pacific, and I told my scornful guide through the medium of pidgin English that I saw perfectly well the differences between those three men in a row over there. The first one had jet-black skin and frizzy hair, the second had much lighter skin and frizzy hair, and the third had straighter hair and more slanty eyes. The only thing the matter with me was that I had no experience of what people from each particular Solomon island looked like. By the end of my first trip through the Solomons, I too could match people to their islands by their skin and hair and eyes.*

Except for some anthropologists, everybody else seems to be able to distinguish people from different parts of the world at a glance by their outward appearance. This, apparently, is also the view of some governmental administrators in countries with programs designed to fight racial discrimination. Obviously, there is a credibility gap on this issue be-

---

*J. Diamond: *The Rise and Fall of the Third Chimpanzee*, p.95, Radius, London, 1991.

tween some anthropologists on one side and the public, as well as the governments of some countries, on the other.

One way to settle the arguments among anthropologists and to reconcile anthropologists with the public might be to move away from physical characters and focus on the genes. If races are real, they should have a genetic basis separable from environmental and cultural influences. Genetics might help to resolve the issue in two ways. First, the physical characters used in the classification of races are controlled by genes which could therefore be used to determine whether there is discreteness at the genotypic level where there appears to be continuity at the phenotypic level. The differences should then reduce to the presence or absence of genes responsible for the particular character and so provide a quantitative measure of racial differentiation. Unfortunately, none of the genes controlling skin color, hair color and texture, or lip and nose shape have been identified. These characters are determined by multiple, interacting genes, so their identification is not easy. But in the near future, the genes will undoubtedly become known and it will then be possible to establish whether there is a correspondence between their distribution and any of the classification schemes that anthropologists have designed for the human species.

There are in essence two mechanisms by which physical differences among the human population may have arisen – natural selection or sexual selection. The former is the traditional explanation based on the argument that each of the physical characters differentiating groups of individuals represents an adaptation to the environment in which these individuals live. Take skin color as an example. There is a certain, though by no means perfect, correlation between the darkness of people's skin and the amount of sunlight to which they are exposed (Fig. 12.4). People living closest to the equator and so receiving the highest intensity of sunlight are more darkly pigmented and generally, the further from the equator, the lighter the skin. One possible explanation for this correlation is the link between the ultraviolet radiation of the sunlight and the synthesis of vitamin D in the human body. Vitamin D is needed for calcium fixation and bone growth, but if present in excess it leads to the calcification of arterial walls and to kidney disease. Humans synthesize vitamin D in the layer of skin beneath the main pigmented layer using precursors obtained from certain types of food in a reaction that requires the energy of ultraviolet radiation. In people living near the equator the heavily pigmented skin absorbs most of the ultraviolet radiation and thus prevents overproduction of vitamin D. By contrast, in people of the northerly latitudes, the paucity of the dark pigment in the deep layers of their skin allows the ultraviolet light to reach the sites of vitamin D synthesis and thus to protect against rickets caused by too little of the vitamin. Other explanations of the correlation between latitude and skin pigmentation include protection against sunburn and skin cancer, protection of internal organs against overheating, protection against sudden drops in temperature in tropical regions, protective coloration in the jungle provided by the dark skin, protection by pale skin against frost-bite, protection by dark skin against beryllium poisoning in the tropics, and avoidance of folic acid deficiency in tropical regions. None of these explanations is fully satisfactory, however, because none explains the many exceptions from the correlation. Satisfactory explanations are also not available for hair color and texture, eye color, and other external differences between human races.

Charles Darwin grappled with this problem over almost 900 pages of his book *The Descent of Man* (1871) and came to the conclusion that the explanation must not be sought in natural selection but in sexual preferences of different groups of people. He pointed out that we select our mates according to their skin color, eye color and shape, hair color and texture, breast size and shape, and other characters that distinguish human races. He further drew attention to the observation that the definition of the character state considered

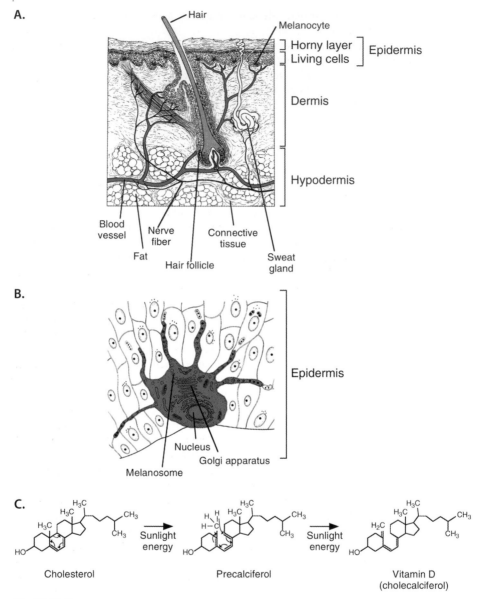

Fig. 12.4A-C
The connection between skin color, sunlight, and vitamin D. **A.** Structure of human skin (cross section). The uppermost layer of horny, dead cells, together with the lower level of living cells, forms the epidermis. In the deepest stratum of the epidermis are the pigmented cells (melanocytes). **B.** From their long, fingerlike extension, melanocytes pinch off granules (melanosomes) containing the pigment melanin which is responsible for the variation in skin color. Melanosomes are deposited around the nuclei inside the epithelial cells of the epidermis. **C.** Vitamin D is synthesized from cholesterol in a reaction that relies on the energy of the ultraviolet rays in the sunlight. The amount of vitamin D synthesized depends on the amount of ultraviolet light that manages to pass through the pigmented layer of the skin. (**A** modified from Spence & Mason 1987; **B** from Singer & Hilgard 1978.)

to be most beautiful varies among people in different parts of the world. Imprinting beauty standards in individuals growing up in a particular community leads to inbreeding among people with the same standards and so to racial differentiation. *Sexual selection* thus favors characters that may otherwise be useless for their bearers, their only function being to attract a mate. Darwin's explanation failed initially to convince most anthropologists, but more recently, after its validity has been demonstrated in many non-human species, it is being considered seriously either as an alternative to or in combination with natural selection as a theory for the origin of racial characters. The selection – natural, sexual, or both – can be expected to have "fixed" different race-specific genes in different races. Genetic studies could therefore resolve the question of whether races exist by demonstrating fixed genetic differences between morphologically defined races.

The second way in which genetics may contribute to the resolution of the question is by the analysis of genes that are not involved in the control of the physical differences between races. In the preceding chapters we described several examples of genetic variation within the human species. In addition to the systems mentioned there, numerous other variable systems are known, probably none of which has anything to do with human morphological diversity. Nevertheless, all these systems can be used to test the hypothesis of genetic racial differentiation. Provided that the races separated a long time ago, random genetic drift should have diversified their genetic composition even in the absence of selection. It can be expected that the longer ago the races diverged, the greater the differences between them will be. Even if there has not been enough time to "fix" different alleles in distinct races, at least differences in gene frequencies should have been generated. If, on the other hand, races are classification artifacts or totally arbitrary categories, this should be reflected in the pattern of genetic variation in the human species. An ideal way of testing these assumptions would be to determine the genetic variation of the entire human species and check whether it sorts out into groups that correspond to the morphologically or geographically defined races. Since, however, the testing involves gene frequencies and the latter can only be obtained by comparing individuals from predefined groups, this approach is not feasible. Instead, the starting point of the genetic test must be the groups defined by anthropologists as races. If the comparison of genetic variation within each race with that between races reveals no significant differences, the whole concept of the human species being genetically subdivided into groups would be seriously undermined.

As far as we can tell, the first geneticist to measure the genetic differences between human races was L.L. Cavalli-Sforza. In a study published in 1966, he summarized genetic distance data for 12 loci (most of them coding for blood group antigens) in seven major human populations and reached the conclusion that of the total genetic variation observed in the human species, less than 15 percent accounts for differences between races. Cavalli-Sforza apparently found this observation unremarkable, or at least unworthy of a comment. The geneticist who did make a big story out of it was Richard C. Lewontin in 1972. He divided the human species into seven races (Caucasians, Black Africans, Mongoloids, South Asian Aborigines, Amerinds, Oceanians, and Australian Aborigines), and each race into a series of populations (Caucasians into Arabs, Armenians, Basques, Bulgarians, Czechs, etc; Black Africans into Abyssinians, Bantu, Burundi, Batutsi, etc; Mongoloids into Ainu, Bhutanese, Bruneians, Buriats, Chinese, etc; etc) delineated by morphological, linguistic, historical, and cultural information. He then compiled gene frequency data for nine blood group systems, four serum proteins, and four soluble enzymes obtained by typing of the individual populations by immunological (antibody-based) and protein electrophoresis methods.

To quantitate genetic variation, Lewontin used an esoteric measure related to *heterozygosity*. Earlier, we defined heterozygosity as the presence of two different alleles at a locus in an individual. In population genetics, however, the words "homozygosity" and "heterozygosity" have acquired slightly different meanings which are best explained in terms of probabilities (see also Chapter Eleven). Consider a locus with two alleles, *a* and *b*, and frequencies p_a and p_b, respectively. We draw two genes from a gene pool at random and ask: what are the probabilities that the genes will be identical or different? The probability of choosing the *a* allele in the first draw is p_a (probability in this case being equal to frequency), as is the probability that *a* will also be chosen in the second draw. Hence the probability of ending up with two *a* alleles is $p_a p_a = p_a^2$. Similarly, the probability of ending up with two *b* alleles is p_b^2 and the total probability of obtaining two identical alleles (homozygosity, either *a* or *b*) is $p_a^2 + p_b^2$. The probability of choosing the *a* allele in the first draw and the *b* allele in the second is $p_a p_b$, so that the total probability of choosing two different alleles (heterozygosity) in the two draws is $p_a p_b + p_a p_b = 2p_a p_b$. More generally, if we designate any allele as *i* and there are *n* alleles at a locus in population *j*, then the homozygosity of population *j* is

$$H_j = \sum_{i=1}^{n} p_{ij}^2$$

and the heterozygosity of the same population is $h_j = 1 - H_j$ since $H_j + h_j = 1$. A low heterozygosity is indicative of high homogeneity (homozygosity) of population *j*, and vice versa. In this sense, heterozygosity is a measure of genetic variation of a population and can therefore be used not only to assess the genetic diversity within a species, but also to compare diversity within and between subdivisions of a species, if the latter is indeed genetically partitioned into populations and races.

The assessment consists of comparing genes sampled separately from individual populations (races) with those sampled randomly from the whole species. The main consequence of the subdivision is that individuals no longer mate randomly with each other within the species; they now tend to mate within each population. As a result, individuals within each population become increasingly more similar to one another genetically but more and more different from individuals of other populations, and the proportion of heterozygotes in the populations decreases through extinction of alleles. The probability that two genes sampled at random will be different is lower if the genes are taken from the same population than if they are taken at random from the entire population comprising the species. Working against these expectations is the gene flow, the exchange of genes between populations. Gene flow increases heterozygosity within populations and decreases it between populations by spreading alleles through the entire species and so homogenizing the species genetically.

Lewontin's study revealed that 85.4 percent of the genetic variation that exists within the human species is contained within the individual populations and that the remaining 14.6 percent is accounted for by differences between human groups. Of the 14.6 percent, 8.3 percent accounts for differences between populations, as defined by Lewontin, and 6.3 percent for differences between races. Of course, since Lewontin's distinction between populations and races is arbitrary, some anthropologists might want to call some or all of his "populations" races.

Lewontin's analysis was later repeated by several investigators using ever larger compilations of gene frequencies and different methods of assessing genetic variation within and between groups. Most of the measures of variation were related to the F_{ST} *statistic* which was introduced by Sewall Wright to determine the effect of fragmentation of a population on its genetic make-up. Wright called the measure *fixation index* (hence the letter

F) because it gauges the increase in fixation of alleles within a subpopulation as a result of the fragmentation. There are several ways of defining the F_{ST}, the most common of which is based on heterozygosity. In a population divided into *s* subpopulations, the average gene frequency of allele *i* ($\bar{p}_i$) is obtained by computing its frequencies in individual subpopulations, summing them up, and dividing the total by *s*, the number of subpopulations:

$$\bar{p}_i = \sum_{j=1}^{s} p_{ij} / s$$

The heterozygosity of the total population (h_T), obtained by sampling individuals of the entire population at random (i.e., disregarding its division into subpopulations), is given by

$$h_T = 1 - \sum_{i=1}^{n} (\bar{p}_i)^2$$

where *n* is the number of alleles at a locus. The h_T is contrasted with h_S, the average heterozygosity of *s* subpopulations, which is obtained by computing the heterozygosity of each subpopulation (h_j), summing up the individual values, and dividing the total by *s*:

$$h_S = h_s = \sum_{j=1}^{s} h_j / s$$

(It is from h_S and h_T that the F_{ST} symbol derives its two other letters.) The F_{ST} statistic is then defined as $F_{ST} = (h_T - h_S)/h_T$. It measures variation of gene frequencies in populations and as such can be rewritten for a single allele *i* as $F_{STi} = Var(p_i)/[\bar{p}_i(1 - \bar{p}_i)]$, where

$$Var(p_i) = \sum_{i=1}^{s} (p_i - \bar{p}_i)^2 / (s-1)$$

(Recall that variance, which measures the spread of distribution of a variable about the mean, is the square of the standard deviation; see Appendix One. Here the mean, $\bar{p}_i$, is indicated by placing a bar over the letter *p*.) Earlier we mentioned that genetic variation is influenced by two opposing processes: random genetic drift within a subpopulation and gene flow between subpopulations. When an equilibrium is reached between the processes, F_{ST} measures the number of migrants between the subpopulations.

Regardless of the method used in the analyses, all investigators reached estimates very close to that obtained by Lewontin: the differences observed between the subdivisions (populations, groups of populations, races) represented 10 to 15 percent of the total genetic variation found in the human species. Formally, these findings demonstrate, first, that the species is indeed subdivided into genetically definable groups of individuals and, second, that at least some of these groups correspond to those defined by anthropologists as races on the basis of physical characters. They do not, however, settle the arguments regarding the merits of the racial classification. Unfortunately, Lewontin did not specify before initiating his analysis how large the difference has to be in order to call the groups "races". Consequently, the results of the studies have led population geneticists to two diametrically opposite conclusions. Lewontin called the observed differences trivial and proclaimed that "racial classification is now seen to be of no genetic or taxonomic significance" so that "no justification can be offered for its continuance." This view is echoed by most authors of similar studies, who seem to be surprised that genetic variation within populations is greater than that between them. By contrast, Sewall Wright, who can hardly be taken for a dilettante in questions of population genetics, has stated emphatically that if differences of this magnitude were observed in any other species, the groups they distinguish would be called subspecies.

One can extend Wright's argument even further. The more than 200 species of hap-lochromine fishes in Lake Victoria differ from each other much less than the human races in their neutral genes, although they are presumably distinguished by genes that control differences in their external appearances. The same can be said about at least some of the currently recognized species of Darwin's finches and about other examples of recent adaptive radiations. In all these cases, reproductively isolated groups are impossible to tell apart by the methods used to measure differences between human races. Obviously, human races are not reproductively isolated (interracial marriages are common and the progenies of such marriages are fully fertile) but the external differences between them are comparable to those between the cichlid fishes and Darwin's finches. Under these circumstances, to claim that the genetic differences between the human races are trivial is more a political statement than a scientific argument. Trivial by what criterion? How much difference would Lewontin and those who side with him consider nontrivial?

By mixing science with politics, geneticists and anthropologists are committing the same infraction of which they are accusing other scientists, whom they themselves label as racist. Even worse, by dismissing the genetic differences as insignificant, they play into the hands of genuine racists who can easily demolish this claim and so further their own agenda. It is intellectually more honest to acknowledge the differences and then point out that they by no means imply supremacy of one race over others. This can be done by demonstrating that the differences are in genes that cannot be linked to any features that would be required for the preeminence of a particular race.

As for the notion of continuity in genetic variation across the entire human species, it must not be confused with the hierarchical structure of the variation. There are indeed different levels of grouping within *H. sapiens*, which are the result of population history, genealogical patterns, geography, cultural differentiation, and other factors. Smaller groups can be clustered into larger ones and these into larger ones still, as is apparent from any phylogenetic tree drawn for human populations. The differentiation and clustering is more pronounced for populations inhabiting Siberia, the Amazon basin, or other such areas in which the traditional organization of human societies has not yet disappeared completely. It is far less apparent in areas of large population expansion and mixing, in which the frequencies of some genes can be shown to form *clines* – relatively smooth gradients decreasing from a center of distribution to the periphery. But even in Europe, with its long tradition of intermarriages and easy opportunities for mixing, Guido Barbujani and Robert R. Sokal could uncover a patchy distribution of allele frequencies and zones of sharp changes in genetic variation, and attribute them to physical and cultural barriers to gene flow. The hierarchical nature of the groupings, of course, begs for an answer to the question: which of the groups should be called races? It could be: all, none, or any, according to one's preference. The name is not important. What is important is to acknowledge the existence of differentiation and its significance for the reconstruction of human history.

A related question is how old the differentiation may be. Here the answer depends on whether one is a multiregionalist or a uniregionalist and if the latter, to what version of the uniregional hypothesis one subscribes. Multiregionalists have no problem in explaining the 10 to 15 percent difference between the human groups. Since they assume that the differentiation began up to 2 my ago, when *H. erectus* established founding populations in the different regions, there has been sufficient time to accumulate the differences. Uniregionalists, who assume that the differentiation into groups began after the exodus of *H. sapiens* from Africa, are at a disadvantage, because calculations indicate that only under highly unrealistic assumptions (e.g., no gene flow between populations) would the time interval suffice for the origin of the observed differences. Henry C. Harpending and his colleagues therefore proposed in 1993 that the ancestral human population was already subdivided before its expansion from Africa so that the race differences are older than the date of the exodus. An alternative considered by Harpending and Alan R. Rogers is the possible colonization of new environments after the exodus took place by small bands which then expanded locally. The small sizes of the founding populations could account for the local losses of neutral genetic variation and thus for the relatively large differences between the populations.

Homo intolerans

Lest the reader may have been confused by the technicality of the arguments presented in the preceding section, the following "Ten Commandments" should make it clear where we stand:

1. All humans belong to the same species, *Homo sapiens.*
2. There are no savages, there are only different cultures. The real barbarians are those ignorant white men who are unable to fathom other races and who exploit them.
3. Different characters are the results of social and environmental factors.
4. There are no inferior races.
5. In each race there are individuals who are more gifted than others.
6. Human races are well adapted to the environment in which they live.
7. Mixing of language and physical characters occurred secondarily.
8. Differences in intelligence and morality between races are no greater than those between individuals of the same race.
9. Inter-racial hybrids are not inferior.
10. Exploitation and depreciation of races must stop.

In case you think that the Commandments have been formulated by a contemporary left-wing activist, you are wrong. They are taken from the work *Rassen, Völker und Sprachen* (Races, Nations, and Languages) published in 1922 in Berlin by the German anthropologist Felix von Luschan. They not only present a list of creeds that every decent person should be prepared to endorse, they also document that even in a nation tainted by the most despicable, racially-motivated crimes against humanity, there have been upright anthropologists. This fact acquires special significance in our own time, when in the same nation racism, xenophobia, and intolerance are on the rise again.

Racial intolerance is probably as old as races themselves, and it is as widely distributed now as it has ever been in the past. Is there any hope that it may be wiped out in the future? Based on recent developments, it can be expected that the existing genetic and hence also physical differences between races will gradually fade and ultimately disappear altogether. The trend toward homogenization of the human species is apparent in all but the most xenophobic countries. Isolated tribal groups are either dying out or are being integrated into larger populations. People are traveling and changing their countries of residence more than ever before. Many aspects of culture are being globalized and many language barriers are falling. Racial integration is actively promoted by governments in countries such as the United States. Inter-racial marriages are no longer a taboo in most countries and are on the rise all over the world.

All these developments intimate that at some time in the not too distant future a counterpart of Diamond's guide will no longer be able to tell the Oceanic island from which a person originates merely by physical appearance. Nor will anybody be able to identify the continent of a person's origin. When this happens, the problem of race and racism should disappear. The spectrum of variability in appearance may become large within any particular group, but since it will not be associated with origin or any cultural characteristic, it will not be possible to use it for discrimination.

Unfortunately, if this happens, it will not be the end of intolerance, of which racism is only one example. The bigots among us will have plenty of pretexts for hating others, religion and nationality being the most common. In Northern Ireland, all that distinguishes Protestants from Catholics is a minor difference in their religion, yet it is a sufficient pretext for the two groups to kill and maim each other. On the Balkan peninsula, Serbs and Croats, who are distinguished only by their slightly different histories, coexisted with each other as long as Tito held an iron grip over them, but as soon as the dictator died, they went to war. We could, in this manner, move around the globe and find examples of people everywhere who are at each other's throats because of differences in religion, customs, tradition, or simply because they have always been at each other's throats. Racism is only one manifestation of the intolerance that characterizes the human species, so much so that a fitting alternative to the designation *H. sapiens* could also be *Homo intolerans*.

Although the conflicts are local, there is always the potential that one of them will draw in other nations and escalate into a global war. And since the military means now available even to some small nations can easily lead to a real Armageddon, intolerance is a major threat to the survival and well-being of the human species. To eliminate the threat, it might be necessary to reconsider the concepts of patriotism and nationalism, and to make tolerance of other faiths the basic principle of all religions.

Finding a nondestructive relationship to the biosphere and eliminating intolerance are two of the three most pressing problems facing the human species. The third, which has a great impact on the first two, is finding a just social system for people to live in, a system that protects individuality and guarantees personal freedom, but stifles bigotry; one that ensures equality but promotes competition, a system without social evils such as poverty, unemployment, drug addiction, hunger, and grossly uneven distribution of wealth. None of the systems currently in existence comes anywhere near this ideal. Communism is, by Marx's definition, a form of dictatorship which greatly limits personal freedom, leads to horrible abuses of power, and drives economies into ruins. At the other extreme, capitalism of the ruthless type results in a grossly inequitable distribution of wealth, rampant consumerism, and class struggle. The middle road taken by many European countries discourages creativity, promotes idleness with consequent economical repercussions, and seems to beget various forms of intolerance. Designing a system that will solve all these problems is a major challenge for humanity; but an even greater challenge will be implementing the system once it has been found.

Toward Universal Mingling

Let us be optimistic for a moment and assume that *H. sapiens* does solve the three paramount problems facing it and also overcomes many other – by no means minor – exigencies such as acquiring new sources of energy after the exhaustion of coal, natural gas, and oil supplies. What will happen to the human species biologically, specifically in regard to its further evolution? Will it transform itself into a new species?

Because its underlying processes constitute random events, evolution is unpredictable. To second-guess the specific direction human evolution will take is futile, an undertaking better suited to science-fiction writers than to scientists. Instead, we consider the general character of human evolution in the future, as it can be deduced from the idiosyncrasies of the human population. Specifically, we shall ponder possible changes to the three principal components of evolution – mutation, random genetic drift, and natural selection.

Mutations will undoubtedly continue to occur in the human genome either at their present rate or at an accelerated speed if the level of mutation-inducing agents (mutagens) in the environment increases dramatically as a result of human activities. Although scientists of the future might be able to design methods of reducing mutation rates or preventing mutations altogether, it would be risky and unwise to apply these abilities: risky because of the unforeseeable long-term consequences of such an intervention and foolish because the absence of mutations would deprive humans of the genetic variation every species depends on to meet the challenges of the changing environment. Humans might, of course, develop the ability of selectively introducing mutations into their genomes that will cope with the changes, but even so the environment will always be one unpredictable step ahead of the genetic engineers and will ultimately catch them unprepared. But let us set these fictional scenarios aside and assume instead that the human genome will continue to change at approximately the same rate as it does at present: what will be the fate of these mutations?

The answer depends to a large degree on the size and the structure of the human population in the future. As far as size is concerned, we must, as before, distinguish between N and N_e, and in the latter case, between N_e calculated from genetic variation and N_e represented by the number of breeding individuals. The census population size, N, includes individuals of pre- and postreproductive age. Taking the reproductive age as extending from 15 to 40 years, it has been estimated that the number of breeding individuals comprises approximately 34 percent of the census population size, hence about two billion individuals of the current world population of six billion. This value is very different from the effective population size estimate of $N_e = 10,000$ individuals based on genetic variation. For our present purpose, it will be the number of breeding individuals rather than the number of individuals needed to account for the observed genetic variation upon which our considerations will be based.

As for its structure, although the current population is divided into races and local populations, the tendency is toward homogenization and loss of subdivision. Ultimately, the trend can be expected to result in a population in which each individual has the same probability of mating with any other individual of the opposite sex. Population geneticists call such a population *panmictic, panmixia* being a technical term for random mating (from Greek *pan*, all; *mixis*, a mingling). You may object that a complete panmixia is an unrealistic assumption: regardless of the extent that human travel and migration will increase in the future, it is unreasonable to assume that every person will have the same opportunity of meeting each of the one billion individuals of the opposite sex. If nothing else, the surface area the traveler would have to cover is simply too large to satisfy the condition of random mating. However, population geneticists, led by Sewall Wright, have been pointing out that to attain panmixia, all that is necessary is the exchange of a few migrants between local populations. The exact number depends on the particular population model. If one assumes that the global population consists of a large number of partially isolated populations, each of which is equally likely to exchange genes with any other partial isolate (= *island model*; Fig. 12.5A), then two migrants per generation suffice to render the global population effectively panmictic. In a population organized like stepping stones in a traditional Japanese garden (= *stepping-stone model*; Fig. 12.5B and C), in which gene exchange always occurs between neighboring isolates, the number of migrants required to attain

A.

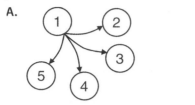

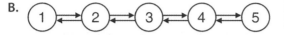

B.

Fig. 12.5A-C
Genetic models of migrations between populations. **A.** The island model: migrants from one subpopulation (e.g., number 1) have an equal probability of ending up (arrows) in any of the other subpopulations (numbers 2 through 5). **B.** The stepping stone model: migrants are exchanged between neighboring populations. **C.** A Japanese garden with stepping stones: Garden of the Heian Jingu shrine.

C.

panmixia is somewhat larger, but still small. Thus in both models only a small amount of gene flow is necessary to erase or to prevent any local genetic differentiation. These considerations apply to neutral genes; the dispersion of genes responsible for adaptations to local conditions requires an increased movement of migrants between neighboring groups. The exact number depends on the intensity of natural selection and is larger, the stronger the selection. Generally, however, as long as neighboring groups are exchanging genes, the whole population can attain panmixia.

The extent of local differentiation is measured by the F_{ST}; the relationship between N and N_e is given by a formula derived by Sewall Wright in 1943: $N_e = N/(1 - F_{ST})$. From this formula it can be seen that if F_{ST} is much smaller than 1, N_e approaches N and the population can be regarded as effectively panmictic. The F_{ST} value of 0.1 (or 10 percent) estimated for the present human population is the result of differentiation into partially isolated groups of individuals in the past. Using this value, we can calculate the $F_{ST}(t)$ value at t generations into the future from the relationship

$$F_{ST}(t) = F_{ST}(\text{future}) + [F_{ST}(\text{present}) - F_{ST}(\text{future})]e^{-\left(\frac{1}{2N_g} + 2m\right)t}$$

where $F_{ST}(\text{future})$ is the value of F_{ST} at the future equilibrium between random genetic drift and migration, N_g is the average number of breeding individuals in a group, and m is the migration rate. If we take $F_{ST}(\text{present})$ to be 0.1 and assume that $F_{ST}(\text{future})$ equals zero and N_g is very large, then the above relationship reduces to approximately $F_{ST}(t) = 0.1e^{-2mt}$. The equation indicates that $F_{ST}(t)$ approaches zero after $t = 5/m$ generations because $F_{ST}(5/m) = 0.1e^{-10} = 4.5 \times 10^{-6}$. If N_g is much larger than 100 and a group of individuals exchanges one gene every 100 generations with a neighboring group, then the differentiation of the human population will disappear within 500 generations and the effective population size will equal the census population size.

These considerations have far-reaching consequences for the fate of neutral mutations arising in the human population. Since it takes on average $4N_e$ generations to fix a neutral mutation, in a population consisting of one billion breeding individuals who have a generation time of 20 years, the time until fixation becomes 20 billion years. This value is much larger than the age of the earth and of the same magnitude as the age of the universe since the Big Bang. In other words, under these conditions, fixation of neutral mutations by random genetic drift can no longer occur on a biologically meaningful time scale. Hence, if the human population size stays at its present level and the population becomes fully panmictic, mutations and random genetic drift will continue to operate in it normally, but fixations of neutral alleles will be suspended. Instead, most new mutations will be lost, although some will retain high frequency without ever reaching the 100 percent mark. The composition of neutral alleles in the global population will change with time as new mutations replace old ones, but very few, if any, of the new mutations will be shared by all the individuals constituting the human species. Technically, the continuing fluctuations of the neutral allele frequencies in the human population will represent evolution, but it will be evolution of polymorphism rather than accumulation of species-specific substitutions.

The suspension of fixation should not, however, apply to advantageous mutations subject to natural selection. Provided that the selection intensity is high enough to shorten the average time to fixation significantly, these mutations would be expected to spread through the large gene pool and to replace the old, less opportune alleles. There is, however, one factor that may work against the fixation even of advantageous mutations – human intervention. The essence of the selection process is that while one allele is promoted, the other is demoted: when a mutation appears that provides the bearer with a selective advantage relative to the carriers of the original alleles, the latter automatically becomes disadvantageous. (Only in certain special forms of selection can several alleles be advantageous simultaneously.) The promotion of one allele and the demotion of the other takes place through differential survival or differential fertility of the bearers. Humans, of course, perceive the possession of the disadvantageous alleles as a genetic defect which modern medicine has the moral obligation to treat and so not only support the survival of the affected individual, but also provide the same opportunity to reproduce as those without the defect. When successful, the treatment works against natural selection by artificially maintaining the disadvantaged gene in the human gene pool and preventing the fixation of the advantageous alleles. This type of influence on the human gene pool is already being exerted via the treatment of several genetic defects and as the nature of more and more genetic diseases becomes known, the practice will probably become widespread.

There is, however, the possibility that the trend will be reversed through genetic engineering – the ability to manipulate the genome by introducing genetic material into it artificially. Once gene therapies become routine, it will become possible to exchange the disadvantageous allele for the advantageous one, just as surgeons replace diseased organs. By intervening, the disadvantageous allele can be eliminated from the gene pool artificially and the fixation of the advantageous allele will be accelerated. The practice would thus amount to humans taking their evolution into their own hands – with unforeseeable consequences.

Homo futurus

It follows from the foregoing discussion that evolution in the human species has not stopped and will never do so. Compared to other species, evolution in humans may assume a somewhat different form in the way mutations are incorporated into the gene pool, but otherwise, like all organisms on this planet, we will remain firmly in the grip of the evolutionary process. If so, does this mean that we will evolve into another species? Will we diverge into more than one species?

Earlier, on several occasions, we distinguished two modes by which a species may lose its identity and turn into another species – cladogenesis and anagenesis. Cladogenesis requires the erection of a barrier that splits a single population into two or more which subsequently evolve in isolation from each other. As far as the human population is concerned, the trend seems to be more in the opposite direction, toward fusion rather than splitting of populations. Despite persisting racism and nationalism, enough inter-racial and inter-national marriages take place to drive the species toward panmixia, and if anything, this trend can be expected to intensify in the future. Seen from this perspective, cladogenesis does not appear to be in the script for the human species – a realistic script, that is. One can, namely, come up with a variety of science-fiction scenarios in which situations favorable to cladogenesis may arise. Of these, we mention only two: a group of humans becomes marooned on a distant planet and separated from its parental population evolves into a different, reproductively isolated species; or a mad dictator orders his genetic engineers to introduce a gene or an entire artificial chromosome into the genomes of his followers which isolates the group reproductively from the rest of humanity. Would *Homo sapiens,* or its descendant *Homo futurus,* tolerate another human and competing species? The history of the hominid lineage suggests that it would not.

Anagenesis does not require the separation of a population in space to produce a new species: here the accumulation of mutations in time is assumed to change a species and isolate it reproductively from its distant ancestors. The ancestors and the descendants, if they could meet, would then represent two different species. Whether human egg and sperm cells can remain viable for as long as it would take to test this assumption is not known. If, however, they can and the experiment is carried out one day in the far geological future, we would expect *H. futurus* to regard *H. sapiens* as a distinct species.

How long can *H. sapiens* or its successor *H. futurus* be expected to survive? One message that the fossil record conveys to us clearly and unambiguously is that there is no permanence in species existence. Some species cease to exist by evolving into new species, but the majority simply become extinct without issue. Which factors determine whether a species will suffer the former or the latter fate is not known. The extinction of a species could simply be the result of bad luck if it finds itself in the wrong place at the wrong time. If, for example, circumstances change and the species suddenly encounters a predator which it cannot escape or a competitor it cannot outcompete, it may be wiped out before it has had time to find an evolutionary solution. Alternatively, in the genetic game of poker, a species may have been dealt a bad hand in the form of a set of genes that do not allow it to adapt adequately to a changed environment. Whatever the reasons for extinction, paleontologists have documented innumerous examples of species that first loiter backstage and then burst onto the scene, flourishing for a while, only to fizzle out again unceremoniously. The apes among the primates are clearly in the league of lineages that are being phased out. Of the 13 to 16 extant ape species (the uncertainty about the actual number reflects a disagreement among taxonomists as to the classification of gibbons into species and subspecies), most are considered to be endangered; only one ape is doing well – *H. sapiens*. The reason why humans are flourishing while the other apes are perishing is very likely to be found in the genes responsible for the enlargement of the brain, increased intelligence, and the development of culture in the former.

Obliteration of a species is as inevitable as the ultimate demise of an individual, and just as individuals have different life spans, so do species. Since our only source of information about the duration of species' life span is the notoriously unreliable fossil record, we can estimate their tenure only approximately. Keeping this qualification in mind, we reckon that the mean life span of species in the hominid lineage (excluding *H. sapiens*) was slightly more than half a million years. The estimated life spans of the individual species are these: *Australopithecus anamensis*, 0.3 my; *A. afarensis*, 0.9 my; *A. africanus*, 0.6 my; *A. aethiopicus*, 0.5 my; *Paranthropus boisei*, 0.9 my; *P. robustus*, 0.5 my; *H. habilis*, 0.2 my; *H. rudolfensis*, 0.8 my; *H. ergaster*, 0.2 my; *H. erectus*, 1.6 my; *H. heidelbergensis*, 0.5 my; *H. neanderthalensis*, 0.3 my. The present age of *H. sapiens* is 200,000–500,000 years according to one version of the uniregional hypothesis and 1-2 my according to the proponents of the multiregional hypothesis. The latter estimate would make *H. sapiens* the longest living hominid on record and so well overdue for extinction.

In contrast to the life span of an individual, which is fixed genetically, that of a species is determined by circumstance and so can vary widely. Albeit, the long ages of the so-called living fossils often cited in textbooks are probably highly exaggerated. Thus the best known of the living fossils, the coelacanth (*Latimeria chalumnae*) is claimed to represent a 380 my-old lineage, which was believed to have become extinct 70 my ago until it was caught off the coast of South Africa in 1937. Since then some 180 specimens have been reported and live coelacanths have been photographed in their natural habitat. Most recently, a second species of *Latimeria* was caught in the Indian ocean. Other examples of living fossils – species purported to have survived unchanged for hundreds of millions years – include the horseshoe crab, cockroaches, sharks, and the gingko tree.

But are all these species really the same as their fossilized ancestors that lived many millions of years ago? After having discussed the presumed long life spans of the living fossils at length, the paleontologist David M. Raup concludes: "I suspect that nearly everything I have written in the previous paragraph is bunk."* Why this skepticism? For one thing, living species are never exactly the same morphologically as the fossil species. The horseshoe crabs that you encounter along the east coast of the United States certainly resemble those that lived in the Jurassic seas, but on closer examination they differ from them in many details. Similarly, none of the extant shark species is identical to any of the extinct species. What the living forms have in common with their fossilized ancestors is a general morphology of their bodies, but this does not make them the same species. If, and this is the

---

* D.M. Raup: *Extinctions: Bad Genes or Bad Luck?* p. 41, W. W. Norton, New York 1991.

second important point, one could compare the DNA or the proteins of the extant and extinct coelacanth species, they would undoubtedly be found to differ as much as two extant species that had a most recent common ancestor 380 my ago. Even if coelacanths evolved exclusively by anagenesis, without splitting – an assumption contradicted by the discovery of the second *Latimeria* species – the ancestral and the extant species would still be expected to differ molecularly by the substitutions accumulated in the intervening 380 my. Similarly, if the human species does continue to exist for several million years and does evolve anagenetically during the entire period without changing morphologically to any great extent, present-day *H. sapiens* and *H. futurus* millions of years from now will be different species molecularly who, if they were to meet, would discover that they are reproductively isolated.

The Oracles Are Dumb *

We humans have such a sense of self-importance that we are often unable to reconcile ourselves with the notion that the world was not created for us, that it existed without us and that it will continue existing long after the last representative of the human lineage has gone the way of all flesh. Yet, nothing is so certain as the fact that we, as individuals, will pass away one day, and that we, the human race, will come to an end without issue at some point in the future. If oracles tell you something different, they are dumb. Some of us may prefer to cling to myths in which the world begins with our creation and ends with our demise. But this is a delusion contradicted by everything that we know about the universe and about life on earth. Others dream about interstellar, intergalactic travel, about colonies in space, Star Wars, Star Trek, wormholes, and parallel universes – about personal and species immortality. But the cold reality is that we are firmly tethered to our solar system and when the system one day perishes, we will either long be gone, or – if our descendants by some miracle stick it out until then – will perish with it.

We live currently in an interglacial, a warm spell between two cold periods. In the past, interglacials varied in length from 20,000 to 70,000 years, so the next glacial can be expected in less than 10,000 to 60,000 years. Although the greenhouse effect may delay the onset of the glacial, it is not likely to prevent it because soon, humans will exhaust the supply of fossil fuels and will thus stop pumping carbon dioxide into the atmosphere (but not before inflicting considerable damage on the biosphere and themselves). The surplus CO_2 will then be gradually absorbed by the oceans, long before the time comes for the earth to plunge into another Ice Age. Assuming that the human species survives until then, it will have to make profound adjustments in its distribution, population size, and survival strategies.

Major adjustments will also be necessary further into the future in the event that *H. sapiens/futurus* makes it through the glaciation period, which could last a good 100,000 years. The slow movement of the plates bearing the continents can be expected to continue for at least as long as it has already been going on – for over three billion years. The movement will wrinkle flatlands into mountains, submerge some parts of the continents and open new seaways through others, and move the continents to different latitudes and into different constellations. Some of the hither and thither of the landmasses will change the climate dramatically, as it did in the past. Ultimately, however, the main source of heat

---

* John Milton: *On the Morning of Christ's Nativity*, 1629.

in the earth's interior, the decay of the radioactive elements, will gradually diminish to the extent that it will not be able to maintain the molten state of the mantle on which the plates float, and the currents in the mantle will first slow down and then stop flowing altogether. Earth's interior will cool down and solidify, the surface water will gradually evaporate without replenishment from the deeper layers, the cooler core will stop functioning as a generator of the earth's magnetic field, in the absence of the magnetic field the upper layers of the atmosphere will disperse and expose the earth's surface to harmful radiation from space, and the planet will die both geologically and biologically.

In the meantime an even greater danger to earth will arise from events taking place in the aging sun. In the sun's dense interior, hydrogen atoms are smashed together and fused into atoms of heavier elements in nuclear reactions that are accompanied by the release of huge amounts of energy. Some three billion years into the future, the supply of hydrogen atoms in the sun's central core will begin to run out, the nuclear reactions will spread outward toward the surface, and the heated surface layers will begin to expand. The sun will grow brighter and larger, gradually entering the phase of stellar evolution that astronomers refer to as the *red giant* stage. At the same time, deprived of its own source of energy, the sun's core will be unable to resist the force of gravity and will begin to collapse. The heat generated in the process will then ignite nuclear reactions among the heavier helium atoms and the surge of heat from the reactions will cause the sun's surface to expand even further, perhaps as far as the earth's orbit. If this happens, the earth might be engulfed by the sun's atmosphere, and drawn by the pull of the sun's gravity might spiral into its interior. The glory of the solar system, and who knows, perhaps even of the universe, will thus end in the oven of its inception. The cremation will be a dignified end to a planet that once resounded with voices singing *The Ode to Joy*.

But perhaps the sun's expansion will not reach this far. The sun's hot breath may merely scorch the earth and melt its surface into the sea of lava that it once was. Some six billion years from now, the red giant will run out of fuel, shrink into a feebly glowing *white dwarf*, and allow the temperature on the surface of the dimly lit earth to drop to –217°C. The declining force of the sun's gravity may not be strong enough to hold its planets in orbit and the earth may drift off. A lonely, dark, barren body will then wander aimlessly through the immensity of space and nothing will betray the fact that it was once inhabited by creatures who had the ambition of reaching for the stars and who were able to ask themselves: *Where do we come from? Who are we? Where are we going?*

A P P E N D I C E S

Appendix One:
Binomial Probability Distribution and Basic Terms in Statistics

An example of a situation to which the binomial probability distribution applies is provided by coin tossing. Here, the distribution describes the probabilities of heads (H) or tails (T) coming up in a sequence of successive coin tosses (or alternatively of a single toss of multiple coins). In statistics, the act of a tryout such as a flip of a coin is generally referred to as a *trial* and any process leading to a collection of data is an *experiment*. Here, we shall regard a single coin toss as a trial and a series of two or more successive coin tosses as an experiment. A trial (experiment) has a certain *outcome*, one of several possible. If only two outcomes are possible or of interest, they are referred to as *success* and *failure*. In the coin flipping trials, heads coming up in a trial can be regarded as success and heads *not* coming (i.e., tails coming up) as failure. It is customary to designate the probability of a success in a trial as p and the probability of a failure as $q = 1 - p$ (because $p + q = 1$ by definition). The quantity that takes values in a certain range with probabilities specified by a probability distribution is a *random variable*. In the coin tossing experiment, the random variable k is the number of heads that appear when a coin is flipped 1, 2, 3, ... n number of times.

Thus, when we toss a coin once, there are two equally likely outcomes, H or T, so that each outcome has the probability of 1/2 and the probability distribution in the two categories is

$$
\begin{array}{cc}
1H & 1T \\
1(1/2) & 1(1/2)
\end{array}
$$

If a coin is tossed twice, there are four equally likely outcomes (HH, HT, TH, TT; here, the first letter signifies the outcome of the first toss and the second letter the outcome of the second toss) and since in each toss the probability of H is always 1/2 as is that of T, the probability of H in both tosses is $(1/2)(1/2)$ or $(1/2)^2$; that of 1H and 1T is $(1/2)(1/2)$ for HT and $(1/2)(1/2)$ for TH or together $2[(1/2)(1/2)]$; and that of T in both tosses $(1/2)(1/2)$ or $(1/2)^2$. The probability distribution is then

$$
\begin{array}{ccc}
HH & HT \text{ or } TH & TT \\
1(1/2)^2 & 2[(1/2)(1/2)] & 1(1/2)^2
\end{array}
$$

If the appearance of heads is regarded as a "success" and a success is the only outcome of a trial that interests us, we can rewrite the distribution thus:

$$
\begin{array}{ccc}
2H & 1H & 0H \\
1(1/2)^2 & 2[(1/2)(1/2)] & 1(1/2)^2
\end{array}
$$

For a coin tossed three times, there are eight equally likely outcomes: HHH, HHT, HTH, THH, HTT, THT, TTH, TTT. The probability of H coming up in all three tosses is $(1/2)(1/2)(1/2)$ or $(1/2)^3$; the probability of H coming up in two tosses and T in one is $3[(1/2)^2(1/2)]$ [since there are three such outcomes possible, HHT, HTH, and THH, and in each the probability of H coming up twice is $(1/2)(1/2)$ or $(1/2)^2$ and that of T coming up once is 1/2] and so on, giving a probability distribution

3H	2H	1H	0H
$1(1/2)^3$	$3[(1/2)^2(1/2)]$	$3[(1/2)(1/2)^2]$	$1(1/2)^3$

We could continue in this fashion indefinitely. If we now generalize the situation and use the p and q notations for probabilities of H coming up in each toss, we can present the probability distributions for n number of trials as follows:

$$(p + q)^1 = 1p + 1q$$

$$(p + q)^2 = 1p^2 + 2pq + 1q^2$$

$$(p + q)^3 = 1p^3 + 3p^2q + 3pq^2 + 1q^3$$

.
.
.

$$(p + q)^n = \binom{n}{n}p^n + \binom{n}{n-1}p^{n-1}q + \binom{n}{n-2}p^{n-2}q^2 + \ldots + \binom{n}{2}p^2q^{n-2} + \binom{n}{1}pq^{n-1} + \binom{n}{0}q^n$$

where $\binom{n}{k}$ is a shorthand notation for

$$\binom{n}{k} = \frac{n!}{k!\,(n-k)!}$$

Here $n!$ stands for "n factorial" and instructs the user to multiply together a string of numbers, starting with n, and continuing with $(n-1)$, $(n-2)$, etc, the last number in the sequence being 1. Thus, 4! factorial, for example, equals $(4)(3)(2)(1) = 24$. The expression $\binom{n}{k}$ is the *combination rule*, which specifies the number of ways of selecting a subset of k elements from a set of n elements, where the order in which the elements are chosen is of no importance. The expanded version of the $(p + q)^n$ expression can be shortened to the formula:

$$B(k;n,p) = \binom{n}{k}p^k q^{n-k} \qquad (k = 0, 1, 2, 3, \ldots n)$$

where $B(k;n,p)$ is the distribution of the probabilities of the various numbers of successes (from 0 to n, e.g., the number of heads) when n trials (e.g., coin tosses) are carried out; n is the number of trials (number of times a coin has been flipped); k is the number of successes in n trials (e.g., the number of heads that can be observed in a given trial series) – the random variable of the binomial distribution; and p is the probability of success in a single trial (e.g., heads coming up in a single coin toss). The $\binom{n}{k}$ is the *binomial coefficient* which gives the number of ways in which a particular outcome (e.g., number of heads) in a given number of trials (experiment) can be arrived at. It represents the coefficients (a numerical factor in an algebraic expression) in the expansion $(p + q)^n$. The coefficients can be obtained from *Pascal's triangle,* named after the seventeenth century French mathematician and philosopher Blaise Pascal:

Row No.										
1:						1				
2:					1		1			
3:				1		2		1		
4:			1		3		3		1	
5:		1		4		6		4		1

etc

Each row of the triangle begins and ends with 1; each other number in the row is the sum of the two nearest numbers in the row immediately above. Rows number 1, 2, 3, 4, etc are the coefficients of the expansions of $(p + q)^0$, $(p + q)^1$, $(p + q)^2$, $(p + q)^3$, etc, respectively. Since p in the binomial formula is the probability of success in a single trial (e.g., the probability of heads appearing in one coin toss) and the $q = (1 - p)$ is the probability of a failure (e.g., probability of tails appearing in a single coin toss), the product $p^k q^{n-k}$ is the probability assigned to each outcome that has k successes and $(n - k)$ failures. The $(p + q)^n$ is the *binomial theorem*, whereby a *theorem* is a mathematical statement that can be proven true by deductive reasoning. It indicates that $(p + q)$ is to be raised to the nth power –multiplied repeatedly by itself for a total of n-times. This operation results in the *expansion* of the theorem, here the *binomial expansion*. The expansion represents mathematical *series* –a succession of numbers or *terms* so related that each can be derived from one or more of the preceding terms in accordance with some fixed rule. Although a purely mathematical construct, the series happens to reflect the probabilities of certain events in the real world, here for example the probabilities of the number of heads coming up in a succession of coin tosses.

To illustrate the use of the binomial formula, we describe a simple experiment consisting of two trials ($n = 2$): a coin that has been tossed twice. There are three possible outcomes in terms of heads (H) coming up (number of successes, k):

1. H does not appear in either of the two tosses: TT; $k = 0$
2. H comes up in one toss, but not in the other: either HT or TH; $k = 1$
3. H appears in both tosses: HH; $k = 2$.

To determine the probability distribution of these various outcomes ($k = 0, k = 1, k = 2$), $B(k;n,p)$, we use the formula

$$B(k;n,p) = \binom{n}{k} p^k q^{n-k}$$

and applying its expanded version to the case under study, we obtain for $p = 1/2, q = 1/2, n = 2, k = 0, 1, 2$:

$$B(k;n,p) = \underbrace{\left[\binom{2}{0}\left(\frac{1}{2}\right)^2\right]}_{P(k=0)} + \underbrace{\left[\binom{2}{1}\left(\frac{1}{2}\right)^1\left(\frac{1}{2}\right)\right]}_{P(k=1)} + \underbrace{\left[\binom{2}{2}\left(\frac{1}{2}\right)^2\right]}_{P(k=2)}$$

$B(k;n,p) = 1/4 + 1/2 + 1/4$

which is the probability distribution of the experiment. We can depict it as

$$B(k = 0; 2, 1/2) = P(TT) = 1/4$$
$$B(k = 1; 2, 1/2) = P(TH) + P(HT) = 1/4 + 1/4 = 1/2$$
$$B(k = 2; 2, 1/2) = P(HH) = 1/4$$

or in a tabular form as

k	$B(k;n,p)$
0	0.25
1	0.50
2	0.25

or graphically as

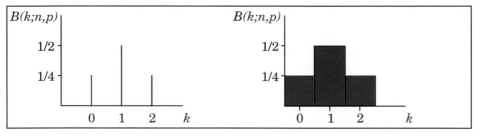

In statistics, any set of items for which information is sought is a *population*. A subset of items taken from the set is a *sample*. A school of fish that darts by is a population. The few specimens caught in the net placed in their path is a sample of that population. The items comprising a heterogeneous population (and the sample) differ from each other (e.g., the fish are of different size) – they vary. Any characteristic of the items in the population and the sample that can be measured and can assume a set of values is a *variable*. If the variable is observed as part of the sampling process (e.g., you are interested in the variation in size) or as part of a statistical experiment in which each outcome has a definite probability of occurrence (e.g., a set of coin-tossing trials), it is referred to as a *random variable*. A count of the number of items having specific values of the random variable is a *frequency distribution*. A model describing the relative frequencies (= probabilities) of the values that the random variable can take on is a *probability distribution* (e.g., the binomial distribution is a model describing the distribution of the number of heads falling face-up in a coin-tossing experiment). Each population and each probability distribution is characterized by certain numerical values called *parameters*. They include mean, variance, and standard deviation.

The *arithmetic mean* (average) of a random variable is the sum of its individual values divided by the total number of values. Hence, if a random variable assumes values $k_1, k_2, k_3, ..., k_n$, the mean, $\bar{k}$, is given by the formula

$$\bar{k} = \frac{k_1 + k_2 + k_3 + ... + k_n}{n} = \frac{\sum_{i=1}^{n} k_i}{n}$$

where the Greek capital letter sigma, Σ, stands for summation (i.e., $k_1 + k_2 + k_3 + ...+ k_n$); $i = 1$ at the bottom indicates that one must start adding where i equals 1; n at the top means one must stop adding where i equals n; and k_i indicates the values to be added.

Variance is a measure of dispersion; it is a measure of the degree to which the values in the data set are spread out. The scatter is determined by measuring the distance of each value from the mean. Population variance, *Var*, or σ^2 (sigma squared), is given by the formula

$$Var = \sigma^2 = \frac{\left(k_1 - \bar{k}\right)^2 + \left(k_2 - \bar{k}\right)^2 + \left(k_3 - \bar{k}\right)^2 + ... + \left(k_n - \bar{k}\right)^2}{n}$$

where $(k_i - \bar{k})^2$ are the squared differences of individual values from the mean (squaring is necessary to turn negative numbers into positive; if this were not done, the numbers would cancel out). Using the summation notation, this formula can be abbreviated to

$$Var = \sigma^2 = \frac{\sum_{i=1}^{n} \left(k_i - \bar{k}\right)^2}{n}$$

The square root of variance is the *standard deviation*, symbolized by σ:

$$\sigma = \sqrt{\frac{\sum_{i=1}^{n} \left(k_i - \bar{k}\right)^2}{n}}$$

Ideally, we would like to obtain these three population parameters directly, by assessing all the items of a population (e.g., all the fish of a school). This is often not possible, however, for technical and other reasons. In such cases, we take samples from the population, assess all items in each sample, and estimate the population parameters from the values thus obtained. An estimate of an unknown numerical quantity is called a *statistic* and a statistic used to estimate a population parameter is an *estimator*. To estimate the population mean, we calculate the mean of the sample in the same way as described above for the former. To distinguish the two, the population mean is conventionally designated by the Greek letter μ, (mu), whereas the sample mean is denoted by the Latin letter m or by a bar ($\bar{m}$) or "hat" ($\hat{m}$) placed above a letter symbol. By repeating the sampling several times, we obtain individual values of the sample mean which we can then treat as if they themselves were a random variable and thus calculate their variances and standard deviations. Such a process, however, is impractical and so these two statistics are commonly calculated from a single sample. The justification for this is the mathematical demonstration that a sample mean $\bar{m}$ of n independent observations has a variance equal to σ^2/n and standard deviation equal to $\sigma/\sqrt{n}$. The standard deviation of the estimator is then referred to as the *standard error* (s.e.). From the formula σ^2/n it is apparent that the larger the sample size (n), the smaller the variance and the standard error of the sample mean.

It is important to keep in mind that the sample mean $\bar{m}$ is merely an estimate of the population mean μ. Under these circumstances one would like to know just how reliable the estimate is; specifically, one would like to define the length of the interval from $\bar{m}-c$ to $\bar{m}+c$ (where c is an unspecified value) for which there is a high probability (usually set at 95 percent) that it contains m. The $\bar{m}-c$ and $\bar{m}+c$ values are the *confidence limits*, the segment between them the *confidence interval*, and the probability value of 0.95 the *confidence level*. The length of the confidence interval is obtained by solving the equation $P(\bar{m}-c < \mu < \bar{m}+c) = 0.95$. The solution is $c = 1.96\sigma/\sqrt{n}$ for the 95 percent confidence level.

Returning to the binomial theorem and to the coin-flipping experiment we note that here the number of successes constitutes a frequency distribution which is predicted by the binomial probability distribution. Like the frequency distribution, the probability distribution, too, can be characterized by two parameters, the mean, μ, and variance, Var (= σ^2). The mean of a probability distribution is the *expected value* of the random variable k denoted as $\mu = E(k)$. To obtain the expected value of the binomial distribution, we note that each term of the distribution consists of two parts: the $\binom{n}{k}$ part, which is the frequency of successes (in the coin-flipping experiment, the number of heads appearing in an experiment); and the $p^k q^{n-k}$ part, which is the probability of occurrence of that value. Each term is thus the product of frequency and probability, and since probability is the number of successful outcomes divided by the total number of outcomes, the summation of the individual terms gives the expected value:

$$\mu = E(k) = \sum_{\text{all } k} kB(k; n, p)$$

Since a probability distribution can be viewed as a representation of a population, population variance is used to measure its variability. The population variance σ^2 is the average of the squared distances of k from the population mean, μ. Since k is a random variable, the squared distance $(k - \mu)^2$ is also a random variable. The variance of the binomial distribution is therefore obtained by multiplying all possible values of $(k - \mu)^2$ by $B(k;n,p)$ and then summing up the results thus obtained:

$$\sigma^2 = E[(k - \mu)^2] = \sum_{all\,k} (k - \mu)^2 \, B(k;n,p)$$

The standard deviation is then the square root of the variance σ^2.

The mean and the variance of the distribution of probabilities in the two-trial (coin flipping) experiment with values

k	$B(k;n=2,p=0.5)$
0	1/4
1	1/2
2	1/4

are calculated as follows:

$$\mu = E(k) = \sum kB(k;n,p) = 0(1/4) + 1(1/2) + 2(1/4) = 1$$
$$\sigma^2 = E[(k - \mu)^2] = \sum(k - \mu)^2 B(k;n,p)$$

Squared distances

$$= (0 - 1)^2(1/4) + (1 - 1)^2(1/2) + (2 - 1)^2(1/4) = 1/2 = 0.5$$

Probability of k number
of heads coming up

Appendix Two:
The Poisson Probability Distribution and the Derivation of the Jukes-Cantor Formula

In Chapter Five we described the Poisson process underlying the Poisson probability distribution. Here we derive the latter from the binomial probability distribution. To use the same example as in Appendix One, we would have to assume the existence of an extremely biased coin which falls tails up in almost every toss. In such a situation the frequency of success (heads-up) would be close to zero. A more realistic example is that of a bag filled with 100 marbles ($n = 100$), 99 of them white and one black. We draw a marble randomly from the bag, record its color, place it back in the bag, and repeat this procedure one hundred times. In this experiment, the probability of drawing a white marble in each single draw is 99/100 ($q = 0.99$); that of drawing the black marble is 1/100 ($p = 0.01$). The probabilities of drawing the black marble 0, 1, 2, 3, ... or generally k times in the 100 draws are specified by the terms of the binomial distribution with $n = 100$ and $p = 0.01$:

$$B(k;n,p) = \binom{n}{k} p^k q^{n-k}$$

Using this formula we calculate the probabilities for $k = 0$, $k = 1$, $k = 2$, $k = 3$, $k = 4$, and $k = 5$ as 0.366, 0.370, 0.185, 0.061, 0.015, and 0.003, respectively. Increasing the number of marbles in the bag to 1000 or more makes the calculations correspondingly more tedious because the length of the expansion grows with each increase. In such a situation, it is more expeditious to use the Poisson distribution for the calculations. Since the product $np = \lambda$ remains the same as n increases and p decreases [e.g. for $n = 100$ and $p = 0.01$, $np = (100)(0.01) = 1$; for $n = 1000$ and $p = 0.001$, $np = (1000)(0.001) = 1$], the application of the Poisson distribution to this situation is justified.

To derive the Poisson probability distribution from the binomial distribution, let us first take the case for $k = 0$. The probability specified by the binomial formula for this case is

$$B(0;\ n,\ p) = \binom{n}{k} p^0 q^{n-0}$$

Here, the binomial coefficient is

$$\binom{n}{0} = \frac{n!}{0!(n-0)!} = 1$$

(because $0! = 1$); p^0 is also 1; and q^{n-0} can be rewritten as $(1 - p)^n$. Hence $B(0; n, p) = (1 - p)^n$. By rearranging the equation $\lambda = np$, we obtain $p = \lambda/n$ and so $B(0; n, p) = (1 - \lambda/n)^n$. The right-hand side of this equation is in fact a variant of the binomial formula $(a - b)^n$ which in turn is a variant of the formula $(a + b)^n$ described in Appendix One. The variants of the type $(1 \pm \lambda/n)^n$ have certain special properties which we must now describe. We begin with $(1 + 1/n)^n$.

By expanding this formula (see Appendix One) we obtain

$$\left(1 + \frac{1}{n}\right)^n = 1 + n\left(\frac{1}{n}\right) + \frac{n(n-1)}{2!}\left(\frac{1}{n}\right)^2 + \frac{n(n-1)(n-2)}{3!}\left(\frac{1}{n}\right)^3 + \ldots + \left(\frac{1}{n}\right)^n$$

which we can manipulate to give

$$\left(1+\frac{1}{n}\right)^n =1+1+\frac{\left(1-\frac{1}{n}\right)}{2!}+\frac{\left(1-\frac{1}{n}\right)\left(1-\frac{2}{n}\right)}{3!}+\ldots+\frac{1}{n^n}$$

Consider now what happens when n increases without bound toward infinity ($n \to \infty$). The expansion will then have more and more terms, while at the same time the expression within the parentheses will get closer and closer to 1. The limit of $(1 + 1/n)^n$ as $n \to \infty$ (see Appendix Four) will become

$$\lim_{n\to\infty}\left(1+\frac{1}{n}\right)^n =1+1+\frac{1}{2!}+\frac{1}{3!}+\ldots$$

The expression continues *ad infinitum*: it is an example of an *infinite series*. The limit of this particular series is denoted by the letter e, a notation introduced by the eighteenth century Swiss mathematician Leonhard Euler because it is the first letter of the word *exponential*, according to some historians, or because it is the next letter after a, b, c, d, the four letters widely used by mathematicians in Euler's time. It is easier and faster to compute the individual terms of the series separately and add up the partial sums than to calculate them from the expression $(1 + 1/n)^n$ directly. The values for the first seven terms are these:

$2 =$	2
$2 + 1/2 =$	2.5
$2 + 1/2 + 1/6 =$	2.666...
$2 + 1/2 + 1/6 + 1/24 =$	2.708333...
$2 + 1/2 + 1/6 + 1/24 + 1/120 =$	2.716666...
$2 + 1/2 + 1/6 + 1/24 + 1/120 + 1/720 =$	2.7180555...
$2 + 1/2 + 1/6 + 1/24 + 1/120 + 1/720 + 1/5{,}040 =$	2.718253968...

The approximate value of e is therefore 2.71828.

By expanding instead of $(1 + 1/n)^n$ the $(1 + \lambda/n)^n$ form and subjecting the expansion to the same manipulations as that of the former variant, we obtain

$$\lim_{n\to\infty}\left(1+\frac{\lambda}{n}\right)^n =1+\frac{\lambda}{1!}+\frac{\lambda^2}{2!}+\frac{\lambda^3}{3!}+\ldots= e^{\lambda}$$

The limit of $(1 - \lambda/n)^n$ as n tends toward infinity is

$$\lim_{n\to\infty}\left(1-\frac{\lambda}{n}\right)^n = 1-\frac{\lambda}{1!}+\frac{\lambda^2}{2!}-\frac{\lambda^3}{3!}+\ldots= e^{-\lambda}$$

Hence the first term of the Poisson probability distribution is $P(0; n, p) \approx e^{-\lambda}$, where the $\approx$ sign indicates approximate equality.

To obtain the second term of the Poisson probability distribution, we resort to a bit of algebraic jugglery. We determine what it takes mathematically to get from one term to the next and then use the result to multiply the term of the preceding step by it. If we start with the k step, next is the $k + 1$ step. Therefore by dividing the term for $B(k + 1; n,p)$ by the term $B(k; n,p)$, we obtain the increment necessary for the multiplication:

$$\frac{B(k+1;n,p)}{B(k;n,p)} = \frac{\dfrac{n!}{(k+1)!(n-k-1)!}\,p^{k+1}q^{n-k-1}}{\dfrac{n!}{k!(n-k)!}\,p^{k}q^{n-k}} = \frac{n!}{n!}\,\frac{k!}{(k+1)!}\,\frac{(n-k)!}{(n-k-1)!}\,\frac{p^{k+1}}{p^{k}}\,\frac{q^{n-k-1}}{q^{n-k}} = \frac{(n-k)p}{(k+1)q}$$

$$= \frac{np}{k+1}\left[\left(1-\frac{k}{n}\right)\frac{1}{q}\right]$$

For k which is much smaller than n and p which is much smaller than 1, the term in the square brackets is almost unity so that

$$\frac{B(k+1;n,p)}{B(k;n,p)} \approx \frac{np}{k+1} = \frac{\lambda}{k+1}$$

Using this increment we can then produce the successive terms of the Poisson probability distribution, $P(k;\lambda)$ for the values $k = 0, 1, 2, 3$, etc:

$$P(0;\lambda) = e^{-\lambda}$$

$$P(1;\lambda) = \frac{\lambda}{1}P(0;\lambda) = \lambda e^{-\lambda}$$

$$P(2;\lambda) = \frac{\lambda}{2}P(1;\lambda) = \left(\frac{\lambda}{2}\right)\left(\lambda e^{-\lambda}\right) = \frac{\lambda^2}{2}e^{-\lambda}$$

$$P(3;\lambda) = \frac{\lambda}{3}P(2;\lambda) = \left(\frac{\lambda}{3}\right)\left(\frac{\lambda^2}{2}e^{-\lambda}\right) = \frac{\lambda^3}{(3)(2)(1)}e^{-\lambda} = \frac{\lambda^3}{3!}e^{-\lambda}$$

.
.
.

$$P(k;\lambda) = \frac{\lambda^k}{k!}e^{-\lambda}$$

This last expression is then the general formula for the Poisson approximation to the binomial probability distribution.

For the example of 99 white and 1 black marble, the probabilities of drawing 0, 1, 2, 3, …, k black marbles in 100 trials ($n = 100$) as determined by the Poisson and the binomial distributions are these:

k	Poisson	Binomial
0	0.366 032	0.367 879
1	0.369 730	0.367 879
2	0.184 865	0.183 940
3	0.060 999	0.061 313
4	0.014 942	0.015 328

The mean and the variance of the Poisson probability distribution are $E(k) = \lambda$ and $\sigma^2 = \lambda$, respectively, where k is the random variable, as before.

One of the processes that can be described by the Poisson probability distribution is the accumulation of substitutions during molecular evolution. There are several models of molecular evolution in which the distribution is the underlying assumption and these are used to correct the number of observed differences for the hidden substitutions (see Chapter Five). The simplest of these models postulates that each nucleotide has an equal probability to change into any of the three other nucleotides. The probability of the change can be calculated by using the Jukes-Cantor formula, whose derivation we now describe.

Consider a particular nucleotide site which has a certain probability of undergoing 0, 1, 2, 3, ...k... substitutions during a given time interval. The probability of k substitutions is specified by the Poisson distribution: $P(k) = \lambda^k/k! \, e^{-\lambda}$ (see Chapter Five). Let us assume that at the onset of this interval, the site was occupied by a certain nucleotide, say an A, and let us designate as $I(k)$ the probability that after k substitutions at the end of the interval, the site will again be occupied by an A. Similarly, let us denote as $D(k)$ the probability that after k substitutions the site will carry a different nucleotide: G, C, or T, whereby $I(k) + D(k) = 1$ so that $D(k) = 1 - I(k)$. Now let us consider what can happen when the next $(k + 1)$ substitution occurs. The corresponding probabilities are then $I(k + 1)$ and $D(k + 1)$. If after k substitutions the site was occupied by an A, then after $k + 1$ substitutions the nucleotide at this site cannot be an A. If after k substitutions the site was occupied by a C, then after one additional substitution the probability of recovering the A is 1/3, and the same is true when the site carries a G or a T after k substitutions, as seen from the following diagram:

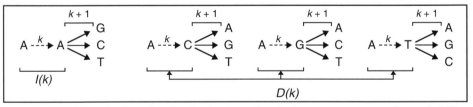

Hence, regardless whether after k substitutions the site became C, T, or G, the probability that it will become A again after $k + 1$ substitutions is 1/3. Since the total probability of A changing to C, G, or T after k substitutions is $D(k)$ and since if it did change to C, G, or T, there is a 1:3 chance that it will change to A again with the new substitution, we can write:

$$I(k+1) = \frac{1}{3} D(k)$$

If we now replace $D(k)$ by $1 - I(k)$, we obtain:

$$I(k+1) = \frac{1}{3}[1 - I(k)]$$

We note that if the site was originally occupied by an A and if no substitution occurred ($k = 0$), then the site remains to be the A. For this reason, we define $I(0) = 1$. For $I(1)$, we obtain

$$I(0+1) = \frac{1}{3}[1 - I(0)] = \frac{1}{3}[1 - 1] = 0$$

To obtain $I(2)$, we write $I(1 + 1) = 1/3(1 - 0)$ or $I(2) = 1/3$. If we repeat this process, we can obtain $I(k)$ and therefore $D(k)$ for all nonnegative integers. When k becomes very large, the difference between $I(k)$ and $I(k + 1)$ is negligible and under these circumstances we can replace both expressions by a common symbol b and re-write the equation $I(k + 1) = 1/3[1 -$

$I(k)$] as $b = 1/3(1 - b)$, which becomes $b = 1/3 - 1/3b$, then $1/3b + b = 1/3$, then $(4b)/3 = 1/3$, and after multiplication of both sides by $3/4$ and cancellations finally $b = 1/4$.

Let us now make $I'(k) = I(k) - b$ [so that $I(k) = I'(k) + b$] and $I'(k + 1) = I(k + 1) - b$. Let us subtract b from both sides of the equation $I(k + 1) = 1/3[1 - I(k)]$ derived above. We can then write:

$$I(k + 1) - b = \frac{1}{3}[1 - I(k)] - b$$

$$= \frac{1}{3} - \frac{1}{3}I(k) - b$$

$$= \frac{1}{3} - \frac{1}{3}[I'(k) + b] - b \quad \text{[here we replaced } I(k) \text{ by } I'(k) + b; \text{ see above]}$$

$$= \frac{1}{3} - \frac{1}{3}I'(k) - \frac{1}{3}b - b$$

$$= \frac{1}{3} - \frac{1}{3}I'(k) - \frac{4}{3}b$$

$$= \frac{1}{3} - \frac{1}{3}I'(k) - \left(\frac{4}{3}\right)\left(\frac{1}{4}\right) \quad \text{(since } b = 1/4)$$

$$= \frac{1}{3} - \frac{1}{3}I'(k) - \frac{1}{3}$$

$$= -\frac{1}{3}I'(k)$$

and since $I(k + 1) - b = I'(k + 1)$, we have

$$I'(k + 1) = -\frac{1}{3}I'(k)$$

We can therefore write

$$I'(0) = I(0) - b$$

$$= 1 - \frac{1}{4} = \frac{3}{4}$$

$$I'(1) = I'(0)\left(-\frac{1}{3}\right)$$

$$= \left(\frac{3}{4}\right)\left(-\frac{1}{3}\right)$$

$$I'(2) = \left(-\frac{1}{3}\right)I'(1)$$

$$= \left(-\frac{1}{3}\right)\left[\left(-\frac{1}{3}\right)I'(0)\right]$$

$$= \left(-\frac{1}{3}\right)^2 I'(0)$$

$$I'(k) = \left(-\frac{1}{3}\right)^k I'(0)$$

$$= \left(-\frac{1}{3}\right)^k \frac{3}{4} \qquad (\text{since } I'(0) = \frac{3}{4})$$

Now take the equation $I'(k) = (-1/3)^k I'(0)$ and add b to both sides:

$$I'(k) + b = \left(-\frac{1}{3}\right)^k I'(0) + b$$

Since $I'(k) + b = I(k)$, we have

$$I(k) = \left(-\frac{1}{3}\right)^k I'(0) + b$$

and since $b = 1/4$ and $I'(0) = 3/4$, we obtain

$$I(k) = \frac{1}{4} + \frac{3}{4}\left(-\frac{1}{3}\right)^k$$

Finally, since $D(k) = 1 - I(k)$, we can write

$$D(k) = 1 - \left[\frac{1}{4} + \frac{3}{4}\left(-\frac{1}{3}\right)^k\right]$$

$$= \frac{3}{4} - \frac{3}{4}\left(-\frac{1}{3}\right)^k$$

$$= \frac{3}{4}\left[1 - \left(-\frac{1}{3}\right)^k\right]$$

Up until now we considered individual substitutions one by one and obtained the probability of identity (difference) at individual sites. Now, however, instead of taking individual values of the random variable 0, 1, 2, 3, …k and specifying the probabilities for each individually, we must consider the whole sequence and the identity (difference) of all the sites together. We must assume that k can be any non-negative integer with a certain probability. To do this, we take the sum of the differences and their probabilities to obtain the mean number of substitutions λ which have led to the proportion of differences observed between two aligned sequences. In other words, we take the sum of the product consisting of the probability of difference for the individual values of the variable and multiply it by the proportion of observed differences between the sequences. Denoting the sum as $\bar{D}$ we can write:

Probability of difference   Proportion of sites
for individual values of k with k substitutions

$$\bar{D} = \sum_{k=0}^{n} \overbrace{\frac{3}{4}\left[1-\left(\frac{1}{3}\right)^k\right]}^{} \overbrace{P(k)}^{}$$

$$= \frac{3}{4}\left[1-\sum_{k=0}^{n}\left(-\frac{1}{3}\right)^k P(k)\right] \qquad \text{(We moved the summation inside the brackets)}$$

$$= \frac{3}{4}\left[1-\sum_{k=0}^{n}\frac{\left\{\left(-\frac{1}{3}\right)\lambda\right\}^k}{k!}e^{-\lambda}\right] \qquad \text{[We replaced the } P(k)\text{, the Poisson element of the product,}$$
by the general formula for the Poisson distribution.]

$$= \frac{3}{4}\left[1-e^{-\frac{\lambda}{3}}e^{-\lambda}\right] \qquad \text{[Recall the definition of } e^x \text{ where } x = -\lambda/3.]$$

$$\bar{D} = \frac{3}{4}\left[1-e^{-\frac{4}{3}\lambda}\right]$$

By rearranging, we obtain

$$\frac{4}{3}\bar{D} = 1-e^{-\frac{4}{3}\lambda}$$

$$e^{-\frac{4}{3}\lambda} = 1-\frac{4}{3}\bar{D}$$

and by taking natural logarithms (ln) of both sides and rearranging, we finally obtain

$$-\frac{4}{3}\lambda = ln\left(1-\frac{4}{3}\bar{D}\right)$$

$$\lambda = -\frac{3}{4}ln\left(1-\frac{4}{3}\bar{D}\right)$$

This is the Jukes-Cantor formula for estimating λ, the mean number of substitutions which led to the observed mean proportion of differences between two aligned (orthologous) sequences.

Based on the formula

$$\bar{D} = \frac{3}{4}\left[1-e^{-\frac{4}{3}\lambda}\right]$$

we can also derive the formula for the probability of identity, P_{ii}, for a particular nucleotide, i, or probability of difference, P_{ij}, between a particular pair of different nucleotide i and j. For a particular nucleotide i at a site, there is a probability of 1/3 of i being substituted by a different nucleotide j. Hence we divide $\bar{D}$ by 3 and have P_{ij}:

$$P_{ij} = \frac{1}{4}-\frac{1}{4}e^{-\frac{4}{3}\lambda}$$

And since $P_{ii} = 1 - \bar{D} = 1 - 3P_{ij}$, we have

$$P_{ii} = \frac{1}{4}+\frac{3}{4}e^{-\frac{4}{3}\lambda}$$

Appendix Three:
Geometric Probability Distribution and the Concept of Function

Suppose we perform a series of independent trials, each of which has two possible outcomes, success or failure, with probabilities of p and $q = 1 - p$, respectively (= *Bernoulli trials* after the Swiss mathematician Jacques Bernoulli). Since each trial can be either a success or a failure, there can be by chance several trial runs all of which may end in failure, until we finally score the first success. Let us designate the number of trials required before we reach success as X and then assess the probability $P(X = n)$ that success occurs in a particular n-th trial. For example, from a bag containing nine white marbles and one black one, we draw one marble (= trial No. 1). If it is not black (= failure) we return it to the bag and draw another one, and continue like this until we eventually draw the black marble (= success). What are the probabilities that we draw the black marble in the first, second, third, or n-th trial?

In each trial, the probability of success is given by the probability that all preceding trials were failures multiplied by the probability that this trial is a success. Thus the probability of success in the first trial is p and that of failure q. The probability of X being equal to 2 is the probability (q) of failure in the first trial times the probability of success in the second trial (p), that is qp. The probability of X being equal to 3 is the probability of failure in the first trial q and in the second trial (also q) times the probability p of success in the third trial, that is qqp or q^2p. Thus for the different values of n, the probabilities $P(X = n)$ are:

n	1	2	3	4	5	...	n
$P(X = n)$	p	qp	q^2p	q^3p	q^4p		$q^{n-1}p$

We therefore obtain a probability distribution specified by the general formula
$$P(X = n) = q^{n-1}p.$$
The terms of the distribution represent a numerical sequence such that each successive term may be obtained by multiplying by the same number [in our example, $1, q, q^2\ q^3\ q^4$, etc is always multiplied by p] and the ratio of each term (except the first) to the preceding one is a constant. Such a sequence is called *geometric progression (sequence)* and this particular form of the sequence is referred to as *geometric probability distribution*. The mean of the geometric distribution is $\mu = q/p$ and its variance $Var = q/p^2$. In the context of this book, a more relevant example of the geometric distribution is the distribution of probabilities that two genes chosen at random from a *Go* pool will be found to have a most recent common ancestor in the past $G_1, G_2, G_3,\dots Gn$ generations.

In the two rows of numbers above, a particular number in one row (n) has a corresponding value (P) in the second row. The two sets of numbers are the *variables* – quantities that may assume any one of a set of values. In each pair, the value of one of the numbers (the *dependent variable,* generally y) depends on the value of the other (the *independent variable,* generally x). The dependence, the relationship between the two numbers, is referred to as a *function* which can be represented mathematically by an *equation*, a statement that two mathematical expressions are equal. Take for example the following relation between two variables x and y:

x	0	1	2	3	4	5	6
y	2	4	6	8	10	12	14

It can be represented by the formula $y = 2x + 2$, which specifies the values of the dependent variable y corresponding to whatever value of x (the independent variable) we might choose. If, for example, we choose $x = 15$, then $y = (2)(15) + 2 = 32$. In mathematics, therefore, a function (f) is a correspondence that associates with each number x some other number $f(x)$ (read "f of x"). Thus, instead of $y = 2x + 2$ in our example we can write $f(x) = 2x + 2$.

The geometric probability distribution is an example of a function specified by the formula $P(X=n) = q^{n-1}p$ which can therefore be rewritten as $f(X=n) = q^{n-1}p$. Other examples are the binomial probability distribution, defined as

$$f(k;n,p) = B(k;n,p) = \binom{n}{k} p^k q^{n-k}$$

and the Poisson probability distribution specified by the formula

$$f(k;\lambda) = \frac{\lambda^k}{k!} e^{-\lambda}$$

Since these three functions specify probabilities, they are referred to as *probability functions* or *probability mass functions*. And since probability is in principle a frequency, yet another name for them is *frequency functions*; all these designations are synonyms.

The relationship between the two variables can be depicted not only *algebraically* by numbers, but also graphically or *geometrically*. Underlying the geometric rendition of functions is the realization by the French mathematician and philosopher René Descartes (Cartesius in its latinized form) that the position of any point in space can be specified by a set of numbers, relative to an appropriate reference system. In two-dimensional space, a plane, the *Cartesian reference system* consists of two lines drawn at right angles to each other, the *coordinate axes*. The horizontal line, the *x-axis*, is used as reference for the independent variable x, whereas the vertical line, the *y-axis,* serves as a reference for the dependent variable y. The point of intersection of the two lines is taken as the *origin* (O point) of the coordinate system. Distances to the right from the origin along the x-axis are expressed in positive values, whereas those to the left are given in negative numbers. Similarly, distances above the origin along the y-axis are positive and those below are negative. The position of a point P in the plane is specified by two numbers, its *coordinates, X* and Y, written as $P(X, Y)$. Here the X or *abscissa* gives the distance from the x-axis parallel to the y-axis, whereas the Y or the *ordinate* provides a distance from the y-axis parallel to the x-axis. The coordinates of the origin are O(0, 0).

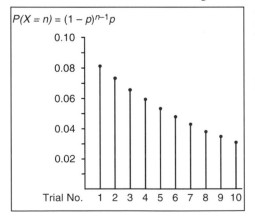

A graphical depiction of the geometric distribution of probabilities of picking the black marble from the bag of ten in the above example is shown on the left. Here the coordinate points (highlighted as enlarged dots) are specified by the number pairs (1; 0.081), (2; 0.073), (3; 0.066), (4; 0.059), (5; 0.053), (6; 0.048), (7; 0.043), (8; 0.038), (9; 0.035), and (10; 0.031). They are placed at the tips of vertical lines, the height of which corresponds to the probabilities that the first success occurs in the first, second, third, ... trial (the first and the second number in each of the coordinate pairs, respectively). The points identified by the individual coordinates and the lines are clearly separated from each other because the values of the independent variable, here the individual trials, are whole numbers (integers); there is no half or any other fraction of a trial. This type of random variable, which takes on isolated, countable values, is referred to as *discrete*. Here the opportunities for success arise only when a trial is conducted and the corresponding discreteness of the probability values is the consequence of the separateness of the independent variable (the individual trials).

One can, however, imagine a situation in which opportunities for an event to occur (for "success") exist continually, without separation by periods in which an event cannot occur. Such a continuum is commonly provided by time. You can divide time into intervals, but you can then subdivide each interval further, and each subdivision further still, *ad infinitum*. A random event can occur in any of these subdivisions and a random variable distributed over such a continuum is said to be *continuous*. The

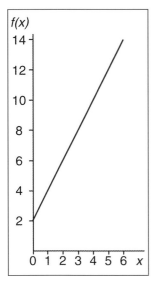

closest graphic approximation to a distribution of this type is a smooth line (or a curve) like the one on the left. The line (= *graph*) represents a graphic depiction of the algebraic equation (function) $y = 2x + 2$ discussed earlier. It can be obtained by choosing x equal not only to 0, 1, 2, 3, 4, 5, 6, ... but also to all the numbers between 0 and 1, to those between 1 and 2, and so on. Hence geometric elements such as lines and curves are described by equations, each equation specifying the coordinates of the element's points. Thus a *straight line*, a geometric element completely determined by two of its points, is described by an equation of the type $y = a + bx$. In this example of a *linear equation*, x and y are the two variables, whereas a and b are constants, the latter being a *coefficient*, a constant that multiplies a variable. The equation represents a function which can be specified for any value of x. The left-hand side of the equation is fully determined by the right-hand side so that the equation can be written without the use of the y as $f(x) = a + bx$. In situations in which mathematicians want to imply the existence of a functional relationship between x and y but, for whatever reason, do not want to (or cannot) specify its type, they may also write $y = f(x)$.

The continuum does not allow us to define probabilities as we did in the case of discrete random variables. In the latter, we can consider the probability that the success will occur in, say, trial No. 4 and obtain a sensible answer by using the geometric distribution. In the case of a continuous random variable, we cannot specify the probability that an event will occur at *exactly* 4 o'clock because "4 o'clock" is in reality a time interval which can be subdivided into an infinite number of subintervals. To which of these subintervals does "4 o'-clock" refer? And even if we could identify the subinterval, we would then face the same problem again since each subinterval could be partitioned even further into sub-subintervals. Obviously, in the case of continuous random variables we must take a different approach to specifying probabilities (see Appendix Four).

Appendix Four:
Exponential Functions and Exponential Probability Distribution

Consider a function $y = b^x$ in which x and y are the independent and dependent variables, respectively, and b is the *base*, a constant. Here the independent variable x is the exponent and so the family of functions of this type are called *exponential*. An *exponent* is a number placed in a superscript position to the right of another number to indicate repeated multiplication (e.g., $b^2 = bb$; $b^3 = bbb$). It can be any real number (i.e., a number that can be represented physically on a scale), either rational (an integer – a whole number – or a number that can be written as a ratio of two integers) or irrational (i.e., a number that cannot be expressed as a ratio of two integers, for example $\sqrt{2}$ or e), positive or negative. If the exponent is negative, then the expression b^{-x} can be written as a reciprocal with a positive value of the exponent: $b^{-x} = 1/b^x$. The b in the exponential function can be any positive number. In the way of an example, let us make the base equal to 2 and consider values of the exponent from –5 to +5. We then obtain the following corresponding values of the dependent variable y:

x	–5	–4	–3	–2	–1	0	1	2	3	4	5
$y = 2^x$	1/32	1/16	1/8	1/4	1/2	1	2	4	8	16	32
	= 0.03	= 0.06	= 0.12	= 0.25	= 0.5						

($b^0 = b^{(x-x)} = b^x/b^x = 1$; $b^1 = b$.) The 11 points placed in the plane relative to the reference system by the chosen integers of the independent variable x merely suggest the shape of the curve described by the function. In reality, an infinite number of points exists in between these points because the function is that of a continuous variable. The graph on the

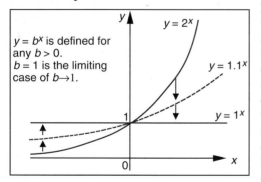

left depicts part of the $y = 2^x$ curve. Note that as x increases in one direction, so does y, slowly at first, but then at an ever faster rate. Similarly, as x decreases in the opposite direction, y decreases with it but at an ever slower rate, coming closer and closer to 0 but never quite reaching it. If there were a number that we could call negative infinity (– ∞), the graph would actually touch the x-axis. Since, however, infinity is not a number but merely a symbol for an increase without an end, there is always a larger number than the largest we can name. All we can say is that if x *could* reach infinity, the corresponding y value would become 0 and hence that 0 is the *limit* of the $y = 2^x$ function as x tends toward negative infinity ($x \rightarrow -\infty$), a statement which in the symbolic language of algebra is written as $\lim_{x \to -\infty} 2^x = 0$. In the opposite direction, as x increases and tends toward positive infinity ($x \rightarrow +\infty$), so does y, and we can write $\lim_{x \to -\infty} 2^x \rightarrow -\infty$.

A characteristic feature of the graph defined by the $y = 2^x$ or any other *exponential equation* is that it changes continuously, always ascending or descending relative to the coordinate system and the rate of change differs depending on the part of the curve we focus on. In some parts the curve ascends or descends steeply, whereas in other parts it changes almost imperceptibly. The problem we now face is to express the rate of change mathematically. Specifically, we would like to determine the rate of change at a particular point P as shown in the diagram below. But what exactly do we mean by "rate of change"?

The word "rate" has different meanings depending on the context. In the present context, it refers to the value obtained by dividing the change in a function of the dependent variable y by the change in the independent variable x. To find the rate of change at point P

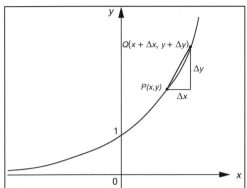

with coordinates (x, y), we choose another point Q on the curve close to P. The horizontal and vertical distances between P and Q are Δx and Δy, respectively, so that the coordinates of point Q are $(x + \Delta x, y + \Delta y)$. Here the capital Greek letter Δ (delta) is used to designate an interval, a distance between two points; when the distance becomes very small, the capital letter is replaced by the small letter δ. Note, however, that Δ and δ are not algebraic quantities and that the expressions Δx and δx do not signify "$\Delta (\delta)$ times x". The Greek letters, in this case, cannot be separated from the x; they simply stand for "small interval in x".

The straight line connecting the points P and Q is the *chord* of the curve and the angle the line makes with the horizontal is its *gradient* (*slope*) – the vertical distance traveled per horizontal distance, here $\Delta y / \Delta x$ or $\delta y / \delta x$. If we now move the point Q along the curve toward P, δx and δy decrease in size and when they get very, very small, they begin to approach 0. They never reach 0, however, so the quotient $\delta y / \delta x$ remains finite. When the interval becomes vanishingly small, the chord PQ becomes indistinguishable from the tangent at point P (a tangent being a line that touches a curve at one point only). When that happens, the gradient of the chord PQ becomes the gradient of the tangent at P. We say that the gradient of the tangent at P is the *limit* of the gradient of the chord PQ as PQ becomes vanishingly small. The gradient of the chord is $\delta y / \delta x$ and the gradient of the tangent is designated as dy/dx. The statement "the limit of the gradient of the chord $\delta y / \delta x$, as δx tends to zero, is the gradient of the tangent dy/dx" is expressed in mathematical shorthand thus:

$$\lim_{\delta x \to 0} \left(\frac{\delta y}{\delta x} \right) = \frac{dy}{dx}$$

The ratio dy/dx, the rate of change of a function with respect to the independent variable is called the *derivative* (*differential coefficient, derived function*). For a function $y = f(x)$, the derivative can also be written as y', $Df(x)$, $D_x y$, or $f'(x)$ (read the last as "f prime of x"). Here again, the d's in the expressions dx and dy are inseparable from x and y and are used to designate vanishingly small intervals.

The process of obtaining a derivative of a function is called *differentiation* and the branch of mathematics concerned with calculating the rate of change of a function with respect to changes in the independent variable is referred to as *differential calculus* (*calculus* being that part of mathematics dealing with rates of change in general, using the idea of a limit). The essence of differentiation is the assessment of small changes in a function and in the independent variable and the determination of the limiting value of the ratio of such changes.

From the specific example $y = 2^x$ we now return to the general form of the exponential function $y = b^x$. Using the diagram above, at point Q we have $y + \delta y = b^{x+\delta x}$. If we subtract $y = b^x$ from $y + \delta y = b^{x+\delta x}$, we obtain $\delta y = (b^{x+\delta x}) - b^x$. By applying the rule of exponentiation that states $a^{m+n} = a^m a^n$, we obtain $\delta y = b^x b^{\delta x} - b^x$ which we can rewrite as $\delta y = b^x (b^{\delta x}$

– 1). The rate of change is then $\delta y/\delta x = b^x(b^{\delta x} - 1)/\delta x$. A common method of finding the limit of a rate of change is to let δx tend to zero. But if δx is made equal to 0 in the last equation, the fraction becomes 0/0 and so indeterminate. Here, therefore, a different method must be used. We replace δx by a single letter h and define the derivative as

$$\frac{dy}{dx} = \lim_{h \to 0} \frac{b^x(b^h - 1)}{h}$$

Because the limit in this equation concerns only the variable h, while x can be regarded as fixed, we can remove b^x from the limit and define the derivative as

$$\frac{dy}{dx} = b^x \lim_{h \to 0} \frac{b^h - 1}{h}$$

If we now denote the value of the limit by the letter k, we reach the important conclusion that if $y = b^x$, then $dy/dx = kb^x = ky$. Expressed in words: the derivative of an exponential function is proportional to the function itself – it equals the function multiplied by the constant of proportionality k.

This conclusion begs the question: is there a situation in which k equals 1, so that the derivative is, in this case, equal to the function itself? In other words, is there a value of b such that $\lim_{h \to 0}(b^h - 1)/h = 1$? To find out, we set $(b^h - 1)/h = 1$ for finite h and seek the value of b that satisfies this equation. From $(b^h - 1)/h = 1$ we obtain $b^h = 1 + h$ so that $b = \sqrt[h]{(1-h)}$ = $(1 + h)^{1/h}$. (In the last step, we replaced the radical sign $\sqrt{}$ with a fractional exponent by applying the rule $\sqrt[a]{b} = b^{1/a}$.) By letting h tend to 0, we obtain $b = \lim_{h \to 0}(1 + h)^{1/h}$ and this, then, should be the value of b that fills the bill in the equation $(b^h - 1)/h = 1$. A quick check confirms this expectation: when b in this equation is replaced by $(1 + h)^{1/h}$, the equation does indeed come to 1. Hence, the expression $\lim_{h \to 0}(b^h - 1)/h$ equals 1 applies when, and only when, b equal to the limit of $(1 + h)^{1/h}$ is chosen. If we now replace $1/h$ by the letter m, then as $h \to 0$, m will tend to infinity and we can write $b = \lim_{m \to \infty}(1 + 1/m)^m$. Finally, we denote the expression $\lim_{m \to \infty}(1 + 1/m)^m$ by the letter e and so come to the conclusion that when the number e is chosen as base, the exponential function is equal to its own derivative – or, if $y = e^x$, then $dy/dx = d(e^x)/dx = e^x$ and $dy/dx = y$. (Equations of this type that contain a derivative are known as *differential equations*.)

But why all this fuss about an exponential function being its own derivative? The reason is that in nature, numerous phenomena are best described by differential equations possessing this property. In all these phenomena the rate of change of a certain quantity is proportional to the quantity itself. Of the examples of such phenomena that we have already encountered in this text, we remind the reader of two – the growth of a population and the decay of a radioactive substance – and use them to illustrate the general characteristics of exponential functions.

Consider a population consisting of 1,000,000 individuals at the beginning of a particular year y_0. We assume that the rate of population growth is 0.01 or 1 percent every year. Therefore during the year y_0, 10,000 individuals are added to the population through births, so that at the beginning of the following year, y_1, the population size has increased to 1,010,000 individuals.* During the year y_1, 1 percent of 1,010,000 individuals or 10,100 individuals are again added to the population so that at the beginning of the year y_2 the population size has increased to 1,020,100 individuals. In this manner the population con-

---

* In this model, deaths of individuals are ignored; it would, however, be a simple matter to incorporate them into it. If, for example, we were to assume that births exceed deaths by 10,000, then 20,000 births and 10,000 deaths per year would give the same result as 10,000 births and no deaths.

tinues to grow, increasing its size every year by 1 percent of the size attained at the end of the preceding year. The population size changes at a rate λ (lambda) which is given by the ratio of the number of births in a given year to the population size at the beginning of that year. Thus in year y_0, the number of births was 10,000 and the population size at the beginning of y_0 was 1,000,000 individuals and this ratio is kept constant for all other years. Denoting the initial population size at the beginning of y_0 as N_0, the population size at the end of given year y as N_y, the number of births during y_0 as B_0, and the number of births during a year y as B_y, we can define both N_y and B_y by similar formulae: $N_y = N_{y-1}(1 + \lambda) = N_0(1 + \lambda)^y$ and $B_y = B_{y-1}(1 + \lambda) = B_0(1 + \lambda)^y$. In our example, the values for the first five years are these:

y	$N_0 (1 + \lambda)^y$	N_y	$B_0(1 + \lambda)^y$	B_y
0	$1{,}000{,}000(1 + 0.01)^0$	1,000,000	$10{,}000(1 + 0.01)^0$	10,000
1	$1{,}000{,}000(1 + 0.01)^1$	1,010,000	$10{,}000(1 + 0.01)^1$	10,100
2	$1{,}000{,}000(1 + 0.01)^2$	1,020,100	$10{,}000(1 + 0.01)^2$	10,201
3	$1{,}000{,}000(1 + 0.01)^3$	1,030,301	$10{,}0001 + 0.01)^3$	10,303
4	$1{,}000{,}000(1 + 0.01)^4$	1,040,604	$10{,}000(1 + 0.01)^4$	10,406

The basic relationship in population growth specifies that if there is a certain number of individuals at a particular time, then the number of individuals at a later point in time is given by the number of individuals at the earlier point multiplied by a constant. When the time unit is one year, the relationship is specified by the equation $N_y = (1 + \lambda)N_{y-1}$, as stated earlier, or $N_{y+1} = (1 + \lambda)N_y$, where $(1 + \lambda)$ is a constant. If we now divide one year into n intervals, each interval having a length of $\Delta t = 1/n$ (so that $n\Delta t = 1$), we can denote λ in units Δt as $m\Delta t$, where m is the particular number of Δt intervals. We can then express the basic relationship in the form $N_{t+\Delta t} = (1 + m\Delta t)N_t$. A modified form of this equation is $N_{t+1} = (1 + m\Delta t)N_{t+1-\Delta t}$, where $N_{t+1-\Delta t}$ is the number of individuals at a point in time preceding that at which the count of N_{t+1} was taken at an interval Δt. We can then replace N_{t+1} in the expression $N_{t+1-\Delta t}$ on the right-hand side of the equation by $(1 + m\Delta t)N_{t+1-\Delta t}$ to obtain $N_{t+1} = (1 + m\Delta t)[(1 + m\Delta t)N_{t+1-2\Delta t}] = (1 + mt)^2 N_{t+1-2\Delta t}$. In this expression, we can again replace N_{t+1} and continue in this manner to obtain a general formula $N_{t+1} = (1 + m\Delta t)^n N_t$. Since we defined $\Delta t = 1/n$, we can rewrite the formula as $N_{t+1} = (1 + m/n)^n N_t$. If n tends toward infinity ($n \to \infty$; i.e., one year is divided into an infinite number of small intervals of length $\Delta t = 1/n$), then $(1 + m/n)^n = e^m$ for a large n and we can rewrite the formula as $N_{t+1} = e^m N_t$. The constant $(1 + \lambda)$ in the equation $N_{y+1} = (1 + \lambda)N_y$ can therefore be replaced by e^m, and because $(e^m)^t = e^{mt}$, the equation $N_y = N_0(1 + \lambda)^y$ can be rewritten as $N_t = N_0 e^{mt}$. When λ is much smaller than 1, so is m and it follows from the equation $1 + \lambda = e^m$ that $\lambda = m$ approximately.

In the case of radioactivity, we have at time t_0 a certain number of radioactive atoms N_0 in the sample. During the time interval from t_0 to some later point t, some of the atoms

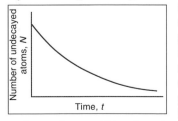

(their nuclei to be more precise) disintegrate (see Chapter Nine) so that the number of undecayed atoms gets progressively smaller and reaches a value of N at time t. We can therefore consider time as the independent continuous variable and the number of undecayed atoms that remain at any specific time point as the dependent variable to obtain the type of curve shown on the left (see also Figure 9.5). We say that the number of undecayed atoms N is a function of time t.

Now we want to describe the relationship between t and N mathematically, specifically to define the rate at which N changes (decreases) with time. The rate is defined as the small number of atoms dN that disintegrate per short time interval dt (= *rate of decay*) or dN/dt. Measurements indicate that the rate decreases proportionately to the number of unde-cayed atoms remaining at a given instant.Mathematically, the proportionality can be ex-

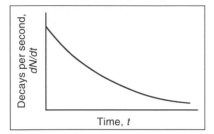

pressed as $dN/dt \propto N$. The proportionality sign "$\propto$" can be replaced by the "equals" sign if a constant of proportionality – call it λ (here the *decay constant*) – is introduced on the right-hand side: $dN/dt = -\lambda N$ (the minus sign indicates that the rate is decreasing). We thus obtain a new function of the type $dN/dt = f(N)$ which makes it explicit that the change in the number of undecayed atoms is a function of the *number of atoms*, not time (in contrast to the earlier function). The plot of this function yields a curve similar to the graph above. The decay constant is different for different isotopes. The greater the λ, the greater the rate of decay and the more radioactive the isotope is said to be.

Let us now look more closely at what happens in the short time interval dt during which atoms are disintegrating one after the other at random. Let us call the time period between two consecutive disintegrations the *decay interval*. The length of this interval depends on the number of undecayed atoms remaining in the sample. If n decays are registered in the time interval $t - t_0$ for a given isotope with a decay constant λ per unit time and the number of radioactive atoms at time t_0 is N_0, then $\lambda = n/(N_0 t)$ [and hence $\lambda t = n/N_0$ and $1/N_0 = (\lambda t)/n$]. The decay interval is constantly lengthening as more and more atoms decay. Let us further take as a *unit decay interval* the interval between t_0 and the first decay and use it to gauge all subsequent decay intervals. After the first decay interval, the probability that a specific atom has decayed is $1/N_0$ and the probability that it has not decayed is $1 - 1/N_0$. At each consecutive unit the probability of non-decay is reduced by $1 - 1/N_0$. Thus, after two units the probability that a given atom has decayed is $2/N_0 - (1/N_0)^2$ and the probability that it has not decayed is $(1 - 1/N_0)^2$. (In the second unit interval there is one atom fewer than before, so the probability of decay is reduced slightly relative to the first decay, but the difference is so small that here it can be ignored.) After n unit intervals, the probability that a given atom has decayed is $1 - (1 - 1/N_0)^n$ and the probability that it has not decayed is $(1 - 1/N_0)^n$. If we now replace $1/N_0$ by $(\lambda t)/n$ (which follows from the definition of the decay constant above), the probability of a given atom remaining undecayed after n unit intervals (which corresponds to the interval from t_0 to t) is $[1 - (\lambda t)/n]^n$. This expression represents a binomial whose limit for large n is $e^{-\lambda t}$. In a population of N_0 radioactive atoms, the number of undecayed atoms at time t will therefore be $N = N_0 e^{-\lambda t}$.

The growth of a population and the decay of radioactive atoms are both examples of processes in which the number of elements, N, (individuals in the former, atoms in the latter) changes continuously with time t. Here time is the independent variable and the number of elements in an instant of time is the dependent variable, whereby the instancy of the determination is expressed as the ratio dN/dt. The behavior of the processes is described by an exponential function which can generally be written as $dN/dt = \lambda N$, where λ is positive for population growth and negative for the radioactive decay and in both cases it is a proportionality constant of the rate of change. The solution of this differential equation under the initial condition of $N = N_0$ at an arbitrary time point t_0 is $N = N_0 e^{\lambda t}$. In this equation, N_0 is a constant specifying that the behavior of the process itself depends on its initial condition (the number of individuals or the number of radioactive atoms) at time t_0. The product λt reflects the fact that the dependent variable N changes during the process at a rate which itself changes continually with time. Finally, e in the solution, the base of the exponent, specifies that in the process the rate of change of the dN/dt with respect to time (more generally, the rate of change of y with respect to x) is equal or proportional to N (generally y). The e itself is defined as the limit of the binomial $(1 + 1/n)^n$. The equation $dN/dt = \lambda N$ represents a special case of the general exponential function $y = b^x$ or $y = ab^{cx}$ [in the former case the intersect of the exponential curve with the y-axis has the coordinates $(0, 1)$, whereas in the latter the coordinates are $(0, a)$ and the rate of change is proportional to c]. It is of such importance that it is often referred to as *the* exponential function.

The exponential function is suited to dealing with phenomena such as probabilities which assume values from 0 to 1. We saw in Appendix Two how the function is used in the Poisson probability distribution which deals with discrete variables (number of events per time interval). The function is, however, particularly useful in dealing with continuous variables such as time. The Poisson distribution describes one aspect of the Poisson process – the probability of an event happening or not happening in a specified time interval. But the process also has another aspect which is revealed in the diagram below:

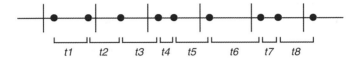

$$t1 \quad t2 \quad t3 \quad t4 \quad t5 \quad t6 \quad t7 \quad t8$$

where the dots represent events, the vertical lines time units, and $t1$ through $t8$ intervals between successive events (compare the diagram with a similar one that appears in Chapter Five). This other aspect concerns the length of time t from the moment an event takes place until the moment the next event happens, an interval referred to as the *waiting time*. In the Poisson process, the length of time spent waiting for the next event is exponentially distributed. If events occur at a rate λ per unit time, the intervals between successive events are exponentially distributed with the mean of $1/\lambda$.

The formula for the exponential probability distribution can be derived from the geometric distribution. The latter deals with discrete variables and the probability values can therefore be represented as the heights of vertical lines (see diagram in Appendix Three).

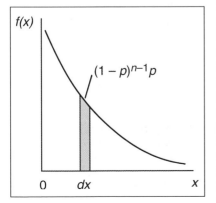

The exponential probability distribution describes the behavior of continuous variables and here the probability values are represented as the area under the curve (as shown in the diagram on the left). Areas are given by the product of two values, height and width. For the area under an exponential curve, the unit of width is the increment of x, the dx, extending from x to $x + \Delta x$. The exponential distribution represents the continuous limit of the geometrical distribution given by the function $P(x = n) = q^{n-1}p = (1 - p)^{n-1}p$. To derive the formula, we replace p by λdx to obtain $(1 - \lambda dx)^{n-1}\lambda dx$ and then replace dx by $x/(n-1)$ to obtain $[1 - (\lambda x)/(n-1)]^{n-1}$. As $n-1$ approaches infinity, we get $= e^{-\lambda x}$ and hence $f(x) = \lambda e^{-\lambda x}dx$, which is the formula for the exponential probability distribution.

Appendix Five: List of Symbols and Abbreviations Used in the Text

A:
in a DNA or RNA sequence, a nucleotide containing the nitrogenous base adenine.

ABO:
human blood group system.

A.D.:
anno Domini, after year of the common era.

Alu:
transposable element in primate genome originally defined by using a restriction enzyme isolated from *Arthrobacter luteus*.

b:
branch length.

B(k;n,p):
binomial distribution.

B.C.:
before common era (Christ).

BIF:
banded-iron formation.

BP:
before present (i.e., before 1950).

by:
billion years.

C:
in a DNA or RNA sequence, a nucleotide containing the nitrogenous base cytosine.

CA:
common ancestor.

CFC:
chlorofluorocarbon.

CMP-NeuAc:
cytidine monophosphate-N-acetyl neuraminic acid.

d:
evolutionary distance; actual number of nucleotide substitutions per site (corrected for hidden substitutions); $d = 2t\mu$.

d:
proportion of nucleotide differences per site corrected for multiple hits = *genetic distance* (sequence divergence).

D:
standard genetic distance; Nei's standard genetic distance; $D = -ln(I)$.

D (loop):
displacement loop in mtDNA replication.

D-:
dextro; see L-.

DNA:
deoxyribonucleic acid; chain of nucleotides in which the sugar residue is deoxyribose.

dsDNA:
double-stranded DNA.

dx, dy (δx, δy):
derivate; the rate of change of function $y = f(x)$ with respect to the independent variable (see Appendix Four).

e:
the limit of $(1 + 1/n)^n$ as $n \to \infty$; the sum of the infinite series
$1 + \dfrac{1}{1!} + \dfrac{1}{2!} + \dfrac{1}{3!} + ...$ the base of natural logarithms. Its value is 2.718281828...

E(t):
expected mean coalescence time; $E(t) = 2N_e$ generations for two randomly selected nuclear genes; $E(t) = 4N_e$ for all genes of a nuclear locus in a population.

E(x):
an expected outcome of x.

EF:
elongation factor.

f:
probability of fixation of an allele.

F (F$^+$,F$^-$):
fertility plasmid of bacteria.

f(x):
function as in $y = f(x)$, a rule that assigns to every element x of a set X a unique element y of a set Y.

f_s:
$$= \frac{\sqrt{S}}{2M}\left[ln\left\{\frac{S}{16\pi M^2}\right\}\right]^{-\frac{3}{2}}$$
where $S = 2N_e s$ and $M = N_e uj$; scaling factor to convert neutral gene genealogy to genealogy of genes evolving unter balancing selection.

F_{ST}: fixation index; F_{ST} statistic; measure of variation in gene frequencies in a population; $F_{ST} = (h_T - h_S)/h_T$.

Fuc: fucose (sugar).

G: generation; period from one point of a life cycle to the corresponding point of the following cycle.

G_0, G_1, G_2, etc: generation 0, 1, 2, etc.

G: in a DNA or RNA sequence, a nucleotide containing the nitrogenous base guanine.

Gal: D-galactose (sugar).

Gal-NAc: N-acetyl-D-galactosamine.

HBB: gene coding for hemoglobin β-subunit.

HFE: hereditary hemochromatosis gene involved in regulation of iron (Fe) metabolism.

h_j: heterozygosity of a population j; the probability of drawing two different alleles (i and j) from a population at random; $h_j = 1 - H_j$.

h_j: heterozygosity of a single subpopulation j.

H_j: homozygosity of a population j; the probability of drawing two identical alleles from a population at random;

$$H_j = \sum_{i=1}^{n} p_{ij}^2$$

h_S: the average heterozygosity of a set of subpopulations;

$$h_S = \sum_{j=1}^{s} h_j / s$$

h_T: heterozygosity of a total population obtained by sampling the entire population at random (disregarding its division into subpopulations).

HVRI and II: hypervariable regions I and II in the control region of mtDNA.

I: genetic identity between two populations;

$$I = J_{XY} / \sqrt{J_X J_Y}$$

i (j): general designation for any allele.

$\bar{p}_i$: the average gene frequency of allele i in a population divided into subpopulations;

$$\bar{p}_i = \sum_{j=1}^{s} p_{ij} / s$$

J_X (J_Y): the probability that by sampling a population X (Y), we choose the same allele.

J_{XY}: the probability that by sampling populations X and Y we choose the same allele from both populations.

k: number in a data set.

k: probability of fixation of a neutral allele; $k = u$.

k: the number of nucleotide differences between two alleles.

kb: kilobase pairs (= 1000 nucleotide pairs).

K-T: Cretaceous-Tertiary (event, extinction).

λ (lambda): constant; the mean (and variance) in the Poisson probability distribution.

L: likelihood; estimate of a parameter from observed data under a specified model.

L:	the number of nucleotide sites compared.
L-:	levo, a certain configuration of atoms around a carbon atom in an organic compound (e.g., amino acids) responsible for the rotation of polarized light in a certain direction.
$L(p;k,n)$:	likelihood function.
lg:	logarithm to base 2 (see *log*).
ln:	natural logarithm (see *log*).
log:	for a positive number n, the logarithm of n (written as log n) is the exponent that indicates the power to which a number b must be raised to produce the number n. The abbreviation is generally used for the base of the logarithm equal to 10. Such logarithms are called *common*.
μ (mu):	mutation rate per site per generation.
μ (mu):	arithmetic mean of a population.
m:	migration rate.
$\bar{m}$:	mean of a sample.
ME:	minimum evolution; a method of phylogenetic reconstruction striving to obtain a phylogenetic tree with the smallest sum of branch lengths.
Mhc:	major histocompatibility complex.
ML:	maximum likelihood; a criterion for estimating a parameter from observed data under a particular model.
MP:	maximum parsimony; a principle of minimizing the number of events needed to explain the observed data.
MRCA:	most recent common ancestor.
mRNA:	messenger ribonucleic acid; RNA whose nucleotide sequence is translated into an amino acid sequence of a protein.
mtDNA:	mitochondrial deoxyribonucleic acid.
my:	million years.
n:	number of trials in a statistical experiment.
n:	number of items (e.g., nucleotide sites) compared or analyzed.
N:	census population size.
N_b:	size of a population in a bottleneck phase.
n_d:	number of nucleotide (amino acid) differences between two sequences.
N_e:	effective population size; the number of reproducing individuals in a population.
NeuAc:	*N*-acetyl-neuraminic acid.
NeuGc:	*N*-glycolyl-neuraminic acid.
N_f:	effective female population size.
N_g:	average number of breeding individuals in a group.
NJ:	neighbor joining; a method of phylogenetic reconstruction based on distance data.
$\binom{n}{k}$:	binomial coefficient, equal to $\dfrac{n!}{k!(n-k)!}$
numt:	a fragment of mitochondrial DNA inserted into the nuclear genome.
π (pi):	average nucleotide diversity; the mean number of nucleotide differences per site between two sequences chosen randomly from a population (gene pool);

$$\pi = \sum_{ij}^{q} d_{ij} x_i x_j$$

p:	probability of success in a single Bernoulli trial such as a single toss of a coin.
p:	proportion of different nucleotides (amino acid residues) between two sequences uncorrected for multiple hits; $p = n_d/n = p$-distance (evolutionary distance).
P:	probability (relative frequency); the ratio of the number of outcomes to the total number of possible outcomes.
P(k;λ):	Poisson probability distribution.
PAR:	pseudoautosomal region of the Y chromosome.
PCR:	polymerase chain reaction; a process for amplifying a target DNA sequence manifold.
PDHA1:	pyruvate dehydrogenase E1-alpha polypeptide 1.
$p_i(j)$:	frequency of allele *i* (*j*) in a subpopulation.
$P_{ii}(t)$:	probability that nucleotide *i* at time t_o is the same as the nucleotide at time *t*.
$P_{ij}(t)$:	probability that nucleotide *i* at time t_o changes to nucleotide *j* at time *t*.
PMRCA:	place of existence of the most recent common ancestor.
q:	probability of failure in a single Bernoulli trial such as a single toss of a coin.
r:	rate of nucleotide (amino acid) substitutions per site per year; $r = d/(2t)$.
R:	rate of population growth.
rDNA:	deoxyribonucleic acid coding for ribosomal ribonucleic acid.
RFLP:	restriction fragment length polymorphism.
RNA:	ribonucleic acid; a chain of nucleotides in which the sugar is ribose.
rRNA:	ribosomal ribonucleic acid; a component of ribosomes, which functions in the translation of mRNA into protein.
rt:	mean number of nucleotide (amino acid) substitutions per site (position) during a time interval *t*.
Σ (sigma):	summation sign; indicates a summation of a sequence of numbers.
σ (sigma):	standard deviation (s.d.); the positive square root of the variance.
σ^2:	variance of a population; see *Var(X)*.
$\sigma^2/\sqrt{n}$:	standard error (s.e.).
s:	number of subpopulations.
s:	selection coefficient (intensity); a measure of the intensity of natural selection; calculated as the proportional reduction in gametic contribution of one genotype compared to that of a standard genotype.
S:	Svedberg unit of sedimentation rate, a rate at which a particle settles in a centrifugal field.
s^2:	variance of a sample; see *Var(X)*.
SRY:	sex-determining gene on Y chromosome.
ssDNA:	single-stranded DNA.
θ (theta):	$4N_e\mu$.
t or T:	time in generations; the length of a time interval.
T:	in a DNA sequence, a nucleotide containing the nitrogenous base thymine.
t_b:	duration of a population bottleneck phase.
TF:	transcription factor.
TMRCA:	time period of the existence of the most recent common ancestor.

tRNA:	transfer ribonucleic acid; RNA that associates with specific amino acids during protein synthesis.
u:	mutation rate per gene per generation.
U:	in an RNA sequence, a nucleotide containing the nitrogenous base uracil.
UV:	ultraviolet (light).
UPGMA:	unweighted pair-group method with arithmetic mean.
$Var(X)$:	variance; the square of the standard deviation, s.d.; a measure of dispersion (the spread of a distribution); calculated from

$$Var(X) = \sum_{i=1}^{n} \left(\frac{x_i - \bar{x}}{n} \right)^2$$

where x is a random variable, x_i is the realized value of the variable, $\bar{x}$ is the mean, and n is the total number of items.

w:	fitness; the probability of transmission of a gene from one generation to the next; $w = 1 - s$.
x:	the horizontal coordinate in the Cartesian system.
x:	specific value of a random variable (estimate).
$\bar{x}$:	(arithmetic) mean; the sum of a set of observations divided by the total number of observations.
X:	random variable (estimator).
X:	one of the sex chromosomes.
y:	vertical coordinate in the Cartesian system.
Y:	one of the sex chromosomes (male).
YAP:	Y chromosome *Alu* element, polymorphic.
!:	factorial; a number obtained by multiplying all the positive integers less than or equal to a given positive integer.
∞ :	infinity; in limits of functions such as $y = 1/x$, the symbol $y \to \infty$ stands for "as x tends to zero, y tends to infinity".
$\propto$:	symbol for proportionality.

Sources and Further Reading

Chapter One

Andersen, W. *Gauguin's Paradise Lost.* Viking Press, New York 1971
Cavendish, R. *Man, Myth, and Magic. An Illustrated Encyclopedia of the Supernatural.* Vol. 1-24. Marshal Cavendish, New York 1970
Gauguin, P. *Noa Noa.* (trans. D. F. Theis). Chronicle Books, San Francisco 1994
Leeming, D. and Leeming, M.A. *A Dictionary of Creation Myths.* Oxford University Press, New York 1994
O'Brien, F. (ed.) *Gauguin's Letters From the South Pacific Seas.* (Translated by R. Pielkovo.) New York 1992
Sweetman, D. *Paul Gauguin. A Life.* Simon & Schuster, New York 1995
The Holy Bible. King James Version. CAMEX International, New York 1989

Chapter Two

Dobel, C. (ed.) *Antony van Leeuwenhoek And His „Little Animals".* Dover Publications, New York 1960
Fruton, J.S. *Molecules and Life. Historical Essays on the Interplay of Chemistry.* Wiley-Interscience, New York 1972
Gillispie, C.C. (ed.) *Dictionary of Scientific Biography. Vol. 1-8.* Charles Scribner's Sons, New York 1981
Orel, V. *Gregor Mendel. The First Geneticist.* (Translated by S. Finn.) Oxford University Press, Oxford 1996
Robinson, G. *A Prelude to Genetics. Theories of a Material Substance of Heredity: Darwin to Weismann.* Coronado Press, Lawrence, KS 1979

Chapter Three
Sources

Ajioka, R.S., Jorde, L.B., Gruen, J.R., Yu, P., Dimitrova, D., Barrow, J., Radisky, E., Edwards, C.Q., Griffen, L.M. & Kushner, J.P. Haplotype analysis of hemochromatosis: evaluation of different linkage-disequilibrium approaches and evolution of disease chromosomes. *Am. J. Hum. Genet.* 60:1439-1447, 1997
Andrews, N.C. Disorders of iron metabolism. *N. Engl. J. Med.* 341:1986-1995, 1999
Burgoyne, P.S. The mammalian Y chromosome: a new perspective. *BioEssays* 20:363-366, 1998
Chang, J.T. Recent common ancestors of all present-day individuals. *Adv. App. Prob.* 31:1002-1026, 1999
Feder, J.N., Penny, D.M., Irrinki, A., Lee, V.K., Lebrón, J.A., Watson, N., Tsuchihashi, Z., Sigal, E., Bjorkman, P.J. & Schatzman, R.C. The hemochromatosis gene product complexes with the transferrin receptor and lowers its affinity for ligand binding. *Proc. Natl. Acad. Sci. USA* 95:1472-1477, 1998
Hammer, M.F. & Zegura, S.L. The role of the Y chromosome in human evolutionary studies. *Evol. Anthropol.* 5:116-134, 1996
Kühn, L.C. Iron overload: molecular clues to its cause. *Trends Biochem. Sci.* 24:164-166, 1999

Rochette, J., Pointon, J.J., Fisher, C.A., Perera, G., Arambepola, M., Kodikara Arichchi, D.S., De Silva, S., Vandwalle, J.L., Monti, J.P., Old, J.M., Merryweather-Clarke, A.T., Weatherall, D.J. & Robson, K.J.H. Multicentric origin of hemochromatosis gene (*HFE*) mutations. *Am. J. Hum. Genet.* 64:1056-1062, 1999

Further Reading

Brown, T.A. *Genomes*. BIOS Scientific Publishers, Oxford, England 1999
Griffiths, A.J.F., Miller, J.H., Suzuki, D.T., Lewontin, R.C. & Gilbert, W.M. *An Introduction to Genetic Analysis*. 7th edn. W.H. Freeman, New York 2000
Strachan, T. & Read, A.P. *Human Molecular Genetics*. BIOS Scientific Publishers, Oxford 1996

Chapter Four
Sources

Box, J.F. & Fisher, R.A. *The Life of a Scientist*. John Wiley, New York 1978
Chappell, M.A. & Snyder, L.R. Biochemical and physiological correlates of deer mouse α-chain hemoglobin. *Proc. Natl. Acad. Sci. USA* 81:5484-5488, 1984
Clark, R. *The Life and Works of J.B.S. Haldane*. Oxford University Press, Oxford 1968
Jessen, T.-H., Weber, R.E., Fermi, G., Tame, J. & Braunitzer, G. Adaptation of bird hemoglobins to high altitudes: Demonstration of molecular mechanism by protein engineering. *Proc. Natl. Acad. Sci. USA* 88:6519-6522, 1991
Kimura, M. Evolutionary rate at the molecular level. *Nature* 217:624-626, 1968
Kleinschmidt, T., März, J., Jürgens, K.D. & Braunitzer, G. Interaction of allosteric effectors with α-globin chains and high altitude respiration of mammals. The primary structure of two tylopoda hemoglobins with high oxygen affinity: Vicuna (*Lama vicugna*) and Alapaca (*Lama pacos*). *Z. Biol. Chem. Hoppe-Seyler* 367:153-160, 1986
Oberthür, W., Voelter, W. & Braunitzer, G. Die Sequenz der Hämoglobine von Streifengans (*Anser indicus*) und Strauss (*Struthio camelus*). Inositpentaphosphat als Modulator der Evolutions-Geschwindigkeit: Die überraschende Sequenz α63 (E12) valin. *Z. Physiol. Chem. Hoppe-Seyler* 361:S969-S975, 1980
Perutz, M.F. Species adaptation in a protein molecule. *Mol. Biol. Evol.* 1:1-28, 1983
Provine, W.B. *Sewall Wright and Evolutionary Biology*. University of Chicago Press, Chicago 1986
Zuckerkandl, E. & Pauling, L. Evolutionary divergence and convergence in proteins. In *Evolving Genes and Proteins* (Bryson, V. and Vogel, H.J. eds.) pp. 97-166, Academic, New York 1965

Further Reading

Dennett, D.C. *Darwin's Dangerous Idea. Evolution and the Meaning of Life*. Simon & Schuster, New York 1995
Gillespie, J.H. *The Causes of Molecular Evolution*. Oxford University Press, Oxford 1991
Kimura, M. *The Neutral Theory of Molecular Evolution*. Cambridge University Press, Cambridge, England 1983
King, J.L. & Jukes, T.H. Non-Darwinian evolution. *Science* 64:788-798, 1969
Lewin, R. *Patterns in Evolution. The New Molecular View*. Scientific American Library, New York 1997
Li, W.-H. & Graur, D. *Fundamentals of Molecular Evolution*. Sinauer Press, Sunderland, MA 1991
Li, W.-H. *Molecular Evolution*. Sinauer Associates, Sunderland, MA 1997
Page, R.D.M. & Holmes, E.C. *Molecular Evolution. A Phylogenetic Approach*. Blackwell Science, Oxford 1998
Ridley, M. *Evolution*. 2nd edn. Blackwell Science, Cambridge, MA 1996
Strickberger, M.W. *Evolution*. 3rd edn. Jones and Bartlett Publ., Sudbury, MA 2000

Chapter Five
Sources

Efron, B. Bootstrap methods: another look at the jackknife. *Ann. Statist.* 7:1-26, 1979

Felsenstein, J. Confidence limits on phylogenies: An approach using the bootstrap. *Evolution* 39:783-791, 1985

Jukes, T.H. & Cantor, C.R. Evolution of protein molecules. In *Mammalian Protein Metabolism* (Munro, H.N. ed.) pp. 21-132, Academic Press, New York 1969

Kimura, M. & Ohta, T. The average number of generations until fixation of a mutant gene in a finite population. *Genetics* 61:763-771, 1969

Kimura, M. The length of time required for a selectively neutral mutant to reach fixation through random frequency drift in a finite population. *Genet. Res.* 15:131-133, 1970

Saitou, N. & Nei, M. The neighbor-joining method: a new method for reconstructing phylogenetic trees. *Mol. Biol. Evol.* 4:406-425, 1987

Further Reading

Hillis, D.M., Moritz, C. & Mable, B.K. (eds.) *Molecular Systematics*. 2nd edn. Sinauer Associates, Sunderland, MA 1996

Maynard Smith, J. *Evolutionary Genetics*. Oxford University Press, Oxford 1989

Nei, M. & Kumar, S. *Molecular Evolution and Phylogenetics*. Oxford University Press, Oxford, England 2000

Nei, M. *Molecular Evolutionary Genetics*. Columbia University Press, New York 1987

See also Li (1997), Li and Graur (2000), and Page and Holmes (1998) in the *Further Reading* section of Chapter Four.

Chapter Six
Sources

Creti, R., Ceccarelli, E., Bocchetta, M., Sanangelantoni, A.M., Tiboni, O., Palm, P. & Cammarano, P. Evolution of translational elongation factor (EF) sequences: reliability of global phylogenies inferred from EF-1α(Tu) and EF-2 (G) proteins. *Proc. Natl. Acad. Sci. USA* 91:3255-3259, 1994

Hacker, J., Blum-Oehler, G., Muhldorfer, I. & Tschape, H. Pathogenicity islands of virulent bacteria: structure, function and impact on microbial evolution. *Mol. Microbiol.* 23:1089-1097, 1997

Iwabe, N., Kuma, K.-I., Hasegawa, M., Osawa, S. & Miyata, T. Evolutionary relationship of archaebacteria, eubacteria, and eukaryotes inferred from phylogenetic trees of duplicated genes. *Proc. Natl. Acad. Sci. USA* 86:9355-9359, 1989

Kroes, I., Lepp, P. & Relman, D. Bacterial diversity within the human subgingival crevice. *Proc. Natl. Acad. Sci. USA* 96:14547-14552, 1999

Lawrence, J.G. & Ochman, H. Molecular archaeology of the *Escherichia coli* genome. *Proc. Natl. Acad. Sci. USA* 95:9413-9417, 1998

Pace, N.R. A molecular view of microbial diversity and the biosphere. *Science* 276:734-740, 1997

Suau, A., Bonnet, R., Sutren, M., Godon, J., Gibson, G., Collins, M. & Dore, J. Direct analysis of genes encoding 16S rRNA from complex communities reveals many novel molecular species within the human gut. *Appl. Environ. Microbiol.* 65:4799-4807, 1999

Wheelis, M.L., Kandler, O. & Woese, C.R. On the nature of global classification. *Proc. Natl. Acad. Sci. USA* 89:2930-2934, 1992

Woese, C.R. & Fox, G.E. Phylogenetic structure of the prokaryotic domains: The primary kingdoms. *Proc. Natl. Acad. Sci. USA* 74:5088-5090, 1977

Woese, C.R. Whither microbiology? Phylogenetic trees. *Curr. Biol.* 6:1060-1063, 1996

Woese, C.R., Kandler, O. & Wheels, M.L. Towards a natural system of organisms: Proposal for the domains Archaea, Bacteria, and Eucarya. *Proc. Natl. Acad. Sci. USA* 87:4576-4579, 1990

Further Reading

Andersson, S.G.E. & Kurland, C.G. Origins of mitochondria and hydrogenosomes. *Curr. Op. Microbiol.* 2:535-541, 1999

de la Cruz, F. & Davie, J. Horizontal gene transfer and the origin of species: lessons from bacteria. *Trends Microbiol.* 128:128-133, 2000

Doolittle, W.F. Phylogenetic classification and the universal tree. *Science* 284:2124-2128, 2000

Doolittle, W.F. Uprooting the Tree of Life. *Sci. Am.* Feb.:72-77, 2000

Doolittle, W.F. You are what you eat: a gene transfer ratchet could account for bacterial genes in eukaryotic nuclear genomes. *Trends Genet.* 14:307-311, 1998

Embley, T.M. & Hirt, R.P. Early branching eukaryotes? *Curr. Op. Genet. Devel.* 8:624-629, 1998

Fortere, P. & Philippe, H. Where is the root of the universal tree of life? *BioEssays* 21:871-879, 1999

Gray, M.W. Evolution of organellar genomes. *Curr. Op. Genet. Dev.* 9:678-687, 1999

Gray, M.W., Burger, G. & Lang, B.F. Mitochondrial evolution. *Science* 283:1476-1481, 1999

Guttman, D.S. Recombination and clonality in natural populations of *Escherichia coli. Trends Ecol. Evol.* 12:16-22, 1997

Lawrence, J.G. Gene transfer, speciation, and the evolution of bacterial genomes. *Curr. Op. Microbiol.* 2:519-523, 1999

Madigan, M.T. & Marrs, B.L. Extremophiles. *Sci. Am.* April:66-71, 1997

Madigan, M.T., Martinko, J.M. & Parker, J. *Brock Biology of Microorganisms.* 8th edn. Prentice Hall International, Upper Saddle River, NJ 1997

Martin, W. Mosaic bacterial chromosomes: a challenge en route to a tree of genomes. *BioEssays* 21:99-104, 1999

Recchia, G.D. & Hall, R.M. Origins of the mobile gene cassettes found in integrons. *Trends Microbiol.* 5:389-394, 1997

Sogin, M.L. Early evolution and the origin of eukaryotes. *Curr. Op. Genet. Dev.* 1:457-463, 1991

Sogin, M.L. History assignment: when was the mitochondrion founded? *Curr. Opin. Genet. Dev.* 7:792-799, 1997

Sogin, M.L. Organelle origins: Energy-producing symbionts in early eukaryotes? *Curr. Biol.* 7:R315-R317, 1997

Stephens, C. Intimate strangers. *Curr. Biol.* 10:R272-R275, 2000

Whitman, W.B., Coleman, D.C. & Wiebe, W.J. Prokaryotes: the unseen majority. *Proc. Natl. Acad. Sci. USA* 95:6578-6583, 1998

Woese, C.R. Interpreting the universal phylogenetic tree. *Proc. Natl. Acad. Sci. USA* 97:8392-8396, 2000

Chapter Seven
Sources

Adoutte, A., Balavoine, G., Lartillot, N., Lespinet, O., Prud'homme, B. & de Rosa, R. The new animal phylogeny: Reliability and implications. *Proc. Natl. Acad. Sci. USA* 97:4453-4456, 2000

Aguinaldo, A.M., Turbeville, J.M., Linford, L.S., Rivera, M.C., Garey, J.R., Raff, R.A. & Lake, J.A. Evidence for a clade of nematodes, arthropods and other moulting animals. *Nature* 387:489-493, 1997

Baldauf, S.L. & Palmer, J.D. Animals and fungi are each other's closest relatives: Congruent evidence from multiple proteins. *Proc. Natl. Acad. Sci. USA* 90:11558-11562, 1993

Cameron, C.B., Garey, J.R. & Swalla, B.J. Evolution of the chordate body plan: New insights from phylogenetic analyses of deuterostome phyla. *Proc. Natl. Acad. Sci. USA* 97:4469-4474, 2000

De Rijk, P., Van de Peer, Y., Van den Broeck, I. & De Wachter, R. Evolution according to large ribosomal subunit RNA. *J. Mol. Evol.* 41:366-367, 1995

Field, K.G., Olsen, G.J., Lane, D.J., Giovannoni, S.J., Ghiselin, M.T., Raff, E.C., Pace, N.R. & Raff, R.A. Molecular phylogeny of the animal kingdom. *Science* 239:748-753, 1988

Turbeville, J.M., Schulz, J.R. & Raff, R.A. Deuterostome phylogeny and the sister group of the chordates: Evidence from molecules and morphology. *Mol. Biol. Evol.* 11:648-655, 1994

Wainright, P.O., Hinkle, G., Sogin, M.L. & Stickel, S.K. Monophyletic origins of the metazoa: An evolutionary link with fungi. *Science* 260:340-342, 1993

Further Reading

Brusca, R.C. & Brusca, G.J. *Invertebrates*. Sinauer Associates, Sunderland, MA 1990

Carroll, R.L. *Patterns and Processes of Vertebrate Evolution*. Cambridge University Press, Cambridge, England 1990

Gerhart, J. & Kirschner, M. *Cells, Embryos, and Evolution. Toward a Cellular and Developmental Understanding of Phenotypic Variation and Evolutionary Adaptability*. Blackwell Science Publications, Malden, MA 1997

Müller, W.A. *Developmental Biology*. Springer, New York 1997

Nielson, C. *Animal Evolution. Interrelationships of the Living Phyla*. Oxford University Press, Oxford 1995

Raff, R.A. *The Shape of Life. Genes, Development, and the Evolution of Animal Form*. The University of Chicago Press, Chicago 1996

Willmer, P. *Invertebrate Relationships. Patterns in Animal Evolution*. Cambridge University Press, Cambridge, England 1990

Wolpert, L., Beddington, R., Brockes, J. & Jessell, T. *Principles of Development*. Oxford University Press, Oxford 1998

Chapter Eight
Sources

Avise, J.C. & Johns, G.C. Proposal for a standardized temporal scheme of biological classification for extant species. *Proc. Natl. Acad. Sci. USA* 96:7358-7363, 1999

Bailey, W., Hayasaka, K., Skinner, C.G., Kehoe, S., Sieu, L.C., Slightom, J.L. & Goodman, M. Reexamination of the African hominoid trichotomy with additional sequences from the primate β-globin gene cluster. *Mol. Phylogenet. Evol.* 1:97-135, 1992

Bailey, W.J. Hominoid trichotomy: A molecular overview. *Evol. Anthropol.* 2:100-108, 1993

Brinkman-Van der Linden, E.C.M., Sjoberg, E.R., Juneja, L.R., Crocker, P.R., Varki, N. & Varki, A. Loss of N-glycolylneuraminic acid in human evolution. *J. Biol. Chem.* 275:8633-8640, 2000

Chen, F.-C. & Li, W.-H. Genomic divergences between human and other hominoids and the effective population size of the common ancestor of human and chimpanzee. *Am. J. Hum. Genet.* 68:444-456, 2001

Figuera, L., Pandolfo, M., Dunne, P.W., Cantú, J.M. & Patel, P.I. Mapping of the congenital generalized hypertrichosis locus to chromosome Xq24-q27.1. *Nat. Genet.* 10:202-207, 1995

Gagneux, P. & Varki, A. Genetic differences between humans and great apes. *Mol. Phylogenet. Evol.* 18:2-13, 2001

Goodman, M., Porter, C.A., Czelusniak, J., Page, S.L., Schneider, H., Shoshani, J., Gunnell, G. & Groves, C.P. Toward a phylogenetic classification of primates based on DNA evidence complemented by fossil evidence. *Mol. Phylogenet. Evol.* 9:585-598, 1998

Goodman, M., Tagle, D.A., Fitch, D.H.A., Bailey, W., Czelusniak, J., Koop, B.F., Benson, P. & Slightom, J.L. Primate evolution at the DNA level and a classification of hominoids. *J. Mol. Evol.* 30:260-266, 1990

Groves, C.P. Systematic of the Great Apes. In *Comparative Primate Biology. Vol. 1. Systematics, Evolution, and Anatomy* (Swindler, D.R. and Erwin, J. eds.) pp. 187-219, Alan R. Liss, New York 1986

Huxley, T.H. *Man's Place in Nature*. University of Michigan Press, Ann Arbor, MI 1959

Irie, A. & Suzuki, A. CMP-N-Acetylneuraminic acid hydroxylase is exclusively inactive in humans. *Biochem. Biophys. Res. Commun.* 248:330-333, 1998

Irie, A., Koyama, S., Kozutsumi, Y., Kawasaki, T. & Suzuki, A. The molecular basis for the absence of N-glycolylneuraminic acid in humans. *J. Biol. Chem.* 273:15866-15871, 1998

Kay, R.F., Ross, C. & Williams, B.A. Anthropoid origins. *Science* 275:797-804, 1997

King, M.-C. & Wilson, A.C. Evolution at two levels in humans and chimpanzees. *Science* 188:107-116, 1975

Koop, B.F., Siemieniak, D., Slightom, J.L., Goodman, M., Dunbar, J., Wright, P.C. & Simons, E.L. Tarsius δ- and β-globin genes: conversions, evolution, and systematic implications. *J. Biol. Chem.* 264:68-79, 1989

Luo, Z.-X., Cifelli, R.L. & Kielan-Jaworowska, Z. Dual origin of tribosphenic mammals. *Nature* 409:53-57, 2001

Madsen, O., Scally, M., Douady, C.J., Kao, D.J., DeBry, R.W., Adkins, R., Amrine, H.M., Stanhope, M.J., de Jong, W.W. & Springer, M.S. Parallel adaptive radiations in two major clades of placental mammals. *Nature* 409:610-614, 2001

Murphy, W.J., Eizirik, E., Johnson, W.E., Zhang, Y.P., Ryder, O.A. & O'Brien, S.J. Molecular phylogenetics and the origins of placental mammals. *Nature* 409:614-618, 2001

O'hUigin, C., Satta, Y., Takahata, N. & Klein, J. Contribution of homoplasy and ancestral polymorphism to the evolution of genes in anthropoid primates. *Mol. Biol. Evol.* 2002

Page, S.L., Chiu, C.-H. & Goodman, M. Molecular phylogeny of Old World monkeys (Cercopithecidae) as inferred from γ-globin DNA sequences. *Mol. Phylogenet. Evol.* 13:348-359, 1999

Porter, C.A., Sampaio, I., Schneider, H., Schneider, M.P.C., Czelusniak, J. & Goodman, M. Evidence on primate phylogeny from ε-globin gene sequences and flanking regions. *J. Mol. Evol.* 40:30-55, 1995

Rogers, J. The phylogenetic relationships among *Homo*, *Pan* and *Gorilla*: a population genetics perspective. *J. Hum. Evol.* 25:201-215, 1993

Satta, Y., Klein, J. & Takahata, N. DNA archives and our nearest relative: The trichotomy problem revisited. *Mol. Phylogenet. Evol.* 14:259-275, 2000

Shosani, J., Groves, C.P., Simons, E.L. & Gunnell, G.F. Primate phylogeny: morphological vs. molecular results. *Mol. Phylogenet. Evol.* 5:102-154, 1996

Springer, M.S., Cleven, G.C., Madsen, O., de Jong, W.W., Waddel, V.G., Amrine, H.M. & Stanhope, M.J. Endemic African mammals shake the phylogenetic tree. *Nature* 388:61-64, 1997

Stanhope, M.J., Waddel, V.G., Madsen, O., de Jong, W.W., Hedges, S.B., Cleven, G.C., Kao, D.J. & Springer, M.S. Molecular evidence for multiple origins of Insectivora and for a new order of endemic African insectivore mammals. *Proc. Natl. Acad. Sci. USA* 95:9967-9972, 1998

Szabó, Z., Levi-Minzi, S.A., Christiano, A.M., Struminger, C., Stoneking, M., Batzer, M.A. & Boyd, C.D. Sequential loss of two neighboring exons of the tropoelastin gene during primate evolution. *J. Mol. Evol.* 49:664-671, 1999

Vallois, H. Ordre des primates (Primates Linne', 1758). In *Traité de Zoologie. Anatomie, Systematique, Biologie. Vol. 172* (Grassé, P.-P. ed.) pp. 1854-2206, Massonet, C., Paris 1955

van Dijk, M.A.M., Madsen, O., Catzeflis, F., Stanhope, M.J., de Jong, W.W. & Pagel, M. Protein sequence signatures support the African clade of mammals. *Proc. Natl. Acad. Sci. USA* 98:188-193, 2001

Winter, H., Langbein, L., Krawczak, M., Cooper, D.N., Jave-Suarez, L.F., Rogers, M.A., Praetzel, S., Heidt, P.J. & Schweizer, J. Human type I hair keratin pseudogene $\phi hH\alpha A$ has functional orthologs in the chimpanzee and gorilla: evidence for recent inactivation of the human gene after the *Pan-Homo* divergence. *Hum. Genet.* 108:37-42, 2001

Further Reading

Corbet, G.B. & Hill, J.E. *A World List of Mammalian Species*. 3rd edn. Oxford University Press, Oxford 1991

Dorit, R.L., Walker, J., W.F. & Barnes, R.D. *Zoology*. Saunders College Publishing, Philadelphia 1991

Martin, R.D. *Primate Origins and Evolution: A Phylogenetic Reconstruction*. Chapman and Hall, London 1990

Napier, J.R. & Napier, P.H. *The Natural History of the Primates*. The MIT Press, Cambridge, MA 1985

Novacek, M.J. Mammalian phylogeny: Shaking the tree. *Nature* 356:121-125, 1992

Passingham, R.E. *The Human Primate*. W.H. Freeman, Oxford 1982

Pough, F.H., Janis, C.M. & Heiser, J.B. *Vertebrate Life*. 5th edn. Prentice Hall, Upper Saddle River, NJ 1998

Wilson, D.E. & Reeder, D.-A.M. (eds.) *Mammal Species of the World. A Taxonomic and Geographic Reference*. 2nd edn. Smithsonian Institution Press, Washington D.C. 1993

Chapter Nine
Sources

Ayala, F.J., Rzhetsky, A. & Ayala, F.J. Origin of the metazoan phyla: Molecular clocks confirm paleontological estimates. *Proc. Natl. Acad. Sci. USA* 95:606-611, 1997

Britten, R.J. Rates of DNA sequence evolution differ between taxonomic groups. *Science* 231:1393-1398, 1986

Doolittle, R.F., Feng, D.-F., Tsang, S., Cho, G. & Little, E. Determining divergence times of the major kingdoms of living organisms with a protein clock. *Science* 271:470-477, 1996

Feng, D.-F., Cho, G. & Doolittle, R.F. Determining divergence times with a protein clock: update and reevaluation. *Proc. Natl. Acad. Sci. USA* 94:13028-13033, 1997

Gu, X. Early metazoan divergence was about 830 million years ago. *J. Mol. Evol.* 47:369-371, 1998

Hailand, W.B., Armstrong, R.L., Cox, A.V., Craig, L.E., Smith, A.G. & Smith, D.G. *A Geologic Time Scale 1989.* Cambridge University Press, Cambridge, England 1990

Kumar, S. & Hedges, S.B. A molecular timescale for vertebrate evolution. *Nature* 392:917-920, 1998

Li, C.-W., Chen, J.-Y. & Hua, T.-E. Precambrian sponges with cellular structures. *Science* 279:879-882, 1998

Li, W.-H., Gojobori, T. & Nei, M. Pseudogenes as a paradigm of neutral evolution. *Nature* 292:237-239, 1981

Martin, A.P. & Palumbi, S.R. Body size, metabolic rate, generation time, and the molecular clock. *Proc. Natl. Acad. Sci. USA* 90:4087-4091, 1993

Mojzsis, S.J., Arrhenius, G., McKeegan, K.D., Harrison, T.M., Nutman, A.P. & Friend, C.R.L. Evidence for life on Earth before 3,800 million years ago. *Nature* 384:55-59, 1996

Runnegar, B. A molecular-clock date for the origin of the animal phyla. *Lethaia* 15:199-205, 1982

Runnegar, B. Molecular palaeontology. *Palaeontology* 29:1-24, 1986

Sarich, V.M. & Wilson, A.C. Immunological time scale for hominid evolution. *Science* 158:1200-1203, 1967

Sarich, V.M. & Wilson, A.C. Quantitative immunochemistry and the evolution of primate albumins: Micro-complement fixation. *Science* 154:1563-1566, 1966

Schopf, J.W. Microfossils of the early archaean apex chart: new evidence of the antiquity of life. *Science* 260:640-646, 1993

Seilacher, A., Bose, P.K. & Pflüger, F. Triploblastic animals more than 1 billion years ago: trace fossil evidence from India. *Science* 282:80-83, 1998

Wang, D.Y.-C., Kumar, S. & Hedges, S.B. Divergence time estimates for the early history of animal phyla and the origin of plants, animals and fungi. *Proc. R. Soc. Lond. B* 266:163-171, 1999

Wray, G.A., Levinton, J.S. & Shapiro, L.H. Molecular evidence for deep precambrian divergences among metazoan phyla. *Science* 274:568-573, 1996

Xiao, S., Zhang, Y. & Knoll, A.H. Three-dimensional preservation of algae and animal embryos in a Neoproterozoic phosphorite. *Nature* 391:553-558, 1998

Further Reading

Conway Morris, S. The fossil record and the early evolution of the Metazoa. *Nature* 361:219-225, 1993

Conway Morris, S. *The Crucible of Creation. The Burgess Shale and the Rise of Animals.* Oxford University Press, Oxford 1998

Doolittle, W.F. Fun with genealogy. *Proc. Natl. Acad. Sci. USA* 94:12751-12753, 1997

Emiliani, C. *Planet Earth. Cosmology, Geology, and the Evolution of Life and Environment.* Cambridge University Press, Cambridge, England 1992

Fortey, R.R., Briggs, D.E. & Wills, M.A. The Cambrian evolutionary „explosion" recalibrated. *BioEssays* 19:429-434, 1997

Knoll, A.H. The early evolution of Eukaryotes: a geological perspective. *Science* 256:622-627, 1992

Lambert, D. & the Diagram Group. *The Field Guide to Prehistoric Life.* Facts on File Publ., New York 1985

Lambert, D. & the Diagram Group *The Field Guide to Geology.* Facts on File Publ., New York 1988

Macdougall, J.D. *A Short History of Planet Earth. Mountains, Mammals, Fire, and Ice.* John Wiley, New York 1996

Schoepf, J.W. *Cradle of Life. The Discovery of Earth's Earliest Fossils.* Princeton University Press, Princeton, NJ 1999

Simpson, G.G. *Fossils and the History of Life.* Scientific American Library, New York 1983

Singer, R. (ed.) *Encyclopedia of Paleontology.* Vol. 1 and 2. Fitzroy Dearborn Publishers, Chicago 1999

Stanley, S.M. *Exploring Earth and Life Through Time.* W.H. Freeman, New York 1993

See also Li (1997) and Li and Graur (2000) in the *Further Reading* section of Chapter Four.

Chapter Ten
Sources

Adcock, G.J., Dennis, E.S., Easteal, S., Huttley, G., Jermiin, L.S., Peacock, W.J. & Thorne, A. Mitochondrial DNA sequences in ancient Australians: implications for modern human origins. *Proc. Natl. Acad. Sci. USA* 98:537-542, 2001

Andrews, P. Evolution and environment in the Hominoidea. *Nature* 360:641-646, 1992

Austin, J.J., Ross, A.J., Smith, A.B., Fortey, R.A. & Thomas, R.H. Problems of reproducibility – does geologically ancient DNA survive in amber-preserved insects? *Proc. R. Soc. Lond. B* 264:467-474, 1997

Bowcock, A.M., Ruiz-Linares, A., Tomfohrde, J., Minch, E., Kidd, J.R. & Cavalli-Sforza, L.L. High resolution of human evolutionary trees with polymorphic microsatellites. *Nature* 368:455-457, 1994

Brown, W.M. Polymorphism in mitochondrial DNA of humans as revealed by restriction endonuclease analysis. *Proc. Natl. Acad. Sci. USA* 77:3605-3609, 1980

Cann, R.L., Stoneking, M. & Wilson, A.C. Mitochondrial DNA and human evolution. *Nature* 325:31-36, 1987

Cano, R.J., Poinar, H.N., Pieniazek, N.J., Acra, A. & Poinar, O. Jr. Amplification and sequencing of DNA from a 120-135-million-year-old weevil. *Nature* 363:536-538, 1993

Chen, Y.-S., Torroni, A., Excoffier, L., Santachiara-Benerecetti, A.S. & Wallace, D.C. Analysis of mtDNA variation in African populations reveals the most ancient of all human continent-specific haplogroups. *Am. J. Hum. Genet.* 57:133-149, 1995

Chen, Y.-S., Olckers, A., Schurr, T.G., Kogelnik, A.M., Huoponen, K. & Wallace, D.C. mtDNA variation in the South African Kung and Khwe - and their genetic relationships to other African populations. *Am. J. Hum. Genet.* 66:1362-1383, 2000

Dorit, R.L., Akashi, H. & Gilbert, W. Absence of polymorphism at the *ZFY* locus on the human Y chromosome. *Science* 268:1183-1185, 1995

Fleagle, J.G. *Primate Adaptation and Evolution.* Academic Press, San Diego, CA 1988

Golenberg, E.M., Giannasi, D.E., Clegg, M.T., Smiley, C.J., Durbin, M., Henderson, D. & Zurawski, G. Chloroplast DNA sequence from a Miocene magnolia species. *Nature* 344:656-658, 1990

Hammer, M.F., Karafet, T., Rasanayagam, A., Wood, E.T., Altheide, T.K., Jenkins, T., Griffiths, R.C., Templeton, A.R. & Zegura, S.L. Out of Africa and back again: nested cladistic analysis of human Y chromosome variation. *Mol. Biol. Evol.* 15:427-441, 1998

Harding, R.M., Fullerton, S.M., Griffiths, R.C., Bond, J., Cox, M.J., Schneider, J.A., Moulin, D.S. & Clegg, J.B. Archaic African and Asian lineages in the genetic ancestry of modern humans. *Am. J. Hum. Genet.* 60:772-789, 1997

Harris, E.E. & Hey, J. X chromosome evidence for ancient human histories. *Proc. Natl. Acad. Sci. USA* 96:3320-3324, 1999

Hawks, J. & Wolpoff, M.H. Paleoanthropology and the population genetics of ancient genes. *Am. J. Phys. Anthropol.* 114:269-272, 2001

Higuchi, R., Bowman, B., Freiberger, M., Ryder, O.A. & Wilson, A.C. DNA sequences from the quagga, an extinct member of the horse family. *Nature* 312:282-284, 1984

Horai, S., Hayasaka, K., Kondo, R., Tsugane, K. & Takahata, N. Recent African origin of modern humans revealed by complete sequences of hominoid mitochondrial DNAs. *Proc. Natl. Acad. Sci. USA* 92:532-536, 1995

Ingman, M., Kaessmann, H., Pääbo, S. & Gyllensten, U. Mitochondrial genome variation and the origin of modern humans. *Nature* 408:708-713, 2000

Klein, R.G. Archeology and the evolution of human behavior. *Evol. Anthropol.* 9:17-36, 2000

Kocher, T.D. & Wilson, A.C. Sequence evolution of mitochondrial DNA in humans and chimpanzees: Control region and a protein-coding region. In *Evolution of Life: Fossils, Molecules, and Culture* (Osawa, S. and Honjo, T. eds.) pp. 391-413, Springer-Verlag, New York 1991

Krings, M., Stone, A., Schmitz, R.W., Krainitzki, H., Stoneking, M. & Pääbo, S. Neandertal DNA sequences and the origin of modern humans. *Cell* 90:19-30, 1997

Krings, M., Geisert, H., Schmitz, R.W., Krainitzki, H. & Pääbo, S. DNA sequence of the mitochondrial hypervariable region II from the Neandertal type specimen. *Proc. Natl. Acad. Sci. USA* 96:5581-5585, 1999

Krings, M., Capelli, C., Tschentscher, F., Geisert, H., Meyer, S., von Haeseler, A., Grossschmidt, K., Possnert, G., Paunovic, M. & Pääbo, S. A view of Neandertal genetic diversity. *Nat. Genet.* 26:144-146, 2000

Lindahl, T. Instability and decay of the primary structure of DNA. *Nature* 362:709-715, 1993

Lindahl, T. Facts and artifacts of ancient DNA. *Cell* 90:1-3, 1997

Nei, M. Evolution of human races at the gene level. In *Human Genetics, Part A: The Unfolding Genome* (Bonne-Tamir, B., Cohen, T. and Goodman, R.M. eds.) pp. 167-181, Alan R. Liss, Inc., New York 1982

Nei, M. & Roychoudhury, A.K. Genetic relationship and evolution of human races. In *Evolutionary Biology* Vol. 14. (Hecht, M.K., Wallace, B. and Prance, C.T. eds.) pp. 1-59, Plenum Publishing Corporation, New York 1982

Nei, M. & Roychoudhury, A.K. Evolutionary relationships of human populations on a global scale. *Mol. Biol. Evol.* 10:927-943, 1993

Ohno, S. & Ohno, M. The all pervasive principle of repetitious recurrence governs not only coding sequence construction but also human endeavor in musical composition. *Immunogenetics* 24:71-78, 1986

Ovchinnikov, I.V., Götherström, A., Romanova, G.P., Kharitonov, V.M., Lidén, K. & Goodwin, W. Molecular analysis of Neanderthal DNA from the northern Caucasus. *Nature* 404:490-493, 2000

Pääbo, S. Molecular cloning of ancient Egyptian mummy DNA. *Nature* 314:644-645, 1985

Poinar, H.N., Höss, M., Bada, J.L. & Pääbo, S. Amino acid racemization and the preservation of ancient DNA. *Science* 272:864-866, 1996

Roychoudhury, A.K. & Nei, M. *Human Polymorphic Genes. World Distribution.* Oxford University Press, New York 1988

Ruvolo, M., Zehr, S., von Dornum, M., Pan, D., Chang, B. & Lin, J. Mitochondrial CO II sequences and modern human origins. *Mol. Biol. Evol.* 10:1115-1135, 1993

Simons, E.L. *Primate Evolution: An Introduction to Man's Place in Nature.* Macmillan, New York 1972

Strait, D.S., Grine, F.E. & Moniz, M.A. A reappraisal of early hominid phylogeny. *J. Hum. Evol.* 32:17-82, 1997

Takahata, N., Lee, S.-H. & Satta, Y. Testing multiregionality of modern human origins. *Mol. Biol. Evol.* 18:172-183, 2001

Vigilant, L., Pennington, R., Harpending, H., Kocher, T.D. & Wilson, A.C. Mitochondrial DNA sequences in single hairs from a southern African population. *Proc. Natl. Acad. Sci. USA* 86:9350-9354, 1989

Vigilant, L., Wilson, A.C. & Harpending, H. African populations and the evolution of human mitochondrial DNA. *Science* 253:1503-1507, 1991

Watson, E., Forster, P., Richards, M. & Bandelt, H.-J. Mitochondrial footprints of human expansions in Africa. *Am. J. Hum. Genet.* 61:691-704, 1997

Wolpoff, M.H., Wu, X. & Thorne, A.G. Modern *Homo sapiens* origins: a general theory of hominid evolution involving the fossil evidence from east Asia. In *The Origins of Modern Humans: A World Survey of the Fossil Evidence* (Smith, F.H. and Spencer, F. eds.) pp. 411-483, Alan R. Liss, New York 1984

Woodward, S.R., Weyand, N.J. & Bunnell, M. DNA sequence from Cretaceous period bone fragments. *Science* 266:1229-1232, 1994

Yu, N. & Li, W.-H. No fixed nucleotide difference between Africans and non-Africans at the pyruvate dehydrogenase E1 α-subunit locus. *Genetics* 155:1481-1483, 2000

Further Reading

Angela, P. & Angela, A. *The Extraordinary Story of Human Origins.* (Translated by G. Tonne.) Prometheus Books, Buffalo, NY 1993

Boaz, N.T. *Ecce Homo. How the Human Being Emerged from the Cataclysmic History of the Earth.* Basic Books, New York 1997

Brown, M.H. *The Search for Eve.* Harper & Row Publishers, New York 1990

Brown, T.A. & Brown, K.A. Ancient DNA: Using molecular biology to explore the past. *BioEssays* 16:719-726, 1994

Cavalli-Sforza, L.L. & Cavalli-Sforza, F. *The Great Human Diasporas. The History of Diversity and Evolution.* Addison-Wesley Publishing, Reading, MA 1995

Cavalli-Sforza, L.L. *Genes, Peoples, and Languages.* (Translated by M. Seielstad.) North Point Press, New York 2000

Cavalli-Sforza, L.L., Menozzi, P. & Piazza, A. *The History and Geography of Human Genes.* Princeton University Press, Princeton, NJ 1994

Foley, R. *Another Unique Species. Patterns in Human Evolutionary Ecology.* Longman Scientific & Technical, Burnt Hill, England 1987

Gould, S.J. *The Book of Life.* W.W. Norton, New York 1993

Groves, C.P. *A Theory of Human and Primate Evolution.* Clarendon Press, Oxford 1989

Jones, S., Martin, R. & Pilbeam, D. (eds.) *The Cambridge Encyclopedia of Human Evolution.* Cambridge University Press, Cambridge, England 1992

Klein, R.G. *The Human Career: Human Biological and Cultural Origins.* 2nd edn. Chicago University Press, Chicago 1999

Lambert, D. & and the Diagram Group. *The Cambridge Guide to Prehistoric Man.* Cambridge University Press, Cambridge, England 1987

Lewin, R. *Bones of Contention.* Simon and Schuster, New York 1988

Lewin, R. *The Origin of Modern Humans.* Scientific American Library, New York 1993

Martin, R.D. *Primate Origins and Evolution: A Phylogenetic Reconstruction.* Chapman and Hall, London 1990

Pääbo, S. Ancient DNA. *Sci. Amer.* November:60-66, 1993

Parker, S. *The Dawn of Man.* Crescent Books, New York 1992

Poinar, G. & Poinar, G. *The Quest for Life in Amber.* Addison-Wesley Publishing, Reading, MA 1999

Shreeve, J. *The Neanderthal Enigma. Solving the Mystery of Modern Human Origins.* William Morrow, New York 1995

Stringer, C. & Gamble, C. *In Search of the Neanderthals. Solving the Puzzle of Human Origins.* Thames and Hudson, London 1993

Stringer, C. & McKie, R. *African Exodus. The Origins of Modern Humanity.* Jonathon Cape, London 1996

Tattersall, I. *The Fossil Trail. How We Know What We Think We Know about Human Evolution.* Oxford University Press, New York 1995

Tattersall, I. & Delson, E. Primates. In *Encyclopedia of Paleontology.* Vol. 2 (Singer, R. ed.) pp. 949-959, Fitzroy Dearborn Publishers, Chicago 1999

Tattersall, I., Delson, E. & Van Couvering, J. (eds.) *Encyclopedia of Human Evolution and Prehistory.* 2nd edn. Garland Publishing, New York 2000

Wenke, R.J. *Patterns in Prehistory. Humankind's First Three Million Years.* 3rd edn. Oxford University Press, New York 1990

Wolpoff, M.H. *Paleoanthropology.* 2nd edn. McGraw-Hill, Boston 1999

Wolpoff, M.H. & Caspari, R. *Race and Human Evolution.* Simon & Schuster, New York 1997

Chapter Eleven
Sources

Ayala, F.J., Escalante, A., O'hUigin, C. & Klein, J. Molecular genetics of speciation and human origins. *Proc. Natl. Acad. Sci. USA* 91:6787-6794, 1994

Figueroa, F., Günther, E. & Klein, J. MHC polymorphism pre-dating speciation. *Nature* 335:265-267, 1988

Grant, P.R. *Ecology and Evolution of Darwin's Finches.* Princeton University Press, Princeton, NJ 1986

Grant, B.R. & Grant, P.R. Evolution of Darwin's Finches caused by a rare climatic event. *Proc. Roy. Soc. Lond. B* 251:111-117, 1993

Greenwood, P.H. *The Haplochromine Fishes of the East African Lakes.* Cornell University Press, Ithaca, NY 1981

Hawks, J., Hunley, K., Lee, S.-H. & Wolpoff, M.H. Population bottlenecks and Pleistocene human evolution. *Mol. Biol. Evol.* 17:2-22, 2000

Johnson, T.C., Scholz, C.A., Talbot, M.R., Kelts, K., Ricketts, R.D., Ngobi, G., Beuning, K., Ssemmanda, I. & McGill, J.W. Late Pleistocene desiccation of Lake Victoria and rapid evolution of cichlid fishes. *Science* 273:1091-1093, 1996

Kaessmann, H., Wiebe, V. & Pääbo, S. Extensive nuclear DNA sequence diversity among chimpanzees. *Science* 286:1159-1162, 1999

Kimura, M. & Ohta, T. The average number of generations until fixation of a mutant gene in a finite population. *Genetics* 61:763-771, 1969

Kingman, J.F.C. *Mathematics of genetic diversity*. CMBS-NSF Regional Conference Series in Applied Mathematics, No. 34. Society for Industrial and Applied Mathematics, Philadelphia 1980

Klein, J. Generation of diversity at MHC loci: Implications for T-cell receptor repertoires. In *Immunology 80* (Fougereau, M. and Dausset, J. eds.) pp. 239-253, Academic Press, London 1980

Klein, J., Gutknecht, J. & Fischer, N. The major histocompatibility complex and human evolution. *Trends in Genetics* 6:7-11, 1990

Lawlor, D.A., Ward, F.E., Ennis, P.D., Jackson, A.P. & Parham, P. *HLA-A* and *B* polymorphism predate the divergence of humans and chimpanzees. *Nature* 335:268-271, 1988

Li, W.-H. & Sadler, L.A. Low nucleotide diversity in man. *Genetics* 129:513-523, 1991

Mayer, W.E., Jonker, M., Klein, D., Ivanyi, P., van Seventer, G. & Klein, J. Nucleotide sequences of chimpanzee *Mhc* class I alleles: evidence for trans-species mode of evolution. *EMBO J.* 7:2765-2774, 1988

McConnell, T.J., Talbot, W.S., McIndoe, R.A. & Wakeland, E.K. The origin of *Mhc* class II gene polymorphism within the genus *Mus*. *Nature* 332:651-654, 1988

Meyer, A., Kocher, T.D., Basasibwaki, P. & Wilson, A.C. Monophyletic origin of Lake Victoria cichlid fishes suggested by mitochondrial DNA sequences. *Nature* 347:550-553, 1990

Nagl, S., Tichy, H., Mayer, W.E., Takezaki, N., Takahata, N. & Klein, J. The origin and age of the haplochromine species flock in Lake Victoria. *Proc. R. Soc. Lond. B* 267:1049-1061, 2000

Nei, M. & Graur, D. Extent of protein polymorphism and the neutral mutation theory. *Evol. Biol.* 17:73-118, 1984

Sato, A., O'hUigin, C., Figueroa, F., Grant, P.R., Grant, B.R., Tichy, H. & Klein, J. Phylogeny of Darwin's finches as revealed by mtDNA sequences. *Proc. Natl. Acad. Sci. USA* 96:5101-5106, 1999

Sato, A., Tichy, H., O'hUigin, C., Grant, P.R., Grant, B.R. & Klein, J. On the origin of Darwin's Finches. *Mol. Biol. Evol.* 18:299-311, 2000

Satta, Y., Takahata, N., Schönbach, C., Gutknecht, J. & Klein, J. Calibrating evolutionary rates at major histocompatibility complex loci. In *Molecular Evolution of the Major Histocompatibility Complex* H 59. (Klein, J. and Klein, D. eds.) pp. 51-62, Springer-Verlag, Heidelberg 1991

Tajima, F. Evolutionary relationship of DNA sequences in finite populations. *Genetics* 105:437-460, 1983

Takahata, N. A simple genealogical structure of strongly balanced allelic lines and trans-species evolution of polymorphism. *Proc. Natl. Acad. Sci. USA* 87:2419-2423, 1990

Takahata, N., Satta, Y. & Klein, J. Divergence time and population size in the lineage leading to modern humans. *Theoret. Pop. Biol.* 48:198-218, 1995

Vincek, V., O'hUigin, C., Satta, Y., Takahata, N., Boag, P.T., Grant, P.R., Grant, B.R. & Klein, J. How large was the founding population of Darwin's finches? *Proc. R. Soc. Lond. B* 264:111-118, 1997

Watterson, G.A. Allele frequencies after bottleneck. *Theoret. Pop. Biol.* 26:387-407, 1984

Further Reading

Deevey, E.S. Jr. The human population. *Sci. Amer.* March:195-204, 1960

Diamond, J. *Guns, Gems, and Steel. The Fates of Human Societies*. W.W. Norton, New York 1997

Gamble, C. *Time Walkers. The Prehistory of Global Colonization*. Alan Suton, Phoenix Mill, England 1993

Klein, J. *Natural History of the Major Histocompatibility Complex*. John Wiley, New York 1986

Klein, J. & Sato, A. The *HLA* system. Parts I and II. *N. Eng. J. Med.* 343:702-709, 782-786, 2000

Lack, D. *Darwin's Finches*. Cambridge University Press, Cambridge, England 1983

Li, W.-H. *Molecular Evolution*. Sinauer Associates, Sunderland, MA 1997

McEvedy, C. & Jones, R. *Atlas of World Population History*. Penguin Books, Harmondsworth, England 1978

Meyer, A. Phylogenetic relationships and evolutionary processes in East African cichlid fishes. *Trends Ecol. Evol.* 8:279-284, 1993

Mitheu, S. *The Prehistory of the Mind. The Cognitive Origins of Art, Religion and Science*. Thames and Hudson, London 1996

Tanner, N.M. *On Becoming Human*. Cambridge University Press, Cambridge, England 1981

Weiner, J. *The Beak of the Finch. A Story of Evolution in Our Time*. Alfred A. Knopf, New York 1994

Chapter Twelve
Sources

Barbujani, G. & Sokal, R.R. The zones of sharp genetic change in Europe are also language boundaries. *Proc. Natl. Acad. Sci. USA* 87:1816-1819, 1990

Cavalli-Sforza, L.L. Population structure and human evolution. *Proc. R. Soc. Lond. B* 164:362-379, 1966

Darwin, C. *The Descent of Man, and Selection in Relation to Sex*. John Murray, London 1871

Harpending, H. & Rogers, A. Genetic perspectives on human origins and differentiation. *Annu. Rev. Genomics Hum. Genet.* 1:361-385, 2000

Harpending, H., Sherry, S.T., Rogers, A. & Stoneking, M. The genetic structure of ancient human populations. *Curr. Anthropol.* 34:483-496, 1993

Harrison, P. & Pearce, F. *AAAS Atlas of Population and Environment*. University of California Press, Berkeley, CA 2000

Houghton, J., Filho, L.G.M. & Callander, B. (eds.) *Climate Change 1995: The Science of Climate Change*. Cambridge Univ. Press, Cambridge 1996

Lewontin, R.C. The apportionment of human diversity. *Evol. Biol.* 6:381-398, 1972

Mayr, E. & Ashlock, P.D. *Principles of Systematic Zoology*. 2nd edn. McGraw Hill, New York 1991

Wright, S. Isolation by distance. *Genetics* 28:114-138, 1943

Further Reading

Milner, R. *The Encyclopedia of Evolution. Humanity's Search for Its Origins*. Henry Holt, New York 1990

Raup, D.M. *Extinctions. Bad Genes or Bad Luck?* W.W. Norton, New York 1991

Stanley, S.M. *Extinction*. Scientific American Books, New York 1987

Thomson, K.S. *Living Fossil: The Story of the Coelacanth*. W.W. Norton, New York 1992

Ward, P.D. *On Methuselah's Trail. Living Fossils and the Great Extinctions*. W.H. Freeman, New York 1992

Wilson, E.O. *Biophilia*. Harvard University Press, Cambridge, MA 1984

Wilson, E.O. *The Diversity of Life*. Harvard University Press, Cambridge, MA 1992

Appendices One through Four
Further Reading and Sources

Causton, D.R. *A Biologist's Mathematics*. Edward Arnold, London 1977

Downing, D. & Clark, J. *Statistics The Easy Way*. 3rd edn. Barron's, New York 1997

Elston, R.C. & Johnson, W.D. *Essentials of Biostatistics*. 2nd edn. F.A. Davis, Philadelphia 1994

Jacobs, H.R. *Mathematics. A Human Endeavor*. 3rd edn. W.H. Freeman, San Francisco 1982

Maor, E. *e The Story of a Number*. Princeton University Press, Princeton, NJ 1994

Moroney, M.J. *Facts from Figures*. 3rd edn. Penguin Books, Harmondsworth, Middlesex, England 1975

Nelson, D. (ed.) *The Penguin Dictionary of Mathematics*. 2nd edn. Penguin Books, London 1998

Porkess, R. *The Harper Collins Dictionary of Statistics*. Harper Collins Publishers, New York 1991

Sawyer, W.W. *Mathematician's Delight*. Penguin Books, London 1973

Simon, W. *Mathematical Techniques for Biology and Medicine*. The MIT Press, Cambridge, MA 1977

Titchmarsh, E.C. *Mathematics for the General Reader*. Doubleday, Garden City, NY 1981

Triola, M.F. *Mathematics and the Modern World*. Cummings Publishing, Menlo Park, CA 1973

Illustration and Quotation Credits

Adoutte, A., Balavoine, G., Lartillot, N., Lespinet, O., Prud'homme, B. & de Rosa, R. The new animal phylogeny: Reliability and implications. *Proc. Natl. Acad. Sci. USA* 97:4453-4456, 2000

Aeschylus. *The Oresteia.* (Translated by R. Fagles.) Viking Penguin, New York 1966

Alberts, B., Bray, D., Lewis, J., Raff, M., Roberts, K. & Watson, J.D. *Molecular Biology of the Cell.* 2nd edn. Garland Publishing, New York 1989

Alighieri, D. *The Divine Comedy. Paradiso.* (Translated by J. Ciardi.) W.W. Norton, New York 1970

Anderson, S., Bankier, A.T., Barrell, B.G., de Bruijn, M.H., Coulson, A.R., Drouin, J., Eperon, I.C., Nierlich, D.P., Roe, B.A., Sanger, F., Schreier, P., Smith, A.J., Staden, R. & Young, I.G. Sequence and organization of the human mitochondrial genome. *Nature* 290:457-465, 1981

Andrews, P. & Stringer, C. The Primates Progress. In *The Book of Life. An Illustrated History of the Evolution of Life on Earth* (Gould, J. ed.) pp. 219-251, W.W. Norton, New York 1993

Ankel, F. *Einführung in die Primatenkunde.* Gustav Fischer Verlag, Stuttgart 1970

Apollinaire, G. And everything has changed so much in me. In *Calligrammes. Poems of Peace and War (1913-1916)* (Translated by A.H. Greet.) University of California Press, Berkeley 1980

Ayala, F.J., Escalante, A., O'hUigin, C. & Klein, J. Molecular genetics of speciation and human origins. *Proc. Natl. Acad. Sci.* 91:6787-6794, 1994

Ball, P. *Designing the Molecular World. Chemistry at the Frontier.* Princeton University Press, Princeton 1994

Basho. *The Narrow Road to the Deep North and Other Travel Sketches.* (Translated by N. Yuasa.) Penguin Books, Harmondsworth, England 1966

Beck, W.S., Liem, K.F. & Simpson, G.G. *Life: An Introduction to Biology.* 3rd edn. Harper Collins, New York 1991

Beckett, S.B. *Waiting for Godot. A tragicomedy in two acts.* Grove Press, New York 1954

Beiser, A. & the Editors of Time-Life Books *The Earth.* Time-Life Books, New York 1963

Beneš, F. *člověk.* Mladá fronta, Praha 1994

Bodmer, W.F. & Cavalli-Sforza, L.L. *Genetics, Evolution, and Man.* W.H. Freeman, San Francisco 1976

Boethius, A.M.S. *The Consolation of Philosophy.* (Translated with introduction and notes by R. Green.) The Bobbs Merrill Co. Publishers, Indianapolis 1962

Bowcock, A.M., Ruiz-Linares, A., Tomfohrde, J., Minch, E., Kidd, J.R. & Cavalli-Sforza, L.L. High resolution of human evolutionary trees with polymorphic microsatellites. *Nature* 368:455-457, 1994

Campbell, B.G. *Humankind Emerging.* Harper Collins, New York 1988

Darnell, J., Lodish, H. & Baltimore, D. *Molecular Cell Biology.* 2nd edn. Scientific American Books, New York 1990

Doolittle, W.F. Phylogenetic classification and the universal tree. *Science* 284:2124-2128, 1999

Dorit, R.L., Walker, J., W.F. & Barnes, R.D. *Zoology.* Saunders College Publishing, Philadelphia 1991

Durrell, L. *Tunc.* French & European Pubns, New York 1979

Edwards, A.W.F. & Cavalli-Sforza, L.L. Reconstruction of evolutionary trees. In *Phenetic and Phylogenetic Classification* (Heywood, V.E. & McNeill, J. eds.) pp. 67-76, The Systematics Association, London 1964

Feng, D.-F., Cho, G. & Doolittle, R.F. Determining divergence times with a protein clock: update and reevaluation. *Proc. Natl. Acad. Sci. USA* 94:13028-13033, 1997

Fioroni, P. *Allgemeine und vergleichende Embryologie der Tiere.* 2nd edn. Springer-Verlag, Berlin 1992

Forberg, G. *Gustav Doré. Das grafische Werk. Vol. 1.* Manfred Pawlak Verlagsgesellschaft, Herrsching, Germany

Gilbert, S.F. *Developmental Biology.* 5th edn. Sinauer Associates, Sunderland, MA. 1997

Hammer, M.F., Karafet, T., Rasanayagam, A., Wood, E.T., Altheide, T.K., Jenkins, T., Griffiths, R.C., Templeton, A.R. & Zegura, S.L. Out of Africa and back again: nested cladistic analysis of human Y chromosome variation. *Mol. Biol. Evol.* 15:427-441, 1998

Harding, R.M., Fullerton, S.M., Griffiths, R.C., Bond, J., Cox, M.J., Schneider, J.A., Moulin, D.S. & Clegg, J.B. Archaic African and Asian lineages in the genetic ancestry of modern humans. *Am. J. Hum. Genet.* 60:772-789, 1997

Harris, E.E. & Hey, J. X chromosome evidence for ancient human histories. *Proc. Natl. Acad. Sci. USA* 96:3320-3324, 1999

Holub, M. *Hominization*. Mladá fronta, Prague 1986

Kent, G.C. & Miller, L. *Comparative Anatomy of the Vertebrates*. 8th edn. Wm.C. Brown, Dubuque, IA 1997

Kimura, M. *The Neutral Theory of Molecular Evolution*. Cambridge University Press, Cambridge, U.K. 1983

Klein, J. & Hořejší, V. *Immunology*. 2nd edn. Blackwell Scientific Publishers, Oxford 1998

Klein, J. & Patten, E. Blood group antigens. In *Medical Microbiology* (Baron, S. ed.) pp. 150-156, Addison-Wesley Publishing, Menlo Park, CA 1986

Klein, J. *Immunology. The Science of Self-Nonself Discrimination*. John Wiley, New York 1982

Klein, J., Gutknecht, J. & Fischer, N. The major histocompatibility complex and human evolution. *Trends Genet.* 6:7-11, 1990

Klein, J., Takahata, N. & Ayala, F.J. MHC polymorphism and human origins. *Sci. Am.* December:46-51, 1993

Klein, R.G. Archeology and the evolution of human behavior. *Evol. Anthropol.* 9:17-36, 2000

Koop, B.F., Siemieniak, D., Slightom, J.L., Goodman, M., Dunbar, J., Wright, P.C. & Simons, E.L. Tarsius δ- and β-globin genes: conversions, evolution, and systematic implications. *J. Biol. Chem.* 264:68-79, 1989

Krings, M., Capelli, C., Tschentscher, F., Geisert, H., Meyer, S., von Haeseler, A., Grossschmidt, K., Possnert, G., Paunovic, M. & Pääbo, S. A view of Neandertal genetic diversity. *Nat. Genet.* 26:144-146, 2000

Krings, M., Stone, A., Schmitz, R.W., Krainitzki, H., Stoneking, M. & Pääbo, S. Neandertal DNA sequences and the origin of modern humans. *Cell* 90:19-30, 1997

Kumar, S. & Hedges, S.B. A molecular timescale for vertebrate evolution. *Nature* 392:917-920, 1998

Le Gros Clark, W.E. *The Antecedents of Man. An Introduction to the Evolution of the Primates*. Edinburgh University Press, Edinburgh 1959

Linder, H. *Biologie*. 19th edn. J.B. Metzlersche Verlagsbuchhandlung, Stuttgart 1983

Madigan, M.T., Martinko, J.M. & Parker, J. *Brock Biology of Microorganisms*. 8th edn. Prentice Hall International, Upper Saddle River, NJ 1997

Madsen, O., Scally, M., Douady, C.J., Kao, D.J., DeBry, R.W., Adkins, R., Amrine, H.M., Stanhope, M.J., de Jong, W.W. & Springer, M.S. Parallel adaptive radiations in two major clades of placental mammals. *Nature* 409:610-614, 2001

Mazák, V. *Jak vznikl člověk*. Edice Kotva, Prague 1977

Micklos, D.A. & Freyer, G.A. *DNA Science. A First Course in Recombinant DNA Technology*. Cold Spring Harbor Laboratory Press, Cold Spring Harbor, NY 1990

Murphy, W.J., Eizirik, E., Johnson, W.E., Zhang, Y.P., Ryder, O.A. & O'Brien, S.J. Molecular phylogenetics and the origins of placental mammals. *Nature* 409:614-618, 2001

Napier, J.R. & Napier, P.H. *The Natural History of the Primates*. The MIT Press, Cambridge, MA 1985

Nei, M. & Roychoudhury, A.K. Genetic relationship and evolution of human races. *Evol. Biol.* 14:1-59, 1982

Nims, J.F. *The Harper Anthropology of Poetry*. pp. 760, Harper & Row, Publ., New York 1981

Novacek, M.J. Mammalian phylogeny: Shaking the tree. *Nature* 356:121-125, 1992

Oakley, K.P. *Man the Tool-Maker*. University of Chicago Press, Chicago 1959

Ovchinnikov, I.V., Götherström, A., Romanova, G.P., Kharitonov, V.M., Lidén, K. & Goodwin, W. Molecular analysis of Neanderthal DNA from the northern Caucasus. *Nature* 404:490-493, 2000

Page, R.D.M. & Holmes, E.C. *Molecular Evolution. A Phylogenetic Approach*. Blackwell Science, Oxford 1998

Passingham, R.E. *The Human Primate*. W.H. Freeman, Oxford 1982

Porkess, R. *The Harper Collins Dictionary of Statistics*. Harper Collins, New York 1991

Porter, C.A., Sampaio, I., Schneider, H., Schneider, M.P.C., Czelusniak, J. & Goodman, M. Evidence on primate phylogeny form ε-globin gene sequences and flanking regions. *J. Mol. Evol.* 40:30-55, 1995

Portman, A. *Einführung in die vergleichende Morphologie der Wirbeltiere*. Beno Schwabe, Basel 1948

Pough, F.H., Heiser, J.B. & McFarland, W.N. *Vertebrate Life*. 3rd edn. Macmillan Publishing, New York 1989

Purves, W.K., Orians, G.H. & Heller, H.C. *Life: The Science of Biology*. 3rd edn. Sinauer Associates, Sunderland, MA 1992

Rabelais, F. *Pantagruel*. In *Les Cinq Livres f. Rabelais*. Tome quatrième. (Flammarian, T. ed.) Librairie des bibliophiles, Paris 1924

Roeder, G. S. Meiotic chromosomes: It takes two to tango. *Genes Develop.* 11:2600-2621, 1997

Sato, A., O'hUigin, C., Figueroa, F., Grant, P.R., Grant, B.R., Tichy, H. & Klein, J. Phylogeny of Darwin's finches as revealed by mtDNA sequences. *Proc. Natl. Acad. Sci. USA* 96:5101-5106, 1999

Sato, A., Tichy, H., O'hUigin, C., Grant, P.R., Grant, B.R. & Klein, J. On the origin of Darwin's finches. *Mol. Biol. Evol.* 18:299-311, 2001

Satta, Y., Takahata, N., Schönbach, C., Gutknecht, J. & Klein, J. Calibrating evolutionary rates at major histocompatibility complex loci. In *Molecular Evolution of the Major Histocompatibility Complex* H 59. (Klein, J. & Klein, D. eds.) pp. 51-62, Springer-Verlag, Heidelberg 1991

Schmitt, M. *Wie sich das Leben entwickelte. Die faszinierende Geschichte der Evolution.* Mosaik Verlag, München, Germany 1994

Scientific American. Title page. October 1995

Shelley, P.B. Time. In *The Complete Poetical Works of Percy Bysshe Shelley* (Hutchinson, T. ed.) Oxford University Press, London 1960

Simpson, G.G. & Beck, W.S. *Life. An Introduction to Biology*. 2nd edn. Harcourt, Brace & World, New York 1965

Singer, S. & Hilgard, H.R. *The Biology of People*. W.H. Freeman, San Francisco 1978

Sogin, M.L. Early evolution and the origin of eukaryotes. *Curr. Op. Genet. Dev.* 1:457-463, 1991

Spence, A.P. & Mason, E.B. *Human Anatomy and Physiology*. 3rd edn. Benjamin/Cummings Publishing, Menlo Park, CA 1987

Stokstad, E. Tooth theory revises history of mammals. *Science* 291:26-26, 2001

Strachan, T. & Read, A.P. *Human Molecular Genetics*. BIOS Scientific Publishers, Oxford 1996

Strait, D.S., Grine, F.E. & Moniz, M.A. A reappraisal of early hominid phylogeny. *J. Hum. Evol.* 32:17-82, 1997

Strickberger, M.W. *Evolution*. 2nd edn. Jones & Bartlett Publ., Sudbury, MA 1996

Stringer, C. & McKie, R. *African Exodus. The Origins of Modern Humanity*. Jonathon Cape, London 1996

Stringer, C.B. The emergence of modern humans. *Sci. Am.* December:68-74, 1990

Sweetman, D. *Paul Gauguin. A Life*. Simon & Schuster, New York 1995

Takahata, N., Lee, S.-H. & Satta, Y. Testing multiregionality of modern human origins. *Mol. Biol. Evol.* 18:172-183, 2001

Terry, C.S. (ed.) *Masterworks of Japanese Art*. Charles E. Tuttle, Rutland, VT 1957

The Venerable Bede. *Ecclesiastical History of the English People*. (Translated by L. Sherley-Price.) Penguin Books, London 1990

The World Book Encyclopedia. Vol. 16. World Book, Inc., Chicago 1996

Tortore, G.J. *Principles of Human Anatomy*. 4th edn. Harper & Row Publ., New York 1986

Vallois, H. Ordre des primates (Primates Linne', 1758). In *Traité de Zoologie. Anatomie, Systematique, Biologie*. Vol. 172 (Grassé, P.-P. ed.) pp. 1854-2206, Massonet et Cie, Paris 1955

Waetzoldt, W. *Dürer und seine Zeit*. George Allen & Unwin, London 1938

Wang, D.Y.-C., Kumar, S. & Hedges, S.B. Divergence time estimates for the early history of animal phyla and the origin of plants, animals and fungi. *Proc. R. Soc. Lond. B* 266:163-171, 1999

Whitman, W.B. Leaves of Grass. In *Leaves of Grass and Selected Prose* (Bradley, S. ed.) Rinehart, Toronto 1951

Williams, P.L. & Warwick, R. (eds.) *Gray's Anatomy*. 36th edn. Churchill Livingston, Edinburgh 1980

Wilson, E.B. *The Cell in Development and Inheritance*. Macmillan, New York 1896

Wolf, J. *Menschen der Urzeit*. Artia Praha/Verlag Dausien, Hanau, Germany 1989

I N D E X

G

Printing and binding:
Mercedes-Druck GmbH, Berlin